VOLUME ONE HUNDRED AND FIFTY FOUR

INTERNATIONAL REVIEW OF NEUROBIOLOGY

Metabolic and Bioenergetic Drivers of Neurodegenerative Disease: Neurodegenerative Disease Research and Commonalities with Metabolic Diseases

INTERNATIONAL REVIEW OF NEUROBIOLOGY

VOLUME 154

VOLUME ONE HUNDRED AND FIFTY FOUR

INTERNATIONAL REVIEW OF NEUROBIOLOGY

Metabolic and Bioenergetic Drivers of Neurodegenerative Disease: Neurodegenerative Disease Research and Commonalities with Metabolic Diseases

Edited by

GRAŻYNA SÖDERBOM
Klipspringer AB, Sävedalen, Sweden

RUSSELL ESTERLINE
AstraZeneca, Gaithersburg MD, United States

JAN OSCARSSON
AstraZeneca, Gothenburg, Sweden

MARK P. MATTSON
Department of Neuroscience, Johns Hopkins University School of Medicine, Baltimore, MD, United States

ELSEVIER

ACADEMIC PRESS
An imprint of Elsevier

Academic Press is an imprint of Elsevier
50 Hampshire Street, 5th Floor, Cambridge, MA 02139, United States
525 B Street, Suite 1650, San Diego, CA 92101, United States
The Boulevard, Langford Lane, Kidlington, Oxford OX5 1GB, United Kingdom
125 London Wall, London, EC2Y 5AS, United Kingdom

First edition 2020

ISBN: 978-0-12-820076-6
ISSN: 0074-7742

For information on all Academic Press publications
visit our website at https://www.elsevier.com/books-and-journals

Publisher: Zoe Kruze
Acquisitions Editor: Sam Mahfoudh
Editorial Project Manager: Shellie Bryant
Production Project Manager: Abdulla Sait
Cover Designer: Greg Harris

Contents

Contributors *xi*
Preface *xv*

Section 1
Overview of neurodegenerative research and commonalities with metabolic diseases **1**

1. Status and future directions of clinical trials in Alzheimer's disease **3**

Germán Plascencia-Villa and George Perry

1. Introduction 5
2. Current status of amyloid-β clinical trials 10
3. Current status of clinical trials evaluating secretases 19
4. Current status of non-amyloid clinical trials 23
5. Alternative approaches or future directions to target Alzheimer's disease 32
6. Conclusions 42

Acknowledgments 42
References 42

2. Tracking the potential involvement of metabolic disease in Alzheimer's disease—Biomarkers and beyond **51**

José A. Luchsinger and Henrik Zetterberg

1. Introduction 53
2. Alzheimer's disease neuropathology as an outcome 54
3. Summary of studies relating type 2 diabetes mellitus and glycemia to Alzheimer's disease biomarkers 59
4. Summary of studies relating adiposity, insulin resistance, and metabolic syndrome to Alzheimer's disease biomarkers 61
5. The promise of the field of metabolomics in the study of Alzheimer's disease biomarkers 66
6. Caveats, recommendations, and future directions 70

Acknowledgments 71
Disclosures 71
References 71

3. Brain glucose and ketone utilization in brain aging and neurodegenerative diseases **79**

Dimitrios Kapogiannis and Konstantinos I. Avgerinos

1. Introduction 81
2. Brain glucose metabolism in aging, Alzheimer's disease and other neurodegenerative diseases 82
3. Brain ketones metabolism in aging, Alzheimer's disease and other neurodegenerative diseases 94
4. Interplay between glucose and ketone brain metabolism: Insights and therapeutic strategies for Alzheimer's disease and other neurodegenerative diseases 97
5. Conclusion 101

Acknowledgment 101

References 101

4. Disturbances in brain energy metabolism in insulin resistance and diabetes and Alzheimer's disease — Learnings from brain imaging biomarkers **111**

Kerstin Heurling, Edvin Johansson, and Antoine Leuzy

1. Introduction to medical imaging modalities 112
2. Imaging cerebral glucose metabolism 113
3. Imaging cerebral lipid metabolism 118
4. Magnetic resonance imaging 119
5. Imaging metabolic changes with proton magnetic resonance spectroscopy 121
6. Conclusions 123

References 124

5. APOE and metabolic dysfunction in Alzheimer's disease **131**

Lance A. Johnson

1. *APOE* and Alzheimer's disease risk 132
2. APOE, insulin resistance, and Alzheimer's disease 133
3. APOE and cerebral glucose and fatty acid metabolism 137
4. APOE and the PPP: A potential link between glucose metabolism and oxidative stress in Alzheimer's disease 139
5. APOE and the inverse Warburg hypothesis 141

References 143

6. Status and future directions of clinical trials in Parkinson's disease **153**

Grażyna Söderbom

1. Parkinson's disease is a complex, multifactorial, clinical syndrome 155
2. Risk factors and pathology associated with Parkinson's disease 156
3. Current status of disease-modifying therapies and repurposing strategies 158
4. Parkinson's disease has many unmet needs: Diagnosis, current clinical therapies, drug development and precision medicine 168
5. Biomarkers for diagnostics, drug development and precision medicine 172
6. Future considerations and conclusion 174

Acknowledgments 177

References 177

Section 2 **Metabolic approaches to neurodegenerative disease** **189**

7. Involvement of GABAergic interneuron dysfunction and neuronal network hyperexcitability in Alzheimer's disease: Amelioration by metabolic switching **191**

Mark P. Mattson

1. Neuronal network hyperexcitability in Alzheimer's disease 192
2. Their high metabolic demand renders GABAergic interneurons vulnerable to Alzheimer's disease 196
3. Intermittent metabolic challenges, ketones and NAD^+ enhance the plasticity and resilience of neuronal networks 197
4. Conclusions 201

References 201

8. The mitochondrial hypothesis: Dysfunction, bioenergetic defects, and the metabolic link to Alzheimer's disease **207**

Russell H. Swerdlow

1. Introduction 209
2. Normal mitochondrial and metabolism function 210
3. Alzheimer's disease-relevant alterations in mitochondrial and metabolism function 215
4. Metabolic endophenotypes 218
5. Relationship to proteostasis 219
6. Genetic contributions 220

7. A mitochondrial cascade hypothesis 224
8. Conclusions and implications 226
Acknowledgments 228
Conflict of interests 228
References 228

9. MAM and C99, key players in the pathogenesis of Alzheimer's disease 235

Marta Pera, Jorge Montesinos, Delfina Larrea, Rishi R. Agrawal, Kevin R. Velasco, Irina G. Stavrovskaya, Taekyung D. Yun, and Estela Area-Gomez

1. Introduction 237
2. C99 in Alzheimer's disease animal models and patients 240
3. Mitochondria-associated endoplasmic reticulum membranes in Alzheimer's disease 244
4. Lipid dyshomeostasis in Alzheimer's disease is MAM-related 250
5. Mitochondrial alterations in Alzheimer's disease and MAM 255
6. Concluding remarks 260
References 262

10. Experimental studies of mitochondrial and lysosomal function in *in vitro* and *in vivo* models relevant to Parkinson's disease genetic risk 279

Ria Thomas, Penelope J. Hallett, and Ole Isacson

1. Introduction 280
2. Mitochondrial dysfunction in Parkinson's disease 281
3. Lysosomal dysfunction in aging and Parkinson's disease 291
4. Future considerations for using cell culture models for studying disease-relevant mechanisms *in vitro* 295
5. Conclusions and future directions 296
References 296

11. Endosomal-lysosomal dysfunction in metabolic diseases and Alzheimer's disease 303

Michael F. Almeida, Ben A. Bahr, and Stephen T. Kinsey

1. Introduction 305
2. Metabolic energy balance 306
3. Alzheimer's disease and related dementias 311

4. α-Synucleinopathies 315
5. Conclusion 318
References 318

12. Neuroinflammation, microglial activation, and glucose metabolism in neurodegenerative diseases 325
Paul Edison

1. Introduction 327
2. Microglia 327
3. Type 2 diabetes mellitus 329
4. Obesity 331
5. Hypercholesterolemia 334
6. Role of systemic inflammation in central inflammation 335
7. Blood-brain barrier and systemic inflammation 337
8. Neuroinflammation and its major players 339
9. Conclusion 340
Acknowledgment and conflicts 340
References 341

13. The NLRP3 inflammasome as a bridge between neuro-inflammation in metabolic and neurodegenerative diseases 345
Grażyna Söderbom and Bai-Yun Zeng

1. Introduction 348
2. The NLRP3 inflammasome and peripheral metabolic disease 348
3. The NLRP3 inflammasome, chronic inflammation and endothelial dysfunction 354
4. Neuro-inflammation in Alzheimer's disease 355
5. Neuro-inflammation in Parkinson's disease 363
6. The NLRP3 inflammasome: A target for novel therapeutic approaches in Alzheimer's and Parkinson's diseases? 369
7. Conclusion 372
References 375

14. Impact of circadian and diurnal rhythms on cellular metabolic function and neurodegenerative diseases 393
S. Kendall Smith and Erik S. Musiek

1. Introduction to the circadian clock 394
2. Evidence of sleep and circadian disruption in neurodegenerative diseases 396

3. Does circadian disruption influence the pathogenesis of neurodegenerative diseases? 398
4. Circadian rhythms orchestrate cellular metabolism and bioenergetics in the periphery 400
5. Circadian rhythms orchestrate cellular metabolism and bioenergetics in the CNS 402
6. Conclusion and future directions 403
References 404

15. The impact of neurovascular, blood-brain barrier, and glymphatic dysfunction in neurodegenerative and metabolic diseases 413

Molly Braun and Jeffrey J. Iliff

1. Introduction 415
2. The cerebral vasculature and blood-brain barrier 415
3. Cerebrospinal fluid barriers 418
4. Perivascular exchange and the glymphatic system 420
5. Aging 424
6. Neurodegenerative diseases 425
7. Metabolic disease 428
8. Conclusion 429
References 430

Contributors

Rishi R. Agrawal
Institute of Human Nutrition, Columbia University Irving Medical Center, New York, NY, United States

Michael F. Almeida
Biotechnology Research and Training Center, University of North Carolina—Pembroke, Pembroke, NC, United States

Estela Area-Gomez
Department of Neurology; Taub Institute for Research on Alzheimer's Disease and the Aging Brain; Institute of Human Nutrition, Columbia University Irving Medical Center, New York, NY, United States

Konstantinos I. Avgerinos
Laboratory of Clinical Investigation, Intramural Research Program, National Institute on Aging, National Institutes of Health, Baltimore, MD, United States

Ben A. Bahr
Biotechnology Research and Training Center, University of North Carolina—Pembroke, Pembroke, NC, United States

Molly Braun
Department of Psychiatry and Behavioral Sciences, University of Washington School of Medicine; VISN 20 Mental Illness Research, Education and Clinical Center (MIRECC), VA Puget Sound Health Care System, Seattle, WA, United States

Paul Edison
Department of Brain Sciences, Faculty of Medicine, Imperial College London, London; Cardiff University, Cardiff, United Kingdom

Penelope J. Hallett
Neuroregeneration Research Institute, Harvard Medical School/McLean Hospital, Belmont, MA, United States

Kerstin Heurling
Antaros Medical AB, Mölndal, Sweden

Jeffrey J. Iliff
Department of Psychiatry and Behavioral Sciences, University of Washington School of Medicine; VISN 20 Mental Illness Research, Education and Clinical Center (MIRECC), VA Puget Sound Health Care System; Department of Neurology, University of Washington School of Medicine, Seattle, WA, United States

Ole Isacson
Neuroregeneration Research Institute, Harvard Medical School/McLean Hospital, Belmont, MA, United States

Edvin Johansson
Antaros Medical AB, Mölndal, Sweden

Lance A. Johnson
Department of Physiology; Sanders-Brown Center on Aging, University of Kentucky College of Medicine, Lexington, KY, United States

Dimitrios Kapogiannis
Laboratory of Clinical Investigation, Intramural Research Program, National Institute on Aging, National Institutes of Health, Baltimore, MD, United States

Stephen T. Kinsey
Department of Biology and Marine Biology, University of North Carolina—Wilmington, Wilmington, NC, United States

Delfina Larrea
Department of Neurology, Columbia University Irving Medical Center, New York, NY, United States

Antoine Leuzy
Department of Clinical Sciences, Clinical Memory Research Unit, Lund University, Malmö, Sweden

José A. Luchsinger
Departments of Medicine and Epidemiology, Columbia University Irving Medical Center, New York, NY, United States

Mark P. Mattson
Department of Neuroscience, Johns Hopkins University School of Medicine, Baltimore, MD, United States

Jorge Montesinos
Department of Neurology; Taub Institute for Research on Alzheimer's Disease and the Aging Brain, Columbia University Irving Medical Center, New York, NY, United States

Erik S. Musiek
Department of Neurology, Washington University School of Medicine, St. Louis, MO, United States

Marta Pera
Department of Neurology, Columbia University Irving Medical Center, New York, NY, United States; Basic Sciences Department, Faculty of Medicine and Health Sciences, Universitat Internacional de Catalunya, Sant Cugat del Vallés, Barcelona, Spain

George Perry
Department of Biology and Neurosciences Institute, The University of Texas at San Antonio (UTSA), 1 UTSA Circle, San Antonio, TX, United States

Germán Plascencia-Villa
Department of Biology and Neurosciences Institute, The University of Texas at San Antonio (UTSA), 1 UTSA Circle, San Antonio, TX, United States

S. Kendall Smith
Department of Anesthesiology, Washington University School of Medicine, St. Louis, MO, United States

Grażyna Söderbom
Klipspringer AB, Sävedalen, Sweden

Irina G. Stavrovskaya
Department of Neurology, Columbia University Irving Medical Center, New York, NY, United States

Russell H. Swerdlow
University of Kansas Alzheimer's Disease Center, University of Kansas Medical Center, Kansas City, KS, United States

Ria Thomas
Neuroregeneration Research Institute, Harvard Medical School/McLean Hospital, Belmont, MA, United States

Kevin R. Velasco
Department of Neurology, Columbia University Irving Medical Center, New York, NY, United States

Taekyung D. Yun
Department of Neurology, Columbia University Irving Medical Center, New York, NY, United States

Bai-Yun Zeng
Institute of Pharmaceutical Science, Faculty of Life Sciences & Medicine, King's College London, London, United Kingdom

Henrik Zetterberg
Department of Psychiatry and Neurochemistry, Institute of Neuroscience & Physiology, The Sahlgrenska Academy at the University of Gothenburg; Clinical Neurochemistry Laboratory, Sahlgrenska University Hospital, Mölndal, Sweden; Department of Neurodegenerative Disease, UCL Institute of Neurology, Queen Square; UK Dementia Research Institute at UCL, London, United Kingdom

Preface

Currently, there are no approved pharmacological therapies that retard or reverse the two most common neurodegenerative diseases (NDDs), Alzheimer's disease (AD) and Parkinson's disease (PD). Their global prevalence is expected to increase, primarily driven by an aging population (Dorsey et al., 2018; Hebert et al., 2013), but also due to an increased prevalence of obesity and type 2 diabetes mellitus (T2DM) (De Pablo-Fernandez et al., 2018; Kivipelto et al., 2018; Li et al., 2015). Obesity and T2DM are metabolic, systemic disorders that affect a multitude of body functions both in the periphery and in the brain. Alzheimer's disease results in the progressive loss of memory functions and an inability to move in its final stages. Parkinson's disease is traditionally considered to be a movement disorder, although this disease can also result in cognitive impairment.

The search for new therapeutics for AD and PD has focused on abnormal aggregation of specific proteins that are diagnostic for the diseases (Brundin et al., 2017; Selkoe & Hardy, 2016). Alzheimer's disease is characterized by extracellular amyloid β (Aβ) plaque that contains aggregated Aβ peptide and intraneuronal neurofibrillary tangles, while PD is characterized by aggregated α-synuclein in the in the brain, particularly the substantia nigra. To date, pharmacological approaches to halt or reverse AD by primarily targeting these aggregated proteins have experienced a huge number of failures in the clinical phase of development (Yiannopoulou et al., 2019). In PD, these approaches continue to be evaluated, and in the case of glucocerebrosidase (GBA)-associated PD have shown a positive clinical impact in Phase 2 (Mullin et al., 2020).

Type 2 diabetes mellitus is a risk factor for both AD and PD, and the metabolic derangements seen in T2DM largely overlap in AD and PD, including disturbed glucose and lipid homeostasis, insulin resistance, mitochondrial dysfunction, oxidative damage, and chronic inflammation (Anandhan et al., 2017; de la Monte & Tong, 2014). Type 2 diabetes mellitus is also associated with cognitive dysfunction. Importantly, T2DM and diabetic complications, as well as AD and PD, are associated with aging and regarded as conditions of accelerated aging. Interestingly, exercise may prevent and retard the development of T2DM, and especially AD, while also positively impacting PD. It could well be, therefore, that the consequences of over-nutrition caused by increased food intake and a sedentary life style causes disease by similar and overlapping mechanisms, while genetics and epigenetics determine which tissues are mainly affected in different individuals (Mattson, 2019).

This book focuses on the metabolic derangements associated with NDDs, especially AD and PD, the overlap between the drivers of T2DM, insulin resistance and NDDs, and the extent to which metabolic disturbances could explain the accumulation of protein aggregates and progression of NDDs.

The first part focuses on NDD research and commonalities with metabolic diseases. Germán Plascencia-Villa and George Perry review the status and future directions of clinical trials in AD (Chapter 1). The overall conclusion is that the last 25 years have seen the failure of a large number of trials targeting the "Aβ cascade" so alternative therapeutic areas, including those focusing on energy metabolism intervention, may provide alternative leads in the hunt for disease-modifying agents for AD.

Early diagnosis of AD will enhance the possibility to treat the disease. Thus, development of AD-specific biomarkers associated with increased risk for disease progression is vital to enable early intervention. Investigation of biomarkers in clinical trials evaluating drugs that correct metabolism would also expand our understanding of the importance of metabolic disturbances in the development of NDDs and present new opportunities to develop therapeutics for NDDs. In Chapter 2, José A. Luchsinger and Henrik Zetterberg review the potential involvement of metabolic disease in AD, which may occur decades before clinical manifestation of AD. They also discuss the importance of clinical biomarkers of AD and their application to research focusing on the metabolic contributions to AD, as well as the increasing importance of metabolomics in AD research.

In situations of age- and NDD-associated glucose hypometabolism, the brain meets its high energy demands using additional fuels, such as ketones, and Dimitrios Kapogiannis and Konstantinos I. Avgerinos discuss this and the potential of ketone supplementation as a future approach to treating AD and other NDDs (Chapter 3). Kerstin Heurling, Edvin Johansson, and Antoine Leuzy review the development of imaging techniques that reveal glucose hypometabolism in similar brain regions in T2DM and AD, indicating overlapping pathology (Chapter 4).

Regarding AD, the strongest genetic risk factor for development of AD is polymorphism of apolipoprotein E (apoE). Individuals with the *APOE* ε4 allele also have higher levels of LDL-cholesterol and increased risk for cardiovascular disease compared to individuals with the other isoforms of apoE. Interestingly, apoE4 may change neuronal glucose metabolism and the risk for oxidative stress, and Lance A. Johnson suggests that we need a better understanding of the metabolic mechanisms that may underlie *APOE* ε4-associated cognitive dysfunction and AD risk in Chapter 5.

In Chapter 6, Grażyna Söderbom discusses the metabolic aspects of some of the risk factors and consequent pathology associated with PD and the status of novel therapies that target these pathologies (e.g., mitochondrial and lysosomal dysfunction, oxidative stress, and neuroinflammation), as well as the importance of biomarkers for future research in PD and its clinical management.

The second part of this book focuses on the mechanistic approaches to NDDs based on metabolic hypotheses, and our current understanding of the metabolic causes and consequences of AD and PD. Mark P. Mattson reviews the evidence that suggests early GABAergic neuron degeneration and consequent neuronal network hyperexcitability is involved in AD pathogenesis, and supports the case for pivotal roles for ketones, NAD^+ and the mitochondrial protein deacetylase sirtuin 3 (SIRT3) in protecting against hyperexcitability in AD (Chapter 7).

In contrast to the Aβ hypothesis that argues for amyloid plaque and tau tangle formation as the initiating event in AD, the mitochondrial cascade hypothesis argues for loss of mitochondrial function as the initiating event in the disease process, while the endosomal-lysosomal dysfunction hypothesis indicates that defective protein/organelle turnover is the earliest cytopathology in AD. Russell H. Swerdlow (Chapter 8) presents evidence for mitochondrial dysfunction in AD, suggesting that mitochondria are more relevant to the etiology of AD than its consequences and that a mitochondrial cascade features prominently in AD. Marta Pera, Jorge Montesinos, Delfina Larrea, Rishi R. Agrawal, Kevin R. Velasco, Irina G. Stavrovskaya, Taekyung D. Yun, and Estella Area-Gomez (Chapter 9) present recent data indicating that increased levels of uncleaved C99 levels at the mitochondria-associated ER membranes are associated with loss of lipid homeostasis and mitochondrial dysfunction, and discuss how this impacts AD neuropathology.

Regarding endosomal-lysosomal dysfunction as an early pathological process associated with NDDs, Ria Thomas, Penelope J. Hallett, and Ole Isacson (Chapter 10) discuss their recent investigations on mitophagy and mitochondrial dynamics in fibroblasts from people with *LRRK2* mutations and those with sporadic PD, and also review the changes in and relevance of lysosomal enzymes and resultant lipid dishomeostasis in both PD and aging in human tissue and preclinical models of PD. Michael F. Almeida, Ben A. Bahr, and Stephen T. Kinsey (Chapter 11) review the endosomal-lysosomal pathways and related autophagic processes fundamental to cell signaling, metabolism and quality control, and their impairment in metabolic diseases, aging and NDDs. They discuss how exercise may positively impact autophagy-lysosomal degradation and thus the importance of functional lysosomal-related systems for healthy cognitive aging.

The association between T2DM, insulin resistance and susceptibility for AD and PD has led to the hypothesis that defective insulin receptor signaling in the brain is an early event in the disease process. Other pathological commonalities of T2DM and NDDs like AD and PD are local inflammation/neuroinflammation and chronic, low-grade (systemic) inflammation. Inflammation is also exacerbated by aging, and contributes to the complex, mechanistically intertwined, pathological processes underlying these diseases. Paul Edison (Chapter 12) discusses how chronic inflammation associated with metabolic disorders, including T2DM, influences microglial activation, the relationship between microglial activation and glucose metabolism, and how microglial activation influences neurodegeneration in AD. Grażyna Söderbom and Bai-Yun Zeng focus on the relationship between inflammation associated with T2DM, AD and PD from the perspective of the proinflammatory signaling of the nucleotide-binding oligomerization domain (NOD)-, leucine-rich repeat (LRR)-, and pyrin domain-containing protein 3 (NLRP3) inflammasome complex (Chapter 13). They also consider how approaches targeting neuro-inflammation may provide novel therapeutics for AD and PD and suggest that viewing AD and PD as systemic diseases may advance the discovery of therapeutics.

Other important aspects of NDDs and possible early events in their pathology include a loss of blood-brain barrier integrity and blood-CSF barrier function that may be associated with chronic inflammation, as well as disturbed glymphatic function and circadian rhythms. S. Kendall Smith and Erik S. Musiek (Chapter 14) describe how metabolic pathways are regulated by the circadian clock. They discuss evidence that implicates circadian and sleep disruption in NDDs like AD and PD, and how optimizing circadian function with clock-targeted drugs theoretically could positively influence brain metabolism, inflammation, redox defenses, proteostasis and synaptic function, all of which are affected in NDDs.

In Chapter 15, Molly Braun and Jeffrey J. Iliff review ground-breaking research demonstrating the vital role of the glymphatic system in maintaining brain homeostasis, together with the cerebral vasculature, the blood-brain barrier and the cerebrospinal fluid. With particular focus on AD, they explain the impact of dysfunction in these structures and suggest that cellular and molecular changes occurring within and around the cerebral vasculature are one of the key drivers of NDD pathogenesis.

For the first time in this series, a renowned group of scientists have come together to review recent developments in NDD intervention trials, particularly in AD and PD, and the growing research on the impact of metabolic dysfunction and T2DM on these diseases. In light of recent disappointments in the search for novel, disease-modifying therapies for NDDs and particularly AD, looking beyond the impact of aberrant proteins can stimulate novel directions of research that may contribute to the development of innovative therapeutic strategies in the future.

We hope that this book will inspire scientists in Academia and Industry to further develop our understanding of how metabolic disturbances could contribute to the pathology of NDDs, including AD and PD, and to explore metabolic targets in the search for clinical treatments that could halt, retard or reverse these devastating diseases.

We thank our team at Elsevier for their support in putting together this volume, particularly Sam Mahfoudh, Peter J. Llewellyn, Shellie Bryant, Priyadarshini Varadarajan, and Abdulla Sait, the serial editors, Peter Jenner and Patricia Janak for this opportunity to present many novel areas of research, and to the authors who contributed their time and ideas to review this rapidly developing field of NDD research.

Grażyna Söderbom
Klipspringer AB, Sävedalen, Sweden
Russell Esterline
AstraZeneca, Gaithersburg MD, United States
Jan Oscarsson
AstraZeneca, Gothenburg, Sweden
Mark P. Mattson
Department of Neuroscience, Johns Hopkins
University School of Medicine, Baltimore, MD, United States

References

Anandhan, A., Jacome, M. S., Lei, S., Hernandez-Franco, P., Pappa, A., Panayiotidis, M. I., et al. (2017). Metabolic dysfunction in Parkinson's disease: Bioenergetics, redox homeostasis and central carbon metabolism. *Brain Research Bulletin*, *133*, 12–30.

Brundin, P., Dave, K. D., & Kordower, J. H. (2017). Therapeutic approaches to target alpha-synuclein pathology. *Experimental Neurology*, *298*(Pt B), 225–235.

de la Monte, S. M., & Tong, M. (2014). Brain metabolic dysfunction at the core of Alzheimer's disease. *Biochemical Pharmacology, 88*(4), 548–559.

De Pablo-Fernandez, E., Goldacre, R., Pakpoor, J., Noyce, A. J., & Warner, T. T. (2018). Association between diabetes and subsequent Parkinson disease: A record-linkage cohort study. *Neurology, 91*(2), e139–e142.

Dorsey, E. R., Sherer, T., Okun, M. S., & Bloem, B. R. (2018). The emerging evidence of the Parkinson pandemic. *Journal of Parkinson's Disease, 8*(s1), S3–S8.

Hebert, L. E., Weuve, J., Scherr, P. A., & Evans, D. A. (2013). Alzheimer disease in the United States (2010-2050) estimated using the 2010 census. *Neurology, 80*(19), 1778–1783.

Kivipelto, M., Mangialasche, F., & Ngandu, T. (2018). Lifestyle interventions to prevent cognitive impairment, dementia and Alzheimer disease. *Nature Reviews Neurology, 14*(11), 653–666.

Li, X., Song, D., & Leng, X. S. (2015). Link between type 2 diabetes and Alzheimer's disease: From epidemiology to mechanism and treatment. *Clinical Interventions in Aging, 10*, 549–560.

Mattson, M. P. (2019). An evolutionary perspective on why food overconsumption impairs cognition. *Trends in Cognitive Sciences, 23*(3), 200–212.

Mullin, S., Smith, L., Lee, K., D'Souza, G., Woodgate, P., Elflein, J., et al. (2020). Ambroxol for the treatment of patients with Parkinson disease with and without glucocerebrosidase gene mutations: A nonrandomized, noncontrolled trial. *JAMA Neurology*, 77, 427–434.

Selkoe, D. J., & Hardy, J. (2016 Jun 1). The amyloid hypothesis of Alzheimer's disease at 25 years. *EMBO Molecular Medicine, 8*(6), 595–608.

Yiannopoulou, K. G., Anastasiou, A. I., Zachariou, V., & Pelidou, S. H. (2019). Reasons for failed trials of disease-modifying treatments for Alzheimer disease and their contribution in recent research. *Biomedicine*, 7(4), pii: E97.

SECTION 1

Overview of neurodegenerative research and commonalities with metabolic diseases

CHAPTER ONE

Status and future directions of clinical trials in Alzheimer's disease

Germán Plascencia-Villa, George Perry*
Department of Biology and Neurosciences Institute, The University of Texas at San Antonio (UTSA), 1 UTSA Circle, San Antonio, TX, United States
*Corresponding author: e-mail address: george.perry@utsa.edu

Contents

1. Introduction 5
2. Current status of amyloid-β clinical trials 10
   2.1 Anti-amyloid-β immunotherapy 10
   2.2 Amyloid-β vaccines 17
   2.3 Inhibition of amyloid-β aggregation 18
3. Current status of clinical trials evaluating secretases 19
   3.1 α-Secretase 19
   3.2 β-Secretase 21
   3.3 γ-Secretase 23
4. Current status of non-amyloid clinical trials 23
   4.1 Tau 23
   4.2 Neuroinflammation 26
   4.3 Neuroprotective compounds 27
   4.4 Brain chemistry and neurotransmitters 30
5. Alternative approaches or future directions to target Alzheimer's disease 32
   5.1 Apolipoprotein E 33
   5.2 Gene therapy 37
   5.3 Cell therapy 37
   5.4 Antimicrobials 38
   5.5 Energy metabolism-based interventions 39
6. Conclusions 42
Acknowledgments 42
References 42

Abstract

Amyloid-β (Aβ) senile plaques and neurofibrillary tangles of tau are generally recognized as the culprits of Alzheimer's disease (AD) and related dementia. About 25 years ago, the amyloid cascade hypotheses postulated a direct correlation of plaques with the development of AD, and it has been the dominant theory since then. In this period, more than

International Review of Neurobiology, Volume 154
ISSN 0074-7742
https://doi.org/10.1016/bs.irn.2020.03.022

200 clinical trials focused mainly on targeting components of the Aβ cascade have dramatically failed, some of them in Phase III. With a greater than 99.6% failure rate at a cost of several billion from governments, industry, and private funders, therapeutic strategies targeting amyloid and tau are now under scrutiny. Therefore, it is time to reevaluate alternatives to targeting Aβ and tau as effective therapeutic strategies for AD. The diagnosis of AD is currently based on medical examination of symptoms including tests to assess memory impairment, attention, language, and other thinking skills. This is complemented with brain scans, such as computed tomography, magnetic resonance imaging, or positron emission tomography with the help of imaging probes targeting Aβ or tau deposits. This approach has contributed to the tunnel vision focus on Aβ and tau as the main culprits of AD. However, events upstream of these proteopathies (age-related impaired neuronal bioenergetics, lysosome function, neurotrophic signaling, and neuroinflammation, among others) are almost surely where the development of alternative therapeutic interventions should be targeted. Here, we present the current status of therapeutic candidates targeting diverse mechanisms and strategies including Aβ and tau, proteins involved in Aβ production and trafficking (ApoE, α/β/γ-secretases), neuroinflammation, neurotransmitters, neuroprotective agents antimicrobials, and gene and stem cell therapy. There are currently around 33 compounds in Phase III, 78 in Phase II, and 32 more in Phase I trials. With the current world health crisis of increased dementia in a rapidly aging population, effective AD therapies are desperately needed.

Abbreviations

AAV	adenoassociated virus
ABCA1	ATP-binding cassette transporter 1
AChE	acetylcholinesterase
AD	Alzheimer's disease
ADAM-10	a desintegrin and metalloproteinase domain-containing protein 10
ADAS-Cog	Alzheimer's Disease Assessment Scale cognitive subscale
ADCS-ADL	Alzheimer's Disease Cooperative Study-activities of daily living scale
ADHD	attention deficit hyperactivity disorder
AdMSC	adipose tissue-derived, mesenchymal stem cells
Akt	protein kinase B
APH-1	anterior pharynx-defective 1
ApoE	apolipoprotein E
APOE	apolipoprotein E gene
APOEε4	apolipoprotein E allele ε4
ARB	angiotensin II receptor blocker
Aβ	amyloid-β peptide
AβPP	amyloid-β precursor protein
BACE	β-secretase, β-site amyloid precursor protein cleaving enzyme 1
CDR-SB	Clinical Dementia Rating Sum of Boxes
CMAI	Cohen-Mansfield Agitation Inventory
CNS	central nervous system
Cox2	cyclooxygenase 2
CSF	cerebrospinal fluid
DHA	docosahexaenoic acid

DPP4	dipeptidyl peptidase-4
EPA	eicosapentaenoic acid
GLP1	glucagon-like peptide 1
HDL	high-density lipoprotein
HGT/MET	hepatocyte growth factors (HGT)/MET receptor system
HHV-6A	human herpes virus 6A
HHV-7	human herpes virus 7
hUCB-MSCs	human umbilical cord-derived, mesenchymal stem cells
IgG	immunoglobulin G
IL-1β	Interleukin-1β
KLH	keyhole limpet hemocyanin
LDL	low-density lipoprotein
LXR	liver X receptors
mAb	monoclonal antibody
MADCS-CGIC	the modified Alzheimer's Disease Cooperative Study-Clinical Global Impression of Change
MAPT	microtubule-associated protein tau
MCI	mild cognitive impairment
MMSE	Mini-Mental State Examination
NFT	neurofibrillary tangles
NGF	nerve growth factor
NMDA	*N*-methyl D-aspartate
PD	Parkinson's disease
PI3K	phosphatidylinositol 3-kinase
PPAR-γ	peroxisome proliferator-activated receptor γ
PSEN1	presenilin 1
PSEN2	presenilin 2
PSP	progressive supranuclear palsy
RAGE	advanced glycation end product-specific receptor
RXR	retinoid X receptors
SSRI	selective serotonin reuptake inhibitor
T2DM	type 2 diabetes mellitus
TNF-α	tumor necrosis factor-α
VLDL	very low-density lipoprotein

1. Introduction

Alzheimer's disease (AD) is a chronic, progressive, and irreversible brain disorder that mainly affects the cerebral cortex and hippocampus causing synaptic loss and neuronal failure. Alzheimer's disease is the most common cause of dementia among the elderly, accounting for around 60–80% of dementia cases. Consequently, patients affected by AD and related dementia present a continuous decline in cognitive functioning, thinking, memory, behavioral, and social skills (Alzheimer's Association, 2019).

In 2020, close to 6 million Americans suffer from dementia, a number that will increase to 8.4 million by 2030 and up to 13.8 million by 2050. These estimates contemplate that the overall population age of 65 years and older will continue to increase in the upcoming years in most countries (Alzheimer's Association, 2019; Pakpoor and Goldacre, 2017). Today, AD and related dementia is considered the sixth leading cause of death in the United States, but it could be categorized only behind cancer and heart disease as a cause of death in the elderly. Aging of our population will not only increase the number of AD patients, but also will have a tremendous impact on public and private health institutions with a burgeoning economic cost resulting from the substantial support required to address the disability, hospitalizations, and constant personal care demented patients require to meet their basic needs. If this trend continues over the coming years, we will face a global pandemic of chronic neurological diseases with social and economic disruption.

The development of AD occurs as a combination of aging, genetics, lifestyle, and environmental conditions that trigger complex pathogenic mechanisms in the brain, even though the specific or real cause of AD is not completely understood. The course of the disease is accompanied by neuropathological changes or histopathological hallmarks, notably accumulation of abnormal protein deposits of amyloid-β (Aβ) senile plaques, neurofibrillary tangles (NFT) of hyperphosphorylated tau, neuroinflammation, synaptic loss, and neuronal cell death. Currently available treatments for AD only ameliorate or attempt to maintain neuronal functions, manage symptoms, or slow down memory loss by regulating neurotransmitters. The cholinesterase inhibitors donepezil, rivastigmine, and galantamine are prescribed to treat mild-to-moderate AD, whereas memantine, a *N*-methyl D-aspartate (NMDA) antagonist, and the combination of memantine and donepezil are prescribed to treat moderate-to-severe AD (NIA, 2020). However, none of these drugs can stop, reverse, or prevent neurodegeneration. A nonregulated dietary supplement, Souvenaid, is also marketed to improve synaptic function with protective effects against Aβ in mild cognitive impairment (MCI) and AD (Cummings et al., 2019b). This product contains a mixture of omega fatty acids, antioxidants, cofactors, vitamins, and minerals.

The therapeutic compounds tested for AD and MCI in the last decades have targeted Aβ, secretases, tau, neuroinflammation, neurotransmitters, and neuroprotection. Unfortunately, around 106 drug candidates tested in more than 200 preclinical and clinical trials have all failed to achieve improved cognitive outcomes or to demonstrate significant benefits (Table 1 and Fig. 1).

Table 1 Therapeutic compounds approved andtrials terminated for Alzheimer's disease.

Approved by FDA	Preclinical/clinical trials terminated	
Neurotransmission • Donepezil • Galantamine • Rivastigmine • Memantine • Memantine + Donapezil • Tacrine (discontinued in 1996) **Not regulated Neuroprotective/ metabolic** • Souvenaid	**Aβ immunotherapy** • AAB-003 (PF-05236812) • AN-1792 (AIP 001) • Amilomotide (CAD106) • Bapineuzumab (AAB-001) • GSK933776 • Gammagard (intravenous IVIg) • LY2599666 • Ponezumab • Venutidecrudificar (ACC-001, PF-05236806) **Aβ aggregation** • Alzhemed (Tramiprosate, Vivimind, NC-531, Homotaurine) • Clioquinol (PBT-1, iodochlorhydroxyquin) • ELND005 (AZD-103, scylloinositol) **Tau immunotherapy** • RG7345 **Tau aggregation** • Davunetide (NAP, AL-108) • Epothilone D (BMS-241027) • Methylene blue (Rember TM, TRx-0014) • TPI 287 • Tideglusib (NP031112, Nypta, Zentylor) **β-Secretase (BACE)** • Atabecestat (JNJ-54861911) • BI 1181181 (VTP 37948) • Elenbecestat (E2609) • LY2886721 • LY3202626 • Lanabecestat (AZD3293, LY3314814) • PF-06751979 • RG7129 • Verubecestat (MK-8931, MK-8931-009)	**Neurotransmission** • ABT-418 • ABT-089 • ABT-288 • AF 102B • AZD1446 (TC-6683) • AZD3480 (TC-1734) • Atomoxetine (ATX, Strattera) • BI 409306 (SUB 166499) • Besipirdine • CX516 • Dimebon (Dimebolin, Latrepirdine, Pf-01913539) • Encenicline (EVP-6124, MT-4666) • Eptastigmine (MF 201) • GSK239512 • Huperzine A (ZT-1, DEBIO 9902) • Idalopirdine • Intepirdine (RVT-101, SB 742457, GSK 742457) • LU25-109 • Ladostigil (TV3326) • MK-7622 • MKC-231 • Metrifonate • Milameline • NGX267 • NS2330 • Nefiracetam • Nelonicline • Neramexane • PF-05212377 • PF-06852231 • PRX-03140 • Phenserine • Physostigmine salicylate • RG3487 • S38093 • S47445 (CX1632) • SB202026 (Memric)

Continued

Table 1 Therapeutic compounds approved andtrials terminated for Alzheimer's disease.—cont'd

Approved by FDA	Preclinical/clinical trials terminated	
	γ-Secretase • Avagacestat (BMS-708163) • CHF 5074 • Flurizan (tarenflurbil, r-flurbiprofen, MPC-7869) • PF-06648671 • Samagacestat (LY450139) **Inflammation** • Celecoxib (Celebrex) • Dapsone (Avlosulfon, DDS, diamino-diphenyl sulfone) • HF0220 • Ibuprofen • Lornoxicam • Minocycline (Solodyn, Arestin, Minocin, Dynacin) • Naproxen • Prednisone • Rofecoxib • Young Plasma **Other** • CERE-110 (Gene therapy) • COGNIShunt (CSF shunts)	• SGS-742 (CGP-36742, DVD-742) • Sabeluzole (R58735) • Sembragiline • Suritozole • Valproate • Varenicline • Xaliproden **Neuroprotective/ metabolic** • ABT-384 • Acetyl-L-carnitine • Atorvastatin (Lipitor, Zarator, Sortis, Tahor) • Edonerpic (T-817 MA, T-817) • Estrogen (Premarin) • Exendin-4 (Exenatide, Byetta, Bydureon) • Idebenone (Catene, Sovrima) • Leuprolide • Linopirdine • MEM 1003 (BAY Z 4406) • NeoTrofin (AIT-082, leteprinim) • Nilvadipine (Nilvad, Nivadil, ARC029) • PF-05251749 • Propentofylline • Rosiglitazone (Avandia) • ST101 (ZSET1446) • Saracatinib (AZD0530)

Some of the drug candidates showed considerable side effects in AD and MCI patients and were subsequently abandoned. Nine of these compounds targeted Aβ removal and three inhibited its aggregation. Six compounds targeted tau, nine compounds were directed to β-secretase (β-site amyloid

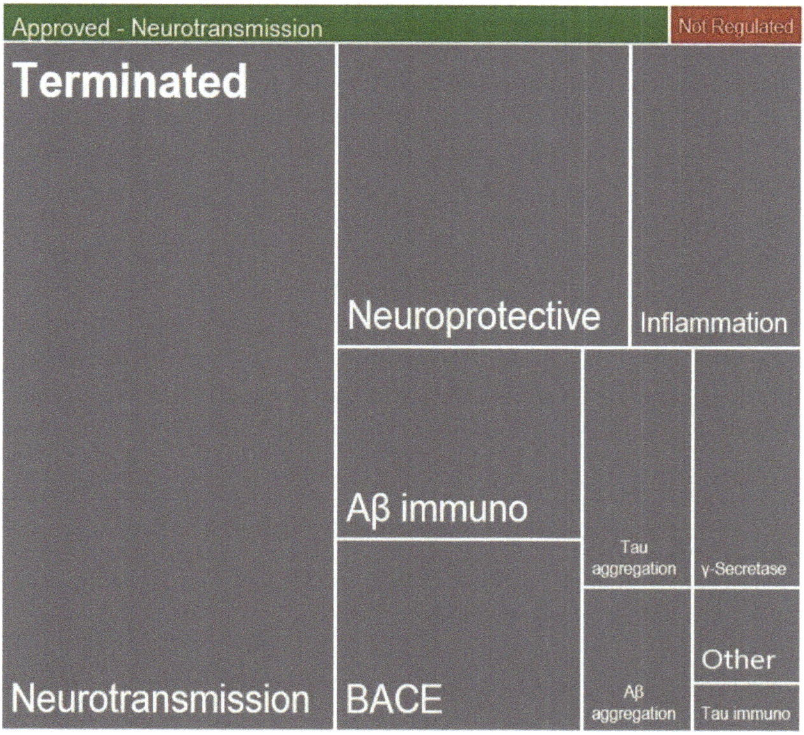

Fig. 1 Treemap of AD drugs approved and preclinical/clinical trials terminated. Hierarchical display of AD therapeutic compounds approved, not regulated, and terminated, classified as Aβ immunotherapy, Aβ aggregation, tau immunotherapy, tau aggregation, α-secretase, β-secretase (BACE), γ-secretase, neurotransmission, neuroprotective/metabolic, inflammation, antimicrobial, and other. Data from Table 1.

precursor protein cleaving enzyme 1, BACE), and five to γ-secretase. These two enzymes are critical in cleavage of amyloid-β precursor protein (AβPP) to produce the neurotoxic Aβ peptide. In all, 10 formulations were tested for neuroinflammation, 17 compounds were examined as neuroprotective agents, but the majority of these trials (44 compounds) focused on improving neurotransmission. Other approaches explored included one gene therapy and one cerebrospinal (CSF) shunt trial (Table 1 and Fig. 1). The combined failure of these efforts represent decades of research and several billion dollars of investment from public and private organizations. As our population ages, the incidence of AD and related dementia will grow further to exacerbate the crisis of chronic neurological diseases.

Herein, we present the current status and directions of therapeutic compounds that target AD that are currently being tested in Phase III, II, and I clinical trials. Overall, at least 147 therapeutic compounds are in clinical trials for AD (Table 2) retrieved from current literature (Cummings et al., 2019a; Morgan, 2019) and databases (alzforum.org and clinicaltrials.gov). Of these, 33 are in Phase III, 82 in Phase II, and 32 in Phase I (Table 2). Fig. 2 shows the relative hierarchical distribution of the AD trials classified as Aβ immunotherapy, Aβ aggregation, tau immunotherapy, tau aggregation, α-secretase, β-secretase (BACE), γ-secretase, neurotransmission, neuroprotective/metabolic, inflammation, antimicrobial, and other. Overall, the main therapeutic targets are neuroprotective/metabolic (47 compounds), neurotransmission (23 drugs) and Aβ (27 compounds), followed by tau (14), neuroinflammation (12), and finally secretases (5).

2. Current status of amyloid-β clinical trials

The accumulation and deposition of Aβ aggregates as the primary cause of AD has been the dominant causative theory for almost 30 years (Kametani and Hasegawa, 2018; Selkoe and Hardy, 2016). Amyloid-β is a short peptide of 40–42 amino acids, overproduced by sequential cleavages of transmembrane AβPP by β-secretase (BACE) and γ-secretase complex (presenilin 1, nicastrin, APH-1, and presenilin enhancer 2) and then released outside the neurons. Normally, Aβ is degraded and removed, but in aged, AD-affected subjects, Aβ accumulates to promote the formation of oligomers, fibers, and ultimately pathogenic clumps called amyloid plaques.

Efforts in the development of AD therapeutic compounds have focused mostly on upstream components of the amyloid cascade, with the premise that inhibition or clearance of amyloid plaques will lead to improvement of symptoms and restoration of cognitive functions. Recent results presented from trials evaluating immunotherapies and drugs designed to target Aβ have shown largely disappointing outcomes. In some clinical trials, Aβ aggregates decreased, but no improvement in clinical outcomes measures resulted, and in some cases, even drastic side effects were observed causing cessation of trials.

2.1 Anti-amyloid-β immunotherapy

The immunotherapies that target Aβ peptide are classified as active or passive. Active immunotherapies are monoclonal or polyclonal antibodies designed to target specific regions of Aβ to promote its functional inhibition

Table 2 Therapeutic compounds in clinical trials for Alzheimer's disease.

Phase III	Phase II	Phase I
Aβ immunotherapy	**Aβ immunotherapy**	**Aβ immunotherapy**
• Aducanumab	• ABvac40	• Aducanumab
• BAN2401	• Aducanumab	• Crenezumab
• Crenezumab	• BAN2401	• Lu AF20513
• Gantenerumab	• Crenezumab	• LY3002813
• Solanezumab	• GV1001	• LY3372993
• Plasma exchange	• LY3002813	• NPT088
• CAD106-CNP520	• Octagam 10%	
	• UB-311	**Aβ aggregation**
Tau aggregation		• CT1812
• ANAVEX2-73	**Aβ aggregation**	• Efavirenz
• TRx0237 (methylene blue)	• ANAVEX 2-73	
	• AR1001	**Tau immunotherapy**
	• CT1812	• BIIB076
β-Secretase (BACE)	• Neurostem	• JNJ-63733657
• CNP520	• Nilotinib	• LY33003560
• Elenbecestat (E2609)	• Posiphen	
	• Sargramostim	**Tau aggregation**
Neurotransmission	• Thiethylperazine (TEP)	• TPI-287
• AVP-786		
• AXS-05		**Neurotransmission**
• Bexpiprazole	**Tau immunotherapy**	• Escitalopram-Venlafaxine
• Escitalopram	• AADvac1	
• Guanfacine	• ABBV-8E12	• THN201
• Methylphenidate	• BIIB092	
• Mirtazapine	• BIIB080	**Neuroprotective/metabolic**
• Nabilone	• LY33003560	
• Octohydroaminoacridine succinate	• RO7105705	• Allopregnanolone
		• ALZ-801
• Zolpidem	**Tau aggregation**	• BDPP
	• Methylene blue	• Dabigatran
Neuroprotective/metabolic	• Nicotinamide	• DNL747
		• Intranasalinsulin
• AGB101	**α-Secretase**	• J147
• BHV-4157	• APH-1105	• Telmisartan
• DHA	• ID1201	• Vorinostat
• Donepezil (transdermal)		
• Gingko biloba	**β-Secretase (BACE)**	**Inflammation**
• Icosapent ethyl (IPE)	• Elenbecestat (E2609)	• AL002
		• AL003
Inflammation	**Neurotransmission**	• Salsalate
• ALZT-OP1a	• AD35	
• COR388	• BI425809	

Continued

Table 2 Therapeutic compounds in clinical trials for Alzheimer's disease.—cont'd

Phase III	Phase II	Phase I
• GV-971 (sodium oligo-mannurate) • Losartan-Amlodipine • Masitinib **Antimicrobial** • COR388	• DAOI • Dronabinol • Lemborexant • Lithium • Nicotine • Pimavanserin • Piromelatine • Prazosin • SUVN-502 **Neuroprotective/ metabolic** • AMX0035 • Benfotiamine • Bryostatin • Cardesartan • CERE-110 • Cilostazol • Curcumin • Dapagliflozin • Deferiprone • DHA • DHP1401 • Formoterol • Grapeseedextract • Intranasalinsulin • Levetiracetam • Liraglutide • LM11A-31-BHS • Lupron • MLC901 • MMFS-205-SR • MP-101 • NA-831 • Omega 3 PUFA • ORY-2001 • Perindopril • Rasagiline • Riluzole • RPh201 • Serine • S-Equol • Telmisartan-Perindopril • Xanamem	**Antimicrobial** • COR388 **Other** • AAVrh.10hAPOE2 (Gene therapy) • CDK-355 (undisclosed) • hMSCs (Stem cell therapy) • MK-4334 (undisclosed) • NDX-1017 (Neuroregeneration)

Table 2 Therapeutic compounds in clinical trials for Alzheimer's disease.—cont'd

Phase III	Phase II	Phase I
	Inflammation • BPN14770 • Elderberry juice • GRF6019 • Montelukast • Neflamapimod • PTI-125	
	Antimicrobial • Rifaximin • Valacyclovir	
	Other • AstroStem (stem cell therapy) • BAC (undisclosed) • hUCB-MSCs (stem cell therapy) • GB301 (regulatory T cells)	

The therapeutic compounds or drugs for AD were classified according to their clinical trial status and mechanism of action.

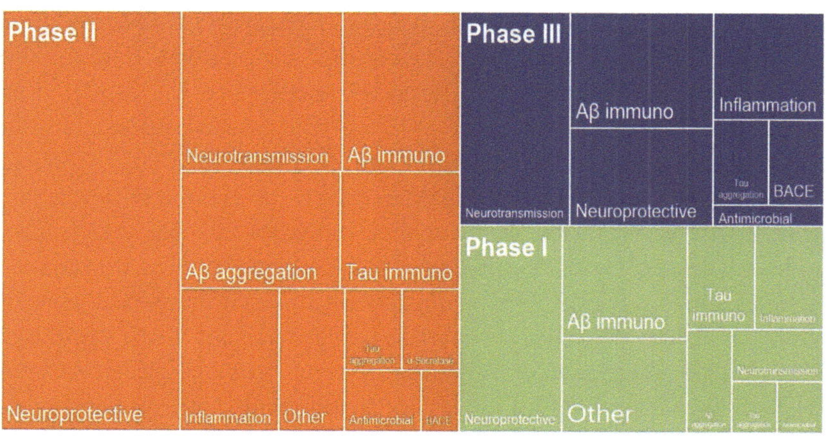

Fig. 2 Treemap of AD clinical trials. Hierarchical display of Phase III, Phase II, and Phase I clinical trials classified as Aβ immunotherapy, Aβ aggregation, tau immunotherapy, tau aggregation, α-secretase, β-secretase (BACE), γ-secretase, neurotransmission, neuroprotective/metabolic, inflammation, antimicrobial, and other. Data from Table 2.

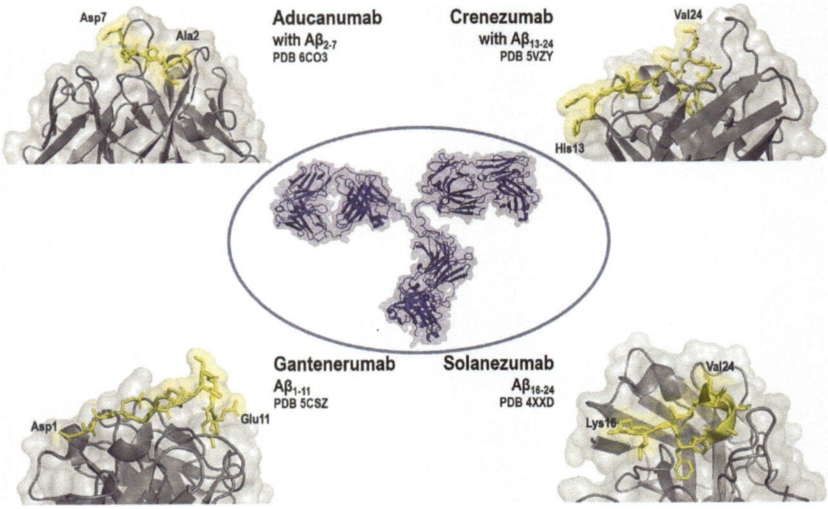

Fig. 3 Immunotherapy anti-Aβ. Structures of therapeutic antibodies that recognize Aβ peptide in different regions. Aducanumab with $A\beta_{2-7}$ (PDB 6CO3), crenezumab with $A\beta_{13-24}$ (PDB 5VZY), gantenerumab with $A\beta_{1-11}$ (PDB 5CSZ), and solanezumab with $A\beta_{16-24}$ (PDB 4XXD).

and clearance (Fig. 3). Currently, there are seven anti-Aβ compounds in Phase III, eight in Phase II, and in another six in Phase I.

Aducanumab (or BIIB037) is an IgG1 monoclonal antibody (mAb) that recognizes aggregated forms of Aβ, including epitopes in soluble and insoluble oligomers (Fig. 3 shows the structure of aducanumab resolved with $A\beta_{2-7}$ fragment, PDB 6CO3) (Budd Haeberlein et al., 2017). Trials showed that intravenous doses of 1–10 mg/day reduced Aβ plaques after 12 months of treatment. Results from Phase III trials (NCT02477800 and NCT02484547) were presented in March 2019 by Biogen indicating that this mAb had failed futility and outcome measures and that the company would end ongoing trials. Surprisingly, in October 2019, it was announced that after reanalysis of data, this company would pursue the FDA approval (Schneider, 2020), as well as an extension of the trial. In July 2020, Biogen completed the submission of a Biologics Licence Application (BLA) to the FDA for approval of aducanumab, as investigational treatment for AD (Biogen News Releases, 2020). Despite the effective reduction of Aβ by aducanumab in AD patients, there are still questions regarding its efficacy on clinical and cognitive effects, especially in *APOEε4* (apolipoprotein E allele ε4) carriers. This gene variant is associated with increased risk for late-onset AD. A complementary Phase II/I trial of aducanumab (BIIB037) in individuals with prodromal and mild AD was terminated based on a futility analysis of Phase III trials (NCT03639987, NCT01677572).

Crenezumab (MABT5102A or RG7412) is a mAb that recognizes aggregated Aβ (Fig. 3 shows the structure of crenezumab resolved with $A\beta_{13-24}$ fragment, PDB 5VZY). Phase II trials evaluating subcutaneous (300 mg every 2 weeks) or intravenous (15 mg/kg every 4 weeks) administration of crenezumab showed a significant decrease in CSF Aβ, but without significant improvement in cognitive assessment as determined by the Alzheimer's Disease Assessment Scale cognitive subscale (ADAS-Cog) and Clinical Dementia Rating Sum of Boxes (CDR-SB) (Cummings et al., 2018; Salloway et al., 2018). High-dose Phase III trials in prodromal-to-mild sporadic AD were discontinued in January 2019 by Roche/Genentech due to a lack of efficacy since primary end points were not met (NCT02670083 and NCT03114657). A Phase II trial evaluating crenezumab in PSEN1 E280A carriers of the familial early-onset mutation is still active (NCT01998841), as is a Phase I trial in mild-to-moderate AD (NCT02353598).

Gantenerumab (RO4909832 or RG1450) is a mAb designed to bind conformational epitopes of aggregated Aβ fibrils in (Fig. 3 shows the structure of gantenerumab resolved with $A\beta_{1-11}$ fragment, PDB 5CSZ). Gantenerumab treatment (105 or 225 mg) every 4 weeks showed a dose-dependent reduction of Aβ deposits after the 6-month trial (Klein et al., 2019). The clearance of Aβ was related to microglia-mediated phagocytosis. The Phase III trial by Hoffmann-La Roche of gantenerumab in 3089 prodromal or mild AD patients was halted early for futility since no differences in end points were observed; however, some dose-dependent effects in AD biomarkers were observed (NCT01224106) (Ostrowitzki et al., 2017). Complementary Phase III trials (NCT03444870 and NCT03443973) are active and recruiting participants.

Solanezumab (LY2062430) is an immunoglobulin G1 (IgG1) that recognizes the mid-domain of soluble Aβ monomers (Fig. 3 shows the structure of solanezumab resolved with $A\beta_{16-24}$ fragment, PDB 4XXD). A Phase II trial showed the safety and tolerability of 100–1600 mg doses of solanezumab administered monthly for 12 weeks. From a total of 2129 patients with mild dementia, 1057 received 400 mg intravenously every 4 weeks for 76 weeks and were tested in the ADAS-cog. The trial showed no significant differences in primary outcomes or effect on cognitive decline (NCT01900665) (Honig et al., 2018). A sporadic adverse effect of cerebral edema or effusion lesions was observed on magnetic resonance imaging (MRI) scans.

The most recent news concerning gantenerumab and solanezumab was from studies in individuals with dominantly inherited familial AD (AβPP or

PSE1 mutations) reported by the media in February 2020. Trials run by Eli Lilly, Roche/Genentech, Washington University in St. Louis, the National Institutes of Health (NIH) and Alzheimer's Association showed that these immunotherapies did not slow down or stop cognitive decline in study participants (NCT01760005). Additional analysis of the results of these trials are ongoing and will be presented in official reports soon.

BAN2401 (mAb158) is a humanized IgG1 that binds to Aβ fibrils. Phase I/II trials with mAb BAN2401 in mild-to-moderate AD showed tolerability of 0.1–10 mg/kg BAN2401 biweekly for 4 months, but no clear effectiveness on AD biomarkers or cognitive benefits. Only the high dose was able to reduce Aβ accumulation (Logovinsky et al., 2016). The Phase III study (NCT03887455) aims to confirm the safety and efficacy of BAN2401 (10 mg/kg biweekly over 18 months); it is currently recruiting up to 1566 individuals with early AD and will continue until 2024.

LY3002813 (Donanemab or N3pG) is a mAb that targets Aβp3–42, an Aβ peptide variant with a residue modification (pyroglutamate) present in amyloid plaques. A Phase I study evaluated the intravenous administration of LY3002813 (0.1–10 mg/kg monthly for 4 months), where the highest dose of LY3002813 was associated with an Aβ reduction. The complementary Phase I trial testing doses of 10–40 mg/kg of LY3002813 confirmed a significant reduction of Aβ deposits. The ongoing Phase II trial from Eli Lilly aims to test safety, tolerability, and efficacy of LY3002813 in 266 early symptomatic AD subjects and is due to complete in 2021 (NCT03367403) (Irizarry et al., 2018). *LY3372993* is another anti-Aβ antibody candidate from Eli Lilly that recently completed a Phase I trial in healthy volunteers aged 18–85 years that evaluated its safety and tolerability (NCT03720548). The results of this study have not been published.

NPT088 is a fusion protein of g3p and IgG_1-Fc that binds to and remodels misfolded amyloid aggregates. A Phase I study sponsored by Proclara Biosciences and the Alzheimer's Association has demonstrated safety and tolerability after administration of NPT088 at four dose levels for 6 months, but no effects on plaques and tangles were observed (NCT03008161) (Michelson et al., 2019).

Passive immunotherapies are intravenous immunoglobulin preparations containing polyclonal antibodies that might recognize Aβ or confer an overall immune response, or albumin formulations administered intravenously that could bind to circulating Aβ, promoting its clearance.

Plasma exchange. A plasmapheresis study by Grifols is evaluating a 5–20% albumin solution with or without 5% immunoglobulin in AD on the premise that this treatment will modify the cognitive, functional, and behavioral aspects of individuals with AD (NCT01561053). Currently in Phase III, the plasma exchange study involves periodically extracting plasma (containing circulating Aβ bound to albumin) and infusing the individuals with a fresh albumin-immunoglobulin solution. This study will evaluate treatment efficacy using neuropsychological tests to measure cognitive function of individuals in the study.

Octagam 10% is an intravenous immunoglobulin formulation designed as a neuroprotective polyclonal Ab against Aβ. A Phase II trial evaluated Octagam 10% in individuals with mild-to-moderate AD aged 50–85 years (NCT00812565). Acceptable safety, but no conclusive evidence of benefit for cognition or function was shown, especially when compared with previous Phase I trials (Dodel et al., 2013). The next trial to assess Octagam 10% will focus on cognitive and functional measures in amnestic MCI and AD patients (NCT01300728).

2.2 Amyloid-β vaccines

CAD106 (or Amilomotide) is an $A\beta_{1-6}$ vaccine. Long-term treatment with CAD106 (150 μg administered as three subcutaneous/intramuscular injections at week 0, 2, and 6, and followed by four injections at 12-week intervals) showed favorable tolerability and production of Aβ-specific antibodies in serum and cerebrospinal fluid (Farlow et al., 2015). Phase II/III trials NCT02565511 of CAD106 and CNP520 (a BACE inhibitor) are assessing the potential of these two therapeutic compounds administered separately [CAD (450 μg) or CNP520 (50 mg)] to slow down the onset and progression of clinical symptoms of AD. Novartis is currently recruiting 480 individuals to this interventional study that is scheduled for completion in 2025.

ABvac40. This vaccine candidate was designed to have multiple repeats of the C-terminal domain of $A\beta_{33-40}$. ABvac40 showed safety and tolerability in 16 patients who received two or three administrations of 200 μg of $A\beta_{33-40}$ peptide coupled to monomeric keyhole limpet hemocyanin (KLH), thereby eliciting a specific immune response to Aβ40 (Lacosta et al., 2018). The follow-up Phase II trial will test the safety and immunogenicity of ABvac40 in MCI and AD patients (NCT03461276).

GV1001 (RIAVAXTM or Tertomotide) is a vaccine peptide based on a peptide fragment of telomerase reverse transcriptase designed to activate the immune system. In AD animal models, this compound blocked Aβ accumulation and exerted antioxidant and neuroprotective properties in primary rat neural stem cells (Park et al., 2014a). A Phase II trial in Korea that tested the safety and efficacy of GV1001 (0.56–1.12 mg, biweekly for 24 weeks) in mild-to-moderate AD showed promising results and a complementary trial is ongoing and recruiting participants in the United States (NCT03184467 and NCT03959553).

UB-311 is a vaccine molecule of $A\beta_{1-14}$ peptide fragment coupled to a T-cell epitope. Preclinical studies in animal models showed that it was associated with production of anti-Aβ antibodies, which neutralized Aβ and induced clearance of amyloid plaques by T-cells (Wang et al., 2007). Its safety, tolerability, and immunogenicity were shown in a small cohort with mild-to-moderate AD subjects. The follow-up Phase II trial will evaluate its impact on cognitive, functional, and neurological outcome measures (NCT02551809).

LuAF20513 consists of three repeats of $A\beta_{1-12}$ coupled with tetanus toxin. Two Phase I trials have evaluated the safety and tolerability of this AD vaccine candidate (NCT03819699 and NCT02388152) and preliminary results indicate that LuAF20513 induces robust T-cell responses and the production of anti-Aβ antibody (Davtyan et al., 2013).

2.3 Inhibition of amyloid-β aggregation

The compounds in this subcategory act by blocking the aggregation of Aβ peptides into dimers, oligomers, fibrils, and finally plaques. Currently, eight compounds are being evaluated in Phase II studies, two in Phase I, and three had been evaluated, but have now been terminated (Tables 1 and 2). Their mechanisms of action are diverse, but generally these compounds are thought to inhibit Aβ aggregation and promote its processing before it can form pathogenic amyloid plaques that subsequently cause synaptic loss and neuronal death.

The small molecule drugs in this category are usually administered orally, for example, *ANAVEX 2–73*, which activates the sigma-1 receptor and *CT1812* (aka Elayta), a sigma-2 receptor activator. Sigma receptors play an important role in the progression of AD by regulating neuronal signal transduction, oxidative stress, neuron plasticity, and neurotransmitter release. *AR1001* is a phosphodiesterase inhibitor, *Nilotinib* (aka Tasigna) is a small molecule drug approved for treating chronic myelogenous leukemia, *Thiethylperazine* is a small molecule antiemetic drug, and *Posiphen* is a small molecule drug acting via inhibition of AβPP translation. *Efavirenz* is a repurposed, antiretroviral agent that may increase cholesterol processing, as well as amyloid processing. The second group of inhibitors or blockers of amyloid aggregation are active proteins, such as *Neurostem*, based on a protein secreted from umbilical cord blood-derived mesenchymal stem cells, and *Sargramostim*, a granulocyte-macrophage colony-stimulating factor, also known as GM-CSF or leukine.

3. Current status of clinical trials evaluating secretases

Secretases are proteases directly involved in the sequential processing of AβPP in the membrane of neurons. These enzymes are critical downstream in the Aβ cascade and direct AβPP into either non-amyloidogenic or amyloidogenic pathways. The non-amyloidogenic pathway involves the cleavage of AβPP by α-secretase followed by γ-secretase, whereas the amyloidogenic pathway involves the cleavage of AβPP first by β-secretase and then by γ-secretase, resulting in Aβ peptide production and its aggregation, fibrillation, and deposition in amyloid plaques (Bozyczko-Coyne and Williams, 2007). Secretase inhibitors or modulators are small drugs designed to selectively target these enzymes with the objective of reducing the amyloidogenic pathway or deviate AβPP into the non-amyloidogenic pathway (Fig. 4).

3.1 α-Secretase

The α-secretase cleaves AβPP between amino acids 16 and 17 of the Aβ peptide region, preventing the production and aggregation of Aβ (De Strooper et al., 2010). Currently, there are only two active trials for α-secretase modulators. The α-secretase modulator *APH-1105* has reached the Phase II study stage in patients with MCI due to AD. An interventional study with 60 individuals with AD will evaluate the safety and tolerability of APH-1105 at doses of 0.5, 1.0, or 2.0 μg administered intranasally, twice weekly for 12 weeks (NCT03806478).

ID1201 is an ethanolic extract of *Meliae toosendan* (the China berry fruit), which has been shown to enhance non-amyloidogenic AβPP processing, resulting from α-secretase activation through the phosphatidylinositol 3-kinase (PI3K)/protein kinase B (Akt) pathway, Aβ reduction, inhibition of the receptor for advanced glycation end products (RAGE) implicated in inflammation, neuroprotection, and anti-inflammatory activities (Park et al., 2014b). A Phase II trial in 116 individuals with mild AD aged 55–85 years evaluating the safety and efficacy of 100, 200, or 400 mg doses of oral ID1201 has been completed, although no results are available yet (NCT03363269).

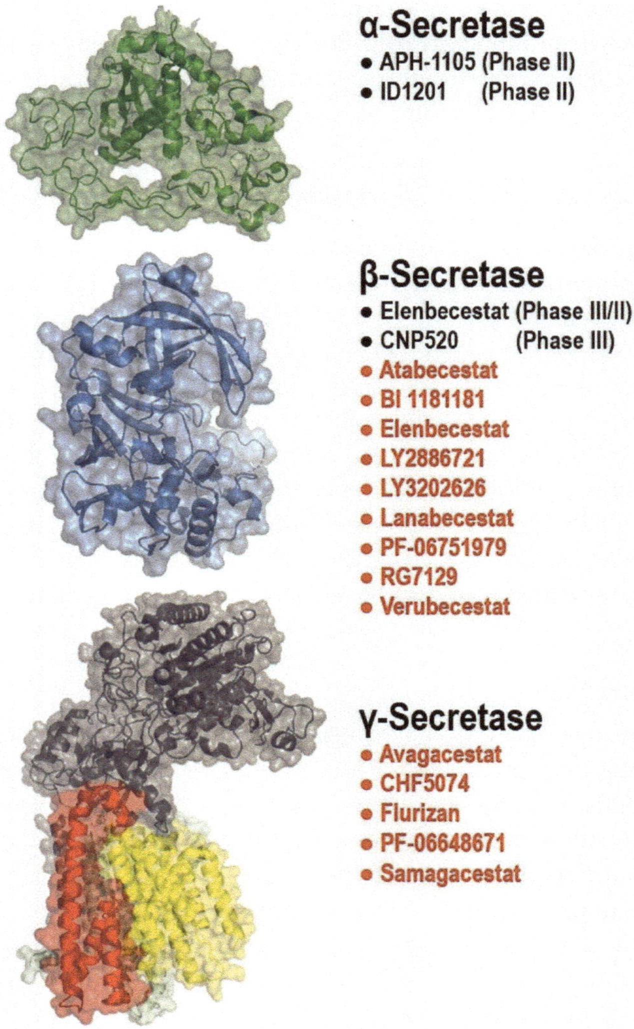

Fig. 4 Therapeutic compounds targeting α/β/γ-secretases. Structures of α/β/γ-secretases and status of corresponding inhibitors/modulators are shown; black text indicates compounds active in clinical trials, whereas red text indicates compounds with clinical trials terminated. α-Secretase ADAM10 extracellular domain (PDB 6Be6), β-secretase (PDB 6EJ2), and γ-secretase complex with PSEN1 domain (yellow) and PSEN2 domain (red) (PDB 5A63).

Acitretin is a retinoic acid receptor agonist. This compound increases the expression of ADAM-10 (disintegrin and metalloprotease), which has α-secretase activity and promotes the non-amyloidogenic processing of AβPP. Acitretin is commercialized as Soriatane, Neotigason, or RO-101670 for psoriasis. *In vitro* assays with neuronal cells overexpressing AβPP695 showed that acitretin stimulated α-secretase ADAM-10 activity, and its administration to APP/PS1–21 mice reduced $A\beta_{40/42}$ (Tippmann et al., 2009). A Phase II study ended in 2011 (NCT01078168) in which individuals with mild-to-moderate AD were treated orally with 30 mg/day for 4 weeks and showed an increase in CSF AβPPs-α levels (Endres et al., 2014). ADAM-10 is expressed in the brain, epethelia, liver, and immune system and its inhibition may result in side effects when using therapeutic compounds targeting its expression or activity.

Etazolate (EHT0202) is an α-secretase activator and modulator of the GABA receptor. In a model of closed-head injury in Swiss mice, a dose of 1, 3, or 10 mg/kg of etazolate was associated with an anti-inflammatory effect, memory improvement, and restoration of sAβPPα levels (Siopi et al., 2013). In a rat model, etazolate (20 nM–2 μM) showed dose-dependent neuroprotection against fibrillar Aβ (Marcade et al., 2008), and in aged rats, etazolate prevented age-related cognitive deficits, including a reduction in memory upon oral administration of up to 2.37 mg/kg (0.5 mL/kg for 7 days) (Drott et al., 2010). A pilot, Phase IIa study with 159 mild-to-moderate individuals with AD evaluated 3-month treatment with a 40 or 80 mg/day dose of etazolate (NCT00880412). Overall, the compound showed acceptable safety and was generally well tolerated (side effects included depression and other alterations in behavior) with no significant efficacy in cognitive tests between treatment groups (Vellas et al., 2011).

Bryostatin-1 also stimulates α-secretase activity, as observed in transgenic APP(V717I) mice, causing a reduction in $A\beta_{40/42}$ and improved behavior outcomes (Etcheberrigaray et al., 2004). Although two Phase II trials with bryostatin-1 have been completed and another has been announced by Neurotrope Bioscience, no Phase III trial has been announced.

3.2 β-Secretase

The β-secretase 1 (BACE) is responsible for cleaving the N-terminus of AβPP between residues 671–672, after which the Aβ peptide is released upon additional processing by the γ-secretase. The rationale to inhibit BACE activity relies upon blocking the initial enzymatic cleavage of

AβPP that leads to the amyloidogenic pathway and deviate AβPP processing to the non-amyloidogenic route. Several natural and designed inhibitors of BACE have been tested as potential therapeutic drugs. In particular, a group of low-molecular-weight, highly potent, and brain-penetrating compounds (analogs of a cyclic amidine, such as isocytosines and dihydroisocytosines) showed good tolerability with long-term dosing (Zhu et al., 2010). Only two BACE modulators are still in clinical trials, elenbecestat in Phase II/III and CNP520 in Phase III.

Elenbecestat (aka E2609) is a small-drug BACE1 inhibitor developed by Biogen and Eisai that has reached Phase III clinical trials. Early preclinical studies with animal models and in pilot clinical trials showed that the administration of elenbecestat caused a reduction in CSF and plasma Aβ levels. Phase II trials evaluated 18-month treatment with elenbecestat at daily doses of 5, 15, or 50 mg in participants with early AD, and showed a decrease in CSF Aβ directly related to the concentration of elenbecestat (Lynch et al., 2018). An accompanying trial tested higher doses (5, 50, and 200 mg) and showed similar pharmacokinetic and pharmacodynamic results. Two multisite Phase III trials recruited a total of 2199 patients with MCI/prodromal AD to evaluate elenbecestat dosing of 50 mg daily for up to 24 months (NCT02956486 and NCT03036280). The clinical trials measuring cognitive and functional outcomes will continue in 2020. Recent publication of additional Phase II studies with elenbecestat characterized the safety, tolerability of the compound along with a reduction of Aβ, and reported improvement in cognitive measures in individuals with MCI or mild AD.

CNP520 (aka Umibecestat) is a designed molecule inhibitor of BACE that has been shown to be safe and well tolerated in animal models. Administration of CNP520 is associated with reduced Aβ in the brain and CSF in Sprague-Dawley rat (5 months old) and Beagle dog (11 months old) models, as well as reduced amyloid plaque formation in APP23 transgenic mice (Neumann et al., 2018). Long-term toxicology studies observed dose-limiting, nonspecific CNS effects in dogs and rats, delayed ovulation and reduced fecundity/fertility, and focal skeletal muscle atrophy in rats, changes that were reversible upon treatment withdrawal (Neumann et al., 2018). Pilot studies demonstrated that the treatment in healthy adults ($\geq$60 years old) was safe, tolerated, and associated with a dose-dependent reduction of CSF Aβ. One trial evaluated the safety and tolerability of CNP520 at a dose of 2, 10, 35, and 85 mg for 13 weeks (NCT02576639). Two pivotal Phase II trials assessed the efficacy and safety of CNP520 in a cognitively unimpaired population at risk for developing AD (*APOE4* carriers,

60–75 years old). Participants received 50 or 15–50 mg capsules (oral intake) for up to 13 weeks (Lopez Lopez et al., 2017, 2019). Ongoing Phase III trials are evaluating CNP520, with outcome measures that include changes in cognitive scales, Aβ and tau biomarkers, and protein deposition measured by brain scans in individuals with MCI or AD (NCT02565511 and NCT02565511).

Another eight BACE modulators have been tested in clinical trials (Table 1), but terminated due to lack of significant activity, not successfully reaching outcome measures, or unacceptable side effect profiles. These compounds are *Atabecestat (JNJ-54861911)*, *BI 1181181 (VTP37948)*, *LY2886721*, *LY3202626*, *Lanabecestat (AZD3293, LY3314814)*, *PF-06751979*, *RG7129*, and *Verubecestat (MK-8931, MK-8931-009)* (Burki, 2018; Panza et al., 2018).

3.3 γ-Secretase

This protein complex formed by presenilin 1 and 2 (PSEN1 and PSEN2) works with BACE to process Aβ peptide from transmembrane AβPP. The PSEN1 and PSEN2 proteins are closely related to AD pathogenesis because several mutations in these genes are associated with a higher risk of developing chronic neurodegeneration. Several drugs have been tested as candidates to modulate or inhibit γ-secretase activity, including *Avagacestat*, *Semagacestat*, *Begacestat*, *EVP-0962*, *PF-06648671*, *CHF-5074*, and *Flurizan*. Unfortunately, these trials have been halted due to the absence of significant effectiveness, not reaching successful outcome measures, or because side effects were observed during the course of treatment. Currently, there are no other γ-secretase inhibitors in advanced trials.

4. Current status of non-amyloid clinical trials

4.1 Tau

Tau is a microtubule-associated protein and its activity is regulated by phosphorylation by numerous protein kinases. Alterations in the normal phosphorylation mechanism result in tau dysfunction (altered microtubule dynamics, axonal transport, and neurite outgrowth), forming intracellular aggregates termed neurofibrillary tangles. In the AD brain, 3R and 4R tau isoforms undergo hyperphosphorylation thereby increasing their tendency to aggregate into NFT, the spread of which correlates with the cognitive and clinical symptoms of neurodegeneration (Kametani and Hasegawa, 2018).

Although NFT and Aβ plaques develop at different times and brain locations, the presence of both is required to confirm a neuropathological diagnosis of AD that indicates their progressive accumulation and spread during AD pathology (Braak and Del Tredici, 2014).

Immunotherapies targeting tau in Phase II include *ABBV-8E12* (aka Tilavonemab or C2N8E12), which is an anti-tau mAb designed to identify aggregated extracellular tau and block its spread in the brain. The initial trial of ABBV-8E12 showed it to be safe and well tolerated at doses of 2.5–50 mg/kg (administered as a solution for intravenous infusion) in 30 patients with progressive supranuclear palsy (PSP), a condition characterized by accumulation of tau in the brain (West et al., 2017). The ongoing Phase II trial (NCT02880956) will continue in a larger cohort of 453 subjects, and additional Phase II trials are in progress.

Gosuranemab (aka BIIB092, BMS-986169, IPN007) is an IgG4 anti-tau targeting the N-terminus of tau. After initial studies demonstrated the safety and tolerability of gosuranemab accompanied by a marked reduction of CSF tau in participants with diagnosis of PSP (Boxer et al., 2019), a Phase II trial by Biogen evaluated an intravenous 50 mg/mL infusion every 4 weeks for 48 weeks in 490 individuals diagnosed with probable or possible PSP (NCT03068468). The company announced in December 2019 that the primary efficacy end point of this Phase II trial was not met and further studies of gosuranemab would be stopped.

Zagotenemab (LY33003560) is an anti-tau mAb targeting soluble tau aggregates. After evaluation in healthy volunteers, a Phase II trial (NCT03518073) has enrolled 285 subjects with AD and results are expected in 2021.

Semorinemab (RO7105705, MTAU9937A, or RG6100) is an IgG4 anti-tau targeting extracellular tau. Phase II studies NCT03289143 and NCT03828747 will evaluate this mAb in patients with prodromal to mild or moderate AD and results are expected in 2022 and 2023, respectively. Three more anti-tau compounds are in Phase I trials (*BIIB076, JNJ-63733657,* and *LY33003560*), and an additional anti-tau mAb has been discontinued (*RG7345*).

AADvac1 is a vaccine candidate consisting of an N-terminal fragment of tau (294–305 residues) coupled to KLH. A Phase I trial evaluated AADvac1 for 72 weeks in mild-to-moderate AD patients, showing an antibody response against tau, but with several patients presenting with side effects (Novak et al., 2018). A Phase II clinical trial was recently completed (NCT02579252), but no results have been published yet.

Similar to the Aβ disruptors, several therapeutic candidates are under evaluation as inhibitors of tau aggregation. Here, the principle relies on avoidance of tau fibrillization into NFT and favoring its clearance pathway. Two compounds, ANAVEZ 2–73 and TRx0237, are in Phase III trials. *ANAVEX 2–73* is a small molecule activator of the sigma-1 receptor that is implicated in protein folding and oxidative stress, and the Phase III trial will end in 2021 (NCT03790709). *TRx0237* (aka leuco-methylthioninium bis-hydromethanesulfonate or LMTM) is a methylene blue formulation that functions as an inhibitor of tau aggregation. Phase II/III trials with 150–250 mg/day of TRx0237 completed in 2014 with inconclusive outcomes regarding effectiveness of this treatment (Wilcock et al., 2018). An additional trial is evaluatingTRx0237 at a dose of 8–16 mg/day twice daily in 450 participants with MCI or probable AD and will complete in 2022 (NCT03446001).

Nicotinamide or *vitamin B3* taken at high doses can reduce the phosphorylation of tau and restore cognition in 3xTg-AD mice (Green et al., 2008). Nicotinamide (750 mg twice daily) will be evaluated in a Phase II trial in 48 individuals with early AD (NCT03061474).

TPI-287 is a small molecule that stabilizes tubulin and tau-containing microtubules intracellularly, thereby improving cognition and neuronal activity (Zempel and Mandelkow, 2014). This compound underwent a Phase I evaluation of its pharmacokinetics and preliminary efficacy in AD (NCT01966666 and NCT02133846). A total of 68 participants [26 diagnosed with AD and 42 diagnosed with 4-repeat tauopathies (TRP)] received 2.0, 6.3 or 20 mg/m^2 of intravenous TPI-287 once every 3 weeks for up to 9 weeks. The results indicated that TPI-287 was less tolerated in individuals with AD than in those with TRP, because only the individuals that showed severe anaphylactoid reactions and dose-related worsening of dementia symptoms had AD (Tsai et al., 2019).

Finally, *BIIB080* (IONIS-MAPT$_{Rx}$, ISIS814907) is an antisense oligonucleotide that binds to mRNA of the microtubule-associated protein tau (MAPT) or tau. A Phase II trial will evaluate its safety, tolerability, pharmacokinetics, and pharmacodynamics when administered intrathecally every 4 weeks in individuals with mild AD (NCT03186989).

In summary, a total of six anti-tau compounds (one immunotherapy and five inhibitors of aggregation) have failed clinical trial evaluation and consequently additional clinical evaluation has been halted (Table 1).

4.2 Neuroinflammation

There are five therapeutic compounds in Phase III clinical trials that target neuroinflammation related to AD and related dementia.

ALZT-OP1 is a combination drug formulation that consists of cromolyn (used to treat asthma) and ibuprofen (a nonsteroidal anti-inflammatory or painkiller). APP Tg2576 mice were treated for 3 months with cromolyn (1–3.15 mg/kg), ibuprofen (0.5–2 mg/kg), or combined cromolyn + ibuprofen (1 + 0.5 mg or 3.15 + 2 mg, respectively). Cromolyn, alone or in combination with ibuprofen, almost abolished aggregation of Aβ40/42 due to microglial activation, but was associated with increased Aβ38 levels (Zhang et al., 2018). An ongoing trial with 620 subjects with evidence of early AD will determine whether the ALZT-OP1 formulation is able to slow down, arrest, or reverse the cognitive and functional decline in participants after 72 weeks of treatment. This trial is scheduled for completion at the end of 2020 (NCT02547818).

GV-971 is a formulation of oligosaccharides (sodium oligomannate) extracted from brown algae with anti-inflammatory properties and with an impact on gut bacteria. This compound received great attention from the scientific community and media because it was approved in China for mild-to-moderate AD at the end of 2019. During AD progression, gut dysbiosis occurs, leading to peripheral accumulation of phenylalanine and isoleucine that promote differentiation and proliferation of pro-inflammatory T helper 1 (TH-1) cells. Oligomannate remodels gut microbiota by suppressing dysbiosis and associated neuroinflammation, and inhibits the progression of AD related to immune cell infiltration and microglial activation in the brain (Wang et al., 2019). Approval was granted after a Phase III clinical trial (NCT02293915) reported that there was a statistical improvement in cognitive function in individuals with mild-to-moderate AD after GV-971 administration. Additional trials in different countries will be initiated to confirm these results and potentially also gain approval.

Losartan and *amlodipine* are both antihypertensive drugs with different mechanisms of action. Losartan is an angiotensin receptor blocker and amlodipine a calcium channel blocker. Although it is still inconclusive whether control of hypertension will reduce Aβ accumulation and improve cognitive function in AD, the aim of this losartan-amlodipine combination formulation is to reduce or control blood pressure, which may subsequently lower brain Aβ accumulation in the cohort group (55–79 years old) with hypertension, but excluding patients with cognitive impairment or dementia. The Phase II/III trial is ongoing (NCT03354143).

Masitinib is a selective, oral tyrosine kinase inhibitor with neuroprotective effects and the ability to modulate inflammation, and is used as an adjunct therapy to standard AD care (Piette et al., 2011). In a Phase III trial, masitinib was administered as add-on therapy to approved treatments (donepezil, rivastigmine, galantamine, and/or memantine) in individuals diagnosed with AD and the results are expected soon (NCT01872598).

Another six therapeutic approaches targeting neuroinflammation are under evaluation in Phase II: *BPN14770*, a small molecule, allosteric inhibitor of phosphodiesterase-4D that is involved in cognition and memory; *elderberry juice*, which has shown significant anti-inflammatory and antioxidant effects in individuals with MCI; *GRF6019*, a proprietary human plasma protein fraction (composed of about 400 proteins) with anti-inflammatory properties; *montelukast*, a corticosteroid that modulates inflammatory responses by inhibiting the leukotriene receptor and indicated for asthma and allergies; *neflamapimod*, an inhibitor of pro-inflammatory kinases, such as tumor necrosis factor (TNF-α) and Interleukin-1β (IL-1β); and *PTI-125*, a small drug that targets misfolded filamin A, which is associated with Aβ-induced tau hyperphosphorylation and synaptic dysfunction. Thus, PTI-125 is used to reduce abnormal Aβ deposition and tau aggregation that trigger neuroinflammation and consequent neurodegeneration.

4.3 Neuroprotective compounds

The course of AD is accompanied by a progressive alteration of, or detriment to, different aspects of brain metabolism. While some of these characteristics could be related to normal aging processes, in individuals with AD and related dementia, the neuroendocrine system, glucose metabolism, antioxidant systems, levels of chelators, and NAD^+ are severely altered. This affects all cognitive function, behavior, and overall brain function. The effective control or restoration to basal levels of brain metabolism has been explored as a therapeutic approach to control or monitor AD progress in the elderly.

The main risk factor for AD is aging, which usually is accompanied by other chronic conditions. Here, we encounter a paradox: do chronic diseases predispose individuals to develop AD or does neurodegeneration subsequently trigger other age-related illnesses? Compounds that target inflammation, oxidative stress, symptoms of chronic neurodegeneration (memory loss, confusion, changes in mood and personality, anxiety, and

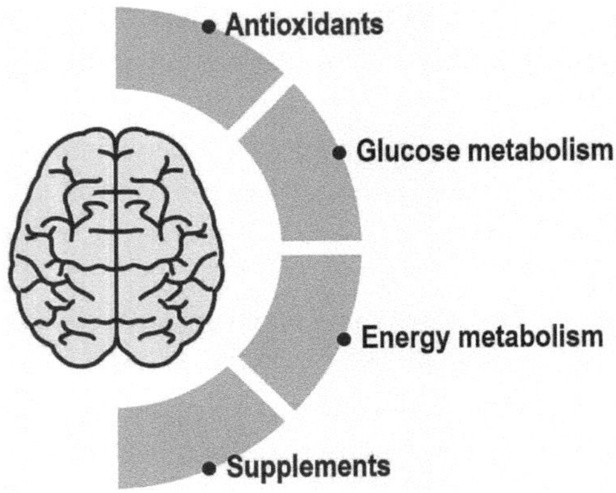

Fig. 5 Neuroprotective compounds targeting Alzheimer's disease. Therapeutic compounds with neuroprotective properties can be antioxidants, modulators of glucose metabolism, modulators of energy metabolism, dietary supplements, chelators, among others.

aggression), and even repurposed drugs have been proposed as non-amyloid, non-tau potential therapies for AD patients (Fig. 5).

AGB101 is a formulation of levetiracetam, an anticonvulsant drug; its mechanism of action has not been fully described. A Phase III trial has been designed to study the efficacy of AGB101 (220 mg administered daily for 78 weeks) on slowing cognitive and functional impairment in 830 individuals with MCI or prodromal AD (NCT03486938).

BHV-4157 (aka Troriluzole) is a prodrug that modulates glutamate by reducing its synaptic levels. Preclinical studies suggest that riluzole, the active metabolite of BHV-4157, may have neuroprotective effects in AD and preserve cognitive function. APP/PS1 mice were kept under early-life stress conditions, affecting synaptic plasticity, learning, memory, and promoting Aβ plaque load. Their treatment with riluzole (4 mg/kg per day in the drinking water) prevented stress-induced cognitive defects and impairments in synaptic plasticity, including reduction in Aβ deposition (Lesuis et al., 2019). An ongoing Phase III study is evaluating the protective effect of BHV-4157 (280 mg daily for up to 48 weeks) on AD-related symptoms and cognitive dysfunction in 336 individuals diagnosed with mild-to-moderate AD (NCT03605667).

Docosahexaenoic acid (DHA) is an omega-3 polyunsaturated fatty acid with antioxidant, anti-inflammatory, and neurotrophic properties. It is administered as a dietary supplement to improve cognitive function in individuals with mild AD. Preliminary studies showed that 204 individuals with moderate AD who received a daily dose of 1.7 g of DHA and 0.6 g of eicosapentaenoic acid (EPA) for 6 months showed no significant effects on oxidative stress (F2-isoprostane and 8-isoprostaglandin F2α) and inflammation (15-keto-dihydro-prostaglandin F2α) biomarkers in comparison in comparison with placebo controls (NCT00211159) (Freund-Levi et al., 2014). A randomized Phase III trial evaluated DHA supplementation (2 g/day for 18 months or placebo) in 295 participants with mild-to-moderate AD (NCT00440050). Supplementation with DHA had no beneficial effect on ADAS-Cog scores and the rate of brain atrophy was not affected by the supplements. Thus, the authors concluded that DHA supplement did not slow the rate of cognitive or functional decline in individuals with AD (Quinn et al., 2010). The effect of long-term omega 3 supplementation has also been tested in 1680 non-demented participants (70 years or older). Participants received 800 mg DHA and 225 mg EPA for 3 years and tested for cognitive decline. This intervention showed no significant effects on prevention or delay of cognitive impairment in elderly individuals (Andrieu et al., 2017). An ongoing Phase III trial in 400 healthy, older adults (70 years and older) receiving 1.53 g of DHA per day for up to 18 months is evaluating the efficacy of omega-3 supplementation on cognitive decline (NCT03691519). A purified form of EPA called *Icosapent ethyl (IPE)* is also under Phase III investigation (NCT02719327), where subjects at high risk of AD receive 4 g daily of IPE for up to 10 months to evaluate its effects on cerebral blood flow, AD biomarkers, and cognitive performance.

Extract of *Gingko biloba* is a dietary supplement with attributed antioxidant properties and taken to improve memory and other age-related loss of cognitive functions (Singh et al., 2019). Several clinical trials with *Gingko biloba* have evaluated its potential benefits on cognition in AD subjects with inconclusive results. Standardized *Gingko biloba* extract is being evaluated for clinical efficacy as a potential treatment in AD, and the Phase III trial will evaluate a dose of 0.15 g of *Gingko biloba* extract three times per day in combination with donepezil (NCT03090516).

At least 32 other, potential neuroprotective compounds are in Phase II trials and nine in Phase I. All these compounds have been tested for potential neuroprotective properties in AD that include compounds with diverse mechanisms of action, including antioxidants, vitamins, dietary supplements, drugs targeting glucose metabolism in the brain or energy metabolism in neurons, and repurposed drugs.

4.4 Brain chemistry and neurotransmitters

The preservation of neurotransmitters important for memory and thinking tasks is also being targeted with novel therapeutic compounds. This group includes the FDA-approved drugs (Table 1), such as donepezil, rivastigmine, galantamine, or memantine. The overall aim of this approach is to maintain brain chemistry and mental function, manage behavioral symptoms (sleeplessness, wandering, agitation, anxiety, aggression, restlessness, and depression), and slow the progression of AD (Fig. 6).

AVP-786 is a dextromethorphan-quinidine formulation that functions as a NMDA antagonist. This formulation is being evaluated for the treatment of agitation in AD and related dementia, which commonly occurs. Following trials that examined the safety, tolerability, and effectiveness of *AVP-786*, the FDA granted a fast-track evaluation of AVP-786 (Garay and Grossberg, 2017). Phase III trials have been conducted (NCT02442765 and NCT02442778) and two more are recruiting participants (NCT03393520 and NCT03896945).

AXS-05 is a combination formulation of the antidepressant bupropion and dextromethorphan, a sigma-1 receptor agonist. These compounds treat neuropsychiatric symptoms, including agitation, aggression, and irritability. The FDA granted a fast-track designation to AXS-05; consequently, a Phase III trial (NCT03226522) will evaluate the effect of AXS-05 on agitation in 435 individuals with AD aged 65–90 years who will receive tablets for 5 weeks, and then be evaluated according to the Cohen-Mansfield Agitation Inventory (CMAI), which is a 29-item scale to systematically assess agitation.

Bexpiprazole is an agonist of the dopamine D2 receptor used to treat the symptoms of schizophrenia and is also being evaluated as an antipsychotic

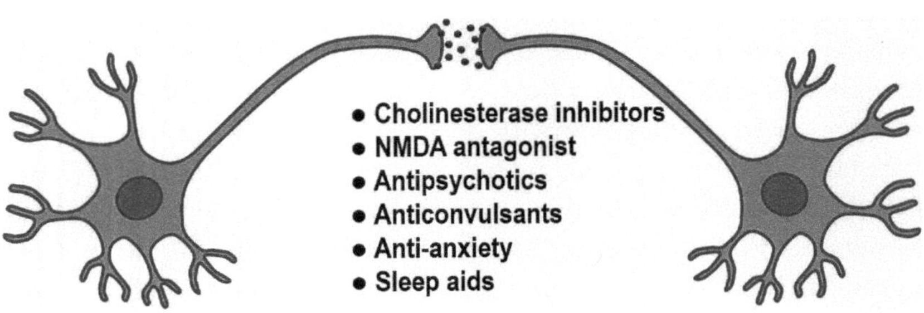

Fig. 6 Therapeutic compounds targeting brain chemistry and neurotransmitters. Symptomatic treatments for AD include cholinesterase inhibitors, NMDA antagonists, antidepressants, anticonvulsants, anti-anxiety, and sleep aid medications.

therapy in AD patients. Phase III trials of bexpiprazole have assessed its efficacy, safety, and tolerability in patients with a diagnosis of probable AD over 12 weeks of administration at a dose of 0.5–2 mg/day to treat symptoms of agitation/aggression associated with dementia (NCT01862640 and NCT01922258) (Grossberg et al., 2020). The high dose of bexpiprazole (2 mg/day) was associated with a statistically significant improvement in CMAI, while the low dose (0.5 mg/day) did not achieve significance. Overall, bexpiprazole showed potential in treating AD symptoms despite some minor side effects (headache, insomnia, dizziness, and urinary infection).

Escitalopram is a selective serotonin reuptake inhibitor (SSRI) used to treat depression and anxiety. This compound is also being tested as a treatment for agitation in AD. A Phase III trial will evaluate the safety and efficacy of escitalopram as a treatment for agitation as a neuropsychiatric symptom in 392 individuals diagnosed with AD who will receive up to 15 mg/day of escitalopram over 12 weeks and subsequently be evaluated with the modified Alzheimer's Disease Cooperative Study-Clinical Global Impression of Change (MADCS-CGIC) (NCT03108846) as the primary outcome (Ehrhardt et al., 2019).

Guanfacine is an agonist of the αA2 adrenergic receptor that decreases sympathetic nervous activity. It is used to treat attention deficit hyperactivity disorder (ADHD) and hypertension. A Phase III trial will evaluate the efficacy of an extended-release guanfacine formulation (2 mg tablet) as add-on therapy in AD for the improvement of cognition and will complete in March 2021 (NCT03116126).

Methylphenidate, commonly known as Ritalin, is prescribed for ADHD and narcolepsy. This compound is used to treat apathy in demented patients, and methylphenidate administered at a target dose of 20 mg per day (range 10–20 mg per day) was shown to significantly improve apathy and global cognition with minimal adverse events in a Phase II trial in individuals diagnosed with possible or probable with AD under the National Institute of Neurological and Communicative Disorders and Stroke-Alzheimer's Disease and Related Disorders Association (NINCDS-ADRDA) criteria, with Mini-Mental State Exam (MMSE) score (Scherer et al., 2018). A Phase III trial (NCT02346201) with 200 participants is scheduled for completion in June 2020.

Mirtazapine is a tricyclic antidepressant with antiserotonergic activity and an α2 receptor blocker. A Phase III trial evaluated the efficacy and safety of mirtazapine in the treatment of sleep disorders in AD, but the 15 mg dose of mirtazapine showed no significant effect on cognitive and functional status

(NCT01867775) (Scoralick et al., 2017). Another Phase III trial is still ongoing with a similar dose of mirtazapine and will be completed in July 2020.

Nabilone possesses antiemetic properties and targets the endocannabinoid system. It was evaluated as a treatment for agitation in AD patients; biomarkers of oxidative stress and neuroinflammation were associated with agitation severity and decreased by nabilone treatment (Ruthirakuhan et al., 2019a, 2019b). A Phase III trial will analyze the safety and efficacy of nabilone for the treatment of agitation in moderate-to-severe AD, comparing a range of doses (0.5–2 mg/day total or placebo) over 6 weeks (NCT02351882) (Ruthirakuhan et al., 2019a, 2019b).

Octohydroaminoacridine succinate is a novel acetylcholinesterase (AChE) inhibitor that has been evaluated for AD. Administered at a dose of 1–12 mg/day for up to 16 weeks, this compound showed a dose-dependent improvement in cognitive function and behavior in patients with mild-to-moderate AD (Xiao et al., 2017). A Phase III trial will continue to evaluate the effectiveness of this acetylcholinesterase inhibitor in AD (NCT03283059).

Zolpidem and *Zopiclone* are non-benzodiazepine sedatives used for sleep disorders. Sleep disorders accelerate with the progression of AD so a Phase III trial will evaluate treatment with zolpidem (10 mg per day, for 14 nights) or zopiclone (7.5 mg per day, for 14 nights) on sleep disorders in AD (NCT03075241).

In addition, a transdermal form of donepezil is in Phase III analysis, with another 11 drug candidates in this group in Phase II trials and two more in Phase I. All are being evaluated for symptomatic management of AD with diverse mechanisms of actions, including cholinesterase inhibitors, NMDA antagonists, antidepressants, anticonvulsants, anti-anxiety, and sleep aid medications.

5. Alternative approaches or future directions to target Alzheimer's disease

Complementary approaches have been proposed beyond those aimed at the amyloid-cascade, including neuroinflammation or agents modulating neuronal metabolism. These strategies are based on apolipoprotein E, gene therapy, stem cell therapies, the use of antimicrobial agents, and energy metabolism-based interventions.

5.1 Apolipoprotein E

Amyloid-β plaques may contain not only $A\beta_{40/42}$ peptides, but also hundreds of different proteins within the abnormal aggregates; at least 26 of these proteins are highly enriched in amyloid plaques, including apolipoprotein E (ApoE) (Liao et al., 2004). Apolipoprotein E is a multifunctional protein with central roles in lipid metabolism [i.e., the transport of high-density lipoprotein (HDL), very low-density lipoprotein (VLDL), and low-density lipoprotein (LDL)] and neuronal homeostasis, including cytoskeleton assembly and disassembly, mitochondrial functions, and dendritic morphology (Huang and Mahley, 2014). Remarkably, ApoE plays a role in facilitating the proteolytic clearance of soluble Aβ and consequently reducing its accumulation and aggregation in the brain. The *APOE* gene presents three allelic variants: ε2, ε3, and ε4. Besides old age, carrying the *APOE*-ε4 gene variant is a great risk factor for late-onset AD and it has been estimated that around 40–65% of all AD cases have one of two copies of this *APOE*-ε4 gene variant. The roles of ApoE4 in AD pathogenesis are to promote Aβ aggregation/deposition and tangle formation, drive synaptic dysfunction, and trigger neuroinflammation. Apolipoprotein E is as a promising target located downstream of the protein aggregation pathways for the development of novel drugs and treatment strategies for AD.

Bexarotene (aka Targretin) is an agonist of retinoid X receptors (RXR) and liver X receptors (LXR), which promotes *APOE* transcription. Oral administration of bexarotene to old Tg2576 (APP/PS1) AD mice for up to 7 days enhanced Aβ clearance, thereby reducing Aβ plaque accumulation by more than 50% and restoring cognitive function (Cramer et al., 2012). Subsequent preclinical analyses have shown an increase of the amount of ApoE in the brain, but with no cognitive benefits in APP/PS1 mice (LaClair et al., 2013; O'Hare et al., 2016). Memory improvement and tau reduction with no significant side effects have been reported in a 68-year-old man (heterozygote *APOE4*) diagnosed with AD who was treated for 6 months with bexarotene (300 mg/day) (Pierrot et al., 2016). Moreover, a Phase II clinical trial with 20 patients treated for 1 month with bexarotene (300 mg/day) showed a significant reduction of brain Aβ in *APOE4* non-carriers, but no reduction in *APOE4* carriers (Cummings et al., 2016). Retinoic acid isomers (all-trans-RA, 9-cis-RA, and 13-cis-RA) increased ApoE in astrocytes through RXR and retinoic acid receptors; the agonist effects were comparable to those of bexarotene and included decreased cellular uptake Aβ in an ApoE-dependent manner (Zhao et al., 2014).

GW3965 is a LRX agonist. In aged Tg2576 mice treated for 4 months with 120 mg/kg (33 mg/kg/day) of GW3965, there was an enhanced expression of ApoE and stimulation of 50% Aβ degradation, reversing contextual memory deficits as measured by the contextual fear conditioning procedure (Jiang et al., 2008). Treatment with 50 mg/kg/day of GW3965 for 6 days in 3xTg-AD mice produced an improvement in cognition, a decrease in methylation in synapse-related genes (*SYP*, *SYN1*, and *DLG3*), proliferation, and neurogenesis-associated genes (*HMGB3* and *RBBP7*) (Sandoval-Hernandez et al., 2016a). Also, preclinical data in 3xTg-AD mice treated with GW3965 showed an increase in the number of astrocytes, but not activation of microglia. In addition, the treatment with GW3965 increased the levels of synapse-related genes (*SYP*, *SYN1,* and *DLG3*) and neurogenesis-associated genes (*HMGB3* and *RBBP7*) in neurons, partially restored microvascular morphology, and decreased Aβ in blood vessels, suggesting that GW3965 has potential therapeutic used in cognitive improvement and synaptic function (Sandoval-Hernandez et al., 2016b).

TO901317. This oxysterol or derivative of cholesterol is another LXR agonist. When administered to APP23 mice (50 mg/kg/day for 6 days) and primary mouse neurons, TO901317 was associated with decreased production of soluble Aβ (Koldamova et al., 2005). When administered to Tg2576 mice at a dose of 50 mg/kg/day for 6 days, TO901317 was associated with reduced accumulation of Aβ in the hippocampus accompanied by increased levels of ABCA1 and ApoE, and a reversal of contextual memory deficiencies (Riddell et al., 2007). Preclinical studies in other age-related diseases, such as Parkinson's disease (PD) have also shown a reduction of neuroinflammation associated with TO901317 (Lei et al., 2017; Paterniti et al., 2017). This is promising since neuroinflammation occurs in diseases, such as AD, PD, and age-related macular degeneration. Despite a promising performance in preclinical animal models that suggest applicability in the treatment of humans, LRX agonists have shown unforeseen side effects, including hepatic steatosis, hypertriglyceridemia, hepatic lipogenesis, and other undisclosed side effects. This has stopped their progression beyond Phase I trials (Fessler, 2018). Nonetheless, this is still an alternative and potentially viable approach and may contribute to future development of novel compounds targeting LXR *per se* or in combination with other AD therapies.

The role of the agonist of peroxisome proliferator-activated receptor γ (PPAR-γ) as a co-regulator of *APOE* transcription is also being evaluated in AD. *Rosiglitazone* was initially developed as an antidiabetic medication for type 2 diabetes mellitus (T2DM). It works as an insulin sensitizer in fat cells and improves glycemic control, while reducing circulating insulin levels. Studies in aged J20 transgenic mice treated with oral rosiglitazone (5 mg/kg/day for 4–12 weeks) was found to promote clearance of Aβ plaques from the hippocampus and cortex and reduce pro-inflammatory markers, such as TNF-α and cyclooxygenase 2 (Cox2). The treatment also reduced amount of neuropil threads containing tau and helped to rescue cognitive and spatial memory functions (Morris water maze test) in these mice (Escribano et al., 2010). Rosiglitazone treatment (5 mg/kg/day for 4 weeks) in the streptozotocin-induced rat model of AD showed a significant reduction of Aβ in the hippocampus (41–54%) and rescued memory impairment (Wang et al., 2017). In a Phase II clinical study with mild-to-moderate AD patients, analysis of cognitive measures demonstrated that only *APOE4* non-carriers showed an improvement in ADAS-Cog in response to rosiglitazone (8 mg/once daily for 24 weeks), whereas *APOE4*+ AD patients showed no improvement or even a decline in ADAS-Cog-related cognitive function (Risner et al., 2006). Phase III studies showed no benefit in cognitive or functional outcomes of rosiglitazone at doses of 2-8 mg/day for 48 weeks even in combination with AChE inhibitor treatment (Harrington et al., 2011). Similarly, a complementary, 6-month Phase III trial in mild-to-moderate AD subjects showed no significant differences in ADAS-Cog between treated and placebo subjects (NCT00428090 and NCT00550420) (Gold et al., 2010). However, some side effects, such as edema and nasopharyngitis, were reported.

Pioglitazone (aka: Actos, Glustin, Piozone) was also developed for T2DM and then tested in AD. A pilot Phase II trial testing the safety of pioglitazone administered at a dose of 45 mg daily for 18 months to non-diabetic AD patients, indicated acceptable tolerability and no adverse events, but showed no clinical benefits as measured with ADAS-Cog (NCT00982202) (Geldmacher et al., 2011). A multisite, Phase III trial started in 2013 to test a biomarker risk algorithm for the prognosis of MCI-AD and evaluate the efficacy of a daily, oral low-dose of pioglitazone (0.8 mg/day) for up to 5 years, was terminated in 2018 due to a lack of efficacy in outcome measures (NCT01931566).

Telmisartan is a partial agonist of PPAR-γ and an angiotensin II receptor blocker (ARB) indicated as an antihypertensive. Preclinical studies in ddY mice with an intracerebroventricular injection of Aβ showed that a single dose of telmisartan alone (100 mg/kg/day) or in combination with

GW9662 (another PPAR-γ antagonist) inhibited cognitive decline after intracerebroventricular injection of Aβ (Mogi et al., 2008). Low-dose telmisartan treatment (3 mg/kg/day for 6–18 months) showed neuroprotective effects in aged rats, reducing Aβ/tau and inflammatory biomarkers (Kurata et al., 2014). A Japanese pilot trial with 20 patients with probable AD were treated with 40–80 mg/day of telmisartan or amlodipine (5–10 mg/day) for 6 months. The telmisartan subgroup showed increased cerebral blood flow, but no changes in cognitive function tests, whereas the amlodipine subgroup showed significantly higher cognitive scores, but no significant increase in cerebral blood flow (Kume et al., 2012). Another pilot trial with 48 patients with probable AD and hypertension received telmisartan (80 mg/day) or amlodipine (10 mg/day) for 6 months. Both groups showed significantly decreased systolic/diastolic blood pressure compared with baseline values. The telmisartan group showed better ADAS-Cog scales scores than the amlodipine group, the changes in cognition correlated with lower levels of TNF-α and IL-1β, and significantly higher plasma Aβ (Li et al., 2012). In a Phase II trial evaluating telmisartan as a cognitive enhancer or neuroprotective agent, 240 patients at risk of AD have been recruited and are being treated with 40–80 mg/day of telmisartan for 6–12 months, and results are expected in 2020 (NCT02085265).

The PPAR-γ agonists mediate Aβ clearance through activation of its transport by ApoE, lipoprotein receptor-related protein 1 (LRP1), and RAGE (Wang et al., 2017). Their impact in AD has been assessed in a meta-analysis that included nine studies involving individuals with AD and mild-to-moderate AD (Cheng et al., 2016). The interventions covered by the analysis included pioglitazone (four studies), telmisartan (one study), and rosiglitazone (four studies) in clinical trials carried out between 2005 and 2012. Pioglitazone was concluded as showing tolerability and some signs of cognitive improvement in AD and mild-to-moderate AD, whereas rosiglitazone exhibited cognitive and functional improvements in *APOE*-ε4 carriers or no effects in *APOE*-ε4 non-carriers (Cheng et al., 2016). It is evident that detailed pharmacogenomic research is necessary to clearly decipher the role of the *APOE*-ε4 genotype in the processes that influence the efficacy and mechanisms of actions of these and other candidate therapeutic compounds to improve cognitive function in individuals with AD.

Immunotherapy with anti-ApoE4 mAbs attempts to counteract pro-Aβ aggregation of ApoE4 and does not inhibit ApoE3. Four anti-ApoE4 antibodies have been developed, among which HJ6.3 IgG2b reduced Aβ deposition by 60–80% and around 30–50% of insoluble $A\beta_{40/42}$ in the cortex and hippocampus when administered intraperitoneally in APPswe/PS1ΔE9 mice weekly for 14 weeks (Kim et al., 2012). Subsequently, treatment for 21 weeks with an intraperitoneal 10 mg/kg/week dose of HJ6.3 IgG2b in APP/PS1 mice only showed mild improvement in cognitive performance yet prevented formation and accumulation of new amyloid deposits (Liao et al., 2014). Another anti-ApoE4 mAb, 9D11, when administered by intraventricular injection in *APOE4* mice prevented Aβ accumulation in the hippocampus. Intraperitoneal administration of 9D11 resulted in a reversal of cognitive impairment, low tau, and under-expression of ApoER2 receptor (Luz et al., 2016).

5.2 Gene therapy

Some cases of AD are caused by the presence of gene mutations. The *APOEε4* gene is associated with a high risk of AD because around 40–65% of cases diagnosed with AD have the *APOEε4* gene, while *APOEε4* is only present in about 20% of the US population. In contrast, *APOE2/3* homozygotes have a significantly lower incidence of AD or even protection from developing AD. Also, mutations in *APP, PSEN1,* or *PSEN2* genes are associated with early-onset familial AD (Alzheimer's Association, 2019). Gene editing offers an alternative to replace those high-risk genes or gene mutations for a functional copy.

AAVrh.10hAPOE2 is a gene therapy that has been tested in a primate model and is currently in Phase I clinical trials with 15 participants (*APOE4* homozygotes) (NCT03634007). This genetically engineered AAV delivers a human *APOE2* gene sequence to the brain cells, and preclinical studies showed that intracisternal delivery of AAVrh.10hAPOE2 was safe and *APOE2* was expressed in the CNS (Rosenberg et al., 2018).

Another AAV gene delivery candidate termed *CERE-110* was evaluated in Phase I (NCT00087789) and Phase II (NCT00876863) trials in 10 and 49 participants with AD, respectively, but did not progress beyond Phase II. The CERE-110 gene therapy was designed to deliver nerve growth factor (NGF), which is a neurotrophic-factor protein with the potential to restore and protect cholinergic neurons (Rafii et al., 2014).

5.3 Cell therapy

The progression of AD is accompanied by a significant reduction in the number of cholinergic neurons. Neuronal networks are also damaged and many regions of the brain begin to shrink. This process called brain atrophy is widespread, being particularly severe in hippocampus and brain cortex causing a significant loss of brain volume. Thus, cell therapies have been proposed as alternative approaches in AD. *AstroStem* consists of autologous, adipose tissue-derived, mesenchymal stem cells (AdMSC), currently in Phase I/II evaluation of safety and efficacy as an investigational stem cell therapy to treat symptoms of AD (NCT03117738). Similarly, *Neurostem* consists of human, umbilical cord-derived, mesenchymal stem cells (hUCB-MSCs), which are administered intraventricularly as a treatment for neuronal loss in AD. In a preclinical model, hUCB-MSCs showed potential with respect to cognitive rescue and restoration of learning and memory functions (Lee et al., 2010). This therapy is being evaluated in a Phase IIa clinical trial (NCT03172117) with 45 individuals with AD who will receive 1×10^7 or 3×10^7 cells in three intraventricular administrations of 2 mL at 4 week intervals, and subsequently be evaluated for changes in cognition with ADAS-Cog after 24 months of first dose.

In contrast, *GB301* is a cell therapy based on regulatory T-cells with amyloid-β-specific CD4(+) responses. A Phase II trial will evaluate the safety and tolerability of multiple intravenous infusion of T-cells in subjects with mild-to-moderate AD (NCT03865017). Finally, *NDX-1017* is a small drug that has been shown to promote proliferation and neuroregeneration via activation of the hepatocyte growth factors (HGT)/MET receptor system (Akimoto et al., 2004). The concept behind *NDX-1017* is that it should promote neuro-regeneration and synaptic plasticity in elderly subjects with mild AD, and it is under Phase I evaluation in healthy young and elderly subjects, as well as elderly subjects with amnestic MCI, mild AD, or mixed dementia with AD (NCT03298672).

5.4 Antimicrobials

COR388 is a small molecule drug with antimicrobial properties against the enzyme gingipain (cysteine protease). This enzyme is produced by the bacteria *Porphyromonas gingivalis* causing chronic periodontitis. This bacteria has been linked to AD because its DNA has been detected in postmortem brains of AD patients (Dominy et al., 2019). The safety and tolerability of COR388 at doses of 25–250 mg twice daily has been demonstrated and the compound has reached the Phase II/III trial stage, where the impact of twice-daily doses of 40–80 mg twice on cognitive outcomes will be evaluated in 573 individuals with AD. It will be run until 2022 (NCT03823404).

Rifaximin is an antibiotic that can lower blood ammonia levels and alter the gut microbiota. It has been hypothesized that rifaximin, by lowering blood ammonia and reducing levels of circulatory, pro-inflammatory cytokines secreted by gut bacteria will improve cognition and function in AD subjects (Kowalski and Mulak, 2019). Rifaximin is currently at the Phase II stage of evaluation in a small cohort of 10 individuals diagnosed with mild-to-moderate AD who will receive an oral dose 550mg twice daily for 3 months (NCT03856359). The study will complete in 2021.

Valacyclovir is an antiviral therapy for herpes virus infections. Human herpes virus 6A (HHV-6A) and 7 (HHV-7) have been detected in RNAs in AD brain samples, although it has not yet been determined if this was a causal contribution to pathology or an opportunistic infection (Readhead et al., 2018). An ongoing Phase II trial is evaluating valacyclovir in 130 participants diagnosed with AD at a dose of 500mg/day for up to 78 weeks (NCT03282916).

Efavirenz is an antiretroviral compound used for HIV/AIDS, but also stimulates cholesterol turnover. Elevated cholesterol levels may increase the risk of AD, and the *APOE4* variant raises levels of cholesterol and LDL. Efavirenz is currently under Phase I evaluation for safety and tolerability at doses of 50–200mg/day administered for 20 weeks to individuals with MCI (NCT03706885).

5.5 Energy metabolism-based interventions

Glucose is the main source of energy for the brain. Alzheimer's disease and MCI are characterized by alterations in glucose metabolism and mitochondrial dysfunction, mainly by oxidative damage of key enzymes involved in ATP biosynthesis, glycolysis, and the tricarboxylic acid pathways (Butterfield and Halliwell, 2019). The consequence of these alterations in glucose metabolism is insulin resistance. Therapeutic approaches in AD that focus on re-establishing glucose metabolism include the administration of biological or small molecule drugs that have been developed for diabetes.

Insulin is a peptide hormone prescribed for type 1 diabetes mellitus. Intranasal insulin is administered for the potential treatment of MCI and AD. A preclinical study tested the effects of 8 weeks of intranasal insulin (4×40 IU/day) in 38 healthy subjects. The treatment showed signs of improved mood and reduced anger, but most importantly the administration of intranasal insulin caused an improvement in memory and attention tests with the absence of systemic side effects (Benedict et al., 2004). A cohort of 25 participants with early AD received intranasal insulin

(20 UI BID) using an electronic atomizer for 21 days. The insulin-treated group showed improvement in cognition (verbal attention and functional status) and a reduction in plasma $A\beta_{40}$ without affecting $A\beta_{42}$ (Reger et al., 2008). A Phase II clinical trial tested the effect of intranasal insulin in memory of patients with MCI or AD (NCT00438568). In all, 90 participants received daily intranasal administration of insulin (10 or 20 IU twice per day) or placebo for 4 months. Treatment with insulin improved delayed memory and cognition as assessed by the ADAS-Cog and Alzheimer's Disease Cooperative Study-activities of daily living (ADCS-ADL) scale. The overall amount of CSF $A\beta_{42}$, $A\beta_{40}$, and tau did not change, but exploratory analysis showed an increase in CSF $A\beta_{42}$ and the ratio of tau/$A\beta_{42}$ (Craft et al., 2012). A Phase II/III clinical trial examined the effects of nasal insulin on cognition, entorhinal cortex and hippocampal atrophy and CSF biomarkers in MCI and AD patients (NCT01767909). Here, 239 participants received 20 IU BID (40 IU daily) for 12 months, and tested for changes in cognition (ADAS-Cog test) at 3, 6, 9, 12, and 15 months. This trial appeared as completed in clinicaltrials.gov, but no peer-reviewed publications with the results are available yet.

Metformin is a common prescription drug used to treat T2DM. This compound increases insulin sensitivity in peripheral tissues and reduces blood glucose. In addition to its activity as activator of AMP-protein kinase, metformin has showed anti-inflammatory properties and reduction of oxidative stress (Campbell et al., 2017). Metformin was tested in a pilot trial in patients with MCI. Overall, this compound was tolerated at doses of 500, 1000, and 1500 mg/day, with no adverse effects. The metformin-treated group showed no changes in ADAS-Cog and the Selective Reminding Test (SRT) (Luchsinger et al., 2016). A Phase II trial (NCT01965756) analyzed the effect of insulin sensitizer metformin. Administration of 2000 mg/day of metformin for 8 weeks to 20 non-diabetic individuals with MCI or AD caused an improvement in executive functioning, including learning/memory and attention tests, an increase in cerebral blood flow, but no change in CSF AD biomarkers (Koenig et al., 2017). The Phase II/III clinical trial NCT04098666 is currently testing long-acting metformin in a cohort of 370 non-diabetic individuals with early and late MCI who are overweight or obese. Metformin will be administered at doses of 500–2000 mg/day every 10 days for up to 24 months. Participants will undergo cognitive testing and brain MRI scans. This trial will be completed in 2024.

Thiazolidinediones are PPARγ agonists that reduce serum glucose level and improve insulin resistance. Compounds in this category are *pioglitazone* and *rosiglitazone* prescribed for diabetes, and previously discussed in Section 5.1.

Glucagon-like peptide 1 (GLP1) is a peptide hormone that targets the GLP1 receptor in pancreatic β-cells to promote insulin release in response to food intake, and is degraded shortly there after by dipeptidyl peptidase-4 (DPP4) (Femminella et al., 2017). Agonists of GLP1 receptor, such as exendin-4, liraglutide, and lixisenatide are drugs approved for diabetes. *Liraglutide* was tested in a pilot trial with 34 patients with AD (NCT01469351). A dose of 1.8 mg daily for 6 months prevented decline of glucose metabolism, but no significant changes in cognitive tests (Femminella et al., 2017; Gejl et al., 2016). Liraglutide is currently in a Phase II, 12-month trial evaluating a dose of 1.8 mg/day by subcutaneous injection (NCT01843075) in 206 participants in order to evaluate its effect on changes in glucose metabolism (via MRI scans), changes in cognitive tests (via ADAS-Cog), and CSF Aβ biomarkers (Femminella et al., 2019). Similarly, *exendin-4* (or exenatide) was evaluated in a Phase II clinical trial testing a dose of 5–10 mcg twice daily in 57 participants with AD over a period of 18 months to determine safety, tolerability, and evidence for improvement in cognitive performance (NCT01255163). Exenatide was safe and well-tolerate, the treatment had no significant effects in clinical and cognitive measures, no differences in AD biomarkers, and only a slight reduction in plasma Aβ. This study was underpowered due to early termination (Mullins et al., 2019).

Ketones. Ketone bodies (β-hydroxybutyrate and acetoacetate) are alternative energy sources to glucose for the brain. A ketogenic diet includes 20–70 g/day of medium-chain triglyceride supplements or a high-fat, low-carbohydrate diet. Pilot studies showed that ketone intake is similar in MCI, AD, and age-matched patients, with modest effects on cognition (Cunnane et al., 2016). Interestingly, the ApoE genotype has an influence on glucose and ketone metabolism as observed in transgenic mice with human *APOE2, APOE3,* and *APOE4* genes. Comparison of gene expression of 43 genes involved in glucose and ketone metabolism showed that *APOE2*-mice had a robust or stable profile, whereas *APOE4*-mice showed deficiencies in genes related to cell uptake and processing of glucose (Wu et al., 2018). In the case of ketone metabolism, *APOE2* and *APOE4* mice showed similar expression profiles and *APOE3* mice had a relatively deficient metabolic profile. The feasibility and efficacy of the ketogenic diet in AD was analyzed in an open-label clinical trial (NCT03690193). In all, 15 individuals with AD received a 3-month intervention of ketogenic diet [increased medium-chain triglyceride (MCT) oil, non-starchy vegetables, butter, eggs, olive oil, avocados, and nuts and seeds with practical elimination of potatoes, grains, red meat, sugar-sweetened beverages, and desserts]. Even though the ketogenic diet showed increased intake of choline, vitamin K, manganese, fiber, and other nutrients, no differences in cognitive performance were reported (Taylor et al., 2019).

6. Conclusions

Here, we have presented the current status of therapeutic compounds and approaches being evaluated in clinical trials for AD, with 33 compounds in Phase III, 78 in Phase II, and 32 in Phase I. The mechanisms and targets cover a wide scope, focusing on amyloid-β, tau, α/β/γ-secretases, neurotransmission, neuroprotection/metabolism, neuroinflammation, antimicrobials, and gene/cell therapies. It is evident from major intervention studies with the once-promising, anti-Aβ immunotherapies and secretases that these compounds have not achieved the long-sought after, clinical outcomes of effectiveness and significance against cognitive measures in AD. A few compounds have been fast tracked for approval by the FDA, highlighting the critical need for effective therapeutics for AD. One compound named GV-971 (oligosaccharide sodium oligomannate) with anti-inflammatory properties and affecting gut bacteria has already been approved in China, but not in other countries. In the near future, we will witness if the compounds and approaches currently in Phases I and II will reach the critical Phase III stage or fail, as has been the case for more than 100 drug candidates over the last two decades.

Acknowledgments

This research project was possible by the generous support from the Alzheimer's Association (AARFD-17-529742), NIH R01AG066749, The Lowe Foundation, and Semmes Foundation.

References

Akimoto, M., Baba, A., Ikeda-Matsuo, Y., Yamada, M. K., Itamura, R., Nishiyama, N., et al. (2004). Hepatocyte growth factor as an enhancer of nmda currents and synaptic plasticity in the hippocampus. *Neuroscience*, *128*(1), 155–162.

Alzheimer's Association. (2019). 2019 Alzheimer's disease facts and figures. *Alzheimer's & Dementia*, *15*(3), 321–387.

Andrieu, S., Guyonnet, S., Coley, N., Cantet, C., Bonnefoy, M., Bordes, S., et al. (2017). Effect of long-term omega 3 polyunsaturated fatty acid supplementation with or without multidomain intervention on cognitive function in elderly adults with memory complaints (MAPT): A randomised, placebo-controlled trial. *Lancet Neurology*, *16*(5), 377–389.

Benedict, C., Hallschmid, M., Hatke, A., Schultes, B., Fehm, H. L., Born, J., et al. (2004). Intranasal insulin improves memory in humans. *Psychoneuroendocrinology*, *29*(10), 1326–1334.
Biogen News Releases. (2020). Biogen completes submission of biologics licence application to FDA for aducanumab as a treatment for Alzheimer's disease. July-8th. https://investors.biogen.com/news-releases/news-release-details/biogen-completes-submission-biologics-license-application-fda.
Boxer, A. L., Qureshi, I., Ahlijanian, M., Grundman, M., Golbe, L. I., Litvan, I., et al. (2019). Safety of the tau-directed monoclonal antibody BIIB092 in progressive supranuclear palsy: A randomised, placebo-controlled, multiple ascending dose phase 1b trial. *Lancet Neurology*, *18*(6), 549–558.
Bozyczko-Coyne, D., & Williams, M. (2007). Neurodegeneration. In J. B. Taylor & D. J. Triggle (Eds.), *Comprehensive medicinal chemistry II* (pp. 193–228). Oxford: Elsevier. 6.08.
Braak, H., & Del Tredici, K. J. A. N. (2014). Are cases with tau pathology occurring in the absence of Aβ deposits part of the AD-related pathological process? *Acta Neuropathologica*, *128*(6), 767–772.
Budd Haeberlein, S., O'Gorman, J., Chiao, P., Bussiere, T., von Rosenstiel, P., Tian, Y., et al. (2017). Clinical development of aducanumab, an anti-Abeta human monoclonal antibody being investigated for the treatment of early Alzheimer's disease. *The Journal of Prevention of Alzheimer's Disease*, *4*(4), 255–263.
Burki, T. (2018). Alzheimer's disease research: The future of BACE inhibitors. *Lancet*, *391*(10139), 2486.
Butterfield, D. A., & Halliwell, B. (2019). Oxidative stress, dysfunctional glucose metabolism and Alzheimer disease. *Nature Reviews. Neuroscience*, *20*(3), 148–160.
Campbell, J. M., Bellman, S. M., Stephenson, M. D., & Lisy, K. (2017). Metformin reduces all-cause mortality and diseases of ageing independent of its effect on diabetes control: A systematic review and meta-analysis. *Ageing Research Reviews*, *40*, 31–44.
Cheng, H., Shang, Y., Jiang, L., Shi, T.-l., & Wang, L. (2016). The peroxisome proliferators activated receptor-gamma agonists as therapeutics for the treatment of Alzheimer's disease and mild-to-moderate Alzheimer's disease: A meta-analysis AU–Cheng, Huawei. *International Journal of Neuroscience*, *126*(4), 299–307.
Craft, S., Baker, L. D., Montine, T. J., Minoshima, S., Watson, G. S., Claxton, A., et al. (2012). Intranasal insulin therapy for Alzheimer disease and amnestic mild cognitive impairment: A pilot clinical trial. *Archives of Neurology*, *69*(1), 29–38.
Cramer, P. E., Cirrito, J. R., Wesson, D. W., Lee, C. Y., Karlo, J. C., Zinn, A. E., et al. (2012). ApoE-directed therapeutics rapidly clear beta-amyloid and reverse deficits in AD mouse models. *Science*, *335*(6075), 1503–1506.
Cummings, J. L., Cohen, S., van Dyck, C. H., Brody, M., Curtis, C., Cho, W., et al. (2018). ABBY: A phase 2 randomized trial of crenezumab in mild to moderate Alzheimer disease. *Neurology*, *90*(21), e1889–e1897.
Cummings, J., Lee, G., Ritter, A., Sabbagh, M., & Zhong, K. (2019a). Alzheimer's disease drug development pipeline: 2019. *Alzheimer's & Dementia (New York, N. Y.)*, *5*, 272–293.
Cummings, J., Passmore, P., McGuinness, B., Mok, V., Chen, C., Engelborghs, S., et al. (2019b). Souvenaid in the management of mild cognitive impairment: An expert consensus opinion. *Alzheimer's Research & Therapy*, *11*(1), 73.
Cummings, J. L., Zhong, K., Kinney, J. W., Heaney, C., Moll-Tudla, J., Joshi, A., et al. (2016). Double-blind, placebo-controlled, proof-of-concept trial of bexarotene Xin moderate Alzheimer's disease. *Alzheimer's Research & Therapy*, *8*, 4.
Cunnane, S. C., Courchesne-Loyer, A., St-Pierre, V., Vandenberghe, C., Pierotti, T., Fortier, M., et al. (2016). Can ketones compensate for deteriorating brain glucose uptake during aging? Implications for the risk and treatment of Alzheimer's disease. *Annals of the New York Academy of Sciences*, *1367*(1), 12–20.

Davtyan, H., Ghochikyan, A., Petrushina, I., Hovakimyan, A., Davtyan, A., Poghosyan, A., et al. (2013). Immunogenicity, efficacy, safety, and mechanism of action of epitope vaccine (Lu AF20513) for Alzheimer's disease: Prelude to a clinical trial. *The Journal of Neuroscience*, *33*(11), 4923–4934.

De Strooper, B., Vassar, R., & Golde, T. (2010). The secretases: Enzymes with therapeutic potential in Alzheimer disease. *Nature Reviews. Neurology*, *6*(2), 99–107.

Dodel, R., Rominger, A., Bartenstein, P., Barkhof, F., Blennow, K., Forster, S., et al. (2013). Intravenous immunoglobulin for treatment of mild-to-moderate Alzheimer's disease: A phase 2, randomised, double-blind, placebo-controlled, dose-finding trial. *Lancet Neurology*, *12*(3), 233–243.

Dominy, S. S., Lynch, C., Ermini, F., Benedyk, M., Marczyk, A., Konradi, A., et al. (2019). Porphyromonas gingivalis in Alzheimer's disease brains: Evidence for disease causation and treatment with small-molecule inhibitors. *Science Advances*, *5*(1)eaau3333.

Drott, J., Desire, L., Drouin, D., Pando, M., & Haun, F. (2010). Etazolate improves performance in a foraging and homing task in aged rats. *European Journal of Pharmacology*, *634*(1–3), 95–100.

Ehrhardt, S., Porsteinsson, A. P., Munro, C. A., Rosenberg, P. B., Pollock, B. G., Devanand, D. P., et al. (2019). Escitalopram for agitation in Alzheimer's disease (S-CitAD): Methods and design of an investigator-initiated, randomized, controlled, multicenter clinical trial. *Alzheimer's & Dementia*, *15*(11), 1427–1436.

Endres, K., Fahrenholz, F., Lotz, J., Hiemke, C., Teipel, S., Lieb, K., et al. (2014). Increased CSF APPs-alpha levels in patients with Alzheimer disease treated with acitretin. *Neurology*, *83*(21), 1930–1935.

Escribano, L., Simón, A.-M., Gimeno, E., Cuadrado-Tejedor, M., López de Maturana, R., García-Osta, A., et al. (2010). Rosiglitazone rescues memory impairment in Alzheimer's transgenic mice: Mechanisms involving a reduced amyloid and tau pathology. *Neuropsychopharmacology: Official Publication of the American College of Neuropsychopharmacology*, *35*(7), 1593–1604.

Etcheberrigaray, R., Tan, M., Dewachter, I., Kuiperi, C., Van der Auwera, I., Wera, S., et al. (2004). Therapeutic effects of PKC activators in Alzheimer's disease transgenic mice. *Proceedings of the National Academy of Sciences of the United States of America*, *101*(30), 11141–11146.

Farlow, M. R., Andreasen, N., Riviere, M. E., Vostiar, I., Vitaliti, A., Sovago, J., et al. (2015). Long-term treatment with active Abeta immunotherapy with CAD106 in mild Alzheimer's disease. *Alzheimer's Research & Therapy*, 7(1), 23.

Femminella, G. D., Bencivenga, L., Petraglia, L., Visaggi, L., Gioia, L., Grieco, F. V., et al. (2017). Antidiabetic drugs in Alzheimer's disease: Mechanisms of action and future perspectives. *Journal of Diabetes Research*, *2017*, 7420796.

Femminella, G. D., Frangou, E., Love, S. B., Busza, G., Holmes, C., Ritchie, C., et al. (2019). Evaluating the effects of the novel GLP-1 analogue liraglutide in Alzheimer's disease: Study protocol for a randomised controlled trial (ELAD study). *Trials*, *20*(1), 191.

Fessler, M. B. (2018). The challenges and promise of targeting the liver X receptors for treatment of inflammatory disease. *Pharmacology & Therapeutics*, *181*, 1–12.

Freund-Levi, Y., Vedin, I., Hjorth, E., Basun, H., Faxen Irving, G., Schultzberg, M., et al. (2014). Effects of supplementation with omega-3 fatty acids on oxidative stress and inflammation in patients with Alzheimer's disease: The OmegAD study. *Journal of Alzheimer's Disease*, *42*(3), 823–831.

Garay, R. P., & Grossberg, G. T. (2017). AVP-786 for the treatment of agitation in dementia of the Alzheimer's type. *Expert Opinion on Investigational Drugs*, *26*(1), 121–132.

Gejl, M., Gjedde, A., Egefjord, L., Moller, A., Hansen, S. B., Vang, K., et al. (2016). In Alzheimer's disease, 6-month treatment with GLP-1 analog prevents decline of brain glucose metabolism: Randomized, placebo-controlled, double-blind clinical trial. *Frontiers in Aging Neuroscience*, *8*, 108.

Geldmacher, D. S., Fritsch, T., McClendon, M. J., & Landreth, G. (2011). A randomized pilot clinical trial of the safety of pioglitazone in treatment of patients with Alzheimer disease. *Archives of Neurology, 68*(1), 45–50.

Gold, M., Alderton, C., Zvartau-Hind, M., Egginton, S., Saunders, A. M., Irizarry, M., et al. (2010). Rosiglitazone monotherapy in mild-to-moderate Alzheimer's disease: Results from a randomized, double-blind, placebo-controlled phase III study. *Dementia and Geriatric Cognitive Disorders, 30*(2), 131–146.

Green, K. N., Steffan, J. S., Martinez-Coria, H., Sun, X., Schreiber, S. S., Thompson, L. M., et al. (2008). Nicotinamide restores cognition in Alzheimer's disease transgenic mice via a mechanism involving sirtuin inhibition and selective reduction of Thr231-phosphotau. *The Journal of Neuroscience, 28*(45), 11500–11510.

Grossberg, G. T., Kohegyi, E., Mergel, V., Josiassen, M. K., Meulien, D., Hobart, M., et al. (2020). Efficacy and safety of brexpiprazole for the treatment of agitation in Alzheimer's dementia: Two 12-week, randomized, double-blind, placebo-controlled trials. *The American Journal of Geriatric Psychiatry, 28*, 383–400.

Harrington, C., Sawchak, S., Chiang, C., Davies, J., Donovan, C., Saunders, A. M., et al. (2011). Rosiglitazone does not improve cognition or global function when used as adjunctive therapy to AChE inhibitors in mild-to-moderate Alzheimer's disease: Two phase 3 studies. *Current Alzheimer Research, 8*(5), 592–606.

Honig, L. S., Vellas, B., Woodward, M., Boada, M., Bullock, R., Borrie, M., et al. (2018). Trial of solanezumab for mild dementia due to Alzheimer's disease. *The New England Journal of Medicine, 378*(4), 321–330.

Huang, Y., & Mahley, R. W. (2014). Apolipoprotein E: Structure and function in lipid metabolism, neurobiology, and Alzheimer's diseases. *Neurobiology of Disease, 72*(Pt. A), 3–12.

Irizarry, M. C., Fleisher, A. S., Hake, A. M., Liu, P., Shcherbinin, S., DeMattos, R. B., et al. (2018). P4–388: Trailblazer-Alz (NCT03367403): A phase 2 disease-modification combination therapy trial targeting multiple mechanisms of action along the amyloid pathway. *Alzheimer's & Dementia, 14*(7S_Part_31), P1622–P1623.

Jiang, Q., Lee, C. Y., Mandrekar, S., Wilkinson, B., Cramer, P., Zelcer, N., et al. (2008). ApoE promotes the proteolytic degradation of Abeta. *Neuron, 58*(5), 681–693.

Kametani, F., & Hasegawa, M. (2018). Reconsideration of amyloid hypothesis and tau hypothesis in Alzheimer's disease. *Frontiers in Neuroscience, 12*, 25.

Kim, J., Eltorai, A. E., Jiang, H., Liao, F., Verghese, P. B., Kim, J., et al. (2012). Anti-apoE immunotherapy inhibits amyloid accumulation in a transgenic mouse model of Abeta amyloidosis. *The Journal of Experimental Medicine, 209*(12), 2149–2156.

Klein, G., Delmar, P., Voyle, N., Rehal, S., Hofmann, C., Abi-Saab, D., et al. (2019). Gantenerumab reduces amyloid-β plaques in patients with prodromal to moderate Alzheimer's disease: A PET substudy interim analysis. *Alzheimer's Research & Therapy, 11*(1), 101.

Koenig, A. M., Mechanic-Hamilton, D., Xie, S. X., Combs, M. F., Cappola, A. R., Xie, L., et al. (2017). Effects of the insulin sensitizer metformin in Alzheimer disease: Pilot data from a randomized placebo-controlled crossover study. *Alzheimer Disease and Associated Disorders, 31*(2), 107–113.

Koldamova, R. P., Lefterov, I. M., Staufenbiel, M., Wolfe, D., Huang, S., Glorioso, J. C., et al. (2005). The liver X receptor ligand T0901317 decreases amyloid beta production in vitro and in a mouse model of Alzheimer's disease. *The Journal of Biological Chemistry, 280*(6), 4079–4088.

Kowalski, K., & Mulak, A. (2019). Brain-gut-microbiota axis in Alzheimer's disease. *Journal of Neurogastroenterology and Motility, 25*(1), 48–60.

Kume, K., Hanyu, H., Sakurai, H., Takada, Y., Onuma, T., & Iwamoto, T. (2012). Effects of telmisartan on cognition and regional cerebral blood flow in hypertensive patients with Alzheimer's disease. *Geriatrics & Gerontology International, 12*(2), 207–214.

Kurata, T., Lukic, V., Kozuki, M., Wada, D., Miyazaki, K., Morimoto, N., et al. (2014). Telmisartan reduces progressive accumulation of cellular amyloid beta and phosphorylated tau with inflammatory responses in aged spontaneously hypertensive stroke resistant rat. *Journal of Stroke and Cerebrovascular Diseases*, *23*(10), 2580–2590.

LaClair, K. D., Manaye, K. F., Lee, D. L., Allard, J. S., Savonenko, A. V., Troncoso, J. C., et al. (2013). Treatment with bexarotene, a compound that increases apolipoprotein-E, provides no cognitive benefit in mutant APP/PS1 mice. *Molecular Neurodegeneration*, *8*, 18.

Lacosta, A. M., Pascual-Lucas, M., Pesini, P., Casabona, D., Perez-Grijalba, V., Marcos-Campos, I., et al. (2018). Safety, tolerability and immunogenicity of an active anti-Abeta40 vaccine (ABvac40) in patients with Alzheimer's disease: A randomised, double-blind, placebo-controlled, phase I trial. *Alzheimer's Research & Therapy*, *10*(1), 12.

Lee, H. J., Lee, J. K., Lee, H., Shin, J. W., Carter, J. E., Sakamoto, T., et al. (2010). The therapeutic potential of human umbilical cord blood-derived mesenchymal stem cells in Alzheimer's disease. *Neuroscience Letters*, *481*(1), 30–35.

Lei, C., Lin, R., Wang, J., Tao, L., Fu, X., Qiu, Y., et al. (2017). Amelioration of amyloid beta-induced retinal inflammatory responses by a LXR agonist TO901317 is associated with inhibition of the NF-kappaB signaling and NLRP3 inflammasome. *Neuroscience*, *360*, 48–60.

Lesuis, S. L., Kaplick, P. M., Lucassen, P. J., & Krugers, H. J. (2019). Treatment with the glutamate modulator riluzole prevents early life stress-induced cognitive deficits and impairments in synaptic plasticity in APPswe/PS1dE9 mice. *Neuropharmacology*, *150*, 175–183.

Li, W., Zhang, J. W., Lu, F., Ma, M. M., Wang, J. Q., Suo, A. Q., et al. (2012). Effects of telmisartan on the level of Abeta1-42, interleukin-1beta, tumor necrosis factor alpha and cognition in hypertensive patients with Alzheimer's disease. *Zhonghua Yi Xue Za Zhi*, *92*(39), 2743–2746.

Liao, L., Cheng, D., Wang, J., Duong, D. M., Losik, T. G., Gearing, M., et al. (2004). Proteomic characterization of postmortem amyloid plaques isolated by laser capture microdissection. *The Journal of Biological Chemistry*, *279*(35), 37061–37068.

Liao, F., Hori, Y., Hudry, E., Bauer, A. Q., Jiang, H., Mahan, T. E., et al. (2014). Anti-ApoE antibody given after plaque onset decreases Abeta accumulation and improves brain function in a mouse model of Abeta amyloidosis. *The Journal of Neuroscience*, *34*(21), 7281–7292.

Logovinsky, V., Satlin, A., Lai, R., Swanson, C., Kaplow, J., Osswald, G., et al. (2016). Safety and tolerability of BAN2401—A clinical study in Alzheimer's disease with a protofibril selective Abeta antibody. *Alzheimer's Research & Therapy*, *8*(1), 14.

Lopez Lopez, C., Caputo, A., Liu, F., Riviere, M. E., Rouzade-Dominguez, M. L., Thomas, R. G., et al. (2017). The Alzheimer's prevention initiative generation program: Evaluating CNP520 efficacy in the prevention of Alzheimer's disease. *The Journal of Prevention of Alzheimer's Disease*, *4*(4), 242–246.

Lopez Lopez, C., Tariot, P. N., Caputo, A., Langbaum, J. B., Liu, F., Riviere, M. E., et al. (2019). The Alzheimer's prevention initiative generation program: Study design of two randomized controlled trials for individuals at risk for clinical onset of Alzheimer's disease. *Alzheimer's & Dementia (New York, N. Y.)*, *5*, 216–227.

Luchsinger, J. A., Perez, T., Chang, H., Mehta, P., Steffener, J., Pradabhan, G., et al. (2016). Metformin in amnestic mild cognitive impairment: Results of a pilot randomized placebo controlled clinical trial. *Journal of Alzheimer's Disease*, *51*(2), 501–514.

Luz, I., Liraz, O., & Michaelson, D. M. (2016). An anti-apoE4 specific monoclonal antibody counteracts the pathological effects of apoE4 in vivo. *Current Alzheimer Research*, *13*(8), 918–929.

Lynch, S. Y., Kaplow, J., Zhao, J., Dhadda, S., Luthman, J., & Albala, B. (2018). P4–389: Elenbecestat, E2609, a bace inhibitor: Results from A phase-2 study in subjects with mild

cognitive impairment and mild-to-moderate dementia due to Alzheimer's disease. *Alzheimer's & Dementia, 14*(7S_Part_31), P1623.
Marcade, M., Bourdin, J., Loiseau, N., Peillon, H., Rayer, A., Drouin, D., et al. (2008). Etazolate, a neuroprotective drug linking GABA(A) receptor pharmacology to amyloid precursor protein processing. *Journal of Neurochemistry, 106*(1), 392–404.
Michelson, D., Grundman, M., Magnuson, K., Fisher, R., Levenson, J. M., Aisen, P., et al. (2019). Randomized, placebo controlled trial of NPT088, a phage-derived, amyloid-targeted treatment for Alzheimer's disease. *The Journal of Prevention of Alzheimer's Disease, 6*(4), 228–231.
Mogi, M., Li, J. M., Tsukuda, K., Iwanami, J., Min, L. J., Sakata, A., et al. (2008). Telmisartan prevented cognitive decline partly due to PPAR-gamma activation. *Biochemical and Biophysical Research Communications, 375*(3), 446–449.
Morgan D, H. D., Rochelle P, Paskus J (2019). "2019 Alzheimer's drug pipeline—The current state of Alzheimer's drug development." UsAgainstAlzheimer's (https://www.usagainstalzheimers.org/sites/default/files/RAAPages_FINAL.pdf).
Mullins, R. J., Mustapic, M., Chia, C. W., Carlson, O., Gulyani, S., Tran, J., et al. (2019). A pilot study of exenatide actions in Alzheimer's disease. *Current Alzheimer Research, 16*(8), 741–752.
Neumann, U., Ufer, M., Jacobson, L. H., Rouzade-Dominguez, M.-L., Huledal, G., Kolly, C., et al. (2018). The BACE-1 inhibitor CNP520 for prevention trials in Alzheimer's disease. *EMBO Molecular Medicine, 10*(11), e9316.
NIA National Institute on Aging (2020). "Treatment of Alzheimer's disease—How is Alzheimer's disease treated?" https://www.nia.nih.gov/health/how-alzheimers-disease-treated.
Novak, P., Schmidt, R., Kontsekova, E., Kovacech, B., Smolek, T., Katina, S., et al. (2018). Fundamant: An interventional 72-week phase 1 follow-up study of AADvac1, an active immunotherapy against tau protein pathology in Alzheimer's disease. *Alzheimer's Research & Therapy, 10*(1), 108.
O'Hare, E., Jeggo, R., Kim, E. M., Barbour, B., Walczak, J. S., Palmer, P., et al. (2016). Lack of support for bexarotene as a treatment for Alzheimer's disease. *Neuropharmacology, 100*, 124–130.
Ostrowitzki, S., Lasser, R. A., Dorflinger, E., Scheltens, P., Barkhof, F., Nikolcheva, T., et al. (2017). A phase III randomized trial of gantenerumab in prodromal Alzheimer's disease. *Alzheimer's Research & Therapy, 9*(1), 95.
Pakpoor, J., & Goldacre, M. (2017). The increasing burden of mortality from neurological diseases. *Nature Reviews. Neurology, 13*(9), 518–519.
Panza, F., Lozupone, M., Solfrizzi, V., Sardone, R., Piccininni, C., Dibello, V., et al. (2018). BACE inhibitors in clinical development for the treatment of Alzheimer's disease. *Expert Review of Neurotherapeutics, 18*(11), 847–857.
Park, H. H., Lee, K. Y., Kim, S., Lee, J. W., Choi, N. Y., Lee, E. H., et al. (2014a). Novel vaccine peptide GV1001 effectively blocks beta-amyloid toxicity by mimicking the extra-telomeric functions of human telomerase reverse transcriptase. *Neurobiology of Aging, 35*(6), 1255–1274.
Park, H., Yoo, J. S., Kim, J. Y., Hwang, B. Y., Han, J. S., Yeon, S. W., et al. (2014b). Anti-amyloidogenic effects of ID1201, the ethanolic extract of the fruits of Melia toosendan, through activation of the phosphatidylinositol 3-kinase/Akt pathway. *Environmental Toxicology and Pharmacology, 37*(2), 513–520.
Paterniti, I., Campolo, M., Siracusa, R., Cordaro, M., Di Paola, R., Calabrese, V., et al. (2017). Liver X receptors activation, through TO901317 binding, reduces neuroinflammation in Parkinson's disease. *PLoS One, 12*(4)e0174470.
Pierrot, N., Lhommel, R., Quenon, L., Hanseeuw, B., Dricot, L., Sindic, C., et al. (2016). Targretin improves cognitive and biological markers in a patient with Alzheimer's disease. *Journal of Alzheimer's Disease, 49*(2), 271–276.

Piette, F., Belmin, J., Vincent, H., Schmidt, N., Pariel, S., Verny, M., et al. (2011). Masitinib as an adjunct therapy for mild-to-moderate Alzheimer's disease: A randomised, placebo-controlled phase 2 trial. *Alzheimer's Research & Therapy*, *3*(2), 16.

Quinn, J. F., Raman, R., Thomas, R. G., Yurko-Mauro, K., Nelson, E. B., Van Dyck, C., et al. (2010). Docosahexaenoic acid supplementation and cognitive decline in Alzheimer disease: A randomized trial. *JAMA*, *304*(17), 1903–1911.

Rafii, M. S., Baumann, T. L., Bakay, R. A., Ostrove, J. M., Siffert, J., Fleisher, A. S., et al. (2014). A phase1 study of stereotactic gene delivery of AAV2-NGF for Alzheimer's disease. *Alzheimer's & Dementia*, *10*(5), 571–581.

Readhead, B., Haure-Mirande, J. V., Funk, C. C., Richards, M. A., Shannon, P., Haroutunian, V., et al. (2018). Multiscale analysis of independent Alzheimer's cohorts finds disruption of molecular, genetic, and clinical networks by human herpesvirus. *Neuron*, *99*(1), 64–82.e67.

Reger, M. A., Watson, G. S., Green, P. S., Wilkinson, C. W., Baker, L. D., Cholerton, B., et al. (2008). Intranasal insulin improves cognition and modulates beta-amyloid in early AD. *Neurology*, *70*(6), 440–448.

Riddell, D. R., Zhou, H., Comery, T. A., Kouranova, E., Lo, C. F., Warwick, H. K., et al. (2007). The LXR agonist TO901317 selectively lowers hippocampal Abeta42 and improves memory in the Tg2576 mouse model of Alzheimer's disease. *Molecular and Cellular Neurosciences*, *34*(4), 621–628.

Risner, M. E., Saunders, A. M., Altman, J. F., Ormandy, G. C., Craft, S., Foley, I. M., et al. (2006). Efficacy of rosiglitazone in a genetically defined population with mild-to-moderate Alzheimer's disease. *The Pharmacogenomics Journal*, *6*(4), 246–254.

Rosenberg, J. B., Kaplitt, M. G., De, B. P., Chen, A., Flagiello, T., Salami, C., et al. (2018). AAVrh.10-mediated APOE2 central nervous system gene therapy for APOE4-associated Alzheimer's disease. *Human Gene Therapy. Clinical Development*, *29*(1), 24–47.

Ruthirakuhan, M., Herrmann, N., Andreazza, A. C., Verhoeff, N., Gallagher, D., Black, S. E., et al. (2019a). Agitation, oxidative stress, and cytokines in Alzheimer disease: Biomarker analyses from a clinical trial with nabilone for agitation. *Journal of Geriatric Psychiatry and Neurology*. 891988719874118.

Ruthirakuhan, M. T., Herrmann, N., Gallagher, D., Andreazza, A. C., Kiss, A., Verhoeff, N., et al. (2019b). Investigating the safety and efficacy of nabilone for the treatment of agitation in patients with moderate-to-severe Alzheimer's disease: Study protocol for a cross-over randomized controlled trial. *Contemporary Clinical Trials Communications*, *15*, 100385.

Salloway, S., Honigberg, L. A., Cho, W., Ward, M., Friesenhahn, M., Brunstein, F., et al. (2018). Amyloid positron emission tomography and cerebrospinal fluid results from a crenezumab anti-amyloid-beta antibody double-blind, placebo-controlled, randomized phase II study in mild-to-moderate Alzheimer's disease (BLAZE). *Alzheimer's Research & Therapy*, *10*(1), 96.

Sandoval-Hernandez, A. G., Hernandez, H. G., Restrepo, A., Munoz, J. I., Bayon, G. F., Fernandez, A. F., et al. (2016a). Liver X receptor agonist modifies the DNA methylation profile of synapse and neurogenesis-related genes in the triple transgenic mouse model of Alzheimer's disease. *Journal of Molecular Neuroscience*, *58*(2), 243–253.

Sandoval-Hernandez, A. G., Restrepo, A., Cardona-Gomez, G. P., & Arboleda, G. (2016b). LXR activation protects hippocampal microvasculature in very old triple transgenic mouse model of Alzheimer's disease. *Neuroscience Letters*, *621*, 15–21.

Scherer, R. W., Drye, L., Mintzer, J., Lanctot, K., Rosenberg, P., Herrmann, N., et al. (2018). The apathy in dementia methylphenidate trial 2 (ADMET 2): Study protocol for a randomized controlled trial. *Trials*, *19*(1), 46.

Schneider, L. (2020). A resurrection of aducanumab for Alzheimer's disease. *Lancet Neurology*, *19*(2), 111–112.

Scoralick, F. M., Louzada, L. L., Quintas, J. L., Naves, J. O., Camargos, E. F., & Nobrega, O. T. (2017). Mirtazapine does not improve sleep disorders in Alzheimer's disease: Results from a double-blind, placebo-controlled pilot study. *Psychogeriatrics*, *17*(2), 89–96.

Selkoe, D. J., & Hardy, J. (2016). The amyloid hypothesis of Alzheimer's disease at 25 years. *EMBO Molecular Medicine*, *8*(6), 595–608.

Singh, S. K., Srivastav, S., Castellani, R. J., Plascencia-Villa, G., & Perry, G. (2019). Neuroprotective and antioxidant effect of Ginkgo biloba extract against AD and other neurological disorders. *Neurotherapeutics*, *16*(3), 666–674.

Siopi, E., Llufriu-Daben, G., Cho, A. H., Vidal-Lletjos, S., Plotkine, M., Marchand-Leroux, C., et al. (2013). Etazolate, an alpha-secretase activator, reduces neuroinflammation and offers persistent neuroprotection following traumatic brain injury in mice. *Neuropharmacology*, *67*, 183–192.

Taylor, M. K., Swerdlow, R. H., Burns, J. M., & Sullivan, D. K. (2019). An experimental ketogenic diet for Alzheimer disease was nutritionally dense and rich in vegetables and avocado. *Current Developments in Nutrition*, *3*(4), nzz003.

Tippmann, F., Hundt, J., Schneider, A., Endres, K., & Fahrenholz, F. (2009). Up-regulation of the alpha-secretase ADAM10 by retinoic acid receptors and acitretin. *The FASEB Journal*, *23*(6), 1643–1654.

Tsai, R. M., Miller, Z., Koestler, M., Rojas, J. C., Ljubenkov, P. A., Rosen, H. J., et al. (2019). Reactions to multiple ascending doses of the microtubule stabilizer TPI-287 in patients with Alzheimer disease, progressive supranuclear palsy, and corticobasal syndrome: A randomized clinical trial. *JAMA Neurology*, *77*(2), 215–224.

Vellas, B., Sol, O., Snyder, P. J., Ousset, P. J., Haddad, R., Maurin, M., et al. (2011). EHT0202 in Alzheimer's disease: A 3-month, randomized, placebo-controlled, double-blind study. *Current Alzheimer Research*, *8*(2), 203–212.

Wang, C. Y., Finstad, C. L., Walfield, A. M., Sia, C., Sokoll, K. K., Chang, T. Y., et al. (2007). Site-specific UBITh amyloid-beta vaccine for immunotherapy of Alzheimer's disease. *Vaccine*, *25*(16), 3041–3052.

Wang, L., Liu, W., Fan, Y., Liu, T., & Yu, C. (2017). Effect of rosiglitazone on amyloid precursor protein processing and Abeta clearance in streptozotocin-induced rat model of Alzheimer's disease. *Iranian Journal of Basic Medical Sciences*, *20*(5), 474–480.

Wang, X., Sun, G., Feng, T., Zhang, J., Huang, X., Wang, T., et al. (2019). Sodium oligomannate therapeutically remodels gut microbiota and suppresses gut bacterial amino acids-shaped neuroinflammation to inhibit Alzheimer's disease progression. *Cell Research*, *29*(10), 787–803.

West, T., Hu, Y., Verghese, P. B., Bateman, R. J., Braunstein, J. B., Fogelman, I., et al. (2017). Preclinical and clinical development of ABBV-8E12, a humanized anti-tau antibody, for treatment of Alzheimer's disease and other tauopathies. *The Journal of Prevention of Alzheimer's Disease*, *4*(4), 236–241.

Wilcock, G. K., Gauthier, S., Frisoni, G. B., Jia, J., Hardlund, J. H., Moebius, H. J., et al. (2018). Potential of low dose leuco-methylthioninium bis(hydromethanesulphonate) (LMTM) monotherapy for treatment of mild Alzheimer's disease: Cohort analysis as modified primary outcome in a phase III clinical trial. *Journal of Alzheimer's Disease*, *61*(1), 435–457.

Wu, L., Zhang, X., & Zhao, L. (2018). Human ApoE isoforms differentially modulate brain glucose and ketone body metabolism: Implications for Alzheimer's disease risk reduction and early intervention. *The Journal of Neuroscience*, *38*(30), 6665–6681.

Xiao, S., Wang, T., Ma, X., Qin, Y., Li, X., Zhao, Z., et al. (2017). Efficacy and safety of a novel acetylcholinesterase inhibitor octohydroaminoacridine in mild-to-moderate Alzheimer's disease: A phase II multicenter randomised controlled trial. *Age and Ageing*, *46*(5), 767–773.

Zempel, H., & Mandelkow, E. (2014). Lost after translation: Missorting of tau protein and consequences for Alzheimer disease. *Trends in Neurosciences*, *37*(12), 721–732.

Zhang, C., Griciuc, A., Hudry, E., Wan, Y., Quinti, L., Ward, J., et al. (2018). Cromolyn reduces levels of the Alzheimer's disease-associated amyloid beta-protein by promoting microglial phagocytosis. *Scientific Reports*, *8*(1), 1144.

Zhao, J., Fu, Y., Liu, C. C., Shinohara, M., Nielsen, H. M., Dong, Q., et al. (2014). Retinoic acid isomers facilitate apolipoprotein E production and lipidation in astrocytes through the retinoid X receptor/retinoic acid receptor pathway. *The Journal of Biological Chemistry*, *289*(16), 11282–11292.

Zhu, H., Young, M. B., Nantermet, P. G., Graham, S. L., Colussi, D., Lai, M. T., et al. (2010). Rapid P1 SAR of brain penetrant tertiary carbinamine derived BACE inhibitors. *Bioorganic & Medicinal Chemistry Letters*, *20*(5), 1779–1782.

CHAPTER TWO

Tracking the potential involvement of metabolic disease in Alzheimer's disease—Biomarkers and beyond

José A. Luchsinger[a,*], Henrik Zetterberg[b,c,d,e]
[a]Departments of Medicine and Epidemiology, Columbia University Irving Medical Center, New York, NY, United States
[b]Department of Psychiatry and Neurochemistry, Institute of Neuroscience & Physiology, The Sahlgrenska Academy at the University of Gothenburg, Mölndal, Sweden
[c]Clinical Neurochemistry Laboratory, Sahlgrenska University Hospital, Mölndal, Sweden
[d]Department of Neurodegenerative Disease, UCL Institute of Neurology, Queen Square, London, United Kingdom
[e]UK Dementia Research Institute at UCL, London, United Kingdom
*Corresponding author: e-mail address: jal94@cumc.columbia.edu

Contents

1. Introduction 53
2. Alzheimer's disease neuropathology as an outcome 54
   2.1 Natural history of Alzheimer's disease in relation to biomarkers 54
   2.2 Alzheimer's disease neuropathology on autopsy 54
   2.3 Alzheimer's disease brain imaging biomarkers 55
   2.4 Alzheimer's disease biomarkers in the cerebrospinal fluid 56
   2.5 Alzheimer's disease blood biomarkers 57
3. Summary of studies relating type 2 diabetes mellitus and glycemia to Alzheimer's disease biomarkers 59
4. Summary of studies relating adiposity, insulin resistance, and metabolic syndrome to Alzheimer's disease biomarkers 61
5. The promise of the field of metabolomics in the study of Alzheimer's disease biomarkers 66
6. Caveats, recommendations, and future directions 70
Acknowledgments 71
Disclosures 71
References 71

Abstract

There is a vast literature linking systemic metabolic conditions to dementia due to Alzheimer's disease (AD). Advances in *in vivo* measurements of AD neuropathology using brain imaging, cerebrospinal fluid (CSF), and/or blood biomarkers have led to research in AD that uses *in vivo* biomarkers as outcomes, focusing primarily on amyloid,

International Review of Neurobiology, Volume 154
ISSN 0074-7742
https://doi.org/10.1016/bs.irn.2020.03.014

tau, and neurodegeneration as constructs. Studies of Type 2 Diabetes Mellitus (T2DM) and AD biomarkers seem to show that T2DM is not related to amyloid deposition, but is related to neurodegeneration and tau deposition. There is a dearth of studies examining adiposity, insulin resistance, and metabolic syndrome in relation to AD biomarkers and the associations in these studies are inconsistent. Metabolomics studies have reported associations of unsaturated fatty acids with AD neuropathology at autopsy, and sphingolipids and glycerophospholipids in relation to neurodegeneration and amyloid and tau. There are other neurodegenerative diseases, such as Lewy body disease that may overlap with AD, and specific biomarkers for these pathologies are being developed and should be integrated into AD biomarker research. More longitudinal studies are needed with concurrent assessment of metabolic factors and AD biomarkers in order to improve the opportunity to assess causality. Ideally, AD biomarkers should be integrated into clinical trials of interventions that affect metabolic factors. Advances in blood-based AD biomarkers, which are less costly and invasive compared with CSF and brain imaging biomarkers, could facilitate widespread implementation of AD biomarkers in studies examining the metabolic contribution to AD.

Abbreviations

AA	Alzheimer's Association
AD	Alzheimer's disease
ADNI	Alzheimer's Disease Neuroimaging Initiative
Aβ	amyloid β
BLSA	Baltimore Longitudinal Study of Aging
CSF	cerebrospinal fluid
DLB	dementia with Lewy bodies
FDG	fluorodeoxyglucose
FTD	frontotemporal dementia
GC	gas chromatography
HOMA-IR	homeostatic model assessment of insulin resistance
IR	insulin resistance
LC	liquid chromatography
MS	mass spectroscopy
MCSA	Mayo clinic study of aging
MRI	magnetic resonance imaging
NFT	neurofibrillary tangles
NP	neuritic plaques
NfL	neurofilament light
NIA	National Institute on Aging
NMR	Nuclear magnetic resonance
PDD	Parkinson's disease dementia
PET	positron emission tomography
P-tau	phosphorylated-tau
TDP-43	transactive response DNA-binding protein 43
T-tau	total-tau
T2DM	type 2 diabetes mellitus

1. Introduction

There is a vast literature linking systemic metabolic conditions of clinical importance, including type 2 diabetes mellitus (T2DM) (Biessels et al., 2014), insulin resistance (Luchsinger, 2010), metabolic syndrome (Muller et al., 2007), and increased adiposity (Luchsinger and Gustafson, 2009) with dementia due to Alzheimer's disease (AD). More recently, the field of metabolomics has sought to elucidate how complex systems of individual metabolites are related to AD, both systemically and in the central nervous system (Kaddurah-Daouk and Krishnan, 2008). There is also increasing interest in the metabolic derangements in the brains of individuals with AD, independent of systemic metabolic derangements, which some have termed brain insulin resistance or type 3 diabetes (Narasimhan et al., 2014). Most of this literature is based on observational studies in which dementia due to AD is defined clinically, without information on AD neuropathology. Alzheimer's disease neuropathology can be examined with the pathologic hallmarks of AD on autopsy, but the availability of large-scale studies with autopsy information is limited. This limitation is being overcome by advances in *in vivo* measurements of AD neuropathology, directly or indirectly, using brain imaging, cerebrospinal fluid (CSF), and/or blood biomarkers (Jack et al., 2013). These advancements have led to research in AD that increasingly uses *in vivo* biomarkers as outcomes, rather than clinical outcomes, which can take decades to develop after the onset of neuropathology. This is particularly important for the study of the metabolic contribution to AD, which may occur decades before clinical manifestations. For example, metabolic changes at midlife may determine the clinical outcomes related to AD in late life. The goal of this review is to provide a summary of the current state of biomarkers of AD, and briefly review the application of these biomarkers to research focusing on the metabolic contributions to AD in representative studies. We emphasize studies examining T2DM and its related conditions, adiposity, insulin resistance, and metabolic syndrome, given that these are the most studied metabolic conditions in AD research and because of their clinical and public health importance. However, we acknowledge the increasing importance of the field of metabolomics in AD research (Kaddurah-Daouk and Krishnan, 2008).

2. Alzheimer's disease neuropathology as an outcome

2.1 Natural history of Alzheimer's disease in relation to biomarkers

Decades of advances in AD research, particularly in CSF and brain imaging biomarkers (Jack et al., 2010, 2013), have led to the dominance of three neuropathological constructs: brain amyloid, brain tau, and neurodegeneration (Jack et al., 2013). Current understanding of the natural history leading to dementia due to AD can be summarized as follows (Jack et al., 2013): the two main proteinopathies underlying AD, amyloid and tau are separate processes, but amyloid deposition accelerates tau deposition; amyloid and tau deposition precede and cause neurodegeneration, which leads to the clinical syndromes of amnestic mild cognitive impairment (MCI) and dementia. The constructs of amyloid, tau, and neurodegeneration feature prominently in the recent National Institute on Aging (NIA)/Alzheimer's Association (AA) 2018 research framework (Jack et al., 2018). This framework proposes to conduct research in which individuals are classified by the presence or absence of evidence of amyloid, tau and neurodegeneration, with or without clinical manifestations, for the purpose of better understanding the mechanisms and sequence of neuropathology. The 2018 research framework has been enabled by the widespread availability of accurate CSF and brain imaging markers of amyloid, tau and neurodegeneration. As compared with cognitively normal individuals, individuals with AD dementia show lower CSF amyloid β 42 (Aβ42) (Olsson et al., 2016) and higher brain amyloid burden on amyloid positron emission tomography (PET) (Ossenkoppele et al., 2015), higher CSF total-tau (T-tau) and phosphorylated-tau (P-tau) (Olsson et al., 2016) and higher brain tau burden on tau PET (James et al., 2015), and lower cortical thickness and brain volumes on magnetic resonance imaging (MRI) (Schwarz et al., 2016). More recently, neurofilament light (NfL) in CSF has also been reported to define neurodegeneration (Olsson et al., 2016). These tools can aid the diagnosis of AD dementia in research and referral settings and are increasingly used in research.

2.2 Alzheimer's disease neuropathology on autopsy

Neuropathologically, AD is characterized by (1) neuronal loss in specific brain regions—notably the medial temporal lobe structures and the temporoparietal association cortices; (2) intraneuronal neurofibrillary tangles, composed of aggregated and often truncated and hyperphosphorylated-tau protein; and (3) extracellular neuritic plaques, which consist of deposits of Aβ peptides, mainly its 42 amino acid isoform, Aβ42 (Lashley et al., 2018).

2.3 Alzheimer's disease brain imaging biomarkers

2.3.1 Amyloid β pathology

There are now several PET ligands that bind to amyloid β (Aβ) plaques in the brain. Amyloid PET has been validated against neuropathology (Cohen et al., 2019), has undergone extensive standardization regarding how to quantify Aβ pathology and how to define cut points for abnormality (Klunk et al., 2015), and has appropriate use criteria (Johnson et al., 2013). Three Aβ ligands (florbetaben, florbetapir, and flutemetamol) have been approved by the Federal Drug Administration (FDA) in the United States for Aβ plaque imaging in patients with cognitive impairment, who are being clinically evaluated for AD (Cohen et al., 2019).

2.3.2 Tau pathology

The recent addition of PET ligands to visualize, map, and quantify tau pathology has contributed information about the temporal and spatial characteristics of tau accumulation in the living brain. To summarize this vast and expanding research field, there are now several tau ligands that reliably detect AD-type tau pathology. The field is still young, lagging behind Aβ imaging, but the results accrued to date corroborate the Braak model of distinct spatial and temporal stages of tau pathology progression in the AD brain (Braak and Braak, 1991), and tau PET is now used in clinical trials to detect pharmacodynamic effects of disease-modifying drug candidates against both Aβ (to monitor downstream tau changes) and tau pathology (Schöll et al., 2019).

2.3.3 Neurodegeneration

Volumetric MRI and fluorodeoxyglucose (FDG) PET have been used commonly as biomarkers for neurodegeneration (Frisoni et al., 2010; Jack et al., 2018). Typical MRI-based measures are derived from T1-weighted imaging and include whole brain atrophy, gray matter atrophy, and regional atrophy (e.g., medial temporal areas, hippocampus, and hippocampal subfields), as well as surface-based cortical thickness (Schwarz et al., 2016). Advances in neuroimaging acquisition and modeling of water diffusion have also facilitated characterization of cortical microstructure, which may provide a sensitive measure of neurodegeneration that precedes overt cell loss (Vogt et al., 2019). FDG PET imaging has long been used clinically for differential diagnosis of AD. Patients with AD dementia show a characteristic pattern of hypometabolism in precuneus, posterior cingulate, parietal cortices, lateral temporal cortex, frontal cortices, as well as medial temporal lobe (Alexander et al., 2002). Hypometabolism likely reflects a combination of synaptic dysfunction, neuronal cell loss, and metabolic dysfunction, in addition to potentially being influenced by astroglial glutamate transport. Despite the complexity of the signal, FDG PET remains an attractive biomarker given its likely sensitivity to synaptic dysfunction, as well as its capability of detecting abnormalities in the preclinical asymptomatic stage of AD.

2.4 Alzheimer's disease biomarkers in the cerebrospinal fluid

2.4.1 Amyloid β pathology

Alzheimer's disease CSF is characterized by reduced (to approximately 50% of normal levels) concentration of the 42 amino acid-long and aggregation-prone form of Aβ (Aβ42). This reduction is caused by the pathological retention of Aβ42 in the brain parenchyma. The diagnostic accuracy for Aβ pathology can be increased by dividing the concentration of aggregation-prone Aβ42 to soluble Aβ40. The CSF Aβ42/Aβ40 ratio, which compensates for interindividual differences in amyloidogenic APP processing (high vs low Aβ producers), is close to 100% concordant with amyloid PET, and discordant cases are often CSF positive and PET negative, turning amyloid PET positive within a few years (Hansson et al., 2019; Lewczuk et al., 2017; Mattsson et al., 2019b).

2.4.2 Tau pathology

Cerebrospinal fluid total-tau (T-tau and phosphorylated-tau (P-tau), together with the CSF Aβ42/Aβ40 ratio, have been proposed as biomarkers that biologically define AD (Jack et al., 2018), and are considered diagnostic in the research criteria for AD (Dubois et al., 2014). Both CSF T-tau and P-tau concentrations are quite AD specific across the neurodegenerative dementias (Itoh et al., 2001; Skillbäck et al., 2015). The most likely explanation for this is that the increased CSF levels of tau (both T-tau and P-tau; in AD, these two markers correlate strongly) are due to increased phosphorylation and secretion of tau from neurons as a neuronal response to Aβ exposure (Maia et al., 2013; Sato et al., 2018). Cerebrospinal fluid T-tau and P-tau may thus be regarded as predictive markers for AD-type neurodegeneration and tangle formation, but not direct markers of these processes (and not markers of non-AD tauopathies, for which we need to continue to look for better biomarkers).

2.4.3 Neurodegeneration

Traditionally, CSF T-tau has been considered a standard biomarker for AD-type neurodegeneration. Although we are revising our interpretation of CSF T-tau in an AD context (a biomarker of increased tau secretion from neurons affected by Aβ pathology and a predictive, but not direct biomarker for neurodegeneration in AD), MRI remains the prototype neurodegeneration biomarker in AD. Recently, NfL has emerged as a general biomarker for neuroaxonal degeneration and injury, irrespective of the cause (Khalil et al., 2018). The biomarker can be measured in both CSF and plasma (or serum), and virtually all CSF findings have been replicated in the blood with sensitive assays. The highest NfL levels are seen in frontotemporal, vascular, and HIV-associated dementias (Bridel et al., 2019). However, the findings in familial AD are also very clear; mutation carriers show a sudden change in their blood NfL levels around a decade before the expected clinical onset, which probably marks the onset of neurodegeneration, and the higher the increase, the more rapid the clinical disease progression (Preische et al., 2019; Weston et al., 2019). In sporadic AD, elevated plasma NfL levels are clearly associated with neurodegeneration, ascertained as brain atrophy on MRI, but also with Aβ positivity and tau positivity, most likely due to the multitude of neurodegenerative changes that may result in NfL increase in people older than 70 years of age (Mattsson et al., 2019a).

2.5 Alzheimer's disease blood biomarkers

Blood-based biomarkers of AD have lagged behind brain imaging and CSF biomarkers, but recent developments are enabling the use of blood-based biomarkers in AD research. The blood-brain barrier is altered in aging and AD (Viggars et al., 2011). The increased permeability between the brain and the periphery makes it possible for blood-based biomarkers to be representative of preclinical changes in AD (Montagne et al., 2015). Extant proteomic methods to measure blood-based biomarkers for AD include mass spectrometry, immunocapture, and aptamer-based techniques. However, issues around the lower limit of detection, depletion of lower molecular weight proteins, and antibody availability have limited the use of these methods in particular (Baird et al., 2015). More recently, ultrasensitive immunoassays coupled with mass spectrometry have shown greater promise (Andreasson et al., 2016). For example, the commercially available single-molecule array (Simoa™) is a novel method to measure Aβ40, Aβ42, tau, and NfL in the plasma (Zetterberg et al., 2011) that is increasingly used in research, whereas CSF and brain imaging biomarkers are relatively well established and standardized in AD research; blood-based biomarkers need further development. Thus, we provide below a brief review of the recent literature for each plasma biomarker.

Using Simoa™ technology in 248 participants with subjective cognitive decline from the SCIENCe project and Amsterdam Dementia Cohort, plasma Aβ42/Aβ40 ratio and plasma Aβ42, but not in plasma total-tau, identified abnormal CSF-amyloid status suggesting that plasma Aβ42/Aβ40 has the potential to be used as a screening measure to identify AD-related neuropathological changes in cognitively normal individuals with subjective cognitive decline (Verberk et al., 2018). In this cohort study, plasma Aβ42 and Aβ40 ascertained with Elecsys immunoassays (Roche Diagnostics) predicted Aβ status, as defined by the cerebrospinal fluid Aβ42/Aβ40 ratio, with an area under the receiver operating curve (AUC) = 0.80, independent of age, *APOE* genotype, or cognitive status (Palmqvist et al., 2019a, 2019b). Plasma Aβ42/Aβ40, measured with Simoa™, was also predictive of cerebral amyloidosis in a sample of 276 cognitively intact individuals with subjective memory complaints from the INSIGHT-preAD study, a French academic university-based cohort that is part of the Alzheimer Precision Medicine Initiative Cohort Program (Vergallo et al., 2019).

Plasma tau ascertained using the Simoa™ assay has been weakly correlated or not correlated with CSF tau levels in both the BioFINDER study and Mayo Clinic Study of Aging (Dage et al., 2016). However, in a sample of patients with AD, MCI and cognitively healthy controls from the Alzheimer's Disease Neuroimaging Initiative (ADNI), higher levels of Simoa™-based plasma tau were observed in AD dementia compared with both amnestic MCI patients and cognitively healthy controls (Mattsson et al., 2016). The Framingham Heart Study examined the use of plasma total-tau, measured with Simoa™, as a blood biomarker for dementia and related endophenotypes. In 1453 participants, a one standard deviation increase in the log of plasma total-tau level was associated with a 35% increase in AD dementia risk. Higher plasma total-tau was also associated with poorer cognition, and smaller hippocampi and more neurofibrillary tangles and microinfarcts at autopsy (Pase et al., 2019). An additional study also reported a significant correlation of plasma phosphorylated-tau and total-tau with brain tau deposition by PET imaging (Park et al., 2019). Mielke et al. (2018) originally demonstrated a correlation between P-tau181, and amyloid and tau PET, which indicates that plasma P-tau181 is a good predictor of brain AD pathology. These findings were replicated in a recent study by Palmqvist et al. (2019a, 2019b), demonstrating that plasma P-tau181 associates with amyloid PET positivity and correlates strongly with CSF P-tau181. Interestingly, the change in plasma P-tau181 became significant before amyloid PET, but after CSF and plasma Aβ42, i.e., already at the sub-PET threshold Aβ pathology (Palmqvist et al., 2019a, 2019b). Thus, plasma P-tau181 might be useful both diagnostically to detect early Aβ-related tau dysmetabolism, as well as for disease staging. Recent large validation studies have shown very similar results (Janelidze et al., 2020; Thijssen et al., 2019), corroborating plasma P-tau as a robust blood biomarker for AD pathology that should be relatively easy to standardize and implement in clinical laboratory practice.

Simoa™-based plasma NfL correlations with CSF NfL levels are high (Gisslén et al., 2016) and reports from ADNI suggest that plasma NFL had high diagnostic accuracy for identifying patients with dementia vs controls (AUC = 0.87) (Mattsson et al., 2017). A study of 1583 participants from ADNI reported the association between clinical diagnoses, CSF biomarkers, imaging measures, and cognition with longitudinal Simoa™-based plasma NfL levels (Mattsson et al., 2019a, 2019b). Levels of plasma NfL increased over 11 years in patients with MCI and AD dementia.

In summary, the data on plasma AD biomarkers is sparse, as compared with CSF and imaging biomarkers, and there is assay variability. Thus, a recent review reported insufficient data to perform a meta-analysis of studies using plasma AD biomarkers (Olsson et al., 2016). However, this is likely to change during the coming years; further validation of the plasma AD biomarkers is underway in both clinical and observational studies to confirm their utility in tracking AD neuropathology.

3. Summary of studies relating type 2 diabetes mellitus and glycemia to Alzheimer's disease biomarkers

Neuropathology studies from Finland (Ahtiluoto et al., 2010; Alafuzoff et al., 2009) and the United States (Arvanitakis et al., 2006; Beeri et al., 2005; Thambisetty et al., 2013) found that T2DM history was not related to higher

AD neuropathology. However, a study of Japanese-Americans found that T2DM is associated with increased AD neuropathology (Peila et al., 2002), and a Japanese autopsy study found that increased glycemia 10 years before autopsy was related to increased AD neuropathology (Matsuzaki et al., 2010).

A study from the Mayo Clinic Study of Aging (MCSA) in the United States among 749 individuals without dementia with a mean age of 79 years reported that T2DM was related to an AD brain metabolism pattern on FDG PET, but not with brain Aβ ascertained by PET (Roberts et al., 2014). The Baltimore Longitudinal Study of Aging (BLSA) (Thambisetty et al., 2013) found among 197 individuals of aged 88 years at autopsy that glucose assessed approximately at age 66 years was not related to AD neuropathology. The same study had data on 53 individuals with Aβ PET at age 79 years and found no evidence of a relation of T2DM and brain Aβ. The Atherosclerosis Risk in Communities (ARIC) Study reported that T2DM was not related to Aβ positivity among 346 participants with a mean age of 75.8 years (Gottesman et al., 2017). A study in the ADNI among 816 individuals with a mean age of 74.3 years reported that T2DM was not associated with *in vivo* Aβ in PET or CSF, but T2DM was associated with higher tau in CSF and lower cortical thickness (Moran et al., 2015). Another analysis of data from 900 participants in ADNI who underwent lumbar puncture with a mean age of 73.54 years (McIntosh and Nation, 2019) showed that individuals with untreated T2DM had higher biomarkers of AD in the CSF (phosphorylated-tau, total-tau, and phosphorylated-tau/Aβ 1–42 levels) than the euglycemic, prediabetes, and treated T2DM groups, and these individuals also had a faster progression to dementia. Another analysis from ADNI relating T2DM status to CSF and cortical Aβ found that individuals with T2DM had lower cerebral Aβ in the anterior cingulate, precuneus, and temporal lobe than individuals with T2DM, and higher levels of CSF Aβ 1-42 (Li et al., 2018).

Fewer studies have examined the relationship of T2DM with plasma biomarkers. Plasma Aβ 1-42 and Aβ 40 levels were compared between 194 individuals with T2DM and 194 matched controls in an Australian cohort (Peters et al., 2017). Median Aβ 40 and Aβ 42 concentrations were significantly lower in those with T2DM. A Korean study compared levels of Aβ autoantibody measured by ELISA in serum samples of T2DM patients ($n = 92$) and age-matched controls ($n = 106$) (Kim et al., 2010). Levels of Aβ autoantibody were increased in T2DM compared with age-matched controls, and this relationship was more apparent in women.

In summary, the studies relating T2DM with AD biomarkers seem to show that T2DM is not related to amyloid deposition, but is related to neurodegeneration ascertained on MRI, with tau-related measures on CSF or in FDG-PET. Table 1 shows further details of the studies summarized above.

4. Summary of studies relating adiposity, insulin resistance, and metabolic syndrome to Alzheimer's disease biomarkers

A study of 942 individuals in the MCSA has found that in individuals in the age range of 70 to 79 years, obesity was not related to amyloid burden on PET or neurodegeneration on MRI (Vemuri et al., 2017). An analysis of two subsamples of BLSA, in which 191 participants underwent autopsy and neuropathological assessment, and 75 non-demented individuals underwent brain amyloid imaging, examined the association between midlife adiposity ascertained as BMI at 50 years of age and AD neuropathology (Willette et al., 2015a). On the whole, higher midlife BMI was associated with greater Braak neurofibrillary, but not CERAD (Consortium to Establish a Registry for Alzheimer's Disease) neuritic plaque scores at autopsy. Higher midlife BMI was also associated with greater fibrillar amyloid on PET, but without statistical significance.

A cross-sectional analysis of the 150 cognitively normal, late middle-aged (mean [SD] age, 60.7 [5.8] years) adults from the Wisconsin Registry for Alzheimer's Prevention (WRAP) study found that insulin resistance defined by the homeostatic model assessment of peripheral insulin resistance (HOMA-IR) was related to higher neurodegeneration, ascertained as lower global glucose metabolism on FDG-PET (Willette et al., 2015a). Another analysis from the same study in 186 individuals with amyloid PET found that higher insulin resistance was associated with higher amyloid burden among individuals with normoglycemia (Willette et al., 2015b). Another analysis from the same study in 113 individuals with CSF biomarkers found that higher insulin resistance among asymptomatic APOEε4 carriers was associated with higher P-Tau181 and T-tau in late middle age (Starks et al., 2015). An analysis of the BLSA in individuals with data on insulin resistance, neuropathology at autopsy, and amyloid PET found no association between insulin resistance and amyloid burden (Thambisetty et al., 2013). A small Finish study in 60 healthy volunteers found that individuals with insulin resistance ascertained by HOMA-IR had higher amyloid burden on amyloid PET (Ekblad et al., 2018). An analysis of the Australian AIBL study in 1264 participants found that insulin

Table 1 Summary of studies relating type 2 diabetes mellitus and glycemia with Alzheimer's disease biomarkers.

Author and year	Study population and design	Metabolic construct and measure	AD biomarker constructs and measures	Summary of results
Ahtiluoto et al. (2010)	Autopsies were performed in 291 persons in the Vantaa 85 study in Vantaa, Finland, who died during the follow-up (48% of total population)	Type 2 diabetes mellitus (T2DM) was assessed according to self-report, medical record, or use of antidiabetic medication	AD neuropathology on autopsy	Individuals with T2DM were less likely to have β-amyloid (hazard ratio [HR] [95% confidence interval (CI)] was 0.48 [0.23–0.98]) and tangles (HR [95% CI] 0.72 [0.39–1.33])
Alafuzoff et al. (2009)	701 subjects who underwent autopsy in Kuopio, Finland	T2DM status defined by medical records	AD neuropathology on autopsy	The loads of Aβ, neuritic plaques (NP), and neurofibrillary tangles (NFT) in the brains were not related to T2DM
Arvanitakis et al. (2006)	Participants were 233 older Catholic clergy in the Religious Orders Study, with brain autopsy at the time of death (mean age 86 years, 45% men)	T2DM was identified by annual direct medication inspection and history	AD neuropathology on autopsy	T2DM (present in 15% patients) was not related to global AD pathology score, or to specific measures of neuritic plaques, diffuse plaques or tangles, or to amyloid burden or tangle density
Beeri et al. (2005)	385 consecutive autopsies of residents of a nursing home, 15.8% diabetics	T2DM status by medical records	AD neuropathology on autopsy	Diabetics had significantly fewer neuritic plaques ($P=0.008$) and NFTs ($P=0.047$) in the cerebral cortex than did non-diabetics. In the hippocampus, diabetics had significantly lower plaque ratings than did non-diabetics ($P=0.019$)

Gottesman et al. (2017)	Longitudinal study of 346 participants without dementia in the Atherosclerosis Risk in Communities (ARIC)-PET Amyloid Imaging Study	T2DM at midlife was defined as fasting glucose $\geq$126 mg/dL, non-fasting glucose $\geq$200 mg/dL, self-report of physician-diagnosed T2DM, or use of oral T2DM medications or insulin	Brain amyloid burden measured with ^{18}F-Florbetapir PET scans	T2DM at midlife not associated with brain amyloid
Kim et al. (2010)	Cross-sectional study of 92 patients with T2DM from Seoul National University Hospital and age-matched controls	T2DM diagnosis not defined	Aβ1-42 antibodies measured with ELISA	Aβ autoantibody levels were increased in T2DM compared with age-matched controls by 45.4 ± 8.1% ($P < 0.001$)
Li et al. (2018)	Cross-sectional analyses of 812 participant of the Alzheimer's disease neuroimaging initiative (ADNI)	T2DM ascertained by medical history	CSF Aβ1-42 and PET with ^{18}F-florbetapir	CSF Aβ1-42 was higher in the T2DM group than the non-diabetic group. Participants with T2DM had a lower cerebral cortical Aβ in the anterior cingulate, precuneus, and temporal lobe than controls
McIntosh and Nation (2019)	Cross-sectional study of 900 participants in ADNI	T2DM category by fasting glucose and medication history	CSF phosphorylated-tau (P-tau), total-tau (T-tau), and β-amyloid 1–42 (Aβ1-42)	The untreated diabetes (UD) group displayed greater P-tau, Ttau, and P-tau/Aβ1-42 levels than the euglycemic (EU), prediabetes (PD), and treated diabetes groups (P values <0.05) and higher T-tau/Aβ1-42 than the EU and PD groups (P values <0.05)
Matsuzaki et al. (2010)	135 persons with autopsies from the town of Hisayama in Fukuoka prefecture in Japan	Fasting glucose, 2-h postload plasma glucose, from 10 years before autopsy	AD neuropathology on autopsy	Higher levels of 2-h postload plasma glucose were associated with increased risk for NPs, but not NFTs. The coexistence of hyperglycemia and APOE ε4 increased the risk for NP formation

Continued

Table 1 Summary of studies relating type 2 diabetes mellitus and glycemia with Alzheimer's disease biomarkers.—cont'd

Author and year	Study population and design	Metabolic construct and measure	AD biomarker constructs and measures	Summary of results
Moran et al. (2015)	Cross-sectional study of 816 participants in ADNI, mean age 75 years, 816 with MRI, 102 with brain amyloid imaging, 415 with CSF	T2DM by glucose levels or T2DM medications or history. There were 124 people with T2DM and 692 without T2DM	Amyloid PET scans with ^{11}C-PIB, CSF amyloid beta 1–42, T-tau, and P-tau 181, hippocampal volume on MRI	T2DM was associated with lower bilateral frontal and parietal cortical thickness (mL) ($\beta = -0.03$, $P = 0.01$). T2DM was not associated with ^{11}C Pittsburgh compound B standardized uptake value ratio (AU) in any brain region or with CSF $A\beta_{42}$ levels (pg/mL). T2DM was associated with greater CSF total-tau (pg/mL) ($\beta = 16.06$, $P = 0.04$) and phosphorylated-tau ($\beta = 5.84$, $P = 0.02$)
Peila et al. (2002)	216 Japanese-American men enrolled in the Honolulu-Asia Aging Study, who underwent autopsy	T2DM was ascertained by interview and direct glucose testing	AD neuropathology on autopsy	Participants with T2DM and the APOEε4 allele had a higher number of hippocampal neuritic plaques (relative risk [RR] 3.0 [CI 1.2–7.3]) and neurofibrillary tangles in the cortex (RR 3.5 [1.6–7.5]) and hippocampus (IRR 2.5 [1.5–3.7]), and they had a higher risk of cerebral amyloid angiopathy (RR 6.6, 1.5–29.6)
Peters et al. (2017)	194 participants with T2DM recruited from the community-based Fremantle T2DM Study Phase II cohort (mean age 71 years, 59% men) and 194 age-, sex-, and APOEε4 allele-matched, control individuals without T2DM from the Australian Imaging, Biomarkers and Lifestyle Study	T2DM status vs controls	Plasma Aβ40 and Aβ42 were measured with multiplex microsphere-based immunoassay	Median Aβ40 and Aβ42 concentrations were significantly lower in those with T2DM (Aβ40 median [interquartile range]: 125.0 [52.6–148.3] vs 149.3 [134.0–165.6] pg/mL; Aβ42: 26.9 [14.5–38.3] vs 33.6 [28.0–38.9] pg/mL, both $P < 0.001$), while the ratio Aβ42:Aβ40 was significantly higher (0.26 [0.23–0.32] vs 0.22 [0.19–0.25], $P < 0.001$)

Roberts et al. (2014)	749 participants without dementia in the Mayo Clinic Study of Aging with a median age of 79 years	T2DM ascertained by medical records, glycemia assessed with HbA1c	Neurodegeneration assessed with flurodeoxyglucose (FDG)-PET. Amyloid assessed with ^{11}C-PIB PET	T2DM and higher HbA1c was related to AD hypometabolism signature on FDG PET, but were not related to amyloid burden
Thambisetty et al. (2013)	Baltimore Longitudinal Study of Aging, 197 participants with brain autopsy, 53 patients with amyloid PET, and with glucose testing two or more times	Glucose intolerance ascertained with oral glucose tolerance test (OGTT), use of T2DM medications	AD neuropathology on autopsy, *in vivo* amyloid burden assessed ^{11}C-PIB	No significant correlations between measures of brain AD pathology or ^{11}C-PiB β-amyloid load and glucose intolerance in those who had a mean (SD) of 6.4 (3.2) OGTT during 22.1 (8.0) years of follow-up. Thirty patients with frank T2DM who received medications also had AD pathology scores that were similar to those of the cohort as a whole

resistance ascertained using HOMA-IR was associated with higher CSF T-tau and P-tau levels, but not with CSF amyloid, amyloid PET, or MRI measures of neurodegeneration (Laws et al., 2017).

Few studies have examined the association of metabolic syndrome with AD biomarkers. A recent analysis of 165 cognitively normal participants with a mean age of 76.4 years from the BLSA has shown evidence of accelerated Aβ accumulation in those who were amyloid positive and had metabolic syndrome followed for an average of 2.6 years (Gomez et al., 2018). Among metabolic syndrome components, only glucose and blood pressure showed associations with some of the outcomes.

In summary, as compared with studies for T2DM, there is a dearth of studies examining adiposity, insulin resistance, and metabolic syndrome in relation to AD biomarkers. The associations in these studies are inconsistent. These inconsistencies may result from the relative lack of studies, from the differences in periods of the lifespan in which the associations are examined, and in the case of longitudinal studies, the lack of concurrent metabolic and AD biomarker measures. Table 2 provides greater details of the studies summarized above.

5. The promise of the field of metabolomics in the study of Alzheimer's disease biomarkers

Recent advances in analytical chemistry, including mass spectroscopy (MS) of different types, liquid chromatography (LC)/MS, gas chromatography (GC)/MS, and nuclear magnetic resonance (NMR) (Anderson and Kodukula, 2014), have led to a relatively new field known as *metabolomics*. *Metabolomics* is the comparative analysis of the endogenous metabolites present in biological systems at a given physiological/pathological state. It has been used in clinical settings, pharmacology, toxicology, and more recently in psychiatric diseases and neurodegenerative disorders (Anderson and Kodukula, 2014; Kaddurah-Daouk and Krishnan, 2008). Metabolomics permits the simultaneous measurement of thousands of metabolites allowing the detection of perturbations in interconnected pathways and networks. It is an interdisciplinary science that combines analytical chemistry, biochemistry, statistics, and bioinformatics (Han et al., 2011; Kaddurah-Daouk et al., 2011, 2013; Kaddurah-Daouk and Krishnan, 2008; Motsinger-Reif et al., 2013). Although the implementation of metabolomics in AD research is more recent than for conditions such as T2DM, it is rapidly increasing. We provide a brief summary of the recent findings here. Metabolomics studies in the

Table 2 Summary of studies relating adiposity, insulin resistance, and metabolic syndrome with Alzheimer's disease biomarkers.

Author and year	Study population and design	Metabolic construct and measure	AD biomarker constructs and measures	Summary of results
Ekblad et al. (2018)	Longitudinal study of 60 elderly volunteers without dementia (mean age at baseline 55.4 and at follow-up 70.9 years, 55.5% women) from the Finnish population-based, nationwide Health2000 study	Insulin resistance (IR) ascertained with homeostatic model assessment of insulin resistance (HOMA-IR)	Amyloid ascertained with ^{11}C-PIB PEtr	An amyloid-positive PET scan was found in 33.3% of the IR negative group and 60.0% of the IR positive group (odds ratio 3.0, 95% confidence interval 1.1–8.9, $P=0.04$). Higher midlife, but not late-life continuous HOMA-IR was associated with a greater brain amyloid burden at follow-up after multivariate adjustments for other cognitive and metabolic risk factors ($\beta=0.11$, 95% confidence interval 0.002–0.22, $P=0.04$)
Gomez et al. (2018)	Longitudinal study of 165 cognitively normal participants of the Baltimore Longitudinal Study of Aging (BLSA) Neuroimaging substudy, aged 55–92 years (mean baseline age = 76.4 years, 85 participants were men)	Metabolic syndrome was defined using the revised Third Adults Treatment Panel of the National Cholesterol Education Program criteria	Amyloid burden assessed with ^{11}C-PIB PET; participants received an average of 2.5 PET-PiB scans over an average interval of 2.6 (3.08 SD) years between first and last visits	Metabolic syndrome was not associated with baseline PiB positivity or concurrent regional Aβ. Metabolic syndrome was associated with increased rates of Aβ accumulation in superior parietal and precuneus regions over time in the PiB + group. Elevated fasting glucose and blood pressure showed individual associations with accelerated Aβ accumulation
Laws et al. (2017)	Cross-sectional study of 1264 participants of the Australian Imaging, Biomarkers and Lifestyle Study, 379 with amyloid PET	IR measured with HOMA-IR, pancreatic β-cell function measured with HOMA-B	Multiple AD biomarkers (CSF Aβ42, T-tau/P-tau, hippocampal volume and neocortical Aβ-amyloid burden with. ^{11}C-PIB PiB, ^{18}F-florebetapir or ^{18}F-flutemetamol	In cognitively normal adults, higher HOMA-IR was associated higher CSF T-tau ($P=0.008$) and P-tau ($P=0.014$) levels. No association was observed with CSF Aβ or imaging modalities. Biomarkers in cognitively normal adults

Continued

Table 2 Summary of studies relating adiposity, insulin resistance, and metabolic syndrome with Alzheimer's disease biomarkers.—cont'd

Author and year	Study population and design	Metabolic construct and measure	AD biomarker constructs and measures	Summary of results
Matsuzaki et al. (2010)	135 persons with autopsies from the town of Hisayama in the Fukuoka prefecture in Japan	Insulin resistance ascertained with HOMA-IR 10 years before autopsy	AD neuropathology on autopsy	Higher levels HOMA-IR were associated with increased risk for NPs after adjustment for age, sex, systolic blood pressure, total cholesterol, body mass index, habitual smoking, regular exercise, and cerebrovascular disease. However, there were no relationships between T2DM-related factors and NFTs. The coexistence of IR and APOE ε4 increased the risk for NP formation
Vemuri et al. (2017)	942 elderly individuals (mean age 79.7 years) with magnetic resonance imaging and Pittsburgh compound B-positron emission tomography scans enrolled in the Mayo Clinic Study of Aging	Obesity defined as body mass index $\geq 30\,k/m^2$ defined at midlife	Amyloid burden measured With ^{11}C-PIB PET; AD pattern neurodegeneration from MRI	Obesity at midlife related to later life neurodegeneration, but not amyloid burden
Willette et al. (2015a)	Cross-sectional study of 150 cognitively normal, late middle-aged adults (mean age = 60 years) from the Wisconsin Registry for Alzheimer's Prevention (WRAP) study, a general community sample enriched for AD parental history	Insulin resistance measured with HOMA-IR	Neurodegeneration measured with FDG-PET	Higher HOMA-IR was associated with lower global glucose metabolism ($\beta = -0.29$; $P < 0.01$) and lower regional glucose metabolism across large portions of the frontal, lateral parietal, lateral temporal, and medial temporal lobes ($P < 0.05$, familywise error corrected). The association was especially robust in the left medial temporal lobe ($R2 = 0.178$)

Willette et al. (2015b)	Cross-sectional study of asymptomatic, 186 late middle-aged adults from the Wisconsin Registry for Alzheimer's Prevention	Insulin resistance measured with HOMA-IR	Amyloid burden measured with ^{11}C-PIB PET	In participants with normoglycemia, but not hyperglycemia, higher IR corresponded to higher PiB uptake in frontal and temporal areas, reflecting increased amyloid deposition
Starks et al. (2015)	113 participants enrolled in the Wisconsin Registry for Alzheimer's Prevention study (mean age = 60.6 years)	Insulin resistance measured with HOMA-IR	Cerebrospinal fluid $A\beta_{42}$, P-Tau181, and T-Tau	No significant main effects of HOMA-IR on P-Tau181, T-Tau, or $A\beta_{42}$ were observed; however, among APOEε4 carriers, higher HOMA-IR was associated with higher P-tau181 and T-tau. Among APOEε4 non-carriers, HOMA-IR was negatively associated with P-tau181 and T-tau. No effects of IR were found on $A\beta_{42}$ levels in CSF
		Resistance (HOMA-IR)		

BLSA (Snowden et al., 2017) have reported associations of unsaturated fatty acids with AD neuropathology at autopsy and sphingolipids and glycerophospholipids (Varma et al., 2018) in relation to MRI measures of AD-related brain atrophy and CSF biomarkers of AD pathology. The Alzheimer's Disease Metabolomics Consortium has conducted studies in multiple cohorts with AD biomarkers, including the ADNI, and has reported associations of metabolites, including phosphatidylcholines and sphingomyelins with Aβ42 positivity on CSF, with T-tau/Aβ42 ratio, and neurodegeneration on MRI (Toledo et al., 2017). Bile acids in serum, which are related to cholesterol metabolism, have also been related to AD biomarkers (Nho et al., 2019), including CSF Aβ42 and CSF P-tau181. Combinations of lipids have also been related to AD biomarkers (Barupal et al., 2019). Given the increasing implementation of metabolomics in AD biomarkers research, a wealth of studies is expected in coming years.

6. Caveats, recommendations, and future directions

Although AD biomarker research currently focuses on amyloid, tau, and neurodegeneration as primary constructs, there are other neurodegenerative diseases that may overlap with AD, such as frontotemporal dementia (FTD), in which inclusions can consist of several different proteins, most typically tau and/or transactive response DNA-binding protein 43 (TDP-43), Parkinson's disease dementia (PDD) and dementia with Lewy bodies (DLB), in which α-synuclein inclusions represent an important part of the pathology. Furthermore, these pathologies are often seen together with classical AD pathology; neuropathological studies show that around 50% of AD patients have Lewy bodies (Robinson et al., 2018), while the corresponding figure for TDP-43 is 30–40% (James et al., 2016). Whether these copathologies contribute to disease progression in AD is not known. The development of specific biomarkers for these pathologies is underway and should be integrated into AD biomarker research.

As shown in our summary of the existing literature, most studies relating metabolic factors to AD biomarkers are cross-sectional, and most studies reported as longitudinal have measurement of metabolic factors and AD biomarkers at discordant points in time. More longitudinal studies are needed with concurrent assessment of metabolic factors and AD biomarkers in order to improve the opportunity to assess causality. Ideally, AD biomarkers should be integrated in clinical trials of interventions that affect metabolic factors. More longitudinal biomarker-based studies are needed to tease out cause-and-effect relationships between metabolic factors and AD-type neuropathologies. Advances in blood-based AD biomarkers, which are less costly and invasive compared with CSF and brain imaging biomarkers, could facilitate their widespread implementation in studies examining the metabolic contribution to AD.

Acknowledgments

JAL is supported by grants from the National Institutes of Health in the United States including K24AG045334, R01AG050440, R01AG062624, R01AG058571, R01AG057898 RF1AG051556, R01AG055299, R01AG058969, and P30AG0593003. HZ is a Wallenberg Scholar supported by grants from the Swedish Research Council (#2018-02532), the European Research Council (#681712), Swedish State Support for Clinical Research (#ALFGBG-720931), and the UK Dementia Research Institute at UCL.

Disclosures

JAL is a consultant to vTv therapeutics and receives a stipend from Wolters-Kluwer as editor-in-chief of the Journal Alzheimer's Disease and Associated Disorders. HZ has served at scientific advisory boards for Roche Diagnostics, Denali, Wave, Samumed and CogRx, has given lectures in symposia sponsored by Alzecure and Biogen, and is a cofounder of Brain Biomarker Solutions in Gothenburg AB, a GU Ventures-based platform company at the University of Gothenburg.

References

Ahtiluoto, S., Polvikoski, T., Peltonen, M., Solomon, A., Tuomilehto, J., Winblad, B., et al. (2010). Diabetes, Alzheimer disease, and vascular dementia: A population-based neuropathologic study. *Neurology*, *75*(13), 1195–1202. https://doi.org/10.1212/WNL.0b013e3181f4d7f8.

Alafuzoff, I., Aho, L., Helisalmi, S., Mannermaa, A., & Soininen, H. (2009). Beta-amyloid deposition in brains of subjects with diabetes. *Neuropathology and Applied Neurobiology*, *35*(1), 60–68. https://doi.org/10.1111/j.1365-2990.2008.00948.x.

Alexander, G. E., Chen, K., Pietrini, P., Rapoport, S. I., & Reiman, E. M. (2002). Longitudinal PET evaluation of cerebral metabolic decline in dementia: A potential outcome measure in Alzheimer's disease treatment studies. *American Journal of Psychiatry*, *159*(5), 738–745. https://doi.org/10.1176/appi.ajp.159.5.738.

Anderson, D. C., & Kodukula, K. (2014). Biomarkers in pharmacology and drug discovery. *Biochemical Pharmacology*, *87*(1), 172–188. https://doi.org/10.1016/j.bcp.2013.08.026.

Andreasson, U., Blennow, K., & Zetterberg, H. (2016). Update on ultrasensitive technologies to facilitate research on blood biomarkers for central nervous system disorders. *Alzheimer's & Dementia: Diagnosis, Assessment & Disease Monitoring*, *3*, 98–102. https://doi.org/10.1016/j.dadm.2016.05.005.

Arvanitakis, Z., Schneider, J. A., Wilson, R. S., Li, Y., Arnold, S. E., Wang, Z., et al. (2006). Diabetes is related to cerebral infarction but not to AD pathology in older persons. *Neurology*, *67*(11), 1960–1965. https://doi.org/10.1212/01.wnl.0000247053.45483.4e.

Baird, A. L., Westwood, S., & Lovestone, S. (2015). Blood-based proteomic biomarkers of Alzheimer's disease pathology. *Frontiers in Neurology*, *6*, 236. https://doi.org/10.3389/fneur.2015.00236.

Barupal, D. K., Baillie, R., Fan, S., Saykin, A. J., Meikle, P. J., Arnold, M., et al. (2019). Sets of coregulated serum lipids are associated with Alzheimer's disease pathophysiology. *Alzheimer's & Dementia: Diagnosis, Assessment & Disease Monitoring, 11*, 619–627. https://doi.org/10.1016/j.dadm.2019.07.002.

Beeri, M. S., Silverman, J. M., Davis, K. L., Marin, D., Grossman, H. Z., Schmeidler, J., et al. (2005). Type 2 diabetes is negatively associated with Alzheimer's disease neuropathology. *The Journals of Gerontology Series A, Biological Sciences and Medical Sciences, 60*(4), 471–475.

Biessels, G. J., Strachan, M. W., Visseren, F. L., Kappelle, L. J., & Whitmer, R. A. (2014). Dementia and cognitive decline in type 2 diabetes and prediabetic stages: Towards targeted interventions. *The Lancet Diabetes and Endocrinology, 2*(3), 246–255. https://doi.org/10.1016/S2213-8587(13)70088-3.

Braak, H., & Braak, E. (1991). Neuropathological stageing of Alzheimer-related changes. *Acta Neuropathologica, 82*(4), 239–259.

Bridel, C., van Wieringen, W. N., Zetterberg, H., Tijms, B. M., Teunissen, C. E., & The N. F. L. Group (2019). Diagnostic value of cerebrospinal fluid neurofilament light protein in neurology: A systematic review and meta-analysis. *JAMA Neurology, 76*(9), 1035–1048. https://doi.org/10.1001/jamaneurol.2019.1534.

Cohen, A. D., Landau, S. M., Snitz, B. E., Klunk, W. E., Blennow, K., & Zetterberg, H. (2019). Fluid and PET biomarkers for amyloid pathology in Alzheimer's disease. *Molecular and Cellular Neuroscience, 97*, 3–17. https://doi.org/10.1016/j.mcn.2018.12.004.

Dage, J. L., Wennberg, A. M. V., Airey, D. C., Hagen, C. E., Knopman, D. S., Machulda, M. M., et al. (2016). Levels of tau protein in plasma are associated with neurodegeneration and cognitive function in a population based elderly cohort. *Alzheimers Dement, 12*(12), 1226–1234. https://doi.org/10.1016/j.jalz.2016.06.001.

Dubois, B., Feldman, H. H., Jacova, C., Hampel, H., Molinuevo, J. L., Blennow, K., et al. (2014). Advancing research diagnostic criteria for Alzheimer's disease: The IWG-2 criteria. *The Lancet Neurology, 13*(6), 614–629. https://doi.org/10.1016/S1474-4422(14)70090-0.

Ekblad, L. L., Johansson, J., Helin, S., Viitanen, M., Laine, H., Puukka, P., et al. (2018). Midlife insulin resistance, APOE genotype, and late-life brain amyloid accumulation. *Neurology, 90*(13), e1150–e1157. https://doi.org/10.1212/WNL.0000000000005214.

Frisoni, G. B., Fox, N. C., Jack, C. R., Jr., Scheltens, P., & Thompson, P. M. (2010). The clinical use of structural MRI in Alzheimer disease. *Nature Reviews. Neurology, 6*(2), 67–77. https://doi.org/10.1038/nrneurol.2009.215.

Gisslén, M., Price, R. W., Andreasson, U., Norgren, N., Nilsson, S., Hagberg, L., et al. (2016). Plasma concentration of the neurofilament light protein (NFL) is a biomarker of CNS injury in HIV infection: A cross-sectional study. *eBioMedicine, 3*, 135–140. https://doi.org/10.1016/j.ebiom.2015.11.036.

Gomez, G., Beason-Held, L. L., Bilgel, M., An, Y., Wong, D. F., Studenski, S., et al. (2018). Metabolic syndrome and amyloid accumulation in the aging brain. *Journal of Alzheimer's Disease, 65*(2), 629–639. https://doi.org/10.3233/JAD-180297.

Gottesman, R. F., Schneider, A. L., Zhou, Y., Coresh, J., Green, E., Gupta, N., et al. (2017). Association between midlife vascular risk factors and estimated brain amyloid deposition. *JAMA, 317*(14), 1443–1450. https://doi.org/10.1001/jama.2017.3090.

Han, X., Rozen, S., Boyle, S. H., Hellegers, C., Cheng, H., Burke, J. R., et al. (2011). Metabolomics in early Alzheimer's disease: Identification of altered plasma sphingolipidome using shotgun lipidomics. *PLoS One, 6*(7). e21643. https://doi.org/10.1371/journal.pone.0021643.

Hansson, O., Lehmann, S., Otto, M., Zetterberg, H., & Lewczuk, P. (2019). Advantages and disadvantages of the use of the CSF amyloid β (Aβ) 42/40 ratio in the diagnosis of Alzheimer's disease. *Alzheimer's Research & Therapy, 11*(1), 34. https://doi.org/10.1186/s13195-019-0485-0.

Itoh, N., Arai, H., Urakami, K., Ishiguro, K., Ohno, H., Hampel, H., et al. (2001). Large-scale, multicenter study of cerebrospinal fluid tau protein phosphorylated at serine 199 for the antemortem diagnosis of Alzheimer's disease. *Annals of Neurology*, *50*(2), 150–156. https://doi.org/10.1002/ana.1054.

Jack, C. R., Bennett, D. A., Blennow, K., Carrillo, M. C., Dunn, B., Haeberlein, S. B., et al. (2018). NIA-AA research framework: Toward a biological definition of Alzheimer's disease. *Alzheimer's Dementia*, *14*(4), 535–562. https://doi.org/10.1016/j.jalz.2018.02.018.

Jack, C. R., Jr., Knopman, D. S., Jagust, W. J., Petersen, R. C., Weiner, M. W., Aisen, P. S., et al. (2013). Tracking pathophysiological processes in Alzheimer's disease: An updated hypothetical model of dynamic biomarkers. *Lancet Neurology*, *12*(2), 207–216. https://doi.org/10.1016/S1474-4422(12)70291-0.

Jack, C. R., Jr., Knopman, D. S., Jagust, W. J., Shaw, L. M., Aisen, P. S., Weiner, M. W., et al. (2010). Hypothetical model of dynamic biomarkers of the Alzheimer's pathological cascade. *Lancet Neurology*, *9*(1), 119–128.

James, O. G., Doraiswamy, P. M., & Borges-Neto, S. (2015). PET imaging of tau pathology in Alzheimer's disease and tauopathies. *Frontiers in Neurology*, *6*, 38. https://doi.org/10.3389/fneur.2015.00038.

James, B. D., Wilson, R. S., Boyle, P. A., Trojanowski, J. Q., Bennett, D. A., & Schneider, J. A. (2016). TDP-43 stage, mixed pathologies, and clinical Alzheimer's-type dementia. *Brain: A Journal of Neurology*, *139*(11), 2983–2993. https://doi.org/10.1093/brain/aww224.

Janelidze, S., Mattsson, N., Palmqvist, S., Smith, R., Beach, T. G., Serrano, G. E., et al. (2020). Plasma P-tau181 in Alzheimer's disease: Relationship to other biomarkers, differential diagnosis, neuropathology and longitudinal progression to Alzheimer's dementia. *Nature Medicine*, *26*(3), 379–386.

Johnson, K. A., Minoshima, S., Bohnen, N. I., Donohoe, K. J., Foster, N. L., Herscovitch, P., et al. (2013). Update on appropriate use criteria for amyloid PET imaging: Dementia experts, mild cognitive impairment, and education. *Journal of Nuclear Medicine*, *54*(7), 1011–1013.

Kaddurah-Daouk, R., & Krishnan, K. R. R. (2008). Metabolomics: A global biochemical approach to the study of central nervous system diseases. *Neuropsychopharmacology*, *34*(1), 173–186.

Kaddurah-Daouk, R., Rozen, S., Matson, W., Han, X., Hulette, C. M., Burke, J. R., et al. (2011). Metabolomic changes in autopsy confirmed Alzheimer's disease. *Alzheimer's Dementia*, 7(3), 309–317. https://doi.org/10.1016/j.jalz.2010.06.001.

Kaddurah-Daouk, R., Zhu, H., Sharma, S., Bogdanov, M., Rozen, S. G., Matson, W., et al. (2013). Alterations in metabolic pathways and networks in Alzheimer's disease. *Translational Psychiatry*, *3*, e244. https://doi.org/10.1038/tp.2013.18 23571809. PubMed Central PMCID: PMC3641405.

Khalil, M., Teunissen, C. E., Otto, M., Piehl, F., Sormani, M. P., Gattringer, T., et al. (2018). Neurofilaments as biomarkers in neurological disorders. *Nature Reviews. Neurology*, *14*(10), 577–589. https://doi.org/10.1038/s41582-018-0058-z.

Kim, I., Lee, J., Hong, H. J., Jung, E. S., Ku, Y. H., Jeong, I. K., et al. (2010). A relationship between Alzheimer's disease and type 2 diabetes mellitus through the measurement of serum amyloid-beta autoantibodies. *Journal of Alzheimer's Disease*, *19*(4), 1371–1376. https://doi.org/10.3233/JAD-2010-1332.

Klunk, W. E., Koeppe, R. A., Price, J. C., Benzinger, T. L., Devous, M. D., Sr., Jagust, W. J., et al. (2015). The Centiloid Project: Standardizing quantitative amyloid plaque estimation by PET. *Alzheimer's Dementia*, *11*(1), 1–15.e154. https://doi.org/10.1016/j.jalz.2014.07.003.

Lashley, T., Schott, J. M., Weston, P., Murray, C. E., Wellington, H., Keshavan, A., et al. (2018). Molecular biomarkers of Alzheimer's disease: Progress and prospects. *Disease Models & Mechanisms, 11*(5). dmm031781. https://doi.org/10.1242/dmm.031781.

Laws, S. M., Gaskin, S., Woodfield, A., Srikanth, V., Bruce, D., Fraser, P. E., et al. (2017). Insulin resistance is associated with reductions in specific cognitive domains and increases in CSF tau in cognitively normal adults. *Scientific Reports*, 7(1), 9766. https://doi.org/10.1038/s41598-017-09577-4.

Lewczuk, P., Matzen, A., Blennow, K., Parnetti, L., Molinuevo, J. L., Eusebi, P., et al. (2017). Cerebrospinal fluid Aβ42/40 corresponds better than Aβ42 to amyloid PET in Alzheimer's disease. *Journal of Alzheimer's Disease: JAD, 55*(2), 813–822. https://doi.org/10.3233/JAD-160722.

Li, W., Risacher, S. L., Gao, S., Boehm, S. L., 2nd, Elmendorf, J. S., & Saykin, A. J. (2018). Type 2 diabetes mellitus and cerebrospinal fluid Alzheimer's disease biomarker amyloid beta1-42 in Alzheimer's disease neuroimaging initiative participants. *Alzheimer's & Dementia, 10*, 94–98. https://doi.org/10.1016/j.dadm.2017.11.002.

Luchsinger, J. A. (2010). Insulin resistance, type 2 diabetes, and AD: Cerebrovascular disease or neurodegeneration? *Neurology, 75*(9), 758–759. https://doi.org/10.1212/WNL.0b013e3181eee287.

Luchsinger, J. A., & Gustafson, D. R. (2009). Adiposity and Alzheimer's disease. *Current Opinion in Clinical Nutrition and Metabolic Care, 12*(1), 15–21. https://doi.org/10.1097/MCO.0b013e32831c8c71.

Maia, L. F., Kaeser, S. A., Reichwald, J., Hruscha, M., Martus, P., Staufenbiel, M., et al. (2013). Changes in amyloid-β and tau in the cerebrospinal fluid of transgenic mice overexpressing amyloid precursor protein. *Science Translational Medicine, 5*(194). 194re192. https://doi.org/10.1126/scitranslmed.3006446.

Matsuzaki, T., Sasaki, K., Tanizaki, Y., Hata, J., Fujimi, K., Matsui, Y., et al. (2010). Insulin resistance is associated with the pathology of Alzheimer's disease: The Hisayama study. *Neurology, 75*, 764–770.

Mattsson, N., Andreasson, U., Zetterberg, H., & Blennow, K. (2017). Association of plasma neurofilament light with neurodegeneration in patients with Alzheimer disease. *JAMA Neurology, 74*(5), 557–566. https://doi.org/10.1001/jamaneurol.2016.6117.

Mattsson, N., Cullen, N. C., Andreasson, U., Zetterberg, H., & Blennow, K. (2019a). Association between longitudinal plasma neurofilament light and neurodegeneration in patients with Alzheimer disease. *JAMA Neurology, 76*, 791–799. https://doi.org/10.1001/jamaneurol.2019.0765.

Mattsson, N., Palmqvist, S., Stomrud, E., Vogel, J., & Hansson, O. (2019b). Staging β-amyloid pathology with amyloid positron emission tomography. *JAMA Neurology, 76*(11), 1319–1329. https://doi.org/10.1001/jamaneurol.2019.2214.

Mattsson, N., Zetterberg, H., Janelidze, S., Insel, P. S., Andreasson, U., Stomrud, E., et al. (2016). Plasma tau in Alzheimer disease. *Neurology, 87*(17), 1827–1835. https://doi.org/10.1212/WNL.0000000000003246.

McIntosh, E. C., & Nation, D. A. (2019). Importance of treatment status in links between type 2 diabetes and Alzheimer's disease. *Diabetes Care, 42*(5), 972. https://doi.org/10.2337/dc18-1399.

Mielke, M. M., Hagen, C. E., Xu, J., Chai, X., Vemuri, P., Lowe, V. J., et al. (2018). Plasma phospho-tau181 increases with Alzheimer's disease clinical severity and is associated with tau- and amyloid-positron emission tomography. *Alzheimer's Dementia, 14*(8), 989–997. https://doi.org/10.1016/j.jalz.2018.02.013.

Montagne, A., Barnes, S. R., Sweeney, M. D., Halliday, M. R., Sagare, A. P., Zhao, Z., et al. (2015). Blood-brain barrier breakdown in the aging human hippocampus. *Neuron, 85*(2), 296–302. https://doi.org/10.1016/j.neuron.2014.12.032.

Moran, C., Beare, R., Phan, T. G., Bruce, D. G., Callisaya, M. L., Srikanth, V., et al. (2015). Type 2 diabetes mellitus and biomarkers of neurodegeneration. *Neurology*, *85*(13), 1123–1130. https://doi.org/10.1212/WNL.0000000000001982.

Motsinger-Reif, A. A., Zhu, H., Kling, M. A., Matson, W., Sharma, S., Fiehn, O., et al. (2013). Comparing metabolomic and pathologic biomarkers alone and in combination for discriminating Alzheimer's disease from normal cognitive aging. *Acta Neuropathologica Communications*, *1*, 28. https://doi.org/10.1186/2051-5960-1-28.

Muller, M., Tang, M. X., Schupf, N., Manly, J. J., Mayeux, R., & Luchsinger, J. A. (2007). Metabolic syndrome and dementia risk in a multiethnic elderly cohort. *Dementia and Geriatric Cognitive Disorders*, *24*(3), 185–192. https://doi.org/10.1159/000105927.

Narasimhan, K., Govindasamy, M., Gauthaman, K., Kamal, M. A., Abuzenadeh, A. M., Al-Qahtani, M., et al. (2014). Diabetes of the brain: Computational approaches and interventional strategies. *CNS & Neurological Disorders Drug Targets*, *13*(3), 408–417. https://doi.org/10.2174/18715273113126660156.

Nho, K., Kueider-Paisley, A., MahmoudianDehkordi, S., Arnold, M., Risacher, S. L., Louie, G., et al. (2019). Altered bile acid profile in mild cognitive impairment and Alzheimer's disease: Relationship to neuroimaging and CSF biomarkers. *Alzheimer's Dementia*, *15*(2), 232–244. https://doi.org/10.1016/j.jalz.2018.08.012.

Olsson, B., Lautner, R., Andreasson, U., Öhrfelt, A., Portelius, E., Bjerke, M., et al. (2016). CSF and blood biomarkers for the diagnosis of Alzheimer's disease: A systematic review and meta-analysis. *The Lancet Neurology*, *15*(7), 673–684. https://doi.org/10.1016/S1474-4422(16)00070-3.

Ossenkoppele, R., Jansen, W. J., Rabinovici, G. D., Knol, D. L., van der Flier, W. M., van Berckel, B. N. M., et al. (2015). Prevalence of amyloid PET positivity in dementia syndromes: A meta-analysis. *JAMA*, *313*(19), 1939–1949. https://doi.org/10.1001/jama.2015.4669.

Palmqvist, S., Insel, P. S., Stomrud, E., Janelidze, S., Zetterberg, H., Brix, B., et al. (2019a). Cerebrospinal fluid and plasma biomarker trajectories with increasing amyloid deposition in Alzheimer's disease. *EMBO Molecular Medicine*, *11*(12). e11170. https://doi.org/10.15252/emmm.201911170.

Palmqvist, S., Janelidze, S., Stomrud, E., Zetterberg, H., Karl, J., Zink, K., et al. (2019b). Performance of fully automated plasma assays as screening tests for Alzheimer disease-related beta-amyloid status. *JAMA Neurology*, *76*, 1060–1069. https://doi.org/10.1001/jamaneurol.2019.1632.

Park, J. C., Han, S. H., Yi, D., Byun, M. S., Lee, J. H., Jang, S., et al. (2019). Plasma tau/amyloid-beta1-42 ratio predicts brain tau deposition and neurodegeneration in Alzheimer's disease. *Brain*, *142*(3), 771–786. https://doi.org/10.1093/brain/awy347.

Pase, M. P., Beiser, A. S., Himali, J. J., Satizabal, C. L., Aparicio, H. J., DeCarli, C., et al. (2019). Assessment of plasma total tau level as a predictive biomarker for dementia and related endophenotypes. *JAMA Neurology*, *76*(5), 598–606. https://doi.org/10.1001/jamaneurol.2018.4666.

Peila, R., Rodriguez, B. L., Launer, L. J., & Honolulu-Asia Aging Study (2002). Type 2 diabetes, APOE gene, and the risk for dementia and related pathologies: The Honolulu-Asia Aging Study. *Diabetes*, *51*(4), 1256–1262.

Peters, K. E., Davis, W. A., Taddei, K., Martins, R. N., Masters, C. L., Davis, T. M., et al. (2017). Plasma amyloid-beta peptides in type 2 diabetes: A matched case-control study. *Journal of Alzheimer's Disease*, *56*(3), 1127–1133. https://doi.org/10.3233/JAD-161050.

Preische, O., Schultz, S. A., Apel, A., Kuhle, J., Kaeser, S. A., Barro, C., et al. (2019). Serum neurofilament dynamics predicts neurodegeneration and clinical progression in presymptomatic Alzheimer's disease. *Nature Medicine*, *25*(2), 277–283. https://doi.org/10.1038/s41591-018-0304-3.

Roberts, R. O., Knopman, D. S., Cha, R. H., Mielke, M. M., Pankratz, V. S., Boeve, B. F., et al. (2014). Diabetes and elevated hemoglobin A1c levels are associated with brain hypometabolism but not amyloid accumulation. *Journal of Nuclear Medicine, 55*(5), 759–764. https://doi.org/10.2967/jnumed.113.132647.

Robinson, J. L., Lee, E. B., Xie, S. X., Rennert, L., Suh, E., Bredenberg, C., et al. (2018). Neurodegenerative disease concomitant proteinopathies are prevalent, age-related and APOE4-associated. *Brain: A Journal of Neurology, 141*(7), 2181–2193. https://doi.org/10.1093/brain/awy146.

Sato, C., Barthélemy, N. R., Mawuenyega, K. G., Patterson, B. W., Gordon, B. A., Jockel-Balsarotti, J., et al. (2018). Tau kinetics in neurons and the human central nervous system. *Neuron, 97*(6), 1284–1298.e1287. https://doi.org/10.1016/j.neuron.2018.02.015.

Schöll, M., Maass, A., Mattsson, N., Ashton, N. J., Blennow, K., Zetterberg, H., et al. (2019). Biomarkers for tau pathology. *Molecular and Cellular Neurosciences, 97*, 18–33. https://doi.org/10.1016/j.mcn.2018.12.001.

Schwarz, C. G., Gunter, J. L., Wiste, H. J., Przybelski, S. A., Weigand, S. D., Ward, C. P., et al. (2016). A large-scale comparison of cortical thickness and volume methods for measuring Alzheimer's disease severity. *NeuroImage Clinical, 11*, 802–812. https://doi.org/10.1016/j.nicl.2016.05.017.

Skillbäck, T., Farahmand, B. Y., Rosén, C., Mattsson, N., Nägga, K., Kilander, L., et al. (2015). Cerebrospinal fluid tau and amyloid-β1-42 in patients with dementia. *Brain, 138*(9), 2716–2731. https://doi.org/10.1093/brain/awv181.

Snowden, S. G., Ebshiana, A. A., Hye, A., An, Y., Pletnikova, O., O'Brien, R., et al. (2017). Association between fatty acid metabolism in the brain and Alzheimer disease neuropathology and cognitive performance: A nontargeted metabolomic study. *PLoS Medicine, 14*(3). e1002266. https://doi.org/10.1371/journal.pmed.1002266.

Starks, E. J., Patrick O'Grady, J., Hoscheidt, S. M., Racine, A. M., Carlsson, C. M., Zetterberg, H., et al. (2015). Insulin resistance is associated with higher cerebrospinal fluid tau levels in asymptomatic APOEvarepsilon4 carriers. *Journal of Alzheimer's Disease, 46*(2), 525–533. https://doi.org/10.3233/JAD-150072.

Thambisetty, M., Jeffrey Metter, E., Yang, A., Dolan, H., Marano, C., Zonderman, A. B., et al. (2013). Glucose intolerance, insulin resistance, and pathological features of Alzheimer disease in the Baltimore Longitudinal Study of Aging. *JAMA Neurology, 70*(9), 1167–1172. https://doi.org/10.1001/jamaneurol.2013.284.

Thijssen, E. H., La Joie, R., Wolf, A., Strom, A., Wang, P., Iaccarino, L., et al. (2019). Diagnostic value of plasma phosphorylated tau181 in Alzheimer's disease and frontotemporal lobar degeneration. *Nature Medicine, 26*(3), 387–397.

Toledo, J. B., Arnold, M., Kastenmuller, G., Chang, R., Baillie, R. A., Han, X., et al. (2017). Metabolic network failures in Alzheimer's disease: A biochemical road map. *Alzheimer's Dementia, 13*(9), 965–984. https://doi.org/10.1016/j.jalz.2017.01.020.

Varma, V. R., Oommen, A. M., Varma, S., Casanova, R., An, Y., Andrews, R. M., et al. (2018). Brain and blood metabolite signatures of pathology and progression in Alzheimer disease: A targeted metabolomics study. *PLoS Medicine, 15*(1). e1002482. https://doi.org/10.1371/journal.pmed.1002482.

Vemuri, P., Knopman, D. S., Lesnick, T. G., Przybelski, S. A., Mielke, M. M., Graff-Radford, J., et al. (2017). Evaluation of amyloid protective factors and Alzheimer disease neurodegeneration protective factors in elderly individuals. *JAMA Neurology, 74*(6), 718–726. https://doi.org/10.1001/jamaneurol.2017.0244.

Verberk, I. M. W., Slot, R. E., Verfaillie, S. C. J., Heijst, H., Prins, N. D., van Berckel, B. N. M., et al. (2018). Plasma amyloid as prescreener for the earliest Alzheimer pathological changes. *Annals of Neurology, 84*(5), 648–658. https://doi.org/10.1002/ana.25334.

Vergallo, A., Megret, L., Lista, S., Cavedo, E., Zetterberg, H., Blennow, K., et al. (2019). Plasma amyloid beta 40/42 ratio predicts cerebral amyloidosis in cognitively normal individuals at risk for Alzheimer's disease. *Alzheimer's Dementia, 15*(6), 764–775. https://doi.org/10.1016/j.jalz.2019.03.009.

Viggars, A. P., Wharton, S. B., Simpson, J. E., Matthews, F. E., Brayne, C., Savva, G. M., et al. (2011). Alterations in the blood brain barrier in ageing cerebral cortex in relationship to Alzheimer-type pathology: A study in the MRC-CFAS population neuropathology cohort. *Neuroscience Letters, 505*(1), 25–30.

Vogt, N. M., Hunt, J. F., Adluru, N., Dean, D. C., Johnson, S. C., Asthana, S., et al. (2019). Cortical Microstructural alterations in mild cognitive impairment and Alzheimer's disease dementia. pii: bhz286. *Cerebral Cortex*. https://doi.org/10.1093/cercor/bhz286. PubMed PMID: 31833550. [Epub ahead of print].

Weston, P. S. J., Poole, T., O'Connor, A., Heslegrave, A., Ryan, N. S., Liang, Y., et al. (2019). Longitudinal measurement of serum neurofilament light in presymptomatic familial Alzheimer's disease. *Alzheimer's Research & Therapy, 11*(1), 19. https://doi.org/10.1186/s13195-019-0472-5.

Willette, A. A., Bendlin, B. B., Starks, E. J., Birdsill, A. C., Johnson, S. C., Christian, B. T., et al. (2015a). Association of insulin resistance with cerebral glucose uptake in late middle-aged adults at risk for Alzheimer disease. *JAMA Neurology, 72*(9), 1013–1020. https://doi.org/10.1001/jamaneurol.2015.0613.

Willette, A. A., Johnson, S. C., Birdsill, A. C., Sager, M. A., Christian, B., Baker, L. D., et al. (2015b). Insulin resistance predicts brain amyloid deposition in late middle-aged adults. *Alzheimer's Dementia, 11*(5), 504–510, e501. https://doi.org/10.1016/j.jalz.2014.03.011.

Zetterberg, H., Mörtberg, E., Song, L., Chang, L., Provuncher, G. K., Patel, P. P., et al. (2011). Hypoxia due to cardiac arrest induces a time-dependent increase in serum amyloid β levels in humans. *PLoS One, 6*(12). e28263. https://doi.org/10.1371/journal.pone.0028263.

CHAPTER THREE

Brain glucose and ketone utilization in brain aging and neurodegenerative diseases

Dimitrios Kapogiannis*, Konstantinos I. Avgerinos
Laboratory of Clinical Investigation, Intramural Research Program, National Institute on Aging, National Institutes of Health, Baltimore, MD, United States
*Corresponding author: e-mail address: kapogiannisd@mail.nih.gov

Contents

1. Introduction 81
2. Brain glucose metabolism in aging, Alzheimer's disease and other neurodegenerative diseases 82
   2.1 Aging 82
   2.2 Alzheimer's disease 88
   2.3 Brain glucose metabolism aberrations in other neurodegenerative diseases 93
3. Brain ketones metabolism in aging, Alzheimer's disease and other neurodegenerative diseases 94
   3.1 Aging 94
   3.2 Alzheimer's disease 95
   3.3 Brain ketone metabolism in other neurodegenerative diseases 96
4. Interplay between glucose and ketone brain metabolism: Insights and therapeutic strategies for Alzheimer's disease and other neurodegenerative diseases 97
   4.1 Ketones are an alternative brain fuel to glucose 97
   4.2 Ketogenic interventions 99
   4.3 The therapeutic potential of the ketone ester in Alzheimer's disease and other neurodegenerative diseases 100
5. Conclusion 101
Acknowledgment 101
References 101

Abstract

To meet its high energy demands, the brain mostly utilizes glucose. However, the brain has evolved to exploit additional fuels, such as ketones, especially during prolonged fasting. With aging and neurodegenerative diseases (NDDs), the brain becomes inefficient at utilizing glucose due to changes in glia and neurons that involve glucose transport, glycolytic and Krebs cycle enzyme activities, and insulin signaling. Positron emission tomography and magnetic resonance spectroscopy studies have identified

International Review of Neurobiology, Volume 154
ISSN 0074-7742
https://doi.org/10.1016/bs.irn.2020.03.015

glucose metabolism abnormalities in aging, Alzheimer's disease (AD) and other NDDs *in vivo*. Despite glucose hypometabolism, brain cells can utilize ketones efficiently, thereby providing a rationale for the development of therapeutic ketogenic interventions in AD and other NDDs. This review compares available ketogenic interventions and discusses the potential of the potent oral Ketone Ester for future therapeutic use in AD and other NDDs characterized by inefficient glucose utilization.

Abbreviations

Aβ	amyloid beta
AD	Alzheimer's disease
ATP	adenosine triphosphate
BBB	blood-brain barrier
BHB	β-hydroxybutyrate
CMR_{AcAc}	cerebral metabolic rate for acetoacetate
CMR_{glc}	cerebral metabolic rate for glucose
DKA	diabetic ketoacidosis
EV	extracellular vesicle
FDG-PET	[^{18}F]-fluorodeoxyglucose positron-emission tomography
FOXO	forkhead box O
FTLD	frontotemporal lobar degenerations
GFAP	glial fibrillary acidic protein
GLAST	glutamate aspartate transporter
GLUTs	glucose transporter proteins
H_2O_2	hydrogen peroxide
HD	Huntington's disease
HIF	hypoxia-inducible factor
HOMA-IR	homeostatic assessment of insulin resistance
IDE	insulin-degrading enzyme
IRS-1	insulin receptor substrate-1
K_{AcAc}	uptake rate constant for acetoacetate
KE	ketone ester
K_{glc}	uptake rate constant for glucose
L1CAM	L1 cell adhesion molecule
MCI	mild cognitive impairment
MCTs	monocarboxylate transporters
MetS	metabolic syndrome
MRS	magnetic resonance spectroscopy
MSA	multiple system atrophy
mTOR	mammalian target of rapamycin
NADH	nicotinamide adenine dinucleotide
NDDs	neurodegenerative diseases
NEVs	neuronal-enriched extracellular vesicles
NFTs	neurofibrillary tangles
NPs	neuritic plaques
PD	Parkinson's disease

PDH pyruvate dehydrogenase
SCOT succinyl-CoA:3-ketoacid coenzyme A transferase
SGLTs sodium-dependent glucose transporters
T2DM type 2 diabetes mellitus

1. Introduction

The brain is a very demanding organ in terms of its energy requirements. While it represents only 2% of the total body mass, it can spend approximately 20–23% of the total body energy, corresponding to ~110–140 g of glucose per day (Cunnane et al., 2016; Owen et al., 1967). It spends this amount of energy maintaining resting potentials, generating action potentials and post-synaptic potentials, regulating presynaptic calcium levels, and recycling neurotransmitters, especially glutamate (Attwell and Laughlin, 2001).

In the "fed state," the brain utilizes almost solely glucose, but during fasting it begins to co-utilize ketone bodies [mainly β-hydroxybutyrate (BHB) and secondarily acetoacetate] (Cahill, 2006). A less recognized energy source for the brain is lactate produced by anaerobic glycolysis and utilized at basal and, most importantly, hyperlactatemic conditions (van Hall et al., 2009).

Aging and neurodegenerative processes are accompanied by alterations in the utilization of energy substrates, especially glucose, by the entire brain and regionally. These alterations could be the cause and/or result of aging and neurodegenerative processes. Understanding the mechanisms implicated in normal and abnormal brain metabolism is crucial for the development of anti-aging strategies and treatments for neurodegenerative diseases (NDDs) (Kapogiannis and Mattson, 2011).

This review investigates the brain metabolism of glucose and ketones, two of the brain's most quantitatively important energy substrates, in the context of aging and neurodegeneration, with a particular focus on Alzheimer's disease (AD). First, we present evidence from observational imaging studies on human brain metabolism and then discuss the alterations that occur at the level of membrane receptors, intracellular cascades, and biochemical pathways in different brain regions and cell types. We believe that the accumulated evidence from preclinical and clinical studies on brain metabolism has reached a critical mass that powerfully motivates the use of metabolic interventions for AD and other NDDs as potential therapeutic strategies (Kapogiannis and Mattson, 2011).

2. Brain glucose metabolism in aging, Alzheimer's disease and other neurodegenerative diseases

2.1 Aging

2.1.1 Brain glucose metabolism in aging: Evidence from imaging studies in humans

[^{18}F]-fluorodeoxyglucose positron-emission tomography (FDG-PET) studies on aging conducted before the year 2000 presented a dichotomy of results; about half of the studies concluded that brain glucose hypometabolism is associated with aging, while the rest found no such associations (Cunnane et al., 2011). The heterogeneity of the age of patients included in these studies, poorly defined age groups, inadequate control of cofounders (e.g., cognitive status and other health parameters), suboptimal PET scan conditions, and no correction for brain atrophy during statistical analysis may partially account for this inconsistency (Cunnane et al., 2011).

More recent studies have consistently shown a decrease in brain glucose metabolism with aging in humans. While most studies report region-specific hypometabolism, some studies report hypometabolism affecting the entire gray matter (Kalpouzos et al., 2009; Nugent et al., 2014a, 2014b, 2016). The frontal lobes, for example, are repeatedly found to be affected in studies comparing ~72-year old with ~25-year old cognitively intact individuals (Kalpouzos et al., 2009; Nugent et al., 2014a, 2014b, 2016). The observed frontal cortex hypometabolism corresponds to the executive dysfunction and frontal-type memory impairment observed in non-demented, older adults (Buckner, 2004). Other brain areas found to display glucose hypometabolism are the temporal lobes (Nugent et al., 2014b), parietal cortex (Nugent et al., 2016), cingulate (Nugent et al., 2014b), and subcortical regions, i.e., the caudate (Nugent et al., 2014a, 2016), putamen (Nugent et al., 2014b), and thalamus (Nugent et al., 2014b). From an evolutionary perspective, it is possible that brain structures that evolved earlier (e.g., the amygdala, hippocampus, thalamus, and cingulate gyrus) are more metabolically resilient, while those that evolved later (e.g., the frontal and parietal cortex) are more vulnerable to aging-associated glucose hypometabolism and atrophy (Grieve et al., 2005).

In addition to FDG-PET scans, we have used magnetic resonance spectroscopy (MRS) to measure glucose concentration (normalized to creatine or water) in the posterior cingulate/precuneus. These brain MRS studies have revealed a stepwise increase in glucose concentrations from younger cognitively normal, to older cognitively normal, to individuals with AD. We interpreted this finding as a manifestation of decreased glucose utilization, which leads to increased concentration of remaining intra- and/or extracellular glucose (Mullins et al., 2018). Data reanalysis from cognitively normal individuals exclusively showed a correlation between age and MRS glucose concentration, where higher glucose concentrations that reflect glucose hypometabolism are associated with aging (Fig. 1).

2.1.2 Brain glucose transport in aging

Glucose transporters are necessary for glucose transfer into brain cells since cell membranes are impermeable to glucose. There are two types of glucose transporters in the brain: the glucose transporter proteins (GLUTs) that transport glucose through facilitative diffusion (a form of passive transport), and sodium-dependent glucose transporters (SGLTs) that use an energy-coupled mechanism (active transport). Given the minor role of SGLTs for brain glucose utilization (Szablewski, 2017), we will focus our review mainly on GLUTs.

The GLUT family consists of at least 14 members (Scheepers et al., 2004). Most GLUTs are expressed by various brain cell types; for example,

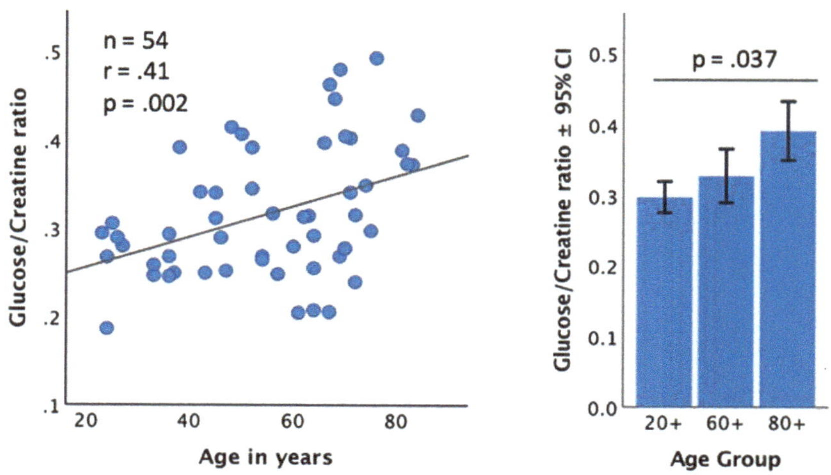

Fig. 1 The stepwise increase in the precuneus glucose to creatine ratio associated with age and by age group. Reanalysis of data included in Mullins et al. (2018).

astrocytes express GLUT-1/2, oligodendrocytes express GLUT-2, microglia express GLUT-5, and neurons express GLUT-1/2/3/4 (Szablewski, 2017). GLUT-3 is considered the principal neuronal glucose transporter. Importantly, GLUT-1 is also expressed by the endothelial cells of the blood-brain barrier (BBB) (Szablewski, 2017). Topographically, GLUTs are expressed in brain areas that are characteristically affected by NDDs (such as the hippocampus and the frontal lobes), but also in brain regions considered resilient to NDDs (such as the cerebellum) (Szablewski, 2017).

GLUT-3 has high affinity for glucose to ensure adequate glucose transport into neurons even when glucose concentration is low (Thorens and Mueckler, 2010). Conversely, GLUT-2 has low affinity for glucose. Interestingly, GLUT-1 is acutely upregulated whenever the concentration of glucose is low (Thorens and Mueckler, 2010). Finally, GLUT-4 is regulated by insulin (Thorens and Mueckler, 2010). The diversity of GLUT transporters likely corresponds to differences in functionality. For example, at the tripartite synapse, glutamate stimulates astrocytic GLUT-1 activity to promote glucose uptake by astrocytes, whereas neuronal GLUT-3 activity is suppressed to decrease glucose influx and promote lactate uptake by adjacent astrocytes (the astrocyte-neuron lactate shuttle hypothesis) (Barros and Deitmer, 2010). It is clear that energy utilization in the brain is a process that requires communication between different cell types, and GLUTs are important components of this cellular "cross-talk." Thus, any alteration in the number and/or function of GLUTs that occurs with aging or disease could have serious implications not only for individual cells, but also for their functional assemblies (i.e., microcircuits and networks) (Kapogiannis and Mattson, 2011).

The expression of GLUTs mRNA and protein in the brain changes with aging (Ding et al., 2013; Mooradian and Shah, 1997). Studies in rodents have revealed age-related decreases in the expression of cerebral and hippocampal GLUT-1 and GLUT-3 (Ding et al., 2013; Mooradian and Shah, 1997; Souza et al., 2015). Increasing age has been associated with decreased expression of GLUT-1 at the BBB (Ding et al., 2013) and in astrocytes (Souza et al., 2015), and importantly with decreased expression of GLUT-3 in rat neurons (Ding et al., 2013). GLUT-3 is the insulin-insensitive, principal neuronal isoform, and its age-associated reduction in the hippocampus of aged rodents (Ding et al., 2013; Fattoretti et al., 2002) is coupled to a decline in brain glucose uptake measured with PET (Ding et al., 2013). In one rat study, there was an increase of the insulin-sensitive GLUT-4 in the membrane despite the age-associated reduction in GLUT-3, although total GLUT-4 levels did not change (Ding et al., 2013). This suggests a potential compensatory mechanism for the observed GLUT-3 reduction. No change in GLUT-5 levels was observed in this study (Ding et al., 2013). The molecular mechanisms responsible for reduced neuronal glucose transport in aging remain to be established, but are likely diverse and may involve amyloid beta (Aβ) peptide induced membrane lipid peroxidation, which may lead to GLUT-3 dysfunction, decreased ATP levels and neuronal death (Mark et al., 1997).

2.1.3 The role of insulin in aging

Glucose utilization in the brain has been considered to be largely independent of the action of insulin (Banks et al., 2012). However, the fact that insulin-sensitive GLUT-4 is expressed in the brain increases the likelihood that glucose metabolism in the brain is at least partially regulated by insulin (Szablewski, 2017). An *in vitro* study has shown that direct infusion of insulin into neuronal cultures promotes translocation of the insulin-insensitive GLUT-3 to the plasma membrane, although it does not promote its fusion to the membrane (Uemura and Greenlee, 2006). However, once the neuron is depolarized, GLUT-3 fuses with the membrane. Moreover, neurons pretreated with insulin express higher levels of GLUT-3 at their membranes (Uemura and Greenlee, 2006), suggesting that insulin may indirectly facilitate glucose uptake through additional GLUTs besides GLUT-4. In the rat brain, GLUT-4 immunoreactive cells co-localize with neurons expressing GLUT-3 (Apelt et al., 1999). This physical proximity of GLUT-3- and GLUT-4-expressing cells further suggests that insulin plays an important role in regulating brain glucose. Finally, there is evidence that GLUT-8, which is expressed by hippocampal and other rat brain neurons, is also insulin-sensitive, but that insulin modulates its function at the subcellular level (i.e., transfer of glucose from intracellular organelles to the cytoplasm) (McEwen and Reagan, 2004). The finding that insulin partially regulates brain glucose uptake through these mechanisms has provided rationale for testing intranasal insulin to treat neurological conditions (Avgerinos et al., 2018).

Aging is associated with chronic hyperinsulinemia combined with reduced expression of insulin receptors and altered insulin signaling cascades (i.e., insulin resistance) in the brain (Steculorum et al., 2014; Zhao et al., 2004). Since glucose transportation in the brain is partially modulated by insulin, it is plausible that insulin resistance may contribute to the altered brain glucose metabolism associated with aging. Interestingly, in aged individuals without cognitive impairment, peripheral insulin resistance (measured by Homeostatic Assessment of Insulin Resistance, HOMA-IR) has been associated with reduced brain glucose metabolism in a regional pattern similar to that of AD (Baker et al., 2011; Willette et al., 2015a).

In the absence of biomarkers for central insulin resistance, earlier investigations have correlated peripheral insulin resistance measures (such as HOMA-IR) with brain function, under the assumption that central and peripheral insulin resistance partly overlap. This has changed since the development of methodologies to derive neuronal-enriched extracellular vesicles (NEVs) from peripheral blood and the introduction of NEV biomarkers. The NEV biomarkers, such as phosphorylated forms of insulin receptor substrate 1 (IRS-1) and downstream mediators, may reflect neuronal insulin signaling. Studies have shown that these biomarkers can differentiate between cognitively normal individuals (with and without diabetes) and individuals with AD (Kapogiannis et al., 2015). They also predict preclinical AD (Kapogiannis et al., 2019) and are associated with gray matter atrophy in AD (Mullins et al., 2017b). Furthermore, they differ between drug-naïve individuals with schizophrenia and control subjects (Goetzl et al., 2019). Importantly, NEVs show a priori-hypothesized and pathway-specific responses to experimental interventions targeting neuronal glucose metabolism, such as (i) dietary protein restriction in cancer (Eitan et al., 2017), (ii) exenatide in Parkinson's disease (PD) (Athauda et al., 2019), and (iii) intranasal insulin in AD (Mustapic et al., 2019). The NEV biomarkers contain signaling molecules related to multiple cellular pathways and may provide a "window to the brain" with a simple blood draw (Mustapic et al., 2017).

2.1.4 Biochemistry of glucose metabolism in aging

Glucose in brain cells can be metabolized to ATP by either oxidative metabolism or non-oxidative metabolism. Oxidative metabolism refers to the final pathway of glucose metabolism via oxidative phosphorylation, after its conversion to pyruvate and passage through the Krebs cycle (Demetrius et al., 2014). On the other hand, non-oxidative metabolism refers to the metabolism of glucose into lactate, and can occur in the presence or absence of oxygen (aerobic or anaerobic glycolysis, respectively) (Demetrius et al., 2014). Although non-oxidative metabolism of glucose produces far less ATP than oxidative phosphorylation, it is often favored by the brain because it also provides substrates for myelination, synaptogenesis, axonal elongation, and produces less reactive oxygen andspecies than oxidative phosphorylation (Bauernfeind et al., 2014; Goyal et al., 2014). Even in the presence of oxygen, glucose is not completely metabolized via the oxidative pathway; some glucose is always being metabolized via the non-oxidative pathway (aerobic glycolysis), which reveals its importance. Notably, PET studies have shown that non-oxidative metabolism of glucose through aerobic glycolysis gradually decreases with aging reaching zero by age 60 (Goyal et al., 2017). This reduction is responsible for the total glucose use reduction associated with aging, while oxidative glucose use remains unaffected (Goyal et al., 2017).

As discussed, glucose handling is a well-orchestrated process between adjacent brain cells of different types. Glial cells take up glucose mainly via GLUT-1 for energy production. During this process, they also produce lactate that can be shuttled into neurons (Jha and Morrison, 2018) and enter the Krebs cycle. Neuronal GLUT-3s also take up glucose, which is used for glycolysis and the pentose phosphate pathway, the latter being necessary for the biosynthesis of molecules that protect neurons against oxidative stress (Benarroch, 2014; Bolanos and Almeida, 2010). In aged mice, the intercellular metabolic "cross-talk" between glia and neurons is disrupted (Drulis-Fajdasz et al., 2018). It is possible that as the lactate shuttle is impaired with aging, neurons rely primarily on their own glycolysis and oxidation for energy (Drulis-Fajdasz et al., 2018) and thus have diminished capacity for energy generation.

In addition to extracellular glucose, brain cells (mainly astrocytes, and to some extent neurons) store glycogen to readily mobilize glucose as an energy source (Magistretti and Allaman, 2007). Acting as an energy reservoir, astrocytic and neuronal glycogen plays an important role in memory formation and learning (Alberini et al., 2018). On the other hand, aging in mice and Drosophila has been associated with increased accumulation of glycogen-based aggregates in the brain. These aggregates resemble the neuronal polyglucosan bodies found in Lafora's disease, an aggressive neurodegenerative disease (Sinadinos et al., 2014). It is possible that the accumulation of these glycogen aggregates with aging may interfere with normal neuronal function (Sinadinos et al., 2014). Interestingly, targeted reduction of glycogen synthesis in aged mice and Drosophila leads to reduced levels of these aggregates and consequently improved neurological function and life extension (Sinadinos et al., 2014). So, although glycogen may play a role in normal cognitive function, its mobilization may be disrupted in aging, resulting in aggregates that exert negative effects on brain function.

2.2 Alzheimer's disease

2.2.1 *Brain glucose metabolism in Alzheimer's disease: Evidence from imaging studies*

Glucose hypometabolism is a consistent finding in FDG-PET studies in AD. The most severely affected areas are the hippocampus (Mosconi et al., 2005), lateral and medial temporal lobes (Castellano et al., 2015; Croteau et al., 2018a), and posterior cingulate/precuneus (Croteau et al., 2018a). These regions are widely known to be vulnerable to AD pathology (Braak and Del Tredici, 2012, 2013; Gefen et al., 2018) and are nodes of the default mode network of the brain (Kapogiannis and Mattson, 2011; Seeley et al., 2009). Regional brain glucose hypometabolism is closely linked to tau deposition and clinical manifestations of AD (Ossenkoppele et al., 2016; Sintini et al., 2019).

Using MRS, we have shown higher glucose concentrations in the precuneus of AD patients compared to younger and older cognitively normal individuals (Mullins et al., 2018) and attributed this to glucose hypometabolism. Our *in vivo* study replicated a post-mortem study that also showed glucose elevation in the brains of AD patients (An et al., 2018). Interestingly, conditions that increase the risk for AD, such as carrying the *ApoE4* allele (Reiman et al., 1996, 2004) or being prediabetic/diabetic and elderly (Baker et al., 2011), cause AD-like reductions of brain glucose metabolism even without any clinical manifestation of the disease. Mild cognitive impairment (MCI) is the prodrome of AD and manifests brain glucose hypometabolism in areas similar to those in AD, but to a lesser degree (Croteau et al., 2018a; Mosconi et al., 2005). We have shown that in AD, as in normal aging, FDG hypometabolism is positively associated with HOMA-IR (Baker et al., 2011; Willette et al., 2015a). Interestingly, insulin resistance in MCI is associated with a paradoxical, transient, and perhaps pathogenic increase in hippocampal glucose metabolism (Willette et al., 2015). Similarly to MCI, individuals with Down Syndrome have an increased risk of developing AD and are characterized by glucose hypometabolism in some areas vulnerable to AD (precuneus), but potentially compensatory hypermetabolism in other areas (inferior temporal/entorhinal cortex) (Haier et al., 2003).

2.2.2 *Brain glucose transport in Alzheimer's disease*

The formation of ATP in the brain is reduced by ~20% in early AD, and an even greater reduction occurs in advanced stages of the disease (Hoyer, 1992). This decrease in energy production can be partially explained by the inadequate cellular uptake of glucose due to a decreased number of astrocytic, BBB and neuronal GLUTs (An et al., 2018; Liu et al., 2008; Simpson et al., 1994). It is not clear, however, if these decreases are the cause or result of AD.

Of the various GLUTs expressed in the brain, isoforms 1–4 have been most studied in the context of AD. Post-mortem studies in individuals with AD have shown reductions in GLUT-1 (expressed by astrocytes and BBB endothelial cells), and especially GLUT-3 (primarily expressed by neurons), which are more evident in frontal, parietal, occipital, temporal cortices, the caudate nucleus and the hippocampus (An et al., 2018; Liu et al., 2008; Simpson et al., 1994). These reductions occur independently of neuronal loss (Liu et al., 2008) and in the same brain regions that appear hypometabolic in FDG-PET scans (Simpson et al., 1994). They are also associated with reduced glycolytic flux (An et al., 2018). Importantly, GLUT reductions strongly correlate to tau-phosphorylation and subsequent increase of the number of hyperphosphorylated tau-containing neurofibrillary tangles (NFTs) (Liu et al., 2008) that are a pathological hallmark of AD.

We have shown that the normal regional expression of GLUTs, as well as insulin signaling mediators, are inversely associated with the density of NFTs and (to a lesser extent) Aβ-containing neuritic plaques (NPs) in AD. In other words, brain areas that normally show a low expression of GLUT-1 and GLUT-4 (as well as IRS-1) are more likely to develop tau pathology in AD (Mullins et al., 2017a) (Fig. 2). Reductions in GLUT-3 levels have also been associated with the formation of plaques in other autopsy studies (An et al., 2018).

In an amyloidogenic mouse model (APP/PS1), hippocampal DG and CA1 areas have been characterized by decreased GLUT-1 levels at 18 months (Hooijmans et al., 2007). This finding supports the hypothesis of an amyloid-driven GLUT-1 decrease (Hooijmans et al., 2007) rather than the possibility that reductions in GLUT-1 lead to increased amyloid burden. In an *in vitro* study, hippocampal neurons from rats were treated with Aβ, resulting in decreased glucose uptake, increased GLUT-3 receptor transcription and translocation, but decreased fusion of GLUT-3 with the plasma membrane (Prapong et al., 2002). It was hypothesized that Aβ induces functional alterations to GLUT-3, which result in decreased glucose uptake despite increased GLUT-3 expression (Prapong et al., 2002). Additionally, it has been shown that Aβ-induced decrease in glucose uptake by cultured hippocampal/cortical rat neurons may be mediated

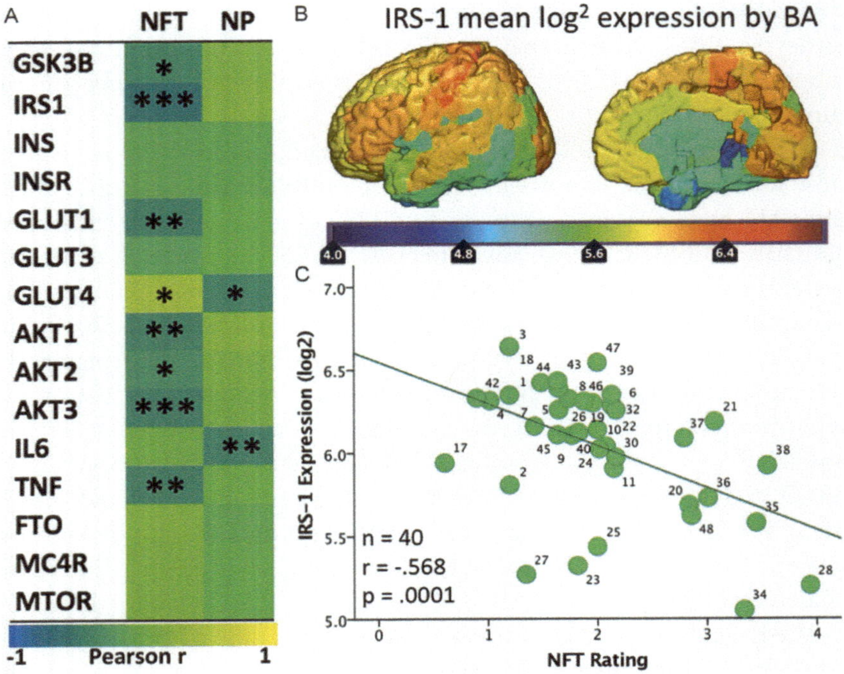

Fig. 2 (A) Heatmap ("jet": red high, blue low) of the spatial correlation between levels of expression of various genes from the Allen Human Brain Atlas and the density of NFTs or NPs in AD brains reported by Arnold et al. (1991). Asterisks (*/**/***) represent *P* values of <0.05/0.01/0.001, respectively. (B) Map of mean IRS-1 log2 expression in the six healthy human specimens included in the Allen Human Brain Atlas. (C) Scatter plot of the mean IRS-1 log2 expression from the Allen Human Brain Atlas and NFT density reported by Arnold et al. (1991). Each of the 40 data points corresponds to a Brodmann Area for which both gene expression levels and tangle density ratings were available. Previously published in Mullins et al. (2017a).

by lipid peroxidation of the neuronal membrane and inactivation of GLUT-3 (Mark et al., 1997).

These findings are generally compatible with the amyloid hypothesis, which posits that an increased Aβ burden results in tau hyperphosphorylation and deposition, and ultimately neurodegeneration. Aβ-induced reduction of GLUT-1,3 number and functionality leads to decreased glucose uptake and decreased protein O-GlcNAcylation; the latter is associated with tau hyperphosphorylation (Liu et al., 2008).

Notably, brain GLUT-2 overexpression, but not GLUT-4, has been reported in AD (Liu et al., 2008). Over-expression of GLUT-2 in AD brain may be due to astrocytic activation, since increased GLUT-2 expression was associated with increased expression of the astrocytic marker, glial fibrillary acidic protein (GFAP). It is possible that the brain attempts to compensate for the energetic deficit due to the neuronal GLUT-1,3 reductions, by inducing reactive astrogliosis to provide additional number of astrocytic GLUTs (Liu et al., 2008), and ultimately shuttle energy to neurons through the astrocytes.

2.2.3 The role of insulin in Alzheimer's disease

Since insulin partially regulates glucose handling in neuronal and glial cells, alterations in the insulin cascades are likely to be implicated in glucose hypometabolism associated with AD. In terms of mechanistic evidence, Aβ oligomers decrease plasma membrane insulin receptors and induce oxidative stress and synaptic spine deterioration in hippocampal neuronal cultures (De Felice et al., 2009). These effects are rescued by insulin (De Felice et al., 2009).

We have associated peripheral insulin resistance with glucose hypometabolism in AD (Willette et al., 2015a, 2015b), as central insulin resistance is believed to partly overlap with peripheral insulin resistance. In studies using novel NEV biomarkers of insulin signaling and insulin resistance, we have been able to differentiate between cognitively normal individuals and those with AD (Kapogiannis et al., 2015), predict preclinical AD (Kapogiannis et al., 2019), and show an association of these biomarkers with cortical atrophy in AD (Mullins et al., 2017b). The areas of atrophy associated with NEV biomarkers of insulin resistance were very similar to the areas of glucose hypometabolism in clinical AD. These areas are also affected in the brains of individuals with a genetic risk of developing AD despite absence of symptoms (Reiman et al., 2004). This reveals a connection between aberrations in central insulin signaling, glucose brain pathology, and glucose underutilization in AD.

An additional parameter is insulin-degrading enzyme (IDE), a protease that degrades insulin and also plays a significant role in Aβ degradation (Kurochkin et al., 2018). Metabolic Syndrome (MetS) and Type 2 Diabetes Mellitus (T2DM) are conditions associated with increased risk of AD. In human and rodent tissues, both elevated glucose and Aβ have been associated with increased S-nitrosylation and subsequent inhibition of IDE, revealing "cross-talk" between MetS/T2DM and AD (Akhtar et al., 2016). An alternative hypothesis is that increased insulin levels in MetS/T2DM may competitively prevent the binding of IDE with Aβ and thereby reducing its degradation (Kurochkin et al., 2018).

2.2.4 Biochemistry of glucose metabolism in Alzheimer's disease

In addition to alterations in GLUT, diminished glycolysis has been shown in AD. Decreased enzymatic activity in several rate-controlling enzymes of glycolysis (An et al., 2018) and in the Krebs cycle occur and constitute a "metabolic block" (Mamelak, 2012). Importantly, these abnormalities in glycolysis are coupled with AD pathology in frontal (Aβ deposition) and temporal areas (tau accumulation) (An et al., 2018). Interestingly, an autopsy study revealed elevation of glycolytic enzymes in AD, although this was attributed to survival bias, i.e., brain cells that survived neurodegeneration may have upregulated rather than downregulated glycolysis (Soucek et al., 2003). This finding corresponds well with *in vitro* studies on Aβ-resistant neurons that show an upregulation of pyruvate kinase and lactate dehydrogenase, the major regulators of aerobic glycolysis (Newington et al., 2011; Soucek et al., 2003).

Upregulation of glycolysis is associated with reduced levels of reactive oxygen species (Newington et al., 2011), while the inhibition of key glycolytic enzymes results in sensitization to Aβ toxicity (Newington et al., 2011). Furthermore, Aβ treatment of cortical neurons in culture leads to the production of nicotinamide adenine dinucleotide (NADH) and hypoxia-inducible factor (HIF)-1, both of which are the result of upregulated glycolysis and may mediate neuroprotection (Soucek et al., 2003). It has been suggested that an increased reliance on glycolysis and suppression of mitochondrial respiration (the Warburg effect) confers increased resistance and survival to neurons (Newington et al., 2011). Upregulation of glycolysis in neurons may act as a compensatory mechanism against AD pathology. Although this compensation might be helpful, it ultimately fails as the disease progresses and the brain manifests prominent glucose hypometabolism, even in preclinical stages of AD.

In addition to autopsy studies showing alterations in glycolytic enzymes, brain imaging studies in patients with AD reveal changes of glucose metabolism *in vivo*. FDG-PET scans reflect glycose utilization as a whole and do not capture the distinction between oxidative phosphorylation and aerobic glycolysis (i.e., the non-oxidative portion of glucose metabolism). In a

FDG-PET study where aerobic glycolysis was calculated by subtracting the oxidative fraction of cerebral metabolic rate of glucose (CMR_{glc}) from total CMR_{glc}, it was found that in individuals with high Aβ burden, reductions in aerobic glycolysis facilitated tau deposition (Vlassenko et al., 2018). Thus, reductions in aerobic glycolysis occur not only in aging (see Section 2.1.4) but also in AD, where they are associated with AD pathology.

2.3 Brain glucose metabolism aberrations in other neurodegenerative diseases

Dysregulated glucose metabolism, including changes in FDG metabolism, alterations in GLUTs, and evidence of central insulin resistance have been observed in other NDDs besides AD. Individuals with PD and intact cognition show FDG hypometabolism in occipital lobes and hypermetabolism in frontal regions (Firbank et al., 2017). In PD with impaired cognition, FDG hypometabolism extends to parietal and temporal cortices (Firbank et al., 2017), predicting atrophy in these regions (Gonzalez-Redondo et al., 2014). At the same time, neuronal insulin resistance is implicated in the pathogenesis of PD and constitutes a target for PD treatment (Athauda and Foltynie, 2016).

In Huntington's disease (HD), glucose hypometabolism in frontal and temporal lobes and the striatum occurs even in asymptomatic disease (Ciarmiello et al., 2006). In HD, the cortex and caudate expresses decreased GLUT-1 and GLUT-3, especially in late disease (Gamberino and Brennan, 1994). Interestingly, in areas of neuronal loss yet seemingly normal glia, glial GLUT-1 and GLUT-3 were under-expressed and may indicate that remaining glia no longer need to support neuronal glucose metabolism (Gamberino and Brennan, 1994). The importance of GLUTs in HD has been also shown in a large human cohort, in which an increased copy number of the gene *SLC2A3* encoding GLUT-3 was associated with delayed HD symptomatology (Vittori et al., 2014).

Brain insulin resistance has also been detected in brain autopsy studies of animal models and humans with Multiple System Atrophy (MSA), localized mostly in putaminal oligodendrocytes bearing α-synuclein aggregates (Bassil et al., 2017).

Evidence that glucose dysmetabolism is present in a spectrum of NDDs reveals the central role of proper glucose handling and metabolism for brain health. It also points toward the possibility of employing similar therapeutic strategies to improve brain energy metabolism across diseases.

3. Brain ketones metabolism in aging, Alzheimer's disease and other neurodegenerative diseases

3.1 Aging

3.1.1 Brain ketones metabolism in aging: Evidence from imaging studies

In a dual tracer PET study comparing glucose and acetoacetate utilization in aged and young rats, the overall cerebral metabolic rate for glucose (CMR_{glc}) and acetoacetate (CMR_{AcAc}) was similar in 21- and 4-month rats (Roy et al., 2012). In a similar study in humans using both glucose and acetoacetate tracers, it was shown that (i) glucose hypometabolism occurred with aging (shown both for the CMR_{glc} and the uptake rate constant for glucose, K_{glc}); and, (ii) there was no difference in CMR_{AcAc} with aging, but K_{AcAc} was reduced at both the whole brain level and regionally (Nugent et al., 2014b). The CMR_{AcAc} varies as a function of peripheral ketone concentration, such that increases in peripheral ketone concentration can increase CMR_{AcAc} (Cunnane et al., 2016). In contrast, the uptake rate constant for AcAc (K_{AcAc}) expresses an inherent property of brain tissue and is independent of peripheral acetoacetate levels (Nugent et al., 2014b). These findings suggest that, although the capacity of the aged brain for ketone uptake is less than that of the young brain, increased peripheral levels of ketones can increase brain ketone metabolism.

3.1.2 Transport and utilization of brain ketones in aging

Monocarboxylate transporters (MCTs) are a family of 14 receptors. Most are expressed on the membrane of brain cells and facilitate passive transport of lactate, pyruvate and ketone bodies into the brain (Perez-Escuredo et al., 2016). In the hippocampus of aged female mice, the expression and glycolytic capacity of GLUTs is decreased, but astrocytic MCT-4 expression is increased with a trend toward increased neuronal MCT-2 expression also (Ding et al., 2013). This rise in MCTs suggests that the aging brain favors ketone metabolism, perhaps as compensatory response to glucose hypometabolism. This is further supported by the observation that the elevation of ketones in aged rat brain coincides with the elevation in markers of mitochondrial dysfunction (Klosinski et al., 2015). It may be that due to the energy deficit, the aged brain tries to meet its energetic needs by utilizing ketones (Klosinski et al., 2015). At the same time, a trend for decreased MCT-1 expression in the BBB (Ding et al., 2013) indicates a diminished capacity for ketone transport from the periphery to the brain with aging.

Under normal circumstances, the liver is the main source of ketones for peripheral organs and the brain. However, brain lipids derived from myelin catabolism can also serve as a source of ketones when peripherally-produced ketones are not available to the brain, as occurs with decreased ketone transport to the brain with aging (Klosinski et al., 2015). The aging brain may compensate for its energy deficit by maladaptively catabolizing lipid-rich myelin to produce fatty acids that can be converted to ketones, inadvertently causing white matter degeneration (Klosinski et al., 2015). Specifically, the age-induced decline in mitochondrial respiration and production of H_2O_2 may activate the $cPLA_2$-sphingomyelinase pathway to catabolize myelin lipids into fatty acids. Those fatty acids may be further catabolized by astrocytes to produce ketone bodies that can enter into neurons via MCTs. This phenomenon may contribute to the demyelination observed in AD and have pathogenic importance (Bartzokis, 2011).

Recently, we have developed extracellular vesicle (EV)-based biomarkers to assess ketone transport by brain cells. We probed MCT-1 and MCT-2 in NEVs and astrocytic-origin EVs expressing L1 cell adhesion molecule (L1CAM) and the astrocytic amino acid transporter (GLAST), respectively. We demonstrated differential expression (enrichment) in these two types of EVs corresponding to their cellular origin, i.e., neurons and astrocytes (Fig. 3) and propose that these novel biomarkers can be used as outcome measures in interventions that aim to boost brain ketone utilization.

3.2 Alzheimer's disease

3.2.1 Brain ketones metabolism in Alzheimer's disease: Evidence from imaging studies

Dual tracer PET studies in MCI and AD individuals have shown that brain regions that are hypometabolic for glucose (i.e., show decreased CMR_{glc}

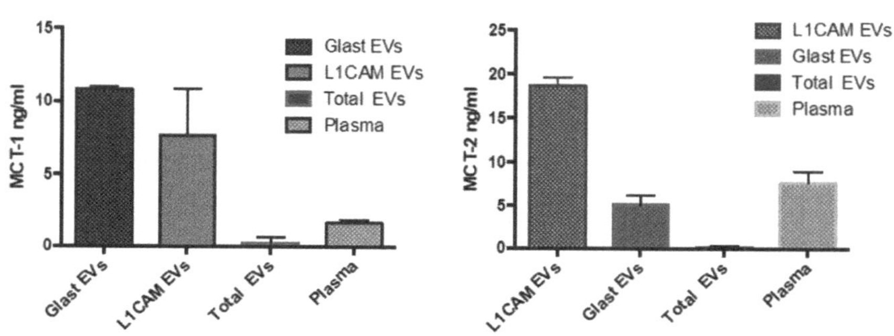

Fig. 3 MCT-1 and MCT-2 in L1CAM and GLAST EVs. Laboratory of Clinical Investigation, NIA, NIH.

and K_{glc}) in parietal, temporal and frontal lobes, and the posterior cingulate, hippocampus, precuneus brain regions, metabolize acetoacetate normally (i.e., unchanged CMR_{AcAc} and K_{AcAc}) (Castellano et al., 2015; Croteau et al., 2018a). Brain ketone uptake is proportional to peripheral levels of et al., ketones; as this ability is conserved in the AD brain, it suggests that normal ketone metabolism occurs in AD brain (Cunnane et al., 2016). Interventions to elevate ketones in periphery, such as exercise and administration of medium-chain triglycerides (ketogenic supplement) result in a direct increase of brain ketone utilization, even in individuals with AD (Castellano et al., 2017; Croteau et al., 2018b).

3.2.2 Brain ketones transport and utilization in Alzheimer's disease

As with aging mice, the development of pathology in AD model mice (3xTgAD) results in alteration of hippocampal MCTs (Ding et al., 2013). In AD, glial MCT-1 and MCT-4 protein expression decreases, but neuronal MCT-2 protein expression increases (Ding et al., 2013). These changes are associated with reductions in GLUT-1 and GLUT-3 protein expression (Ding et al., 2013), suggesting a compensatory mechanism to tackle glucose hypometabolism.

In female 3xTgAD mice, the expression of glycolytic enzymes, such as pyruvate dehydrogenase (PDH), decreases in parallel with increases in enzymes that metabolize ketones, such as succinyl-CoA:3-ketoacid coenzyme A transferase (SCOT) (Yao et al., 2010). Such shifts in enzymatic activity indicate a bioenergetic shift away from glucose and toward ketones as metabolic substrates in AD (Yao et al., 2010). During this bioenergetic shift, the brain may utilize ketones produced by liver ketogenesis. In addition, evidence from mice studies show that myelin may serve as a source of fatty acids that are converted to ketone bodies (especially when peripheral ketone bodies are suppressed or cannot cross the BBB) causing white matter degeneration (Yao et al., 2011). Notably, the latter is an early hallmark of AD (Bartzokis, 2011; Kuczynski et al., 2010). Alternatively, it may be that ketone bodies transported from periphery to the AD brain are mostly spent covering the energetic deficit rather than in the repair of myelin degeneration (Morris, 2005).

3.3 Brain ketone metabolism in other neurodegenerative diseases

The ability of the brain to utilize ketones in other NDDs, such as PD, HD or frontotemporal lobar degenerations (FTLD), has not been studied thoroughly. However, it has been shown in human studies that, as with AD, most NDDs present with brain glucose hypometabolism, albeit with different regional distributions specific to their pathology (Ciarmiello et al., 2006; Firbank et al., 2017; Foster et al., 2007). As with aging and AD, these diseases may be associated with preserved metabolism of ketones, which could compensate for the energy deficit related to glucose hypometabolism. This is supported by the positive effects of ketogenic interventions in experimental models of PD and HD (Chen et al., 2016; Tieu et al., 2003).

4. Interplay between glucose and ketone brain metabolism: Insights and therapeutic strategies for Alzheimer's disease and other neurodegenerative diseases

4.1 Ketones are an alternative brain fuel to glucose

Glucose is the main energy substrate for the brain in the "fed state"; during fasting, however, ketone bodies (primarily BHB and secondarily, acetoacetate and acetone) become important for brain preservation (Cahill, 2006; Cahill and Veech, 2003). In addition, lactate produced by muscle during physical exercise can also serve as a supplemental energy fuel for the brain (Dienel, 2019).

It has been hypothesized that BHB may have conveyed evolutionary advantages that helped large, human brains to survive during periods of prolonged fasting (Cahill, 2006; Cahill and Veech, 2003). β-Hydroxybutyrate is an efficient fuel for brain cells (Cahill, 2006; Cahill and Veech, 2003; Veech et al., 2001). In experiments where the main energy fuel is BHB, the reactants at one step of the mitochondrial chain (NAD+/NADH) become more reduced, while those at the next step (Coenzyme Q/Coenzyme QH_2) become more oxidized. Thus, increased availability of BHB widens the energetic difference between these two steps, and this increased energetic gap is associated with greater ATP production (Pawlosky et al., 2017). In an experiment where BHB was perfused into hearts of rats, there was an increase in the hydraulic work of heart, despite decreased oxygen consumption (Kashiwaya et al., 1994; Veech et al., 2001). Similarly, BHB-treated sperm increased its mobility, again, despite decreased oxygen consumption (Veech et al., 2001). More pertinently, neurons in models of AD and PD, as well as cells under conditions of low oxygen availability, survive longer if supplied with BHB (Cahill, 2006; Kashiwaya et al., 2000). However, it should be emphasized that the brain never relies solely on ketones and that glucose is important not only as an energy substrate, but also plays additional roles for normal brain function, for example, by promoting expression of genes regulating synaptic plasticity (Goyal et al., 2014).

Many clinicians may have underestimated the importance of ketones because they link the term "ketosis" with the dangerous medical condition termed "Diabetic Ketoacidosis" (DKA) (Cahill and Veech, 2003) that occurs only with complete insulin deficiency (Veech et al., 2001). Quantitatively, during the "fed state" ketones cover almost none of the brain's needs, but they can suffice for 60% or more of the energy demands during a state of starvation (Cahill, 2006; Cahill and Veech, 2003; Cunnane et al., 2011). Evidence suggests that in the presence of both glucose and ketones, the brain preferentially metabolizes ketones (Cahill and Veech, 2003).

The PDH complex links glycolysis with the Krebs cycle, and has been found to decrease with aging and AD. Interestingly, preservation of PDH complex activity has been shown to increase life span (Liu et al., 2009). Ketones bypass this block by their direct conversion into acetyl-CoA and thus direct entry into the Krebs cycle. Blockade of PDH in AD also causes decreased levels citrate, which is the precursor of acetylcholine (Veech et al., 2001). An acetylcholine deficit is one of the main neurotransmitter deficits in AD and may be ameliorated by ketones (Veech et al., 2001).

Decreased insulin signaling may result from a metabolic switch from glucose to ketones, leading to increased forkhead box O (FOXO) proteins and decreased activation of the mammalian target of rapamycin (mTOR) pathway (Edwards et al., 2014; Veech et al., 2017). In animal models undergoing ketogenic interventions, these factors have been proposed as mechanisms that mediate life span extension (Edwards et al., 2014; Veech et al., 2017). In a mouse model of AD, ketones blocked the intracellular accumulation of pathogenic Aβ42 protein and rescued the activity of the mitochondrial complex 1 (restoring normal energy production by the mitochondria), which was impaired by Aβ42 (Yin et al., 2016). Furthermore, ketones reduced oxidative stress and improved synaptic plasticity (Yin et al., 2016).

In summary, the brain demonstrates a decreased capacity for glucose utilization, while ketone metabolism is relatively spared in NDDS. However, brain ketone utilization is dependent on plasma ketone levels, which are usually suppressed by normal or even increased insulin levels, in NDDs. The challenge is to effectively elevate brain ketone utilization as a therapeutic strategy for AD and other NDDs.

4.2 Ketogenic interventions

Strategies that elevate brain ketone utilization are directed toward inducing peripheral ketosis. These strategies are collectively known as ketogenic interventions and are relatively well known from their long-standing use in the treatment of epilepsy. From the least to the most robust, ketogenic interventions include: short-term fasting (12–24h) (Castellano et al., 2017); moderate aerobic exercise (Castellano et al., 2017); the intake of medium-chain triglycerides at doses of 20–30g/day while on usual diet (Avgerinos et al., 2019); a low carbohydrate diet (<50g carbs/day) for several weeks (Boden et al., 2005); a high fat (>150g fat/day) combined with low carbohydrate (<20g carbs/day) diet for several weeks (Phillips et al., 2018); and oral intake of ketone salts and esters (Fischer et al., 2018; Stubbs et al., 2017). Oral intake of ketone ester [KE; chemical name: (*R*)-3-hydroxybutyl (*R*)-3-hydroxybutyrate] is considered to be a more robust way to induce ketosis (Soto-Motaand et al., 2019; Stubbs et al., 2017). After its ingestion, KE is rapidly hydrolyzed by gut enzymes into D-β-hydroxybutyrate, which is readily absorbed, and 1,3-butanediol, which is also absorbed and converted to D-β-hydroxybutyrate in the liver (Cox et al., 2016). Absorption of both D-β-hydroxybutyrate and 1,3-butanediol results in a rapid rise in BHB blood concentration, which becomes available to the various tissues and brain (Cox et al., 2016).

A single oral dose of around 25–30g KE can elevate plasma BHB up to 4mM which is comparable to the BHB levels induced after complete starvation for 2 or more weeks (Cahill, 2006; Cahill and Veech, 2003); it is also four times greater than the levels induced after several weeks on a low carbohydrate/high-fat diet (Phillips et al., 2018); 10 times greater than a single high dose (~30g) of oral medium-chain triglycerides (Avgerinos et al., 2019) and 40 times more than the levels reached after overnight fasting (Castellano et al., 2017). In terms of safety, a single dose of KE induces less than half the elevation of BHB seen in DKA (BHB in DKA reaches ~9–10mM) and BHB levels return back to normal within a 4-h time period. In contrast, ketones in DKA continue to rise uncontrollably unless insulin is provided.

Factors that should be considered when selecting a candidate ketogenic intervention to restore brain energy in AD or other NDDs are: (a) its potency, i.e., its ability to induce peripheral and thus central ketosis; (b) its risk profile; and, (c) the convenience of the intervention, which is vital, especially for individuals with cognitive impairment who have difficulty following complex instructions, and for their caregivers. Fortunately, the most robust ketogenic intervention, the oral supplement KE, is associated with minimal side effects and is considered to be convenient for daily intake. A recent 28-day study in healthy individuals taking the KE showed that it produces robust peripheral ketosis and infrequent mild gastrointestinal side effects, mainly nausea (Soto-Mota et al., 2019). Multiple other studies have reported good tolerability and safety of KE supplementation, however no study has been conducted in individuals with MCI or AD so far (Clarke et al., 2012; Cox et al., 2016; Dearlove et al., 2019; Faull et al., 2019; Holdsworth et al., 2017; Myette-Cote et al., 2018; Soto-Mota et al., 2019; Stubbs et al., 2017, 2018, 2019).

In contrast, long-term restrictive (low carbohydrate) and high-fat diets may be very difficult for AD patients to comply with and may be associated with a negative long-term impact on cardiovascular health (e.g., atherogenesis). Supplementation with medium-chain triglycerides is associated with frequent and intolerable side effects, such as diarrhea (Courchesne-Loyer et al., 2017). In addition, both ketogenic diets and medium-chain triglycerides produce less robust ketosis than KE (Avgerinos et al., 2019; Boden et al., 2005; Phillips et al., 2018). Furthermore, ketone salts are also less robust than KE (Stubbs et al., 2017) and linked to more frequent gastrointestinal effects, perhaps due to their salt content (Stubbs et al., 2019).

4.3 The therapeutic potential of the ketone ester in Alzheimer's disease and other neurodegenerative diseases

So far, only medium-chain triglycerides have been extensively studied in AD and while small cognitive benefits have been reported, overall results remain inconsistent (Avgerinos et al., 2019; Henderson et al., 2009). We speculate that the inconsistent cognitive benefit may be due to the mildness of BHB elevation as a result of this intervention (~0.3–0.5 mM) (Avgerinos et al., 2019; Henderson et al., 2009). In comparison, KE administration induces multi-fold higher elevation of peripheral BHB (~3–4 mM) (Soto-Mota et al., 2019). Nevertheless, KE has never been tested for its therapeutic effects in AD despite preclinical evidence to support this approach.

In a study involving healthy rats that consumed KE vs. isocaloric placebo for 5 days as part of their diet (30% of calories), rats in the KE group made more correct decisions on a eight-arm radial maze test, suggesting improved cognition (Murray et al., 2016). Moreover, when chronic KE administration was compared with isocaloric placebo in triple-transgenic AD

(3xTgAD) mice, KE was shown to improve learning and spatial memory while reducing Aβ and hyperphosphorylated tau brain deposits (Kashiwaya et al., 2013). In another study using 3xTgAD mice, KE administration increased hippocampal D-β-hydroxybutyrate and decreased oxidative stress (Pawlosky et al., 2017).

β-Hydroxybutyrate infusion confers resistance to dopaminergic neurodegeneration in animal models of PD, *in vitro* and *in vivo* (Kashiwaya et al., 2000; Tieu et al., 2003). In HD mouse models, BHB improves electrophysiological measures, attenuates motor symptoms, reduces microgliosis and striatal lesions, and extends lifespan (Chen et al., 2016; Lim et al., 2011). These studies suggest that ketones induce multiple beneficial cellular effects besides providing energy as "another fuel," and may even mimic the life extending properties of caloric restriction (Veech et al., 2017).

5. Conclusion

In summary, metabolic dysregulation and especially that associated with glucose is a common feature of NDDs, such as AD, PD, HD, and FTLD. In contrast, the capacity for ketone metabolism in these diseases appears to be relatively preserved. Preclinical and limited clinical data suggest that boosting ketone levels has therapeutic potential for these diseases, although the mechanisms by which this strategy may be effective are not completely understood. Unfortunately, only a small percentage of clinical trials in AD explicitly aim to improve metabolism (Cummings et al., 2019). We view this situation as an unfortunate translational gap from animal to human studies that needs to be addressed, especially since dominant therapeutic strategies that sought to reduce Aβ burden (the amyloid hypothesis) have largely failed, and a new direction may be valuable.

Acknowledgment

This research was supported entirely by the Intramural Research Program of the National Institute on Aging, NIH.

References

Akhtar, M. W., Sanz-Blasco, S., Dolatabadi, N., Parker, J., Chon, K., Lee, M. S., et al. (2016). Elevated glucose and oligomeric beta-amyloid disrupt synapses via a common pathway of aberrant protein S-nitrosylation. *Nature Communications*, 7, 10242. https://doi.org/10.1038/ncomms10242.

Alberini, C. M., Cruz, E., Descalzi, G., Bessieres, B., & Gao, V. (2018). Astrocyte glycogen and lactate: New insights into learning and memory mechanisms. *Glia*, *66*(6), 1244–1262. https://doi.org/10.1002/glia.23250.

An, Y., Varma, V. R., Varma, S., Casanova, R., Dammer, E., Pletnikova, O., et al. (2018). Evidence for brain glucose dysregulation in Alzheimer's disease. *Alzheimer's & Dementia*, *14*(3), 318–329. https://doi.org/10.1016/j.jalz.2017.09.011.

Apelt, J., Mehlhorn, G., & Schliebs, R. (1999). Insulin-sensitive GLUT4 glucose transporters are colocalized with GLUT3-expressing cells and demonstrate a chemically distinct neuron-specific localization in rat brain. *Journal of Neuroscience Research*, *57*(5), 693–705. https://doi.org/10.1002/(SICI)1097-4547(19990901)57:5<693::AID-JNR11 IRN154>3.0.CO;2-X.

Arnold, S. E., Hyman, B. T., Flory, J., Damasio, A. R., & Van Hoesen, G. W. (1991). The topographical and neuroanatomical distribution of neurofibrillary tangles and neuritic plaques in the cerebral cortex of patients with Alzheimer's disease. *Cerebral Cortex*, *1*(1), 103–116.

Athauda, D., & Foltynie, T. (2016). Insulin resistance and Parkinson's disease: A new target for disease modification? *Progress in Neurobiology*, *145–146*, 98–120. https://doi.org/10.1016/j.pneurobio.2016.10.001.

Athauda, D., Gulyani, S., Karnati, H. K., Li, Y., Tweedie, D., Mustapic, M., et al. (2019). Utility of neuronal-derived exosomes to examine molecular mechanisms that affect motor function in patients with Parkinson disease: A secondary analysis of the exenatide-PD trial. *JAMA Neurology*, *76*(4), 420–429. https://doi.org/10.1001/jamaneurol.2018.4304.

Attwell, D., & Laughlin, S. B. (2001). An energy budget for signaling in the grey matter of the brain. *Journal of Cerebral Blood Flow and Metabolism*, *21*(10), 1133–1145. https://doi.org/10.1097/00004647-200110000-00001.

Avgerinos, K. I., Egan, J. M., Mattson, M. P., & Kapogiannis, D. (2019). Medium chain triglycerides induce mild ketosis and may improve cognition in Alzheimer's disease. A systematic review and meta-analysis of human studies. *Ageing Research Reviews*, *58*, 101001. https://doi.org/10.1016/j.arr.2019.101001.

Avgerinos, K. I., Kalaitzidis, G., Malli, A., Kalaitzoglou, D., Myserlis, P. G., & Lioutas, V. A. (2018). Intranasal insulin in Alzheimer's dementia or mild cognitive impairment: A systematic review. *Journal of Neurology*, *265*(7), 1497–1510. https://doi.org/10.1007/s00415-018-8768-0.

Baker, L. D., Cross, D. J., Minoshima, S., Belongia, D., Watson, G. S., & Craft, S. (2011). Insulin resistance and Alzheimer-like reductions in regional cerebral glucose metabolism for cognitively normal adults with prediabetes or early type 2 diabetes. *Archives of Neurology*, *68*(1), 51–57. https://doi.org/10.1001/archneurol.2010.225.

Banks, W. A., Owen, J. B., & Erickson, M. A. (2012). Insulin in the brain: There and back again. *Pharmacology & Therapeutics*, *136*(1), 82–93. https://doi.org/10.1016/j.pharmthera.2012.07.006.

Barros, L. F., & Deitmer, J. W. (2010). Glucose and lactate supply to the synapse. *Brain Research Reviews*, *63*(1–2), 149–159. https://doi.org/10.1016/j.brainresrev.2009.10.002.

Bartzokis, G. (2011). Alzheimer's disease as homeostatic responses to age-related myelin breakdown. *Neurobiology of Aging*, *32*(8), 1341–1371. https://doi.org/10.1016/j.neurobiolaging.2009.08.007.

Bassil, F., Canron, M. H., Vital, A., Bezard, E., Li, Y., Greig, N. H., et al. (2017). Insulin resistance and exendin-4 treatment for multiple system atrophy. *Brain*, *140*(5), 1420–1436. https://doi.org/10.1093/brain/awx044.

Bauernfeind, A. L., Barks, S. K., Duka, T., Grossman, L. I., Hof, P. R., & Sherwood, C. C. (2014). Aerobic glycolysis in the primate brain: Reconsidering the implications for growth and maintenance. *Brain Structure & Function*, *219*(4), 1149–1167. https://doi.org/10.1007/s00429-013-0662-z.

Benarroch, E. E. (2014). Brain glucose transporters: Implications for neurologic disease. *Neurology, 82*(15), 1374–1379. https://doi.org/10.1212/WNL.0000000000000328.

Boden, G., Sargrad, K., Homko, C., Mozzoli, M., & Stein, T. P. (2005). Effect of a low-carbohydrate diet on appetite, blood glucose levels, and insulin resistance in obese patients with type 2 diabetes. *Annals of Internal Medicine, 142*(6), 403–411. https://doi.org/10.7326/0003-4819-142-6-200503150-00006.

Bolanos, J. P., & Almeida, A. (2010). The pentose-phosphate pathway in neuronal survival against nitrosative stress. *IUBMB Life, 62*(1), 14–18. https://doi.org/10.1002/iub.280.

Braak, H., & Del Tredici, K. (2012). Alzheimer's disease: Pathogenesis and prevention. *Alzheimer's & Dementia, 8*(3), 227–233. https://doi.org/10.1016/j.jalz.2012.01.011.

Braak, H., & Del Tredici, K. (2013). Evolutional aspects of Alzheimer's disease pathogenesis. *Journal of Alzheimer's Disease, 33*(Suppl. 1), S155–S161. https://doi.org/10.3233/JAD-2012-129029.

Buckner, R. L. (2004). Memory and executive function in aging and AD: Multiple factors that cause decline and reserve factors that compensate. *Neuron, 44*(1), 195–208. https://doi.org/10.1016/j.neuron.2004.09.006.

Cahill, G. F., Jr. (2006). Fuel metabolism in starvation. *Annual Review of Nutrition, 26*, 1–22. https://doi.org/10.1146/annurev.nutr.26.061505.111258.

Cahill, G. F., Jr., & Veech, R. L. (2003). Ketoacids? Good medicine? *Transactions of the American Clinical and Climatological Association, 114*, 149–161. discussion 162-143.

Castellano, C. A., Nugent, S., Paquet, N., Tremblay, S., Bocti, C., Lacombe, G., et al. (2015). Lower brain 18F-fluorodeoxyglucose uptake but normal 11C-acetoacetate metabolism in mild Alzheimer's disease dementia. *Journal of Alzheimer's Disease, 43*(4), 1343–1353. https://doi.org/10.3233/JAD-141074.

Castellano, C. A., Paquet, N., Dionne, I. J., Imbeault, H., Langlois, F., Croteau, E., et al. (2017). A 3-month aerobic training program improves brain energy metabolism in mild Alzheimer's disease: Preliminary results from a neuroimaging study. *Journal of Alzheimer's Disease, 56*(4), 1459–1468. https://doi.org/10.3233/JAD-161163.

Chen, J. Y., Tran, C., Hwang, L., Deng, G., Jung, M. E., Faull, K. F., et al. (2016). Partial amelioration of peripheral and central symptoms of Huntington's disease via modulation of lipid metabolism. *Journal of Huntingtons Disease, 5*(1), 65–81. https://doi.org/10.3233/JHD-150181.

Ciarmiello, A., Cannella, M., Lastoria, S., Simonelli, M., Frati, L., Rubinsztein, D. C., et al. (2006). Brain white-matter volume loss and glucose hypometabolism precede the clinical symptoms of Huntington's disease. *Journal of Nuclear Medicine, 47*(2), 215–222.

Clarke, K., Tchabanenko, K., Pawlosky, R., Carter, E., Todd King, M., Musa-Veloso, K., et al. (2012). Kinetics, safety and tolerability of (R)-3-hydroxybutyl (R)-3-hydroxybutyrate in healthy adult subjects. *Regulatory Toxicology and Pharmacology, 63*(3), 401–408. https://doi.org/10.1016/j.yrtph.2012.04.008.

Courchesne-Loyer, A., Lowry, C. M., St-Pierre, V., Vandenberghe, C., Fortier, M., Castellano, C. A., et al. (2017). Emulsification increases the acute ketogenic effect and bioavailability of medium-chain triglycerides in humans: Protein, carbohydrate, and fat metabolism. *Current Developments in Nutrition, 1*(7). e000851https://doi.org/10.3945/cdn.117.000851.

Cox, P. J., Kirk, T., Ashmore, T., Willerton, K., Evans, R., Smith, A., et al. (2016). Nutritional ketosis alters fuel preference and thereby endurance performance in athletes. *Cell Metabolism, 24*(2), 256–268. https://doi.org/10.1016/j.cmet.2016.07.010.

Croteau, E., Castellano, C. A., Fortier, M., Bocti, C., Fulop, T., Paquet, N., et al. (2018a). A cross-sectional comparison of brain glucose and ketone metabolism in cognitively healthy older adults, mild cognitive impairment and early Alzheimer's disease. *Experimental Gerontology, 107*, 18–26. https://doi.org/10.1016/j.exger.2017.07.004.

Croteau, E., Castellano, C. A., Richard, M. A., Fortier, M., Nugent, S., Lepage, M., et al. (2018b). Ketogenic medium chain triglycerides increase brain energy metabolism in Alzheimer's disease. *Journal of Alzheimer's Disease, 64*(2), 551–561. https://doi.org/10.3233/JAD-180202.

Cummings, J., Lee, G., Ritter, A., Sabbagh, M., & Zhong, K. (2019). Alzheimer's disease drug development pipeline: 2019. *Alzheimers & Dementia (New York, N. Y.), 5*, 272–293. https://doi.org/10.1016/j.trci.2019.05.008.

Cunnane, S. C., Courchesne-Loyer, A., St-Pierre, V., Vandenberghe, C., Pierotti, T., Fortier, M., et al. (2016). Can ketones compensate for deteriorating brain glucose uptake during aging? Implications for the risk and treatment of Alzheimer's disease. *Annals of the New York Academy of Sciences, 1367*(1), 12–20. https://doi.org/10.1111/nyas.12999.

Cunnane, S., Nugent, S., Roy, M., Courchesne-Loyer, A., Croteau, E., Tremblay, S., et al. (2011). Brain fuel metabolism, aging, and Alzheimer's disease. *Nutrition, 27*(1), 3–20. https://doi.org/10.1016/j.nut.2010.07.021.

De Felice, F. G., Vieira, M. N., Bomfim, T. R., Decker, H., Velasco, P. T., Lambert, M. P., et al. (2009). Protection of synapses against Alzheimer's-linked toxins: Insulin signaling prevents the pathogenic binding of Abeta oligomers. *Proceedings of the National Academy of Sciences of the United States of America, 106*(6), 1971–1976. https://doi.org/10.1073/pnas.0809158106.

Dearlove, D. J., Faull, O. K., Rolls, E., Clarke, K., & Cox, P. J. (2019). Nutritional ketoacidosis during incremental exercise in healthy athletes. *Frontiers in Physiology, 10*, 290. https://doi.org/10.3389/fphys.2019.00290.

Demetrius, L. A., Magistretti, P. J., & Pellerin, L. (2014). Alzheimer's disease: The amyloid hypothesis and the inverse Warburg effect. *Frontiers in Physiology, 5*, 522. https://doi.org/10.3389/fphys.2014.00522.

Dienel, G. A. (2019). Brain glucose metabolism: Integration of energetics with function. *Physiological Reviews, 99*(1), 949–1045. https://doi.org/10.1152/physrev.00062.2017.

Ding, F., Yao, J., Rettberg, J. R., Chen, S., & Brinton, R. D. (2013). Early decline in glucose transport and metabolism precedes shift to ketogenic system in female aging and Alzheimer's mouse brain: Implication for bioenergetic intervention. *PLoS One, 8*(11), e79977. https://doi.org/10.1371/journal.pone.0079977.

Drulis-Fajdasz, D., Gizak, A., Wojtowicz, T., Wisniewski, J. R., & Rakus, D. (2018). Aging-associated changes in hippocampal glycogen metabolism in mice. Evidence for and against astrocyte-to-neuron lactate shuttle. *Glia, 66*(7), 1481–1495. https://doi.org/10.1002/glia.23319.

Edwards, C., Canfield, J., Copes, N., Rehan, M., Lipps, D., & Bradshaw, P. C. (2014). D-beta-hydroxybutyrate extends lifespan in *C. elegans*. *Aging (Albany NY), 6*(8), 621–644. https://doi.org/10.18632/aging.100683.

Eitan, E., Tosti, V., Suire, C. N., Cava, E., Berkowitz, S., Bertozzi, B., et al. (2017). In a randomized trial in prostate cancer patients, dietary protein restriction modifies markers of leptin and insulin signaling in plasma extracellular vesicles. *Aging Cell, 16*(6), 1430–1433. https://doi.org/10.1111/acel.12657.

Fattoretti, P., Bertoni-Freddari, C., Casoli, T., Di Stefano, G., Solazzi, M., & Giorgetti, B. (2002). Decreased expression of glucose transport protein (Glut3) in aging and vitamin E deficiency. *Annals of the New York Academy of Sciences, 973*, 293–296. https://doi.org/10.1111/j.1749-6632.2002.tb04653.x.

Faull, O. K., Dearlove, D. J., Clarke, K., & Cox, P. J. (2019). Beyond RPE: The perception of exercise under normal and ketotic conditions. *Frontiers in Physiology, 10*, 229. https://doi.org/10.3389/fphys.2019.00229.

Firbank, M. J., Yarnall, A. J., Lawson, R. A., Duncan, G. W., Khoo, T. K., Petrides, G. S., et al. (2017). Cerebral glucose metabolism and cognition in newly diagnosed Parkinson's disease: ICICLE-PD study. *Journal of Neurology, Neurosurgery, and Psychiatry, 88*(4), 310–316. https://doi.org/10.1136/jnnp-2016-313918.

Fischer, T., Och, U., Klawon, I., Och, T., Gruneberg, M., Fobker, M., et al. (2018). Effect of a sodium and calcium DL-beta-hydroxybutyrate salt in healthy adults. *Journal of Nutrition and Metabolism, 2018*, 9812806. https://doi.org/10.1155/2018/9812806.

Foster, N. L., Heidebrink, J. L., Clark, C. M., Jagust, W. J., Arnold, S. E., Barbas, N. R., et al. (2007). FDG-PET improves accuracy in distinguishing frontotemporal dementia and Alzheimer's disease. *Brain, 130*(Pt. 10), 2616–2635. https://doi.org/10.1093/brain/awm177.

Gamberino, W. C., & Brennan, W. A., Jr. (1994). Glucose transporter isoform expression in Huntington's disease brain. *Journal of Neurochemistry, 63*(4), 1392–1397. https://doi.org/10.1046/j.1471-4159.1994.63041392.x.

Gefen, T., Papastefan, S. T., Rezvanian, A., Bigio, E. H., Weintraub, S., Rogalski, E., et al. (2018). Von Economo neurons of the anterior cingulate across the lifespan and in Alzheimer's disease. *Cortex, 99*, 69–77. https://doi.org/10.1016/j.cortex.2017.10.015.

Goetzl, E. J., Nogueras-Ortiz, C., Mustapic, M., Mullins, R. J., Abner, E. L., Schwartz, J. B., et al. (2019). Deficient neurotrophic factors of CSPG4-type neural cell exosomes in Alzheimer disease. *The FASEB Journal, 33*(1), 231–238. https://doi.org/10.1096/fj.201801001.

Gonzalez-Redondo, R., Garcia-Garcia, D., Clavero, P., Gasca-Salas, C., Garcia-Eulate, R., Zubieta, J. L., et al. (2014). Grey matter hypometabolism and atrophy in Parkinson's disease with cognitive impairment: A two-step process. *Brain, 137*(Pt. 8), 2356–2367. https://doi.org/10.1093/brain/awu159.

Goyal, M. S., Hawrylycz, M., Miller, J. A., Snyder, A. Z., & Raichle, M. E. (2014). Aerobic glycolysis in the human brain is associated with development and neotenous gene expression. *Cell Metabolism, 19*(1), 49–57. https://doi.org/10.1016/j.cmet.2013.11.020.

Goyal, M. S., Vlassenko, A. G., Blazey, T. M., Su, Y., Couture, L. E., Durbin, T. J., et al. (2017). Loss of brain aerobic glycolysis in normal human aging. *Cell Metabolism, 26*(2). https://doi.org/10.1016/j.cmet.2017.07.010. 353-360.e353.

Grieve, S. M., Clark, C. R., Williams, L. M., Peduto, A. J., & Gordon, E. (2005). Preservation of limbic and paralimbic structures in aging. *Human Brain Mapping, 25*(4), 391–401. https://doi.org/10.1002/hbm.20115.

Haier, R. J., Alkire, M. T., White, N. S., Uncapher, M. R., Head, E., Lott, I. T., et al. (2003). Temporal cortex hypermetabolism in Down syndrome prior to the onset of dementia. *Neurology, 61*(12), 1673–1679.

Henderson, S. T., Vogel, J. L., Barr, L. J., Garvin, F., Jones, J. J., & Costantini, L. C. (2009). Study of the ketogenic agent AC-1202 in mild to moderate Alzheimer's disease: A randomized, double-blind, placebo-controlled, multicenter trial. *Nutrition & Metabolism (London), 6*, 31. https://doi.org/10.1186/1743-7075-6-31.

Holdsworth, D. A., Cox, P. J., Kirk, T., Stradling, H., Impey, S. G., & Clarke, K. (2017). A ketone ester drink increases postexercise muscle glycogen synthesis in humans. *Medicine and Science in Sports and Exercise, 49*(9), 1789–1795. https://doi.org/10.1249/MSS.0000000000001292.

Hooijmans, C. R., Graven, C., Dederen, P. J., Tanila, H., van Groen, T., & Kiliaan, A. J. (2007). Amyloid beta deposition is related to decreased glucose transporter-1 levels and hippocampal atrophy in brains of aged APP/PS1 mice. *Brain Research, 1181*, 93–103. https://doi.org/10.1016/j.brainres.2007.08.063.

Hoyer, S. (1992). Oxidative energy metabolism in Alzheimer brain. Studies in early-onset and late-onset cases. *Molecular and Chemical Neuropathology, 16*(3), 207–224. https://doi.org/10.1007/bf03159971.

Jha, M. K., & Morrison, B. M. (2018). Glia-neuron energy metabolism in health and diseases: New insights into the role of nervous system metabolic transporters. *Experimental Neurology, 309*, 23–31. https://doi.org/10.1016/j.expneurol.2018.07.009.

Kalpouzos, G., Chetelat, G., Baron, J. C., Landeau, B., Mevel, K., Godeau, C., et al. (2009). Voxel-based mapping of brain gray matter volume and glucose metabolism profiles in normal aging. *Neurobiology of Aging, 30*(1), 112–124. https://doi.org/10.1016/j.neurobiolaging.2007.05.019.

Kapogiannis, D., Boxer, A., Schwartz, J. B., Abner, E. L., Biragyn, A., Masharani, U., et al. (2015). Dysfunctionally phosphorylated type 1 insulin receptor substrate in neural-derived blood exosomes of preclinical Alzheimer's disease. *The FASEB Journal, 29*(2), 589–596. https://doi.org/10.1096/fj.14-262048.

Kapogiannis, D., & Mattson, M. P. (2011). Disrupted energy metabolism and neuronal circuit dysfunction in cognitive impairment and Alzheimer's disease. *Lancet Neurology, 10*(2), 187–198. https://doi.org/10.1016/S1474-4422(10)70277-5.

Kapogiannis, D., Mustapic, M., Shardell, M. D., Berkowitz, S. T., Diehl, T. C., Spangler, R. D., et al. (2019). Association of extracellular vesicle biomarkers with Alzheimer disease in the baltimore longitudinal study of aging. *JAMA Neurology, 76*(11), 1340–1351. https://doi.org/10.1001/jamaneurol.2019.2462.

Kashiwaya, Y., Bergman, C., Lee, J. H., Wan, R., King, M. T., Mughal, M. R., et al. (2013). A ketone ester diet exhibits anxiolytic and cognition-sparing properties, and lessens amyloid and tau pathologies in a mouse model of Alzheimer's disease. *Neurobiology of Aging, 34*(6), 1530–1539. https://doi.org/10.1016/j.neurobiolaging.2012.11.023.

Kashiwaya, Y., Sato, K., Tsuchiya, N., Thomas, S., Fell, D. A., Veech, R. L., et al. (1994). Control of glucose utilization in working perfused rat heart. *The Journal of Biological Chemistry, 269*(41), 25502–25514.

Kashiwaya, Y., Takeshima, T., Mori, N., Nakashima, K., Clarke, K., & Veech, R. L. (2000). D-beta-hydroxybutyrate protects neurons in models of Alzheimer's and Parkinson's disease. *Proceedings of the National Academy of Sciences of the United States of America, 97*(10), 5440–5444. https://doi.org/10.1073/pnas.97.10.5440.

Klosinski, L. P., Yao, J., Yin, F., Fonteh, A. N., Harrington, M. G., Christensen, T. A., et al. (2015). White matter lipids as a ketogenic fuel supply in aging female brain: Implications for Alzheimer's disease. *eBioMedicine, 2*(12), 1888–1904. https://doi.org/10.1016/j.ebiom.2015.11.002.

Kuczynski, B., Targan, E., Madison, C., Weiner, M., Zhang, Y., Reed, B., et al. (2010). White matter integrity and cortical metabolic associations in aging and dementia. *Alzheimer's & Dementia, 6*(1), 54–62. https://doi.org/10.1016/j.jalz.2009.04.1228.

Kurochkin, I. V., Guarnera, E., & Berezovsky, I. N. (2018). Insulin-degrading enzyme in the fight against Alzheimer's disease. *Trends in Pharmacological Sciences, 39*(1), 49–58. https://doi.org/10.1016/j.tips.2017.10.008.

Lim, S., Chesser, A. S., Grima, J. C., Rappold, P. M., Blum, D., Przedborski, S., et al. (2011). D-beta-hydroxybutyrate is protective in mouse models of Huntington's disease. *PLoS One, 6*(9), e24620. https://doi.org/10.1371/journal.pone.0024620.

Liu, H. Y., Han, J., Cao, S. Y., Hong, T., Zhuo, D., Shi, J., et al. (2009). Hepatic autophagy is suppressed in the presence of insulin resistance and hyperinsulinemia: Inhibition of FoxO1-dependent expression of key autophagy genes by insulin. *The Journal of Biological Chemistry, 284*(45), 31484–31492. https://doi.org/10.1074/jbc.M109.033936.

Liu, Y., Liu, F., Iqbal, K., Grundke-Iqbal, I., & Gong, C. X. (2008). Decreased glucose transporters correlate to abnormal hyperphosphorylation of tau in Alzheimer disease. *FEBS Letters, 582*(2), 359–364. https://doi.org/10.1016/j.febslet.2007.12.035.

Magistretti, P. J., & Allaman, I. (2007). Glycogen: A Trojan horse for neurons. *Nature Neuroscience, 10*(11), 1341–1342. https://doi.org/10.1038/nn1107-1341.

Mamelak, M. (2012). Sporadic Alzheimer's disease: The starving brain. *Journal of Alzheimer's Disease, 31*(3), 459–474. https://doi.org/10.3233/JAD-2012-120370.

Mark, R. J., Pang, Z., Geddes, J. W., Uchida, K., & Mattson, M. P. (1997). Amyloid beta-peptide impairs glucose transport in hippocampal and cortical neurons: Involvement of membrane lipid peroxidation. *The Journal of Neuroscience, 17*(3), 1046–1054.

McEwen, B. S., & Reagan, L. P. (2004). Glucose transporter expression in the central nervous system: Relationship to synaptic function. *European Journal of Pharmacology, 490*(1–3), 13–24. https://doi.org/10.1016/j.ejphar.2004.02.041.

Mooradian, A. D., & Shah, G. N. (1997). Age-related changes in glucose transporter-one mRNA structure and function. *Proceedings of the Society for Experimental Biology and Medicine, 216*(3), 380–385. https://doi.org/10.3181/00379727-216-44185.

Morris, A. A. (2005). Cerebral ketone body metabolism. *Journal of Inherited Metabolic Disease, 28*(2), 109–121. https://doi.org/10.1007/s10545-005-5518-0.

Mosconi, L., Tsui, W. H., De Santi, S., Li, J., Rusinek, H., Convit, A., et al. (2005). Reduced hippocampal metabolism in MCI and AD: Automated FDG-PET image analysis. *Neurology, 64*(11), 1860–1867. https://doi.org/10.1212/01.WNL.0000163856.13524.08.

Mullins, R. J., Diehl, T. C., Chia, C. W., & Kapogiannis, D. (2017a). Insulin resistance as a link between amyloid-beta and Tau pathologies in Alzheimer's disease. *Frontiers in Aging Neuroscience, 9*, 118. https://doi.org/10.3389/fnagi.2017.00118.

Mullins, R. J., Mustapic, M., Goetzl, E. J., & Kapogiannis, D. (2017b). Exosomal biomarkers of brain insulin resistance associated with regional atrophy in Alzheimer's disease. *Human Brain Mapping, 38*(4), 1933–1940. https://doi.org/10.1002/hbm.23494.

Mullins, R., Reiter, D., & Kapogiannis, D. (2018). Magnetic resonance spectroscopy reveals abnormalities of glucose metabolism in the Alzheimer's brain. *Annals of Clinical Translational Neurology, 5*(3), 262–272. https://doi.org/10.1002/acn3.530.

Murray, A. J., Knight, N. S., Cole, M. A., Cochlin, L. E., Carter, E., Tchabanenko, K., et al. (2016). Novel ketone diet enhances physical and cognitive performance. *The FASEB Journal, 30*(12), 4021–4032. https://doi.org/10.1096/fj.201600773R.

Mustapic, M., Eitan, E., Werner, J. K., Jr., Berkowitz, S. T., Lazaropoulos, M. P., Tran, J., et al. (2017). Plasma extracellular vesicles enriched for neuronal origin: A potential window into brain pathologic processes. *Frontiers in Neuroscience, 11*, 278. https://doi.org/10.3389/fnins.2017.00278.

Mustapic, M., Tran, J., Craft, S., & Kapogiannis, D. (2019). Extracellular vesicle biomarkers track cognitive changes following intranasal insulin in Alzheimer's disease. *Journal of Alzheimer's Disease, 69*(2), 489–498. https://doi.org/10.3233/JAD-180578.

Myette-Cote, E., Neudorf, H., Rafiei, H., Clarke, K., & Little, J. P. (2018). Prior ingestion of exogenous ketone monoester attenuates the glycaemic response to an oral glucose tolerance test in healthy young individuals. *The Journal of Physiology, 596*(8), 1385–1395. https://doi.org/10.1113/JP275709.

Newington, J. T., Pitts, A., Chien, A., Arseneault, R., Schubert, D., & Cumming, R. C. (2011). Amyloid beta resistance in nerve cell lines is mediated by the Warburg effect. *PLoS One, 6*(4), e19191. https://doi.org/10.1371/journal.pone.0019191.

Nugent, S., Castellano, C. A., Bocti, C., Dionne, I., Fulop, T., & Cunnane, S. C. (2016). Relationship of metabolic and endocrine parameters to brain glucose metabolism in older adults: Do cognitively-normal older adults have a particular metabolic phenotype? *Biogerontology, 17*(1), 241–255. https://doi.org/10.1007/s10522-015-9595-7.

Nugent, S., Castellano, C. A., Goffaux, P., Whittingstall, K., Lepage, M., Paquet, N., et al. (2014a). Glucose hypometabolism is highly localized, but lower cortical thickness and brain atrophy are widespread in cognitively normal older adults. *American Journal of Physiology. Endocrinology and Metabolism, 306*(11), E1315–E1321. https://doi.org/10.1152/ajpendo.00067.2014.

Nugent, S., Tremblay, S., Chen, K. W., Ayutyanont, N., Roontiva, A., Castellano, C. A., et al. (2014b). Brain glucose and acetoacetate metabolism: A comparison of young and older adults. *Neurobiology of Aging, 35*(6), 1386–1395. https://doi.org/10.1016/j.neurobiolaging.2013.11.027.

Ossenkoppele, R., Schonhaut, D. R., Scholl, M., Lockhart, S. N., Ayakta, N., Baker, S. L., et al. (2016). Tau PET patterns mirror clinical and neuroanatomical variability in Alzheimer's disease. *Brain, 139*(Pt. 5), 1551–1567. https://doi.org/10.1093/brain/aww027.
Owen, O. E., Morgan, A. P., Kemp, H. G., Sullivan, J. M., Herrera, M. G., & Cahill, G. F., Jr. (1967). Brain metabolism during fasting. *The Journal of Clinical Investigation, 46*(10), 1589–1595. https://doi.org/10.1172/JCI105650.
Pawlosky, R. J., Kemper, M. F., Kashiwaya, Y., King, M. T., Mattson, M. P., & Veech, R. L. (2017). Effects of a dietary ketone ester on hippocampal glycolytic and tricarboxylic acid cycle intermediates and amino acids in a 3xTgAD mouse model of Alzheimer's disease. *Journal of Neurochemistry, 141*(2), 195–207. https://doi.org/10.1111/jnc.13958.
Perez-Escuredo, J., Van Hee, V. F., Sboarina, M., Falces, J., Payen, V. L., Pellerin, L., et al. (2016). Monocarboxylate transporters in the brain and in cancer. *Biochimica et Biophysica Acta, 1863*(10), 2481–2497. https://doi.org/10.1016/j.bbamcr.2016.03.013.
Phillips, M. C. L., Murtagh, D. K. J., Gilbertson, L. J., Asztely, F. J. S., & Lynch, C. D. P. (2018). Low-fat versus ketogenic diet in Parkinson's disease: A pilot randomized controlled trial. *Movement Disorders, 33*(8), 1306–1314. https://doi.org/10.1002/mds.27390.
Prapong, T., Buss, J., Hsu, W. H., Heine, P., West Greenlee, H., & Uemura, E. (2002). Amyloid beta-peptide decreases neuronal glucose uptake despite causing increase in GLUT3 mRNA transcription and GLUT3 translocation to the plasma membrane. *Experimental Neurology, 174*(2), 253–258. https://doi.org/10.1006/exnr.2001.7861.
Reiman, E. M., Caselli, R. J., Yun, L. S., Chen, K., Bandy, D., Minoshima, S., et al. (1996). Preclinical evidence of Alzheimer's disease in persons homozygous for the epsilon 4 allele for apolipoprotein E. *The New England Journal of Medicine, 334*(12), 752–758. https://doi.org/10.1056/NEJM199603213341202.
Reiman, E. M., Chen, K., Alexander, G. E., Caselli, R. J., Bandy, D., Osborne, D., et al. (2004). Functional brain abnormalities in young adults at genetic risk for late-onset Alzheimer's dementia. *Proceedings of the National Academy of Sciences of the United States of America, 101*(1), 284–289. https://doi.org/10.1073/pnas.2635903100.
Roy, M., Nugent, S., Tremblay-Mercier, J., Tremblay, S., Courchesne-Loyer, A., Beaudoin, J. F., et al. (2012). The ketogenic diet increases brain glucose and ketone uptake in aged rats: A dual tracer PET and volumetric MRI study. *Brain Research, 1488*, 14–23. https://doi.org/10.1016/j.brainres.2012.10.008.
Scheepers, A., Joost, H. G., & Schurmann, A. (2004). The glucose transporter families SGLT and GLUT: Molecular basis of normal and aberrant function. *JPEN Journal of Parenteral and Enteral Nutrition, 28*(5), 364–371. https://doi.org/10.1177/0148607104028005364.
Seeley, W. W., Crawford, R. K., Zhou, J., Miller, B. L., & Greicius, M. D. (2009). Neurodegenerative diseases target large-scale human brain networks. *Neuron, 62*(1), 42–52. https://doi.org/10.1016/j.neuron.2009.03.024.
Simpson, I. A., Chundu, K. R., Davies-Hill, T., Honer, W. G., & Davies, P. (1994). Decreased concentrations of GLUT1 and GLUT3 glucose transporters in the brains of patients with Alzheimer's disease. *Annals of Neurology, 35*(5), 546–551. https://doi.org/10.1002/ana.410350507.
Sinadinos, C., Valles-Ortega, J., Boulan, L., Solsona, E., Tevy, M. F., Marquez, M., et al. (2014). Neuronal glycogen synthesis contributes to physiological aging. *Aging Cell, 13*(5), 935–945. https://doi.org/10.1111/acel.12254.
Sintini, I., Schwarz, C. G., Martin, P. R., Graff-Radford, J., Machulda, M. M., Senjem, M. L., et al. (2019). Regional multimodal relationships between tau, hypometabolism, atrophy, and fractional anisotropy in atypical Alzheimer's disease. *Human Brain Mapping, 40*(5), 1618–1631. https://doi.org/10.1002/hbm.24473.

Soto-Mota, A., Vansant, H., Evans, R. D., & Clarke, K. (2019). Safety and tolerability of sustained exogenous ketosis using ketone monoester drinks for 28 days in healthy adults. *Regulatory Toxicology and Pharmacology*, *109*, 104506. https://doi.org/10.1016/j.yrtph.2019.104506.

Soucek, T., Cumming, R., Dargusch, R., Maher, P., & Schubert, D. (2003). The regulation of glucose metabolism by HIF-1 mediates a neuroprotective response to amyloid beta peptide. *Neuron*, *39*(1), 43–56. https://doi.org/10.1016/s0896-6273(03)00367-2.

Souza, D. G., Bellaver, B., Raupp, G. S., Souza, D. O., & Quincozes-Santos, A. (2015). Astrocytes from adult Wistar rats aged in vitro show changes in glial functions. *Neurochemistry International*, *90*, 93–97. https://doi.org/10.1016/j.neuint.2015.07.016.

Steculorum, S. M., Solas, M., & Bruning, J. C. (2014). The paradox of neuronal insulin action and resistance in the development of aging-associated diseases. *Alzheimer's & Dementia*, *10*(1 Suppl), S3–11. https://doi.org/10.1016/j.jalz.2013.12.008.

Stubbs, B. J., Cox, P. J., Evans, R. D., Cyranka, M., Clarke, K., & de Wet, H. (2018). A ketone ester drink lowers human Ghrelin and appetite. *Obesity (Silver Spring)*, *26*(2), 269–273. https://doi.org/10.1002/oby.22051.

Stubbs, B. J., Cox, P. J., Evans, R. D., Santer, P., Miller, J. J., Faull, O. K., et al. (2017). On the metabolism of exogenous ketones in humans. *Frontiers in Physiology*, *8*, 848. https://doi.org/10.3389/fphys.2017.00848.

Stubbs, B. J., Cox, P. J., Kirk, T., Evans, R. D., & Clarke, K. (2019). Gastrointestinal effects of exogenous ketone drinks are infrequent, mild and vary according to ketone compound and dose. *International Journal of Sport Nutrition and Exercise Metabolism*, *29*(6), 1–23. https://doi.org/10.1123/ijsnem.2019-0014.

Szablewski, L. (2017). Glucose transporters in brain: In health and in Alzheimer's disease. *Journal of Alzheimer's Disease*, *55*(4), 1307–1320. https://doi.org/10.3233/JAD-160841.

Thorens, B., & Mueckler, M. (2010). Glucose transporters in the 21st century. *American Journal of Physiology. Endocrinology and Metabolism*, *298*(2), E141–E145. https://doi.org/10.1152/ajpendo.00712.2009.

Tieu, K., Perier, C., Caspersen, C., Teismann, P., Wu, D. C., Yan, S. D., et al. (2003). D-beta-hydroxybutyrate rescues mitochondrial respiration and mitigates features of Parkinson disease. *The Journal of Clinical Investigation*, *112*(6), 892–901. https://doi.org/10.1172/JCI18797.

Uemura, E., & Greenlee, H. W. (2006). Insulin regulates neuronal glucose uptake by promoting translocation of glucose transporter GLUT3. *Experimental Neurology*, *198*(1), 48–53. https://doi.org/10.1016/j.expneurol.2005.10.035.

van Hall, G., Stromstad, M., Rasmussen, P., Jans, O., Zaar, M., Gam, C., et al. (2009). Blood lactate is an important energy source for the human brain. *Journal of Cerebral Blood Flow and Metabolism*, *29*(6), 1121–1129. https://doi.org/10.1038/jcbfm.2009.35.

Veech, R. L., Bradshaw, P. C., Clarke, K., Curtis, W., Pawlosky, R., & King, M. T. (2017). Ketone bodies mimic the life span extending properties of caloric restriction. *IUBMB Life*, *69*(5), 305–314. https://doi.org/10.1002/iub.1627.

Veech, R. L., Chance, B., Kashiwaya, Y., Lardy, H. A., & Cahill, G. F., Jr. (2001). Ketone bodies, potential therapeutic uses. *IUBMB Life*, *51*(4), 241–247. https://doi.org/10.1080/152165401753311780.

Vittori, A., Breda, C., Repici, M., Orth, M., Roos, R. A., Outeiro, T. F., et al. (2014). Copy-number variation of the neuronal glucose transporter gene SLC2A3 and age of onset in Huntington's disease. *Human Molecular Genetics*, *23*(12), 3129–3137. https://doi.org/10.1093/hmg/ddu022.

Vlassenko, A. G., Gordon, B. A., Goyal, M. S., Su, Y., Blazey, T. M., Durbin, T. J., et al. (2018). Aerobic glycolysis and tau deposition in preclinical Alzheimer's disease. *Neurobiology of Aging*, *67*, 95–98. https://doi.org/10.1016/j.neurobiolaging.2018.03.014.

Willette, A. A., Bendlin, B. B., Starks, E. J., Birdsill, A. C., Johnson, S. C., Christian, B. T., et al. (2015a). Association of insulin resistance with cerebral glucose uptake in late middle-aged adults at risk for Alzheimer disease. *JAMA Neurology*, *72*(9), 1013–1020. https://doi.org/10.1001/jamaneurol.2015.0613.

Willette, A. A., Modanlo, N., Kapogiannis, D., & Alzheimer's Disease Neuroimaging Initiative. (2015b). Insulin resistance predicts medial temporal hypermetabolism in mild cognitive impairment conversion to Alzheimer disease. *Diabetes*, *64*(6), 1933–1940. https://doi.org/10.2337/db14-1507.

Yao, J., Hamilton, R. T., Cadenas, E., & Brinton, R. D. (2010). Decline in mitochondrial bioenergetics and shift to ketogenic profile in brain during reproductive senescence. *Biochimica et Biophysica Acta*, *1800*(10), 1121–1126. https://doi.org/10.1016/j.bbagen.2010.06.002.

Yao, J., Rettberg, J. R., Klosinski, L. P., Cadenas, E., & Brinton, R. D. (2011). Shift in brain metabolism in late onset Alzheimer's disease: Implications for biomarkers and therapeutic interventions. *Molecular Aspects of Medicine*, *32*(4–6), 247–257. https://doi.org/10.1016/j.mam.2011.10.005.

Yin, J. X., Maalouf, M., Han, P., Zhao, M., Gao, M., Dharshaun, T., et al. (2016). Ketones block amyloid entry and improve cognition in an Alzheimer's model. *Neurobiology of Aging*, *39*, 25–37. https://doi.org/10.1016/j.neurobiolaging.2015.11.018.

Zhao, W. Q., Chen, H., Quon, M. J., & Alkon, D. L. (2004). Insulin and the insulin receptor in experimental models of learning and memory. *European Journal of Pharmacology*, *490*(1–3), 71–81. https://doi.org/10.1016/j.ejphar.2004.02.045.

CHAPTER FOUR

Disturbances in brain energy metabolism in insulin resistance and diabetes and Alzheimer's disease — Learnings from brain imaging biomarkers

Kerstin Heurling[a,*], Edvin Johansson[a], Antoine Leuzy[b]
[a]Antaros Medical AB, Mölndal, Sweden
[b]Department of Clinical Sciences, Clinical Memory Research Unit, Lund University, Malmö, Sweden
[*]Corresponding author: e-mail address: kerstin.heurling@antarosmedical.com

Contents

1. Introduction to medical imaging modalities 112
2. Imaging cerebral glucose metabolism 113
   2.1 Effects of plasma glucose levels on the brain glucose uptake 114
   2.2 Insulin resistance and the effect of insulin on glucose uptake 115
   2.3 Treatment effects measured using FDG 117
3. Imaging cerebral lipid metabolism 118
4. Magnetic resonance imaging 119
5. Imaging metabolic changes with proton magnetic resonance spectroscopy 121
6. Conclusions 123
References 124

Abstract

Medical imaging techniques, such as structural and functional magnetic resonance imaging and positron emission tomography, have been used to gain a better understanding of the alterations of the metabolic processes in the brain relating to type 2 diabetes melltius, insulin resistance and Alzheimer's disease. These studies have shown that there are several similarities in the effects that these seemingly disparate diseases have on the brain, and that some of the abnormalities are reversed by metabolic interventions. This review provides an overview of the overlap between these diseases using medical imaging, focusing on glucose metabolism, mitochondrial function and lipid metabolism.

International Review of Neurobiology, Volume 154
ISSN 0074-7742
https://doi.org/10.1016/bs.irn.2020.02.011

Abbreviations

AD Alzheimer's disease
BBB blood-brain barrier
CMRglc cerebral metabolic rate of glucose
CNS central nervous system
FDG fluoro-2-deoxy-D-glucose
fMRI functional magnetic resonance imaging
FTHA ^{18}F-fluoro-thia-heptadecanoic acid
GLP-1 glucagon-like peptide-1
HOMA-IR homeostatic model assessment of insulin resistance
MCI mild cognitive impairment
MRI magnetic resonance imaging
MRS magnetic resonance spectroscopy
NAA *N*-acetyl-aspartate
PET positron emission tomography
SGLT2 sodium-glucose cotransporter type 2
T2DM type 2 diabetes mellitus

1. Introduction to medical imaging modalities

The growing attention to the neurological effects of type 2 diabetes mellitus (T2DM), including its association with the development of neurodegenerative diseases and dementia, calls for techniques that allow the *in vivo* assessment of varied brain related processes. Although fluid-based biomarkers (e.g., measurements of protein levels in cerebrospinal fluid) can provide insight into the state of the brain, such measures do not provide spatial information, may be more related to disease stage, and are known to be affected by preanalytical factors.

The majority of knowledge regarding metabolic processes of the brain gathered with imaging techniques is based on positron emission tomography (PET) performed with the fluorine-18 labeled glucose analog fluoro-2-deoxy-D-glucose (FDG). Positron emission tomography is a medical imaging technique utilizing short-lived radioactive isotopes that decay with the emission of positrons, and a scanner consisting of a ring of detectors to record the decays and the position of each event. The most common positron emitting isotope is fluorine-18. The short-lived isotopes used in PET are incorporated into biologically active molecules of interest, for example metabolic substrates, perfusion markers and ligands with affinity to receptors, and the resulting molecules are known as PET tracers. The involvement of a molecule in a given physiological process determines how the radioactivity will distribute in the body. Furthermore, the physiological state of the subject being examined can be altered, meaning that the distribution of the PET tracer is measured in response to a specific stimuli, for instance after administration of a pharmaceutical agent with affinity to the same receptor as the tracer or under metabolic challenge such as occurs during a hyperinsulinemic euglycemic clamp.

Magnetic resonance imaging (MRI), on the other hand, uses static magnetic fields and radio waves to acquire signals within the nuclei of water hydrogen atoms. Based on the MRI technique, high-resolution morphological images of the brain with high soft-tissue contrast can be generated, as well as functional data, such as water diffusion maps and brain activation images. In addition, since different molecules have distinct magnetic resonance properties, maps of several common cerebral metabolites can be determined via a technique known as magnetic resonance spectroscopy (MRS).

This review will introduce the reader to the medical imaging techniques most relevant for studies of brain metabolism and provide an overview of the overlapping findings in insulin resistance, T2DM and Alzheimer's disease (AD).

2. Imaging cerebral glucose metabolism

For the past 40 years, FDG has been the most commonly used PET tracer. The FDG method used to measure the cerebral metabolic rate of glucose (CMRglc) quantitatively with PET in humans is an adaptation of the 2-deoxy-D-[^{14}C]glucose model originally developed by Sokoloff et al. (1977). Fluoro-2-deoxy-D-glucose based measurements of CMRglc reflect regional brain metabolism associated with physiological or pathophysiological conditions and can be viewed as a marker of neuronal integrity (Phelps et al., 1979). Fluoro-2-deoxy-D-glucose and glucose in plasma share and compete for common transporters on the blood-brain barrier (BBB), mainly GLUT1 and GLUT3, for transfer from plasma into the brain; in the brain, FDG and glucose compete for the enzyme hexokinase, which phosphorylates FDG irreversibly to fluoro-deoxy-glucose-6-phosphate and glucose to glucose-6-phosphate. While glucose-6-phosphate is further metabolized, fluoro-deoxy-glucose-6-phosphate is not a substrate for any enzyme and becomes trapped and accumulates in the tissue. In Fig. 1A, example images of the uptake of FDG are shown, where the higher metabolic activity of the gray matter compared to the white matter is evident.

Patients with AD show prominent reductions in FDG uptake in the temporal cortex, posterior cingulate and inferior parietal cortices (Jagust

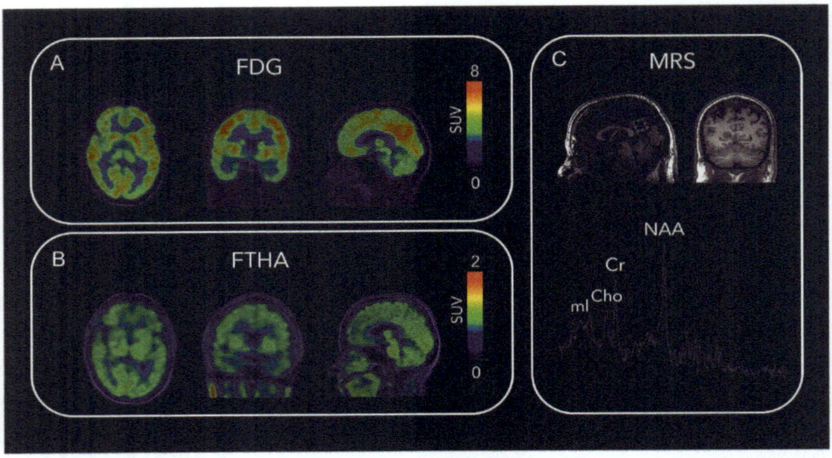

Fig. 1 Example uptake images in the same subject of (A) ^{18}F-fluoro-2-deoxy-D-glucose (FDG) PET and (B) ^{18}F-fluoro-thia-heptadecanoic acid (FTHA) PET, shown with an overlay of a T1 MRI, reflecting glucose and fatty acid uptake, respectively. The difference in maximum standardized uptake value (SUV) on the color scale indicate that while the distribution pattern demonstrates a higher uptake in the cortex for both PET tracers, the absolute uptake of glucose is higher than the uptake of fatty acids. In (C), an example of a voxel placement for magnetic resonance spectroscopy imaging in the precuneus is shown on a T1 MRI , combined with the spectrum (below). The peaks of some key different metabolites, myo-inositol (ml), choline (Cho), creatine (Cr) and *N*-acetyl-aspartate (NAA) are identified in the spectrum.

et al., 2007; Minoshima et al., 1997). Identification of this distribution pattern of FDG as part of the clinical work-up has been demonstrated to increase the sensitivity and specificity of the clinical diagnosis of AD and is included as a supportive biomarker (Sperling et al., 2011). Fluoro-2-deoxy-D-glucose is also of value in the differential diagnosis of neurodegenerative diseases as different conditions present with other characteristic patterns of hypometabolism. The association between T2DM and hypometabolism has been confirmed in a cohort of cognitively healthy elderly subjects using FDG PET, highlighting regions typically presenting as hypometabolic in AD, including the posterior cingulate, temporal and parietal cortex (Roberts et al., 2014).

2.1 Effects of plasma glucose levels on the brain glucose uptake

Since FDG uptake competes with glucose for transport into the brain and hexokinase phosphorylation in all tissues utilizing glucose, it is not surprising that the uptake of FDG is affected by plasma glucose levels.

Quantitative estimates of CMRglc include a correction for this, assuming a linear and uniform behavior of the competition. Studies have shown, however, that increased plasma glucose levels, even in cognitively normal subjects, can alter the cerebral uptake pattern of FDG and reduce the uptake in AD-related regions, resulting in the appearance of an AD-like pattern (Burns et al., 2013; Kawasaki et al., 2008). Furthermore, findings indicate that this hypometabolic pattern can already appear at borderline elevated plasma glucose levels and that it is reversible, i.e., it disappears when returning to a normoglycemic state (Ishibashi et al., 2014, 2015). Interestingly, acute hyperglycemia induced experimentally has been shown to result in a significantly increased uptake of FDG in the white matter, without affecting glucose uptake in cortical and subcortical gray matter structures or the transport capacity of the BBB (Hasselbalch et al., 2001). Taking these aspects into account, it becomes clear that when assessing neurodegeneration with FDG PET, it is important to measure plasma glucose levels carefully, not only for the sake of correcting for the levels, but also because examinations may have to be rescheduled or other measures taken to lower the plasma glucose levels before performing the assessment. The direct impact of plasma glucose levels on the diagnostic accuracy of FDG PET in the context of dementia diseases, however, has been deemed limited (Apostolova et al., 2018).

A key question that remains is whether the pattern change seen with FDG as a consequence of increased plasma glucose levels is in fact a result of a protective mechanism aimed at maintaining normoglycemic glucose tissue utilization in critical regions, with an apparent pattern of hypometabolism caused by an overflow of glucose in surrounding regions. Alternatively, increased serum glucose levels may result in prolonged or chronic hyperglycemia in these regions; this disruption to energy homeostasis could lead to tissue damage and, over time, neurodegeneration and subsequent hypometabolism.

2.2 Insulin resistance and the effect of insulin on glucose uptake

Insulin resistance has long been recognized as a central feature of T2DM, but it has also recently been demonstrated to occur in the brains of individuals with neurodegenerative diseases, such as AD, even in the absence of concurrent T2DM. Brain insulin resistance can be defined as the failure of brain cells to respond to insulin (Mielke et al., 2005). The cause, however, can be a downregulation of the expression of insulin receptors, a developed inability of insulin receptors to bind insulin, defective activation of the insulin-signaling cascade, or a combination of all three factors (Arnold et al., 2018). This may result in impaired neuroplasticity, receptor regulation or neurotransmitter release, as well as impaired processes directly related to insulin metabolism, such as glucose uptake.

Interestingly, acute and chronic intranasal insulin administration has been shown to improve memory and other cognitive functions in cognitively unimpaired adults without obesity or T2DM (Benedict et al., 2004), and in patients with mild cognitive impairment (MCI), AD, Parkinson's disease and multiple system atrophy (Avgerinos et al., 2018; Novak et al., 2019). Similarly, positive effects on memory have also been shown after acute glucose administration (Stollery and Christian, 2016).

Accumulating evidence suggests that insulin resistance is associated with brain changes that may contribute to the pathological features of AD in the preclinical phase. Post-mortem assessments of brains from patients with AD show deficient insulin signaling (Talbot et al., 2012). Studies utilizing FDG PET have shown that more severe insulin resistance is related to lower glucose metabolism which in turn is related to memory function (Willette et al., 2015a). Other studies have also indicated that insulin resistance (Baker et al., 2011) and diabetes mellitus (Roberts et al., 2014) are associated with hypometabolism using FDG PET. In particular, studies evaluating more severe insulin resistance using the homeostatic model assessment of insulin resistance (HOMA-IR) and its relationship with to glucose hypometabolism demonstrate the strongest effects on the AD-related regions such as bilateral prefrontal cortex, temporal cortex, and posteromedial parietal cortex in cognitively healthy subjects (Baker et al., 2011; Willette et al., 2015a). An association between HOMA-IR and hypometabolism has also been confirmed in patients with AD (Willette et al., 2015b).

Brain glucose utilization has been studied under the condition of hyperinsulinemia. The effects of insulin *per se* can be investigated by maintaining physiological glucose levels using a hyperinsulinemic euglycemic clamp and combining this with brain FDG PET (Boersma et al., 2018; Eastman et al., 1990; Hirvonen et al., 2011; Johansson et al., 2018). Eastman et al. (1990) also performed repeat experiments during a sham clamp, indicating not only that the total accumulation of radioactivity in brain tissue was reduced during the hyperinsulinemic euglycemic clamp, but also that radioactivity in the blood was lower, possibly due to a higher removal of FDG from blood into the skeletal muscle. Hirvonen et al. (2011) showed that the glucose uptake increased in the hyperinsulinemic condition compared to fasting only in patients with impaired glucose tolerance but not in healthy controls, establishing that the underlying reason was not changes in cerebral blood flow, which were unaffected by the clamp procedure. Johansson et al. (2018) and Boersma et al. (2018) investigated the relationship between glucose uptake rate and the whole-body insulin sensitivity (M-value) estimated from the clamp itself across the T2DM spectrum, and showed a negative correlation in the brain, opposite to the positive correlation demonstrated in skeletal muscle. Voxel-wise analysis also revealed that this correlation was strongest in the aforementioned AD-related brain regions (Boersma et al., 2018). These studies suggest that the response to insulin, in terms of glucose uptake, may be increased in subjects along the spectrum of T2DM, ranging from obese and insulin resistant, to manifested T2DM.

2.3 Treatment effects measured using FDG

New classes of glucose-lowering agents for the treatment of T2DM, such as glucagon-like peptide-1 (GLP-1) receptor agonists and sodium-glucose cotransporter type 2 (SGLT2) inhibitors, have received attention for their potential use in the treatment of neurodegenerative diseases (NCT03801642 and NCT01843075). Glucagon-like peptide-1 receptor agonists are of interest since GLP-1, as well as the GLP-1 receptors, are also expressed within the central nervous system (CNS) (Gejl et al., 2012). In the hyperglycemic state, infusion of native GLP-1 has been shown to increase the clearance of glucose from the brain, whereas no effects of GLP-1 could be demonstrated in the hypoglycemic state, supporting the neuroprotective effects of GLP-1 during hyperglycemia and suggesting a lack of deleterious effects in the hypoglycemic (Gejl, 2013; Gejl et al., 2012). In a more clinically applicable approach, the same research group also investigated whether long-term treatment with a GLP-1 analog could impact disease progression in patients with AD (Gejl et al., 2016). A 6-month treatment, however, had no effect on the accumulation of β-amyloid or on cognitive outcomes compared with placebo; nevertheless, its use significantly increased cortical glucose consumption, whereas the placebo group showed continued decreased glucose consumption with disease progression (Gejl et al., 2016). Although proof of principle clinical trials are currently ongoing, the pro-amnestic and neuroprotective effects of SGLT2 inhibitors have to date been shown only in animal models (Sa-Nguanmoo et al., 2017; van der Kooij et al., 2018). The usefulness of GLP-1 agonists and SGLT2 inhibitors, however, with respect to preventing or slowing neurodegeneration, will hopefully be demonstrated soon.

3. Imaging cerebral lipid metabolism

While the primary fuel for the brain is glucose under physiological conditions, it is believed that the brain can switch to ketone bodies and fatty acids for energy under circumstances where the supply of glucose is limited (LaManna et al., 2009). Although a controversial topic, it has been suggested that the brain can use fatty acid oxidation as an energy source during conditions of fasting, prolonged starvation or diabetes, especially in glial cells (Romano et al., 2017). Imaging studies investigating fatty acid metabolism in the brain is to date limited. However, imaging using the fatty acid PET tracers ^{18}F-fluoro-thia-heptadecanoic acid (FTHA) and ^{11}C-palmitate in healthy controls and patients with metabolic syndrome has demonstrated that fatty acids are indeed taken up in the brain and likely also metabolized, and that their increased uptake is associated with metabolic syndrome (Karmi et al., 2010). This suggests that changes in fatty acid uptake can occur early in the diabetes continuum. In Fig. 1B, example images of FTHA uptake are shown.

The link between changes in brain lipid metabolism, AD and the risk gene *APOE* ε4 was also highlighted in a study using ^{11}C-docosahexaenoic acid, a polyunsaturated fatty acid PET tracer (Yassine et al., 2017). Docosahexaenoic acid constitutes up to 40% of fatty acids in certain gray matter lipids, is involved in neuronal homeostasis, and is concentrated in the synapses (Lauritzen et al., 2001). The PET study demonstrated an increased uptake of the tracer among subjects carrying the *APOE* ε4 allele, indicating that the genotype influences the metabolism of at least this specific fatty acid, but the finding may also apply to other fatty acids (Yassine et al., 2017).

^{11}C-β-Hydroxybutyrate is a ketone PET tracer that has been used to study the brain uptake of ketone bodies and shown that the transport of ketone bodies across the BBB is irreversible (Blomqvist et al., 1995, 2002). There was no apparent difference in the ketone body uptake during acute hyperketonemia in patients with T2DM compared to healthy controls (Blomqvist et al., 2002). Preclinical studies in rats using FDG and ^{11}C-acetoacetate have shown that uptake of these two energy sources did not differ between young and old rats, but that a ketogenic diet increased the uptake of both the FDG and ^{11}C-acetoacetate (Roy et al., 2012). ^{11}C-acetoacetate has also been used to study the effects of aging and AD in humans together with FDG, showing no significant effects on the ^{11}C-acetoacetate uptake resulting from normal aging compared to healthy young, or AD compared to age-matched controls (Castellano et al., 2014; Nugent et al., 2014). The metabolic shift toward ketone bodies as a preferred energy source has been confirmed, however, using the dual tracer study design, showing that a short-term ketogenic diet results in a decrease in the utilization of glucose and an increase in ketone utilization (Courchesne-Loyer et al., 2016). Interestingly, when a ketogenic drink or placebo was added to the diet of patients with MCI for 6 months and compared to placebo, ketone metabolism was doubled and the treatment group showed improvement in several cognitive domains (Fortier et al., 2019).

While imaging of cerebral glucose metabolism has been extensively utilized during the past four decades, the imaging of cerebral lipid metabolism is still a fairly novel and unexplored field. It holds great potential in expanding our understanding of the metabolic flexibility of the brain, as well as its involvement in the development of brain disorders. Given the relative novelty of and limited access to fatty acid PET compared to FDG PET, applications of the same methodology in both T2DM and neurodegenerative diseases, such as AD, are so far lacking, precluding the possibility of direct comparisons.

4. Magnetic resonance imaging

In its most common clinical application, MRI employs a combination of a strong static magnetic field, weaker time-varying magnetic fields and radio waves to generate images from signal emanating from the ^{1}H nuclei in water molecules. Besides the non-invasive nature of the technique and its high spatial resolution, the key strength of MRI is often the excellent soft-tissue contrast compared to other medical imaging techniques, such as computed tomography. In cerebral applications this can be exemplified by the good separation of gray and white matter. Within the field of neurodegenerative diseases, MRI has contributed to a better understanding of the morphological changes related to disease. Hippocampal atrophy, for instance, can be used to identify AD as the cause underlying the syndrome of MCI (Sperling et al., 2011). In several studies focusing on the association between morphological changes in the brain and T2DM, cortical thickness has been studied. While overall cortical thickness does not appear to be strongly linked to T2DM-related clinical parameters (e.g., HbA1c) certain regions, such as the medial prefrontal cortex, seem to be more affected (Ajilore et al., 2010). Furthermore, cerebrovascular risk factors are negatively correlated with cortical thickness and subcortical volumes

(Cox et al., 2019). Lower hippocampal volumes are more strongly associated with the presence of diabetes, increased disease duration, higher HbA1c and body mass index (Bruehl et al., 2009; Cox et al., 2019; Piao et al., 2002). In AD, gray matter atrophy in temporal, parietal and limbic regions occurs first before spreading into frontal and occipital regions (Thompson et al., 2003). While hippocampal volume is commonly used as an AD biomarker and staging aid (Jack et al., 1999), other morphological landmarks, such as temporal and parietal volumes, have been shown to be of value in identifying cognitively unimpaired individuals at risk of developing dementia (Chiang et al., 2011).

In addition to morphological imaging, MRI can also be used to visualize several functional properties of the brain. Diffusion tensor imaging is sensitive to the effects of tissue microstructure on water diffusion, caused by subtle white matter microstructural changes. Key diffusion tensor imaging parameters are mean diffusivity and fractional anisotropy, where mean diffusivity reflects the magnitude of water diffusion making it suitable for assessing gray matter tissue density, whereas fractional anisotropy measures the degree of directionality of water diffusion. Diffusion tensor imaging may be a useful tool for investigating early microstructural changes indicating a risk for future cognitive impairment; increased mean diffusivity in gray matter and decreased fractional anisotropy in white matter tracts are proposed to reflect microstructural abnormalities, including demyelination and axonal loss (Bozzali and Cherubini, 2007). In studies using diffusion tensor imaging, widespread gray and white matter microstructural abnormalities have been shown in T2DM, primarily in the temporal lobe and fractional anisotropy reduction in the white matter of the left hemisphere (Yau et al., 2009, 2014). One of these studies also showed an association between verbal memory decline and mean diffusivity of the temporal cortex in T2DM (Yau et al., 2014). Similar findings have been shown in a large population of older T2DM patients with reduced fractional anisotropy in the total white matter and increased mean diffusivity in gray matter structures (Falvey et al., 2013). Unsurprisingly, widespread increases in mean diffusivity in both gray and white matter have also been shown in patients with MCI and AD as neuronal damage is a key feature of AD and includes the hyper-phosphorylation of the axonal protein tau (Agosta et al., 2011).

Magnetic resonance imaging can be used to study local brain activation via the technique of functional MRI (fMRI). The activation is assessed indirectly via signal changes that are associated with the different magnetic properties of oxygenated and deoxygenated hemoglobin and the local transient increase of the oxyhemoglobin-to-deoxyhemoglobin ratio during the hemodynamic response related to neuronal activity (Ogawa et al., 1992). Functional MRI is typically used in combination with task-based or stimulus driven paradigms (Pohl et al., 2017), where signals in the baseline state are assessed to identify activated brain regions associated with the task. Alternatively, fMRI can be used to study the intrinsic activity of the brain at rest without any explicit task; this is known as resting state fMRI or functional connectivity MRI. The approach allows for the visualization of simultaneous activations in separate regions to identify resting state networks (Lee et al., 2013). Studies utilizing fMRI have indicated anomalies in the functional connectivity in both patients with T2DM and AD. Higher connectivity has been demonstrated within the default mode network, also correlating with fasting plasma glucose in T2DM (Chen et al., 2015), whereas alterations within the default mode network in AD have been shown to involve both increased and decreased connectivity in different sub-networks (Toussaint et al., 2014).

5. Imaging metabolic changes with proton magnetic resonance spectroscopy

While the signal detected in MRI mainly originates from the ^{1}H nuclei of water molecules, signals from nuclei in other molecules can also be assessed. The resonance frequency of the nuclei varies as a result of its surrounding electron distribution, which shields the magnetic field to a small, but detectable extent. The amount of shielding is distinct for different metabolites, which forms the basis of MRS and enables non-invasive quantification of various metabolites within the tissue. The spatial resolution of MRS is comparatively low, however, due to the low abundance of these metabolites and the correspondingly low signal; despite the low resolution, MRS is unique in providing direct information on endogenous metabolic processes *in vivo*. An example of a voxel placed in the precuneus for the assessment of metabolite concentrations in this region and the corresponding spectrum is shown in Fig. 1C.

While other nuclei besides ^{1}H can also be detected, ^{1}H is the most common variant of MRS. It has been used to study effects of neurodegenerative diseases, such as AD and T2DM, on concentrations of neurochemical metabolites (for example, see meta-analyses by Wang et al., 2015; Wu et al., 2017).

Most studies have focused on four neurometabolites that contribute the strongest signals to *in vivo* spectra: *N*-acetyl-aspartate (NAA), choline, creatine and myo-inositol. Detecting the absolute concentrations from MRS results is relatively difficult, however, such that ratios with metabolites known to be unaffected by pathology are often reported (Helms, 2008).

N-Acetyl-aspartate is a general measure of neuronal density and a biomarker of functional neurons. *N*-acetyl-aspartate has been suggested to be a neuronal osmolyte, and has a neuronal-glial cell specific signaling role important for CNS functions (Baslow, 2002; Mäkimattila et al., 2004). The production of NAA relates to mitochondrial function and is, therefore, expected to be involved in the energy metabolism of neurons (Moffett et al., 2007). A decline in brain NAA reflects axonal or neuronal dysfunction and has been demonstrated in several neurological conditions ranging from MCI to schizophrenia (Kraguljac et al., 2012; Wang et al., 2015; Waraga et al., 2017). In patients suffering from acute traumatic brain injury leading to hypoxia, NAA predicts recovery (De Stefano et al., 1995; Signoretti et al., 2001). Therefore, NAA can be used as a biomarker of neuronal function, density and degeneration, and as a predictor of recovery following neuronal injury. Decreased levels of NAA have also been demonstrated in insulin resistance and T2DM (Sahin et al., 2008; Sinha et al., 2014; Tiehuis et al., 2010; Wu et al., 2017).

Creatine also plays an important role in cellular energy utilization. It is involved in ATP regeneration and is particularly important for maintaining membrane potentials, ion gradients and Ca^{2+} homeostasis (Rackayova et al., 2017). Creatine is relatively easy to detect with proton MRS, the peak represents the total concentration of creatine and phospho-creatine. As the concentrations of creatine are fairly stable under physiological conditions and show little variation throughout the day (Soreni et al., 2006), with age (Saunders et al., 1999) or as a result of AD and T2DM (Kantarci, 2013; Sinha et al., 2014), creatine is often used as an internal reference for other metabolites.

Choline-containing compounds are associated with membrane homeostasis and immobile when incorporated in the membranes, and contribute to the MRS signal only when released, for example, as glycerophosphoryl choline, which is a breakdown product of membrane phospholipids (Saunders et al., 1999). Choline has thus been indicated as a marker of alterations in membrane turnover or changes in cell density, where fluctuating levels of choline can potentially reflect cell damage or inflammation (Chiu et al., 2014; Harris et al., 2014). In T2DM, an increased choline/creatine ratio has been reported in the lenticular nucleus, but not in other regions studied, such as the frontal and occipital lobe (Wu et al., 2017). In AD, a significant increase in choline/creatine ratio has been demonstrated in the posterior cingulate (Wang et al., 2015), but limited overlap in the brain regions studied in T2DM and AD limits the opportunity for direct comparisons between the diseases beyond the indication of elevated choline/creatine ratios in both diseases.

Myo-inositol is another brain metabolite expressed at high concentrations. It is a glucose-like molecule expressed in different cell types, but found mainly in glial cells. Similar to NAA, myo-inositol is a precursor of membrane phospholipids and also involved in cell membrane and myelin sheet structures (Brand et al., 1993). Increased glial expression of myo-inositol can be a result of increased membrane turnover, damage to myelin sheets, glial proliferation and microglial activation (Brand et al., 1993), and increased glial expression has been demonstrated in both T2DM (occipital and parietal lobe) and AD (posterior cingulate and parietal lobe) using MRS (Wang et al., 2015; Wu et al., 2017).

While the changes in metabolites related to AD and T2DM mentioned in the literature typically goes in the same direction (i.e., decreased NAA, increased myo-inositol), the brain regions where these findings have been demonstrated rarely overlap. Although an increased myo-inositol/creatine ratio has been shown in the parietal lobe in both T2DM and AD, a decreased NAA/creatine ratio has been demonstrated in the frontal lobe and lenticular nucleus in T2DM, but not in AD. Conversely, the temporal and parietal lobe have been highlighted as regions showing a decreased NAA/creatine ratio in AD but not T2DM patients (Wang et al., 2015; Wu et al., 2017). This may be because the region of interest needs to be predefined in single-voxel MRS, unlike most medical imaging methodologies. Therefore, regions selected for study are biased by the condition or research topic in question; this may have led to the above described regional mismatches in AD and T2DM. As whole-brain MRS becomes more common, however, comparisons between diseases like AD and T2DM will become more straight-forward.

6. Conclusions

Medical imaging techniques allow brain metabolism to be studied in many ways, focusing on glucose, fatty acid or ketone metabolism using PET or mitochondrial functions using MRS. While glucose metabolism is the most extensively characterized metabolic aspect of the brain studied with medical imaging so far, much remains to be explored in terms of how the brain can alter its energy utilization as a response to external factors and fuel supply. Alterations in brain glucose metabolism in T2DM strongly resemble the pattern of neurodegeneration seen in AD, with signature areas like the posterior cingulate and temporoparietal cortex being affected by hypometabolism. Studies with MRI have also revealed that AD and T2DM are characterized by similar changes in hippocampal volume, cortical thickness, and diffusivity (increased mean diffusivity, accompanied by reduced fractional anisotropy). Although direct comparisons are still limited for other imaging modalities, impaired mitochondrial function is known to occur in both AD and T2DM. This, combined with the observation that both diseases have shown altered fatty acid metabolism, support holistic studies that address the role of metabolism along the AD-T2DM spectrum.

References

Agosta, F., Pievani, M., Sala, S., Geroldi, C., Galluzzi, S., Frisoni, G. B., et al. (2011). White matter damage in Alzheimer disease and its relationship to gray matter atrophy. *Radiology, 258*(3), 853–863. https://doi.org/10.1148/radiol.10101284.

Ajilore, O., Narr, K., Rosenthal, J., Pham, D., Hamilton, L., Watari, K., et al. (2010). Regional cortical gray matter thickness differences associated with type 2 diabetes and major depression. *Psychiatry Research: Neuroimaging, 184*(2), 63–70. https://doi.org/10.1016/j.pscychresns.2010.07.003.

Apostolova, I., Lange, C., Suppa, P., Spies, L., Klutmann, S., Adam, G., et al. (2018). Impact of plasma glucose level on the pattern of brain FDG uptake and the predictive power of FDG PET in mild cognitive impairment. *European Journal of Nuclear Medicine and Molecular Imaging, 45*(8), 1417–1422. https://doi.org/10.1007/s00259-018-3985-4.

Arnold, S. E., Arvanitakis, Z., Macauley-Rambach, S. L., Koenig, A. M., Wang, H.-Y., Ahima, R. S., et al. (2018). Brain insulin resistance in type 2 diabetes and Alzheimer disease: Concepts and conundrums. *Nature Reviews Neurology, 14*(3), 168–181. https://doi.org/10.1038/nrneurol.2017.185.

Avgerinos, K. I., Kalaitzidis, G., Malli, A., Kalaitzoglou, D., Myserlis, P. G., & Lioutas, V.-A. (2018). Intranasal insulin in Alzheimer's dementia or mild cognitive impairment: A systematic review. *Journal of Neurology, 265*(7), 1497–1510. https://doi.org/10.1007/s00415-018-8768-0.

Baker, L. D., Cross, D. J., Minoshima, S., Belongia, D., Watson, G. S., & Craft, S. (2011). Insulin resistance and Alzheimer-like reductions in regional cerebral glucose metabolism for cognitively normal adults with prediabetes or early type 2 diabetes. *Archives of Neurology, 68*(1), 51–57. https://doi.org/10.1001/archneurol.2010.225.

Baslow, M. H. (2002). Evidence supporting a role for N-acetyl-l-aspartate as a molecular water pump in myelinated neurons in the central nervous system: An analytical review. *Neurochemistry International, 40*(4), 295–300. https://doi.org/10.1016/S0197-0186(01)00095-X.

Benedict, C., Hallschmid, M., Hatke, A., Schultes, B., Fehm, H. L., Born, J., et al. (2004). Intranasal insulin improves memory in humans. *Psychoneuroendocrinology, 29*(10), 1326–1334. https://doi.org/10.1016/j.psyneuen.2004.04.003.

Blomqvist, G., Alvarsson, M., Grill, V., Von Heijne, G., Ingvar, M., Thorell, J. O., et al. (2002). Effect of acute hyperketonemia on the cerebral uptake of ketone bodies in nondiabetic subjects and IDDM patients. *American Journal of Physiology-Endocrinology and Metabolism, 283*(1), E20–E28. https://doi.org/10.1152/ajpendo.00294.2001.

Blomqvist, G., Thorell, J. O., Ingvar, M., Grill, V., Widen, L., & Stone-Elander, S. (1995). Use of R-beta-[1-11C]hydroxybutyrate in PET studies of regional cerebral uptake of ketone bodies in humans. *American Journal of Physiology-Endocrinology and Metabolism, 269*(5), E948–E959. https://doi.org/10.1152/ajpendo.1995.269.5.E948.

Boersma, G. J., Johansson, E., Pereira, M. J., Heurling, K., Skrtic, S., Lau, J., et al. (2018). Altered glucose uptake in muscle, visceral adipose tissue, and brain predict whole-body insulin resistance and may contribute to the development of type 2 diabetes: A combined PET/MR study. *Hormone and Metabolic Research, 50*(08), 627–639. https://doi.org/10.1055/a-0831-3620.

Bozzali, M., & Cherubini, A. (2007). Diffusion tensor MRI to investigate dementias: A brief review. *Magnetic Resonance Imaging, 25*(6), 969–977. https://doi.org/10.1016/j.mri.2007.03.017.

Brand, A., Richter-Landsberg, C., & Leibfritz, D. (1993). Multinuclear NMR studies on the energy metabolism of glial and neuronal cells. *Developmental Neuroscience, 15*(3–5), 289–298. https://doi.org/10.1159/000111347.

Bruehl, H., Wolf, O. T., & Convit, A. (2009). A blunted cortisol awakening response and hippocampal atrophy in type 2 diabetes mellitus. *Psychoneuroendocrinology, 34*(6), 815–821. https://doi.org/10.1016/j.psyneuen.2008.12.010.

Burns, C. M., Chen, K., Kaszniak, A. W., Lee, W., Alexander, G. E., Bandy, D., et al. (2013). Higher serum glucose levels are associated with cerebral hypometabolism in Alzheimer regions. *Neurology, 80*(17), 1557–1564. https://doi.org/10.1212/WNL.0b013e31828f17de.

Castellano, C.-A., Nugent, S., Paquet, N., Tremblay, S., Bocti, C., Lacombe, G., et al. (2014). Lower brain 18F-fluorodeoxyglucose uptake but normal 11C-acetoacetate metabolism in mild Alzheimer's disease dementia. *Journal of Alzheimer's Disease, 43*(4), 1343–1353. https://doi.org/10.3233/JAD-141074.

Chen, Y., Liu, Z., Zhang, J., Tian, G., Li, L., Zhang, S., et al. (2015). Selectively disrupted functional connectivity networks in type 2 diabetes mellitus. *Frontiers in Aging Neuroscience*, 7, 233. Retrieved from https://www.frontiersin.org/article/10.3389/fnagi.2015.00233.

Chiang, G. C., Insel, P. S., Tosun, D., Schuff, N., Truran-Sacrey, D., Raptentsetsang, S., et al. (2011). Identifying cognitively healthy elderly individuals with subsequent memory decline by using automated MR temporoparietal volumes. *Radiology, 259*(3), 844–851. https://doi.org/10.1148/radiol.11101637.

Chiu, P.-W., Mak, H. K.-F., Yau, K. K.-W., Chan, Q., Chang, R. C.-C., & Chu, L.-W. (2014). Metabolic changes in the anterior and posterior cingulate cortices of the normal aging brain: Proton magnetic resonance spectroscopy study at 3 T. *Age (Dordrecht, Netherlands), 36*(1), 251–264. https://doi.org/10.1007/s11357-013-9545-8.

Courchesne-Loyer, A., Croteau, E., Castellano, C.-A., St-Pierre, V., Hennebelle, M., & Cunnane, S. C. (2016). Inverse relationship between brain glucose and ketone metabolism in adults during short-term moderate dietary ketosis: A dual tracer quantitative positron emission tomography study. *Journal of Cerebral Blood Flow & Metabolism, 37*(7), 2485–2493. https://doi.org/10.1177/0271678X16669366.

Cox, S. R., Lyall, D. M., Ritchie, S. J., Bastin, M. E., Harris, M. A., Buchanan, C. R., et al. (2019). Associations between vascular risk factors and brain MRI indices in UK Biobank. *European Heart Journal, 40*(28), 2290–2300. https://doi.org/10.1093/eurheartj/ehz100.

De Stefano, N., Matthews, P. M., & Arnold, D. L. (1995). Reversible decreases in N-acetylaspartate after acute brain injury. *Magnetic Resonance in Medicine*, *34*(5), 721–727. https://doi.org/10.1002/mrm.1910340511.

Eastman, R. C., Carson, R. E., Gordon, M. R., Berg, G. W., Lillioja, S., Larson, S. M., et al. (1990). Brain glucose metabolism in noninsulin-dependent diabetes mellitus: A study in Pima Indians using positron emission tomography during hyperinsulinemia with euglycemic glucose clamp. *Journal of Clinical Endocrinology and Metabolism*, *71*(6), 1602–1610.

Falvey, C. M., Rosano, C., Simonsick, E. M., Harris, T., Strotmeyer, E. S., Satterfield, S., et al. (2013). Macro- and microstructural magnetic resonance imaging indices associated with diabetes among community-dwelling older adults. *Diabetes Care*, *36*(3), 677–682. https://doi.org/10.2337/dc12-0814.

Fortier, M., Castellano, C.-A., Croteau, E., Langlois, F., Bocti, C., St-Pierre, V., et al. (2019). A ketogenic drink improves brain energy and some measures of cognition in mild cognitive impairment. (Report). *Alzheimer's & Dementia: The Journal of the Alzheimer's Association*, *15*(5), 625. https://doi.org/10.1016/j.jalz.2018.12.017.

Gejl, M. (2013). Glucagon-like peptide-1 (GLP-1) raises blood-brain glucose transfer capacity and hexokinase activity in human brain. *Frontiers in Neuroenergetics*, *5*(March), 1–9. https://doi.org/10.3389/fnene.2013.00002.

Gejl, M., Egefjord, L., Lerche, S., Vang, K., Bibby, B. M., Holst, J. J., et al. (2012). Glucagon-like peptide-1 decreases intracerebral glucose content by activating hexokinase and changing glucose clearance during hyperglycemia. *Journal of Cerebral Blood Flow and Metabolism*, *32*(12), 2146–2152. https://doi.org/10.1038/jcbfm.2012.118.

Gejl, M., Gjedde, A., Egefjord, L., Møller, A., Hansen, S. B., Vang, K., et al. (2016). In Alzheimer's disease, 6-month treatment with GLP-1 analog prevents decline of brain glucose metabolism: Randomized, placebo-controlled, double-blind clinical trial. *Frontiers in Aging Neuroscience*, *8*(May), 1–10. https://doi.org/10.3389/fnagi.2016.00108.

Harris, J. L., Yeh, H.-W., Swerdlow, R. H., Choi, I.-Y., Lee, P., & Brooks, W. M. (2014). High-field proton magnetic resonance spectroscopy reveals metabolic effects of normal brain aging. *Neurobiology of Aging*, *35*(7), 1686–1694. https://doi.org/10.1016/j.neurobiolaging.2014.01.018.

Hasselbalch, S. G., Knudsen, G. M., Capaldo, B., Postiglione, A., & Paulson, O. B. (2001). Blood-brain barrier transport and brain metabolism of glucose during acute hyperglycemia in humans. *The Journal of Clinical Endocrinology & Metabolism*, *86*(5), 1986–1990. https://doi.org/10.1210/jcem.86.5.7490.

Helms, G. (2008). The principles of quantification applied to in vivo proton MR spectroscopy. *European Journal of Radiology*, *67*(2), 218–229. https://doi.org/10.1016/j.ejrad.2008.02.034.

Hirvonen, J., Virtanen, K. A., Nummenmaa, L., Hannukainen, J. C., Honka, M. J., Bucci, M., et al. (2011). Effects of insulin on brain glucose metabolism in impaired glucose tolerance. *Diabetes*, *60*(2), 443–447. https://doi.org/10.2337/db10-0940.

Ishibashi, K., Kawasaki, K., Ishiwata, K., & Ishii, K. (2015). Reduced uptake of 18F-FDG and 15O-H_2O in Alzheimer's disease-related regions after glucose loading. *Journal of Cerebral Blood Flow & Metabolism*, *35*(8), 1380–1385. https://doi.org/10.1038/jcbfm.2015.127.

Ishibashi, K., Miura, Y., Oda, K., Ishiwata, K., & Ishii, K. (2014). Alzheimer's disease-like pattern of 18F-FDG uptake during a hyperglycemic state and negative 11 C-PiB binding in a patient with mild cognitive impairment. *Journal of Alzheimer's Disease*, *42*, 385–389. https://doi.org/10.3233/JAD-140639.

Jack, C. R., Petersen, R. C., Xu, Y. C., O'Brien, P. C., Smith, G. E., Ivnik, R. J., et al. (1999). Prediction of AD with MRI-based hippocampal volume in mild cognitive impairment. *Neurology*, *52*(7), 1397–1403. https://doi.org/10.1212/WNL.52.7.1397.

Jagust, W., Reed, B., Mungas, D., Ellis, W., & Decarli, C. (2007). What does fluorodeoxyglucose PET imaging add to a clinical diagnosis of dementia? *Neurology*, *69*(9), 871–877. https://doi.org/10.1212/01.wnl.0000269790.05105.16.

Johansson, E., Lubberink, M., Heurling, K., Eriksson, J. W., Skrtic, S., Ahlström, H., et al. (2018). Whole-body imaging of tissue specific insulin sensitivity and body composition by using an integrated PET/MR system: A feasibility study. *Radiology*, *286*(1), 271–278. https://doi.org/10.1148/radiol.2017162949.

Kantarci, K. (2013). Proton MRS in mild cognitive impairment. *Journal of Magnetic Resonance Imaging*, *37*(4), 770–777. https://doi.org/10.1002/jmri.23800.

Karmi, A., Iozzo, P., Viljanen, A., Hirvonen, J., Fielding, B. A., Virtanen, K., et al. (2010). Increased brain fatty acid uptake in metabolic syndrome. *Diabetes*, *59*(9), 2171–2177. https://doi.org/10.2337/db09-0138.

Kawasaki, K., Ishii, K., Saito, Y., Oda, K., Kimura, Y., & Ishiwata, K. (2008). Influence of mild hyperglycemia on cerebral FDG distribution patterns calculated by statistical parametric mapping. *Annals of Nuclear Medicine*, *22*, 191–200. https://doi.org/10.1007/s12149-007-0099-7.

Kraguljac, N. V., Reid, M., White, D., Jones, R., den Hollander, J., Lowman, D., et al. (2012). Neurometabolites in schizophrenia and bipolar disorder—A systematic review and meta-analysis. *Psychiatry Research*, *203*(2–3), 111–125. https://doi.org/10.1016/j.pscychresns.2012.02.003.

LaManna, J. C., Salem, N., Puchowicz, M., Erokwu, B., Koppaka, S., Flask, C., et al. (2009). Ketones suppress brain glucose consumption. *Advances in Experimental Medicine and Biology*, *645*, 301–306. https://doi.org/10.1007/978-0-387-85998-9_45.

Lauritzen, L., Hansen, H. S., Jørgensen, M. H., & Michaelsen, K. F. (2001). The essentiality of long chain n-3 fatty acids in relation to development and function of the brain and retina. *Progress in Lipid Research*, *40*(1), 1–94. https://doi.org/10.1016/S0163-7827(00)00017-5.

Lee, M. H., Smyser, C. D., & Shimony, J. S. (2013). Resting-state fMRI: A review of methods and clinical applications. *American Journal of Neuroradiology*, *34*(10), 1866–1872. https://doi.org/10.3174/ajnr.A3263.

Mäkimattila, S., Malmberg-Cèder, K., Häkkinen, A.-M., Vuori, K., Salonen, O., Summanen, P., et al. (2004). Brain metabolic alterations in patients with type 1 diabetes–hyperglycemia-induced injury. *Journal of Cerebral Blood Flow & Metabolism*, *24*(12), 1393–1399. https://doi.org/10.1097/01.WCB.0000143700.15489.B2.

Mielke, J. G., Taghibiglou, C., Liu, L., Zhang, Y., Jia, Z., Adeli, K., et al. (2005). A biochemical and functional characterization of diet-induced brain insulin resistance. *Journal of Neurochemistry*, *93*(6), 1568–1578. https://doi.org/10.1111/j.1471-4159.2005.03155.x.

Minoshima, S., Giordani, B., Berent, S., Frey, K. A., Foster, N. L., & Kuhl, D. E. (1997). Metabolic reduction in the posterior cingulate cortex in very early Alzheimer's disease. *Annals of Neurology*, *42*(1), 85–94. https://doi.org/10.1002/ana.410420114.

Moffett, J. R., Ross, B., Arun, P., Madhavarao, C. N., & Namboodiri, A. M. A. (2007). N-acetylaspartate in the CNS: From neurodiagnostics to neurobiology. *Progress in Neurobiology*, *81*(2), 89–131. https://doi.org/10.1016/j.pneurobio.2006.12.003.

Novak, P., Pimentel Maldonado, D. A., & Novak, V. (2019). Safety and preliminary efficacy of intranasal insulin for cognitive impairment in Parkinson disease and multiple system atrophy: A double-blinded placebo-controlled pilot study. *PLoS One*, *14*(4), e0214364. Retrieved from https://doi.org/10.1371/journal.pone.0214364.

Nugent, S., Tremblay, S., Chen, K. W., Ayutyanont, N., Roontiva, A., Castellano, C.-A., et al. (2014). Brain glucose and acetoacetate metabolism: A comparison of young and older adults. *Neurobiology of Aging*, *35*(6), 1386–1395. https://doi.org/10.1016/j.neurobiolaging.2013.11.027.

Ogawa, S., Tank, D. W., Menon, R., Ellermann, J. M., Kim, S. G., Merkle, H., et al. (1992). Intrinsic signal changes accompanying sensory stimulation: Functional brain mapping with magnetic resonance imaging. *Proceedings of the National Academy of Sciences of the United States of America, 89*(13), 5951–5955. https://doi.org/10.1073/pnas.89.13.5951.

Phelps, M. E., Huang, S. C., Hoffman, E. J., Selin, C., Sokoloff, L., & Kuhl, D. E. (1979). Tomographic measurement of local cerebral glucose metabolic rate in humans with (F-18)2-fluoro-2-deoxy-D-glucose: Validation of method. *Annals of Neurology, 6*(5), 371–388. https://doi.org/10.1002/ana.410060502.

Piao, Y. S., Hayashi, S., Wakabayashi, K., Kakita, A., Aida, I., Yamada, M., et al. (2002). Cerebellar cortical tau pathology in progressive supranuclear palsy and corticobasal degeneration. *Acta Neuropathologica, 103*(5), 469–474. https://doi.org/10.1007/s00401-001-0488-2.

Pohl, T. M., Tempelmann, C., & Noesselt, T. (2017). How task demands shape brain responses to visual food cues. *Human Brain Mapping, 38*(6), 2897–2912. https://doi.org/10.1002/hbm.23560.

Rackayova, V., Cudalbu, C., Pouwels, P. J. W., & Braissant, O. (2017). Creatine in the central nervous system: From magnetic resonance spectroscopy to creatine deficiencies. *Analytical Biochemistry, 529*, 144–157. https://doi.org/10.1016/j.ab.2016.11.007.

Roberts, R. O., Knopman, D. S., Cha, R. H., Mielke, M. M., Pankratz, V. S., Boeve, B. F., et al. (2014). Diabetes and elevated hemoglobin A1c levels are associated with brain hypometabolism but not amyloid accumulation. *Journal of Nuclear Medicine, 55*(5), 759–764. https://doi.org/10.2967/jnumed.113.132647.

Romano, A., Koczwara, J. B., Gallelli, C. A., Vergara, D., Micioni Di Bonaventura, M. V., Gaetani, S., et al. (2017). Fats for thoughts: An update on brain fatty acid metabolism. *International Journal of Biochemistry and Cell Biology, 84*, 40–45. https://doi.org/10.1016/j.biocel.2016.12.015.

Roy, M., Nugent, S., Tremblay-Mercier, J., Tremblay, S., Courchesne-Loyer, A., Beaudoin, J. F., et al. (2012). The ketogenic diet increases brain glucose and ketone uptake in aged rats: A dual tracer PET and volumetric MRI study. *Brain Research, 1488*, 14–23. https://doi.org/10.1016/j.brainres.2012.10.008.

Sahin, I., Alkan, A., Keskin, L., Cikim, A., Karakas, H. M., Firat, A. K., et al. (2008). Evaluation of in vivo cerebral metabolism on proton magnetic resonance spectroscopy in patients with impaired glucose tolerance and type 2 diabetes mellitus. *Journal of Diabetes and its Complications, 22*(4), 254–260. https://doi.org/10.1016/j.jdiacomp.2007.03.007.

Sa-Nguanmoo, P., Tanajak, P., Kerdphoo, S., Jaiwongkam, T., Pratchayasakul, W., Chattipakorn, N., et al. (2017). SGLT2-inhibitor and DPP-4 inhibitor improve brain function via attenuating mitochondrial dysfunction, insulin resistance, inflammation, and apoptosis in HFD-induced obese rats. *Toxicology and Applied Pharmacology, 333*, 43–50. https://doi.org/10.1016/j.taap.2017.08.005.

Saunders, D. E., Howe, F. A., van den Boogaart, A., Griffiths, J. R., & Brown, M. M. (1999). Aging of the adult human brain: In vivo quantitation of metabolite content with proton magnetic resonance spectroscopy. *Journal of Magnetic Resonance Imaging, 9*(5), 711–716. https://doi.org/10.1002/(SICI)1522-2586(199905)9:5<711::AID-JMRI14>3.0.CO;2-3.

Signoretti, S., Marmarou, A., Tavazzi, B., Lazzarino, G., Beaumont, A., & Vagnozzi, R. (2001). N-acetylaspartate reduction as a measure of injury severity and mitochondrial dysfunction following diffuse traumatic brain injury. *Journal of Neurotrauma, 18*(10), 977–991. https://doi.org/10.1089/08977150152693683.

Sinha, S., Ekka, M., Sharma, U., Raghunandan, P., Pandey, R. M., & Jagannathan, N. R. (2014). Assessment of changes in brain metabolites in Indian patients with type-2 diabetes mellitus using proton magnetic resonance spectroscopy. *BMC Research Notes*, 7(1), 1–7. https://doi.org/10.1186/1756-0500-7-41.

Sokoloff, L., Reivich, M., Kennedy, C., Des Rosiers, M. H., Patlak, C. S., Pettigrew, K. D., et al. (1977). The [14C]deoxyglucose method for the measurement of local cerebral glucose utilization: Theory, procedure, and normal values in the conscious and anesthetized albino rat. *Journal of Neurochemistry*, *28*(5), 897–916. Retrieved from https://www.ncbi.nlm.nih.gov/pubmed/864466.

Soreni, N., Noseworthy, M. D., Cormier, T., Oakden, W. K., Bells, S., & Schachar, R. (2006). Intraindividual variability of striatal 1H-MRS brain metabolite measurements at 3 T. *Magnetic Resonance Imaging*, *24*(2), 187–194. https://doi.org/10.1016/j.mri.2005.10.027.

Sperling, R. A., Aisen, P. S., Beckett, L. A., Bennett, D. A., Craft, S., Fagan, A. M., et al. (2011). Toward defining the preclinical stages of Alzheimer's disease: Recommendations from the National Institute on Aging-Alzheimer's Association workgroups on diagnostic guidelines for Alzheimer's disease. *Alzheimer's & Dementia: The Journal of the Alzheimer's Association*, 7(3), 280–292. https://doi.org/10.1016/j.jalz.2011.03.003.

Stollery, B., & Christian, L. (2016). Glucose improves object-location binding in visual-spatial working memory. *Psychopharmacology*, *233*(3), 529–547. https://doi.org/10.1007/s00213-015-4125-5.

Talbot, K., Wang, H.-Y., Kazi, H., Han, L.-Y., Bakshi, K., Stucky, A., et al. (2012). Demonstrated brain insulin resistance in Alzheimer's disease patients is associated with IGF-1 resistance, IRS-1 dysregulation, and cognitive decline. *Journal of Clinical Investigation*, *122*(4), 1316–1338.

Thompson, P. M., Hayashi, K. M., de Zubicaray, G., Janke, A. L., Rose, S. E., Semple, J., et al. (2003). Dynamics of gray matter loss in Alzheimer's disease. *The Journal of Neuroscience*, *23*(3), 994–1005. https://doi.org/10.1523/JNEUROSCI.23-03-00994.2003.

Tiehuis, A., Van Der Meer, F., Mali, W., Pleizier, M., Biessels, G. J., Kappelle, J., et al. (2010). MR spectroscopy of cerebral white matter in type 2 diabetes; no association with clinical variables and cognitive performance. *Neuroradiology*, *52*(2), 155–161. https://doi.org/10.1007/s00234-009-0598-4.

Toussaint, P.-J., Maiz, S., Coynel, D., Doyon, J., Messé, A., de Souza, L. C., et al. (2014). Characteristics of the default mode functional connectivity in normal ageing and Alzheimer's disease using resting state fMRI with a combined approach of entropy-based and graph theoretical measurements. *NeuroImage*, *101*, 778–786. https://doi.org/10.1016/j.neuroimage.2014.08.003.

van der Kooij, M. A., Jene, T., Treccani, G., Miederer, I., Hasch, A., Voelxen, N., et al. (2018). Chronic social stress-induced hyperglycemia in mice couples individual stress susceptibility to impaired spatial memory. *Proceedings of the National Academy of Sciences of the United States of America*, *115*(43), E10187–E10196. https://doi.org/10.1073/pnas.1804412115.

Wang, H., Tan, L., Wang, H. F., Liu, Y., Yin, R. H., Wang, W. Y., et al. (2015). Magnetic resonance spectroscopy in Alzheimer's disease: Systematic review and meta-analysis. *Journal of Alzheimer's Disease*, *46*(4), 1049–1070. https://doi.org/10.3233/JAD-143225.

Waragai, M., Moriya, M., & Nojo, T. (2017). Decreased N-acetyl aspartate/Myo-inositol ratio in the posterior cingulate cortex shown by magnetic resonance spectroscopy may be one of the risk markers of preclinical Alzheimer's disease: A 7-year follow-up study. *Journal of Alzheimer's Disease*, *60*(4), 1411–1427. https://doi.org/10.3233/JAD-170450.

Willette, A. A., Bendlin, B. B., Starks, E. J., Birdsill, A. C., Johnson, S. C., Christian, B. T., et al. (2015a). Association of insulin resistance with cerebral glucose uptake in late middle-aged adults at risk for Alzheimer disease. *JAMA Neurology*, *72*(9), 1013–1020. https://doi.org/10.1001/jamaneurol.2015.0613.

Willette, A. A., Modanlo, N., & Kapogiannis, D. (2015b). Insulin resistance predicts medial temporal hypermetabolism in mild cognitive impairment conversion to Alzheimer disease. *Diabetes*, *64*(6), 1933–1940. https://doi.org/10.2337/db14-1507.

Wu, G., Zhang, Q., Wu, J., Li, J., Tan, Y., Qiu, T., et al. (2017). Changes in cerebral metabolites in type 2 diabetes mellitus: A meta-analysis of proton magnetic resonance spectroscopy. *Journal of Clinical Neuroscience*, *45*, 9–13. https://doi.org/10.1016/j.jocn.2017.07.017.

Yassine, H. N., Croteau, E., Rawat, V., Hibbeln, J. R., Rapoport, S. I., Cunnane, S. C., et al. (2017). DHA brain uptake and APOE4 status: A PET study with [1-11C]-DHA. *Alzheimer's Research & Therapy*, *9*(1), 23. https://doi.org/10.1186/s13195-017-0250-1.

Yau, P. L., Javier, D., Tsui, W., Sweat, V., Bruehl, H., Borod, J. C., et al. (2009). Emotional and neutral declarative memory impairments and associated white matter microstructural abnormalities in adults with type 2 diabetes. *Psychiatry Research: Neuroimaging*, *174*(3), 223–230. https://doi.org/10.1016/j.pscychresns.2009.04.016.

Yau, P. L., Kluger, A., Borod, J. C., & Convit, A. (2014). Neural substrates of verbal memory impairments in adults with type 2 diabetes mellitus. *Journal of Clinical and Experimental Neuropsychology*, *36*(1), 74–87. https://doi.org/10.1080/13803395.2013.869310.

CHAPTER FIVE

APOE and metabolic dysfunction in Alzheimer's disease

Lance A. Johnson*
Department of Physiology, University of Kentucky College of Medicine, Lexington, KY, United States
Sanders-Brown Center on Aging, University of Kentucky College of Medicine, Lexington, KY, United States
*Corresponding author: e-mail address: Johnson.Lance@uky.edu

Contents

1. *APOE* and Alzheimer's disease risk 132
2. APOE, insulin resistance, and Alzheimer's disease 133
3. APOE and cerebral glucose and fatty acid metabolism 137
4. APOE and the PPP: A potential link between glucose metabolism and oxidative stress in Alzheimer's disease 139
5. APOE and the inverse Warburg hypothesis 141
References 143

Abstract

The strongest genetic risk factor for sporadic Alzheimer's disease (AD) is carriage of the E4 allele of APOE. Metabolic dysfunction also increases risk of dementia and AD. Facing a need for effective therapies and an aging global population, studies aimed at uncovering new therapeutic targets for AD have become critical. Insight into the biology underlying the effects of E4 and metabolic impairment on the brain may lead to novel therapies to reduce AD risk. An understudied hallmark of both AD patients and E4 individuals is a common metabolic impairment—cerebral glucose hypometabolism. This is a robust and replicated finding in humans, and begins decades prior to cognitive decline. Possession of E4 also appears to alter several other aspects of cerebral glucose metabolism, fatty acid metabolism, and management of oxidative stress through the pentose phosphate pathway. A critical knowledge gap in AD is the mechanism by which *APOE* alters cerebral metabolism and clarification as to its relevance to AD risk. Facing a need for effective therapies, studies aimed at uncovering new therapeutic targets have become critical. One such approach is to gain a better understanding of the metabolic mechanisms that may underlie E4-associated cognitive dysfunction and AD risk.

International Review of Neurobiology, Volume 154
ISSN 0074-7742
https://doi.org/10.1016/bs.irn.2020.02.002

Abbreviations

AD Alzheimer's disease
Akt protein kinase B
ApoE 2,3, 4 apolipoprotein, isoforms 2, 3, 4
APOE apolipoprotein E
APOE apolipoprotein E gene
ATP adenosine triphosphate
Aβ amyloid β
BBB blood-brain barrier
CPT carnitine palmitoyltransferase
CSF cerebrospinal fluid
CVD cardiovascular disease
DHA docosahexaenoic acid
FA fatty acids
FDG 18F-deoxyglucose
G6PD glucose-6-phosphate dehydrogenase
GLUT4 glucose transporter 4
GSH glutathione
GSSG glutathione disulfide
HFD high fat diet
IGF1 insulin growth factor 1
IR insulin resistance
IRS insulin receptor substrates
LDL low-density lipoprotein
MCI mild cognitive impairment
NADPH nicotinamide adenine dinucleotide phosphate
PDH pyruvate dehydrogenase
PET positron emission tomography
PFK-1 6-phosphofructo-1-kinase
PPARγ proliferator-activated receptor γ
PPP pentose phosphate pathway
ROS reactive oxygen species
T2DM type 2 diabetes mellitus
TBI traumatic brain injury

1. *APOE* and Alzheimer's disease risk

In humans, the gene Apolipoprotein E (*APOE*) encodes for three common isoforms: E2, E3, and E4 (Mahley and Rall, 2000). As an apolipoprotein, APOE is primarily found on very low density and high density lipoproteins and plays an important role in the metabolism and redistribution of lipoproteins and cholesterol (Eichner et al., 2002; Pendse et al., 2009).

In the periphery, APOE is primarily produced by the liver, but is also expressed by a number of other tissues including adipose tissue (Driscoll and Getz, 1984; Zechner et al., 1991). In the brain, APOE is primarily produced by astrocytes and microglia under normal physiological conditions (Eichner et al., 2002; Mahley and Rall, 2000; Xu et al., 2006). Apolipoprotein E3 is the major isoform expressed in humans, and the effects of E2 and E4 are typically compared to those of E3 to determine relative risk (Phillips, 2014). The E4 isoform is associated with higher LDL cholesterol levels and an increased risk of CVD (Bennet et al., 2007; Wilson et al., 1996). Importantly, E4 is the strongest genetic risk factor for late onset Alzheimer's disease (AD), conferring between a 2 (heterozygous) to 15-fold (homozygous) increase in risk for developing the disease (Bertram et al., 2007; Bertram and Tanzi, 2009; Harold et al., 2009; Raber et al., 2004). Conversely, E2 is protective in regards to AD (Chiang et al., 2010; Corder et al., 1994; Farrer et al., 1997), and has been associated with slower mental decline and increased longevity (Garatachea et al., 2014; Wilson et al., 2002).

2. APOE, insulin resistance, and Alzheimer's disease

Individuals with peripheral metabolic dysfunction, for example those with insulin resistance (IR) and type 2 diabetes mellitus (T2DM), face an increased risk of dementia (Peila et al., 2002). Even in the absence of overt diabetes, obesity alone is associated with an increased risk of developing AD (Craft, 2009; Kaur, 2014). Obesity also negatively affects cognition prior to the onset of dementia, especially in women, and is a strong risk factor for cognitive impairment in later life (Gunstad et al., 2010).

While both obesity and *APOE* genotype are independent risk factors for AD, they paint a complicated picture when together. The *APOE* gene is associated with adiposity in the order of E2 > E3 > E4 (Elosua et al., 2003; Tejedor et al., 2014). Paradoxically, mice and humans with E4 have a substantially higher risk of AD—despite *reduced* adiposity compared to E3 (Bennet et al., 2007; Elosua et al., 2003; Raber et al., 2004; Wilson et al., 1996). Conversely, E2 is associated with a decreased risk of AD—despite *increased* adiposity (Duman et al., 2004; Harold et al., 2009; Raber et al., 2004). Unfortunately, the relatively low frequency of the E2 allele in the general population (~8%) has made it challenging to study the biological mechanisms responsible for this AD protection (Farrer et al., 1997; Raber et al., 2004). The vast majority of clinical studies, including those that examine brain metabolism, have focused on the binary comparison of individuals with or without E4 alleles. Therefore, the influences of the E2 isoform on brain metabolism, and whether they are responsible for the observed neuroprotective effects, remain unknown.

Interestingly, AD shares multiple pathological characteristics with obesity, IR and T2DM, including impairments in vascular function, increases in oxidative stress, and inflammation (Craft, 2009; Walker and Harrison, 2015). The growing elderly demographic in the developed world means that the number of individuals suffering from dementia is expected to rise dramatically over the coming years, and because the incidence of metabolic disorders such as IR and T2DM also increase with age, this rapidly aging demographic effect will be further compounded (Narayan et al., 2006).

Carriers of the E4 gene have been shown to have elevated IR and glucose intolerance compared to non-carriers (Morris et al., 2017). Several other features of IR and T2DM that may contribute to the development and progression of AD are also observed in individuals possessing an E4 allele. For example, reductions in cerebral blood flow are observed in both E4 carriers (Filippini et al., 2011; Thambisetty et al., 2010) and individuals with T2DM (Chung et al., 2015; Pallas and Larson, 1996). Both E4 and T2DM are also associated with cerebral glucose hypometabolism (Brandon et al., 2018). Peripheral IR leads to a decrease in transport of insulin across the blood-brain barrier (BBB) and its subsequent function, and AD patients (regardless of *APOE* genotype) have reduced concentrations of cerebrospinal fluid (CSF) insulin and show signs of brain IR (Talbot et al., 2012).

However, the relationship between E4 status, brain IR and cognitive function is still quite unclear (Craft et al., 2000; Farmer et al., 2019a). Carriage of E4 does appear to affect the correlation between IR and amyloid β, as only non-carriers show a correlation between CSF:plasma glucose ratio and CSF amyloid β levels. Furthermore, only non-carriers showed an improvement in memory after an insulin infusion (Craft et al., 2000). Interestingly, given that the western-style diet is a primary contributor to development of IR and T2DM, E4 carriers also appear to differ in their acute response to a high fat or high sugar meal. For example, non-carriers had a significant increase in plasma amyloid beta (Aβ) following a high-fat meal, while E4 carriers showed both a larger increase in insulin and higher plasma Aβ after a high-carbohydrate meal (Hanson et al., 2015a).

In addition to sharing these characteristics as independent risk factors, E4 and IR appear to synergistically impair cognitive function in humans (Dore et al., 2009) and mouse models (Johnson et al., 2017a, 2017b), and worsen AD neuropathology (Peila et al., 2002; Salameh et al., 2015). This potential interaction has led several groups to directly study the effect of E4 on insulin signaling in mouse and cell models of human APOE. Using female E2, E3 and E4 targeted replacement mice, Keeney et al. (2015) showed that expression of multiple genes involved in insulin signaling and glucose uptake were differentially affected by *APOE* in the mouse hippocampus. This included higher expression of insulin receptor substrates (IRS), insulin growth factor 1 (IGF-11) and glucose transporter 4 (GLUT4) in E2 compared to E3 and E4, and significantly lower peroxisome proliferator-activated receptor γ (PPARγ) and insulin degrading enzyme in E4 compared to E2 and E3 hippocampal tissue (Keeney et al., 2015). The same group then followed up these findings by demonstrating that neuronal cells transfected with ApoE2 had significantly higher glycolytic activity, while those transfected with ApoE4 had significantly lower glycolytic activity, compared to ApoE3 transfected cells (Wu et al., 2018) In another study, Ong et al. (2014) showed that E4 reduced the phosphorylation (and thus activation) of proteins, such as Akt and IRS1. Similarly, when Chan et al. (2015, 2016) introduced the human APOE isoforms into an AD mouse model of amyloid overexpression, mice with E4 showed exaggerated impairments in both insulin signaling and cognitive performance.

Postmortem studies of brain tissue from young cognitively normal adults appear to support the idea that *APOE* alters pathways of glucose metabolism, even decades prior to the onset of disease. An analysis by Perkins et al. (2016) of human autopsy tissue from the posterior cingulate (a region previously shown to exhibit declines in mitochondrial activity strongly associated with AD risk) showed that young adult E4 carriers displayed multiple alterations in glucose and monocarboxylate transporters, as well as in complexes 1, II and IV of the electron transport chain. Recent studies have begun to shed light on the potential biological mechanisms that could explain the observed links between E4 and IR. A study by Zhao et al. (2017) implicated endosomal trapping of the insulin receptor by the E4 protein in neurons, showing that expression of E4 impaired brain insulin signaling in an age-dependent manner. The authors also tied this phenomenon back to peripheral IR and T2DM by showing that E4 also acted synergistically to impair insulin signaling in the brain when mice were fed a high fat diet (HFD) (Zhao et al., 2017).

Furthermore, exploring the relationship between peripheral IR and cognitive health, To et al. (2011) examined the efficacy of the insulin sensitizing drug pioglitazone on E3 and E4 mice with diet-induced obesity and IR, finding that E3 but not E4 mice showed a reduction in tau phosphorylation after treatment. In an attempt to gain more insight into potential interactions between E4 and IR on cognitive function, our group fed E3 and E4 mice a HFD for 6 months followed by a battery of behavioral and metabolic assays. We saw that diet-induced obesity and IR resulted in more pronounced impairments in spatial learning and memory in E4, compared to E3, mice (Johnson et al., 2017a). In parallel with the exaggerated cognitive dysfunction, we observed a number of *APOE*-specific changes in cerebral metabolism in high fat fed mice. Using an unbiased multi-omics approach, we identified several pathways that were significantly altered by *APOE* genotype, in particular the pentose phosphate pathway, purine metabolism, and glutamine-glutamate metabolism (Johnson et al., 2017a). Interestingly, these metabolic pathways were also substantially altered in E4 compared to E3 mice fed the low fat control diet, suggesting that the intrinsic effects of APOE on cerebral metabolism may outweigh those of even a long-term HFD. In a follow up study using a similar paradigm, we saw that high fat fed E4 mice also showed a reduction in cerebral glucose uptake compared to both low fat diet control mice and HFD E3 mice (Johnson et al., 2019). Compared to E3 mice, mice expressing E4 also had decreased cerebral blood flow when mice challenged with a HFD. Paradoxically, despite being slightly more glucose intolerant, these same insulin resistant E4 mice actually benefitted acutely from a bolus of glucose; they showed both increased blood flow and improved performance in a memory task 15 min after a glucose gavage (Johnson et al., 2019).

Given the strong link between IR, brain metabolic dysfunction, and AD, both insulin sensitizing agents and direct administration of insulin via intranasal delivery are being explored as therapies for the disease. Although the results are mixed, many of these treatments have shown efficacy in improving cognitive function and/or memory, but only in E4 non-carriers (Hanson et al., 2015b). Taken together with the results from mouse models of human APOE described above, the blunted therapeutic effects observed in E4 carriers may point toward an inherent E4-associated dysregulation of glucose and insulin pathways—a metabolic reprogramming that is present well before the initiation of disease and resistant to pharmacological intervention.

3. APOE and cerebral glucose and fatty acid metabolism

Glucose hypometabolism, defined here as a decrease in the uptake of ^{18}F-deoxyglucose (FDG) as measured by PET imaging, is now considered one of the hallmarks of AD (Small et al., 2000). FDG-PET is able to differentiate AD from other types of dementia with a high degree of specificity due to specific regional patterns of signal (Laforce and Rabinovici, 2011). Clinical AD symptoms essentially never occur without glucose hypometabolism, and the extent of the metabolic changes are strongly correlated with the severity of clinical symptoms (Blass, 2002; Grady et al., 1986; Haxby et al., 1990). Furthermore, recent evidence suggests that these alterations in glucose metabolism occur very early in the neurodegenerative process (de Leon et al., 2001; Mosconi et al., 2008; Reiman et al., 1996; Small et al., 1995).

Perhaps the most consistent metabolic feature associated with E4 carriage is a pattern of brain glucose hypometabolism similar to that seen in AD (Brandon et al., 2018; Wolf et al., 2013). This E4-associated reduction in glucose uptake via FDG-PET, first described several decades ago, is typically observed in the posterior cingulate, anterior cingulate cortex, and several other parietal, temporal, and prefrontal regions, including many regions within the default mode network (Altmann et al., 2015; Reiman et al., 1996). Even non-demented, cognitively normal E4 carriers demonstrate this pattern of glucose hypometabolism (Reiman et al., 2001). Importantly, these metabolic deficits are present decades in advance of AD onset in E4+ and other at risk individuals, (Benzinger et al., 2013; Reiman et al., 2004), thereby lending support to this being an inherent biological feature of E4, rather than simply a byproduct of dementia (Reiman et al., 2004). In fact, this pattern of hypometabolism in young E4 carriers is considered the earliest brain abnormality described to date in living individuals at risk for AD (Mosconi et al., 2008). While the low frequency of the E2 allele makes it challenging to conduct neuroimaging studies such as those that have shown robust E4-associated metabolic changes (Grothe et al., 2017), mouse models and *in vitro* systems may provide valuable insight into the biological mechanism(s) underlying this phenomenon.

While the brain primarily relies on glucose as an energy substrate, and cerebral glucose hypometabolism is a firmly established feature of E4 carriers, other energy substrates may also contribute to the influence of *APOE* on brain health. The dogma in neuroscience has long been that the brain does not take up or utilize a significant amount of fatty acids (FA). However, recent studies clearly show that the brain does in fact utilize a substantial amount of FA, with some estimates suggesting up to 20% of cerebral ATP is generated from FA (Ebert et al., 2003; Panov et al., 2014). Although transport of FA through the BBB may be too limited to make FA a primary energy substrate in brain compared to glucose, appreciable concentrations of FA cross the BBB and are oxidized (Allweis et al., 1966; Dhopeshwarkar and Mead, 1970; Spitzer, 1973), including FA derived from peripheral lipoproteins (Gao et al., 2017; Lee et al., 2017). In contrast to neurons, astrocytes can readily β-oxidize FA (Auestad et al., 1991; Edmond et al., 1987). However, unlike most cells in the periphery, they primarily oxidize glucose as an energy source. Several reasons have been proposed to explain this brain-specific avoidance of FA oxidation, including: (1) FA β-oxidation is more efficient than glycolysis, but slower; (2) increased FA oxidation requires more oxygen, thereby potentially leading to cerebral hypoxia; and, (3) increased FA oxidation results in increased reactive oxygen species (ROS), and neurons have been shown to be particularly sensitive to oxidative stress (Schonfeld and Reiser, 2013; Vaughn and Deshmukh, 2008; Wang and Michaelis, 2010). Several features related to these arguments have been shown to be affected by *APOE* genotype, specifically increased ROS and myelin breakdown in the E4 brain (Bartzokis et al., 2006, 2007; Glodzik-Sobanska et al., 2009; Lauderback et al., 2002; Montine et al., 1998; Ramassamy et al., 1999, 2000; Tamaoka et al., 2000).

The Randle Cycle, also known as the glucose-FA cycle, is a metabolic process involving the competition of glucose and FA for substrates (Randle et al., 1963). The cycle controls energy substrate selection and adapts the fuel supply and demand in normal tissues. It is well-established in the periphery, describing a metabolic cycling between muscle, blood and adipose tissue. At the molecular level, glucose metabolism is downregulated by FA oxidation via short-term inhibition of several glycolytic processes, primarily at the level of pyruvate dehydrogenase (PDH) and 6-phosphofructo-1-kinase (PFK-1) (Hue and Taegtmeyer, 2009). Conversely, FA oxidation is downregulated by malonyl-CoA inhibition of carnitine palmitoyltransferase (CPT), which controls the entry and oxidation of long chain FA (Hue and Taegtmeyer, 2009). While not specifically described in the brain, all the necessary enzymatic and metabolic components exist. Furthermore, some evidence exists to suggest that this same balance is a critical feature of cerebral metabolism. For example, when astrocytes take up less lipid, they undergo more glycolysis (Gao et al., 2017). Might *APOE* alter this balance?

Multiple studies have shown that *APOE* plays an important role in fat usage and storage in the periphery (Arbones-Mainar et al., 2008, 2010, 2016; Johnson et al., 2011). Although its influence on FA metabolism in the brain remains relatively unexplored, there is some evidence indirectly linking *APOE* to cerebral FA metabolism and dementia. For example, FA concentrations are decreased in the CSF of AD patients (de Wilde et al., 2017; Fonteh et al., 2014). Although it is unclear how *APOE* might affect this association (Mulder et al., 1998; Yassine et al., 2016, 2017), a recent study in E4 mice showed decreases in several FA species in brain tissue (Nuriel et al., 2017). Our group recently found that compared to E3-expressing astrocytes, E4-expressing astrocytes exhibit increased storage and oxidation fatty acids (Farmer et al., 2019b). Lastly, a recent clinical research study showed that E4+ humans β-oxidize more docosahexaenoic acid (DHA) (an omega-3 FA) than E4- individuals (Chouinard-Watkins et al., 2013). Together, these studies strongly suggest an important role for *APOE* in regulating both cerebral glucose and FA metabolism. In particular, E4-associate alterations in energy substrate uptake, storage and utilization may occur early in life and thus potentially contribute to AD pathogenesis over the decades.

4. APOE and the PPP: A potential link between glucose metabolism and oxidative stress in Alzheimer's disease

During the normal aging process, neurons and glia experience changes in redox reactions and are exposed to escalating levels of ROS, often a consequence of perturbed energy homeostasis (increased mitochondrial ROS generation). Increasing evidence suggests that this potential cycle of deficient energy metabolism and increasing oxidative stress contribute to the genesis and progression of neuropathology, synaptic dysfunction and neuronal death in AD (Tramutola et al., 2017). The most prevalent antioxidant in most brain cells is glutathione (GSH), which scavenges ROS and thereby gets oxidized to glutathione disulfide (GSSG) (Fang et al., 2002). The GSSG can then be converted back to reduced GSH by glutathione reductase in a NADPH-dependent reaction. Studies of human brain tissue have shown that the ratio between reduced and oxidized glutathione (GSH/GSSG) is lower in affected brain regions of individuals with AD and mild cognitive impairment (MCI) (Lin et al., 2015). Importantly, neurons are inherently deficient in glutathione, and hence they rely heavily on the recycling of GSH to maintain their redox status and ensure survival (Bolanos et al., 1996).

While E4 has been associated with increased oxidative damage in the AD brain as well as in peripheral systems, an explanation for this effect has been elusive. Numerous studies have reported a decreased antioxidant capacity of the E4 protein relative to E3 in various models of cardiovascular disease and AD (Dietrich et al., 2005; Jolivalt et al., 2000; Talmud et al., 2005; Tamaoka et al., 2000). *In vitro* studies have also shown that recombinant APOE may act as an antioxidant in an isoform-dependent manner, with E2 being the most, and E4 the least protective, in response to Aβ treatment (Filippini et al., 2011). The E4 isoform has also been linked to increased oxidative stress in the human brain, both in AD patients and cognitively normal subjects (Glodzik-Sobanska et al., 2009; Montine et al., 1998; Ramassamy et al., 1999, 2000; Tamaoka et al., 2000). A series of studies have shown that apoE protects synapses from oxidative damage, specifically stress brought on by Aβ (Keller et al., 2000; Lauderback et al., 2001). Furthermore, synaptasomes were protected from oxidative damage induced by Aβ in an APOE isoform-dependent manner; specifically, E4 is less efficient at this protection than E3 and E2, with ROS formation and protein and lipid oxidation always being greatest in E4 synaptosomes (Lauderback et al., 2002).

The pentose phosphate pathway (PPP) is a first-line defense response to oxidative stress, and is critical to cell survival (Kuehne et al., 2015). The primary source of cellular NADPH, and thus regeneration of GSH, is derived from glucose flux through the PPP (Kuehne et al., 2015). Recent work suggests that neurons maintain survival by managing their redox state via the PPP (Vaughn and Deshmukh, 2008). Specifically, glucose acting through the PPP inhibits apoptosis in neurons via redox inactivation of cytochrome *c*, and limits oxidative stress via the generation of NADPH (Vaughn and Deshmukh, 2008). Interestingly, a APOE mimetic peptide reduced apoptotic neurons, and down-regulated cytochrome *c* expression, in a dose-dependent manner in a rat traumatic brain injury (TBI) model (Zhao et al., 2014). While estimates of the rate of glucose entry into the PPP vary substantially depending on biological system, cell type, and method of estimate (Allaman et al., 2010; Hostetler and Landau, 1967; Takahashi et al., 2012), PPP activity is clearly upregulated when the brain is subjected to abnormal levels of oxidative stress, and when the PPP is unable to respond, toxic levels of ROS accumulate and oxidative damage occurs (Kuehne et al., 2015).

Might the observed APOE-specific influences on oxidative balance may be due to altered activity of the PPP? Brain tissue from AD patients show alterations in several metabolites within the PPP, as well as increases in two critical PPP enzymes, glucose-6-phosphate dehydrogenase (G6PD) and 6-phosphogluconate dehydrogenase (Oresic et al., 2011; Palmer, 1999). Our own unpublished measures of G6PD in the AD brain suggest that E3/E4 individuals have increased G6PD enzyme activity compared to E3/E3 individuals. Using untargeted metabolomics, we searched for potential E4-specific alterations in cerebral metabolism in a mouse model of human APOE. Our unbiased approach highlighted the PPP as one of the top two pathways that were differentially affected by *APOE* genotype in the hippocampus (Johnson et al., 2017a). Finally, our recent work examining glucose flux in human APOE expressing astrocytes suggests that cells with E4 diverted a higher fraction of glucose into the PPP compared to cells expressing E3 or E2 (Williams et al., 2019). In summary, studies noting associations between *APOE*, glucose metabolism and management of oxidative stress may implicate glucose flux through the PPP as a potential unifying mechanism.

5. APOE and the inverse Warburg hypothesis

Alzheimer's disease is not the only age-related disease of which the development and progression may be influenced by metabolic reprogramming. In fact, the most famous example of metabolic reprogramming in disease is the characteristic up-regulation of glycolysis in cancer cells—the Warburg effect (Potter et al., 2016). A fascinating theory, the *inverse* Warburg hypothesis provides a novel framework to explain the sporadic forms of both cancer and AD as a result of cell selection based on bioenergetic needs (Demetrius and Simon, 2012). The hypothesis states that while an up-regulation of glycolytic activity in cancer cells provides a proliferative and survival advantage (Warburg effect), an opposite effect underlies AD pathogenesis—whereby a decrease in glycolytic activity and an excessive up-regulation of oxidative phosphorylation drives neurodegeneration (inverse Warburg effect).

There is an inverse comorbidity between cancer and AD. Several large epidemiological studies show that (i) patients with AD have a reduced risk of incident cancer, and (ii) cancer survivors have a lower risk of AD than those without cancer (Shi et al., 2015).

Carriers of E4 have an increased risk of AD, but a decreased risk of cancer. The E4 isoform has been associated with decreased tumor size (Tian et al., 2014), and multiple studies show an effect of *APOE* genotype on risk of colorectal adenomas and carcinomas (Kervinen et al., 1996; Shinomiya et al., 2001; Souza et al., 2009; Watson et al., 2003; Zhongyin et al., 2006). A patient's *APOE* genotype also appears to modify cognitive function in breast cancer and lymphoma (Ahles et al., 2003; Koleck et al., 2014), as well as in patients with brain tumors (Correa et al., 2007, 2014, 2019). Finally, E4 protected against all cancer in the parents and offspring of the Framingham Heart Study participants, a cohort of over 5000 individuals (Kulminski et al., 2013).

Despite clear structural and functional differences, both cancer cells and neurons predominantly metabolize glucose and rely on the PPP for a variety of biological processes. Alzheimer's disease is characterized by neuronal death, and recent work suggests that, like cancer cells, neurons display an adaptive advantage for survival by managing their redox state via the PPP. Specifically, glucose acting through the PPP inhibits apoptosis in both cancer cells and neurons via redox inactivation of cytochrome *c* (Vaughn and Deshmukh, 2008).

Might E4 drive AD risk through metabolic reprogramming similar to that described by the inverse Warburg effect? Neuronal-like N2a cells transfected with E4 had a less robust gene expression profile in glucose uptake and glycolysis and a lower rate of glycolysis compared to those transfected with E3, while E2 expressing cells had the highest rate of glycolysis (Wu et al., 2018). Our own unpublished data shows that both primary astrocytes derived from human APOE mice, as well as immortalized human APOE expressing astrocytes, show a clear stepwise *APOE* effect on glucose uptake, basal glycolytic rate, and maximal glycolytic capacity (E2 > E3 > E4). Thus, perhaps the same metabolic phenotype that predisposes an E4 neuron to cell death also limits the tumorigenic potential of an E4 expressing cancer cell.

References

Ahles, T. A., Saykin, A. J., Noll, W. W., Furstenberg, C. T., Guerin, S., Cole, B., et al. (2003). The relationship of APOE genotype to neuropsychological performance in long-term cancer survivors treated with standard dose chemotherapy. *Psychooncology, 12*(6), 612–619. https://doi.org/10.1002/pon.742.

Allaman, I., Gavillet, M., Belanger, M., Laroche, T., Viertl, D., Lashuel, H. A., et al. (2010). Amyloid-beta aggregates cause alterations of astrocytic metabolic phenotype: Impact on neuronal viability. *The Journal of Neuroscience, 30*(9), 3326–3338. https://doi.org/10.1523/JNEUROSCI.5098-09.2010.

Allweis, C., Landau, T., Abeles, M., & Magnes, J. (1966). The oxidation of uniformly labelled albumin-bound palmitic acid to CO2 by the perfused cat brain. *Journal of Neurochemistry, 13*(9), 795–804.

Altmann, A., Ng, B., Landau, S. M., Jagust, W. J., Greicius, M. D., & Alzheimer's Disease Neuroimaging Initiative. (2015). Regional brain hypometabolism is unrelated to regional amyloid plaque burden. *Brain, 138*(Pt. 12), 3734–3746. https://doi.org/10.1093/brain/awv278.

Arbones-Mainar, J. M., Johnson, L. A., Altenburg, M. K., Kim, H. S., & Maeda, N. (2010). Impaired adipogenic response to thiazolidinediones in mice expressing human apolipoproteinE4. *The FASEB Journal, 24*(10), 3809–3818. https://doi.org/10.1096/fj.10-159517.

Arbones-Mainar, J. M., Johnson, L. A., Altenburg, M. K., & Maeda, N. (2008). Differential modulation of diet-induced obesity and adipocyte functionality by human apolipoprotein E3 and E4 in mice. *International Journal of Obesity, 32*(10), 1595–1605. https://doi.org/10.1038/ijo.2008.143.

Arbones-Mainar, J. M., Johnson, L. A., Torres-Perez, E., Garcia, A. E., Perez-Diaz, S., Raber, J., et al. (2016). Metabolic shifts toward fatty-acid usage and increased thermogenesis are associated with impaired adipogenesis in mice expressing human APOE4. *International Journal of Obesity, 40*(10), 1574–1581. https://doi.org/10.1038/ijo.2016.93.

Auestad, N., Korsak, R. A., Morrow, J. W., & Edmond, J. (1991). Fatty acid oxidation and ketogenesis by astrocytes in primary culture. *Journal of Neurochemistry, 56*(4), 1376–1386.

Bartzokis, G., Lu, P. H., Geschwind, D. H., Edwards, N., Mintz, J., & Cummings, J. L. (2006). Apolipoprotein E genotype and age-related myelin breakdown in healthy individuals: Implications for cognitive decline and dementia. *Archives of General Psychiatry, 63*(1), 63–72. https://doi.org/10.1001/archpsyc.63.1.63.

Bartzokis, G., Lu, P. H., Geschwind, D. H., Tingus, K., Huang, D., Mendez, M. F., et al. (2007). Apolipoprotein E affects both myelin breakdown and cognition: Implications for age-related trajectories of decline into dementia. *Biological Psychiatry, 62*(12), 1380–1387. https://doi.org/10.1016/j.biopsych.2007.03.024.

Bennet, A. M., Di Angelantonio, E., Ye, Z., Wensley, F., Dahlin, A., Ahlbom, A., et al. (2007). Association of apolipoprotein E genotypes with lipid levels and coronary risk. *JAMA, 298*(11), 1300–1311. https://doi.org/10.1001/jama.298.11.1300.

Benzinger, T. L., Blazey, T., Jack, C. R., Jr., Koeppe, R. A., Su, Y., Xiong, C., et al. (2013). Regional variability of imaging biomarkers in autosomal dominant Alzheimer's disease. *Proceedings of the National Academy of Sciences of the United States of America, 110*(47), E4502–E4509. https://doi.org/10.1073/pnas.1317918110.

Bertram, L., McQueen, M. B., Mullin, K., Blacker, D., & Tanzi, R. E. (2007). Systematic meta-analyses of Alzheimer disease genetic association studies: The AlzGene database. *Nature Genetics*, *39*(1), 17–23. https://doi.org/10.1038/ng1934.

Bertram, L., & Tanzi, R. E. (2009). Genome-wide association studies in Alzheimer's disease. *Human Molecular Genetics*, *18*(R2), R137–R145. https://doi.org/10.1093/hmg/ddp406.

Blass, J. P. (2002). Alzheimer's disease and Alzheimer's dementia: Distinct but overlapping entities. *Neurobiology of Aging*, *23*(6), 1077–1084.

Bolanos, J. P., Heales, S. J., Peuchen, S., Barker, J. E., Land, J. M., & Clark, J. B. (1996). Nitric oxide-mediated mitochondrial damage: A potential neuroprotective role for glutathione. *Free Radical Biology & Medicine*, *21*(7), 995–1001.

Brandon, J. A., Farmer, B. C., Williams, H. C., & Johnson, L. A. (2018). APOE and Alzheimer's disease: Neuroimaging of metabolic and cerebrovascular dysfunction. *Frontiers in Aging Neuroscience*, *10*, 180. https://doi.org/10.3389/fnagi.2018.00180.

Chan, E. S., Chen, C., Cole, G. M., & Wong, B. S. (2015). Differential interaction of Apolipoprotein-E isoforms with insulin receptors modulates brain insulin signaling in mutant human amyloid precursor protein transgenic mice. *Scientific Reports*, *5*, 13842. https://doi.org/10.1038/srep13842.

Chan, E. S., Shetty, M. S., Sajikumar, S., Chen, C., Soong, T. W., & Wong, B. S. (2016). ApoE4 expression accelerates hippocampus-dependent cognitive deficits by enhancing Abeta impairment of insulin signaling in an Alzheimer's disease mouse model. *Scientific Reports*, *6*, 26119. https://doi.org/10.1038/srep26119.

Chiang, G. C., Insel, P. S., Tosun, D., Schuff, N., Truran-Sacrey, D., Raptentsetsang, S. T., et al. (2010). Hippocampal atrophy rates and CSF biomarkers in elderly APOE2 normal subjects. *Neurology*, *75*(22), 1976–1981. https://doi.org/10.1212/WNL.0b013e3181ffe4d1.

Chouinard-Watkins, R., Rioux-Perreault, C., Fortier, M., Tremblay-Mercier, J., Zhang, Y., Lawrence, P., et al. (2013). Disturbance in uniformly 13C-labelled DHA metabolism in elderly human subjects carrying the apoE epsilon4 allele. *The British Journal of Nutrition*, *110*(10), 1751–1759. https://doi.org/10.1017/S0007114513001268.

Chung, C. C., Pimentel, D., Jor'dan, A. J., Hao, Y., Milberg, W., & Novak, V. (2015). Inflammation-associated declines in cerebral vasoreactivity and cognition in type 2 diabetes. *Neurology*, *85*(5), 450–458. https://doi.org/10.1212/WNL.0000000000001820.

Corder, E. H., Saunders, A. M., Risch, N. J., Strittmatter, W. J., Schmechel, D. E., Gaskell, P. C., Jr., et al. (1994). Protective effect of apolipoprotein E type 2 allele for late onset Alzheimer disease. *Nature Genetics*, 7(2), 180–184. https://doi.org/10.1038/ng0694-180.

Correa, D. D., DeAngelis, L. M., Shi, W., Thaler, H. T., Lin, M., & Abrey, L. E. (2007). Cognitive functions in low-grade gliomas: Disease and treatment effects. *Journal of Neuro-Oncology*, *81*(2), 175–184. https://doi.org/10.1007/s11060-006-9212-3.

Correa, D. D., Satagopan, J., Baser, R. E., Cheung, K., Richards, E., Lin, M., et al. (2014). APOE polymorphisms and cognitive functions in patients with brain tumors. *Neurology*, *83*(4), 320–327. https://doi.org/10.1212/wnl.0000000000000617.

Correa, D. D., Satagopan, J., Martin, A., Braun, E., Kryza-Lacombe, M., Cheung, K., et al. (2019). Genetic variants and cognitive functions in patients with brain tumors. *Neuro-Oncology*, *21*(10), 1297–1309. https://doi.org/10.1093/neuonc/noz094.

Craft, S. (2009). The role of metabolic disorders in Alzheimer disease and vascular dementia: Two roads converged. *Archives of Neurology*, *66*(3), 300–305. https://doi.org/10.1001/archneurol.2009.27.

Craft, S., Asthana, S., Schellenberg, G., Baker, L., Cherrier, M., Boyt, A. A., et al. (2000). Insulin effects on glucose metabolism, memory, and plasma amyloid precursor protein in Alzheimer's disease differ according to apolipoprotein-E genotype. *Annals of the New York Academy of Sciences*, *903*, 222–228.

de Leon, M. J., Convit, A., Wolf, O. T., Tarshish, C. Y., DeSanti, S., Rusinek, H., et al. (2001). Prediction of cognitive decline in normal elderly subjects with 2-[(18)F] fluoro-2-deoxy-D-glucose/poitron-emission tomography (FDG/PET). *Proceedings of the National Academy of Sciences of the United States of America*, *98*(19), 10966–10971. https://doi.org/10.1073/pnas.191044198.

de Wilde, M. C., Vellas, B., Girault, E., Yavuz, A. C., & Sijben, J. W. (2017). Lower brain and blood nutrient status in Alzheimer's disease: Results from meta-analyses. *Alzheimer's & dementia: Translational Research & Clinical Interventions*, *3*(3), 416–431. https://doi.org/10.1016/j.trci.2017.06.002.

Demetrius, L. A., & Simon, D. K. (2012). An inverse-Warburg effect and the origin of Alzheimer's disease. *Biogerontology*, *13*(6), 583–594. https://doi.org/10.1007/s10522-012-9403-6.

Dhopeshwarkar, G. A., & Mead, J. F. (1970). Fatty acid uptake by the brain. 3. Incorporation of (1-14C)oleic acid into the adult rat brain. *Biochimica et Biophysica Acta*, *210*(2), 250–256.

Dietrich, M., Hu, Y., Block, G., Olano, E., Packer, L., Morrow, J. D., et al. (2005). Associations between apolipoprotein E genotype and circulating F2-isoprostane levels in humans. *Lipids*, *40*(4), 329–334.

Dore, G. A., Elias, M. F., Robbins, M. A., Elias, P. K., & Nagy, Z. (2009). Presence of the APOE epsilon4 allele modifies the relationship between type 2 diabetes and cognitive performance: The Maine-Syracuse Study. *Diabetologia*, *52*(12), 2551–2560. https://doi.org/10.1007/s00125-009-1497-2.

Driscoll, D. M., & Getz, G. S. (1984). Extrahepatic synthesis of apolipoprotein E. *Journal of Lipid Research*, *25*(12), 1368–1379.

Duman, B. S., Ozturk, M., Yilmazer, S., & Hatemi, H. (2004). Apolipoprotein E polymorphism in Turkish subjects with type 2 diabetes mellitus: Allele frequency and relation to serum lipid concentrations. *Diabetes, Nutrition & Metabolism*, *17*(5), 267–274.

Ebert, D., Haller, R. G., & Walton, M. E. (2003). Energy contribution of octanoate to intact rat brain metabolism measured by 13C nuclear magnetic resonance spectroscopy. *The Journal of Neuroscience*, *23*(13), 5928–5935.

Edmond, J., Robbins, R. A., Bergstrom, J. D., Cole, R. A., & de Vellis, J. (1987). Capacity for substrate utilization in oxidative metabolism by neurons, astrocytes, and oligodendrocytes from developing brain in primary culture. *Journal of Neuroscience Research*, *18*(4), 551–561. https://doi.org/10.1002/jnr.490180407.

Eichner, J. E., Dunn, S. T., Perveen, G., Thompson, D. M., Stewart, K. E., & Stroehla, B. C. (2002). Apolipoprotein E polymorphism and cardiovascular disease: A HuGE review. *American Journal of Epidemiology*, *155*(6), 487–495.

Elosua, R., Demissie, S., Cupples, L. A., Meigs, J. B., Wilson, P. W., Schaefer, E. J., et al. (2003). Obesity modulates the association among APOE genotype, insulin, and glucose in men. *Obesity Research*, *11*(12), 1502–1508. https://doi.org/10.1038/oby.2003.201.

Fang, Y. Z., Yang, S., & Wu, G. (2002). Free radicals, antioxidants, and nutrition. *Nutrition*, *18*(10), 872–879.

Farmer, B. C., Johnson, L. A., & Hanson, A. J. (2019a). Effects of apolipoprotein E on nutritional metabolism in dementia. *Current Opinion in Lipidology*, *30*(1), 10–15. https://doi.org/10.1097/MOL.0000000000000566.

Farmer, B. C., Kluemper, J., & Johnson, L. A. (2019b). Apolipoprotein E4 alters astrocyte fatty acid metabolism and lipid droplet formation. *Cell*, *8*(2), 182. https://doi.org/10.3390/cells8020182.

Farrer, L. A., Cupples, L. A., Haines, J. L., Hyman, B., Kukull, W. A., Mayeux, R., et al. (1997). Effects of age, sex, and ethnicity on the association between apolipoprotein E genotype and Alzheimer disease. A meta-analysis. APOE and Alzheimer Disease Meta Analysis Consortium. *JAMA*, *278*(16), 1349–1356.

Filippini, N., Ebmeier, K. P., MacIntosh, B. J., Trachtenberg, A. J., Frisoni, G. B., Wilcock, G. K., et al. (2011). Differential effects of the APOE genotype on brain function across the lifespan. *NeuroImage*, *54*(1), 602–610. https://doi.org/10.1016/j.neuroimage.2010.08.009.
Fonteh, A. N., Cipolla, M., Chiang, J., Arakaki, X., & Harrington, M. G. (2014). Human cerebrospinal fluid fatty acid levels differ between supernatant fluid and brain-derived nanoparticle fractions, and are altered in Alzheimer's disease. *PLoS One*, *9*(6). https://doi.org/10.1371/journal.pone.0100519.
Gao, Y., Layritz, C., Legutko, B., Eichmann, T. O., Laperrousaz, E., Moulle, V. S., et al. (2017). Disruption of lipid uptake in Astroglia exacerbates diet-induced obesity. *Diabetes*, *66*(10), 2555–2563. https://doi.org/10.2337/db16-1278.
Garatachea, N., Emanuele, E., Calero, M., Fuku, N., Arai, Y., Abe, Y., et al. (2014). ApoE gene and exceptional longevity: Insights from three independent cohorts. *Experimental Gerontology*, *53*, 16–23. https://doi.org/10.1016/j.exger.2014.02.004.
Glodzik-Sobanska, L., Pirraglia, E., Brys, M., de Santi, S., Mosconi, L., Rich, K. E., et al. (2009). The effects of normal aging and ApoE genotype on the levels of CSF biomarkers for Alzheimer's disease. *Neurobiology of Aging*, *30*(5), 672–681. https://doi.org/10.1016/j.neurobiolaging.2007.08.019.
Grady, C. L., Haxby, J. V., Schlageter, N. L., Berg, G., & Rapoport, S. I. (1986). Stability of metabolic and neuropsychological asymmetries in dementia of the Alzheimer type. *Neurology*, *36*(10), 1390–1392.
Grothe, M. J., Villeneuve, S., Dyrba, M., Bartrés-Faz, D., Wirth, M., Weiner, M., et al. (2017). Multimodal characterization of older APOE2 carriers reveals selective reduction of amyloid load. *Neurology*, *88*(6), 569–576. https://doi.org/10.1212/wnl.0000000000003585.
Gunstad, J., Lhotsky, A., Wendell, C. R., Ferrucci, L., & Zonderman, A. B. (2010). Longitudinal examination of obesity and cognitive function: Results from the Baltimore longitudinal study of aging. *Neuroepidemiology*, *34*(4), 222–229. https://doi.org/10.1159/000297742.
Hanson, A. J., Bayer, J. L., Baker, L. D., Cholerton, B., VanFossen, B., Trittschuh, E., et al. (2015a). Differential effects of meal challenges on cognition, metabolism, and biomarkers for apolipoprotein E varepsilon4 carriers and adults with mild cognitive impairment. *Journal of Alzheimer's Disease*, *48*(1), 205–218. https://doi.org/10.3233/JAD-150273.
Hanson, A. J., Craft, S., & Banks, W. A. (2015b). The APOE genotype: Modification of therapeutic responses in Alzheimer's disease. *Current Pharmaceutical Design*, *21*(1), 114–120.
Harold, D., Abraham, R., Hollingworth, P., Sims, R., Gerrish, A., Hamshere, M. L., et al. (2009). Genome-wide association study identifies variants at CLU and PICALM associated with Alzheimer's disease. *Nature Genetics*, *41*(10), 1088–1093. https://doi.org/10.1038/ng.440.
Haxby, J. V., Grady, C. L., Koss, E., Horwitz, B., Heston, L., Schapiro, M., et al. (1990). Longitudinal study of cerebral metabolic asymmetries and associated neuropsychological patterns in early dementia of the Alzheimer type. *Archives of Neurology*, *47*(7), 753–760.
Hostetler, K. Y., & Landau, B. R. (1967). Estimation of the pentose cycle contribution to glucose metabolism in tissue in vivo. *Biochemistry*, *6*(10), 2961–2964.
Hue, L., & Taegtmeyer, H. (2009). The Randle cycle revisited: A new head for an old hat. *American Journal of Physiology. Endocrinology and Metabolism*, *297*(3), E578–E591. https://doi.org/10.1152/ajpendo.00093.2009.
Johnson, L. A., Arbones-Mainar, J. M., Fox, R. G., Pendse, A. A., Altenburg, M. K., Kim, H. S., et al. (2011). Apolipoprotein E4 exaggerates diabetic dyslipidemia and atherosclerosis in mice lacking the LDL receptor. *Diabetes*, *60*(9), 2285–2294. https://doi.org/10.2337/db11-0466.

Johnson, L. A., Torres, E. R., Impey, S., Stevens, J. F., & Raber, J. (2017a). Apolipoprotein E4 and insulin resistance interact to impair cognition and alter the epigenome and metabolome. *Scientific Reports*, 7, 43701. https://doi.org/10.1038/srep43701.

Johnson, L. A., Torres, E. R., Weber Boutros, S., Patel, E., Akinyeke, T., Alkayed, N. J., et al. (2017b). Apolipoprotein E4 mediates insulin resistance-associated cerebrovascular dysfunction and the post-prandial response. *Journal of Cerebral Blood Flow and Metabolism*, *39*, 770–781, 271678x17746186. https://doi.org/10.1177/0271678x17746186.

Johnson, L. A., Torres, E. R., Weber Boutros, S., Patel, E., Akinyeke, T., Alkayed, N. J., et al. (2019). Apolipoprotein E4 mediates insulin resistance-associated cerebrovascular dysfunction and the post-prandial response. *Journal of Cerebral Blood Flow and Metabolism*, *39*(5), 770–781. https://doi.org/10.1177/0271678X17746186.

Jolivalt, C., Leininger-Muller, B., Bertrand, P., Herber, R., Christen, Y., & Siest, G. (2000). Differential oxidation of apolipoprotein E isoforms and interaction with phospholipids. *Free Radical Biology & Medicine*, *28*(1), 129–140.

Kaur, J. (2014). A comprehensive review on metabolic syndrome. *Cardiology Research and Practice*, *2014*, 943162. https://doi.org/10.1155/2014/943162.

Keeney, J. T., Ibrahimi, S., & Zhao, L. (2015). Human ApoE isoforms differentially modulate glucose and amyloid metabolic pathways in female brain: Evidence of the mechanism of neuroprotection by ApoE2 and implications for Alzheimer's disease prevention and early intervention. *Journal of Alzheimer's Disease*, *48*(2), 411–424. https://doi.org/10.3233/JAD-150348.

Keller, J. N., Lauderback, C. M., Butterfield, D. A., Kindy, M. S., Yu, J., & Markesbery, W. R. (2000). Amyloid beta-peptide effects on synaptosomes from apolipoprotein E-deficient mice. *Journal of Neurochemistry*, *74*(4), 1579–1586.

Kervinen, K., Sodervik, H., Makela, J., Lehtola, J., Niemi, M., Kairaluoma, M. I., et al. (1996). Is the development of adenoma and carcinoma in proximal colon related to apolipoprotein E phenotype? *Gastroenterology*, *110*(6), 1785–1790. https://doi.org/10.1053/gast.1996.v110.pm8964404.

Koleck, T. A., Bender, C. M., Sereika, S. M., Ahrendt, G., Jankowitz, R. C., McGuire, K. P., et al. (2014). Apolipoprotein E genotype and cognitive function in postmenopausal women with early-stage breast cancer. *Oncology Nursing Forum*, *41*(6), E313–E325. https://doi.org/10.1188/14.Onf.E313-e325.

Kuehne, A., Emmert, H., Soehle, J., Winnefeld, M., Fischer, F., Wenck, H., et al. (2015). Acute activation of oxidative pentose phosphate pathway as first-line response to oxidative stress in human skin cells. *Molecular Cell*, *59*(3), 359–371. https://doi.org/10.1016/j.molcel.2015.06.017.

Kulminski, A. M., Culminskaya, I., Arbeev, K. G., Ukraintseva, S. V., Arbeeva, L., & Yashin, A. I. (2013). Trade-off in the effect of the APOE gene on the ages at onset of cardiocascular disease and cancer across ages, gender, and human generations. *Rejuvenation Research*, *16*(1), 28–34. https://doi.org/10.1089/rej.2012.1362.

Laforce, R., Jr., & Rabinovici, G. D. (2011). Amyloid imaging in the differential diagnosis of dementia: Review and potential clinical applications. *Alzheimer's Research & Therapy*, *3*(6), 31. https://doi.org/10.1186/alzrt93.

Lauderback, C. M., Hackett, J. M., Keller, J. N., Varadarajan, S., Szweda, L., Kindy, M., et al. (2001). Vulnerability of synaptosomes from apoE knock-out mice to structural and oxidative modifications induced by A beta(1-40): Implications for Alzheimer's disease. *Biochemistry*, *40*(8), 2548–2554.

Lauderback, C. M., Kanski, J., Hackett, J. M., Maeda, N., Kindy, M. S., & Butterfield, D. A. (2002). Apolipoprotein E modulates Alzheimer's Abeta(1-42)-induced oxidative damage to synaptosomes in an allele-specific manner. *Brain Research*, *924*(1), 90–97.

Lee, L. L., Aung, H. H., Wilson, D. W., Anderson, S. E., Rutledge, J. C., & Rutkowsky, J. M. (2017). Triglyceride-rich lipoprotein lipolysis products increase blood-brain barrier transfer coefficient and induce astrocyte lipid droplets and cell stress. *American Journal of Physiology. Cell Physiology*, *312*(4), C500–c516. https://doi.org/10.1152/ajpcell.00120.2016.

Lin, K. A., Choudhury, K. R., Rathakrishnan, B. G., Marks, D. M., Petrella, J. R., Doraiswamy, P. M., et al. (2015). Marked gender differences in progression of mild cognitive impairment over 8 years. *Alzheimer's & dementia: Translational Research & Clinical Interventions*, *1*(2), 103–110. https://doi.org/10.1016/j.trci.2015.07.001.

Mahley, R. W., & Rall, S. C., Jr. (2000). Apolipoprotein E: Far more than a lipid transport protein. *Annual Review of Genomics and Human Genetics*, *1*, 507–537. https://doi.org/10.1146/annurev.genom.1.1.507.

Montine, K. S., Reich, E., Neely, M. D., Sidell, K. R., Olson, S. J., Markesbery, W. R., et al. (1998). Distribution of reducible 4-hydroxynonenal adduct immunoreactivity in Alzheimer disease is associated with APOE genotype. *Journal of Neuropathology and Experimental Neurology*, *57*(5), 415–425.

Morris, J. K., Uy, R. A. Z., Vidoni, E. D., Wilkins, H. M., Archer, A. E., Thyfault, J. P., et al. (2017). Effect of APOE epsilon4 genotype on metabolic biomarkers in aging and Alzheimer's disease. *Journal of Alzheimer's Disease*, *58*(4), 1129–1135. https://doi.org/10.3233/JAD-170148.

Mosconi, L., De Santi, S., Li, J., Tsui, W. H., Li, Y., Boppana, M., et al. (2008). Hippocampal hypometabolism predicts cognitive decline from normal aging. *Neurobiology of Aging*, *29*(5), 676–692. https://doi.org/10.1016/j.neurobiolaging.2006.12.008.

Mulder, M., Ravid, R., Swaab, D. F., de Kloet, E. R., Haasdijk, E. D., Julk, J., et al. (1998). Reduced levels of cholesterol, phospholipids, and fatty acids in cerebrospinal fluid of Alzheimer disease patients are not related to apolipoprotein E4. *Alzheimer Disease and Associated Disorders*, *12*(3), 198–203.

Narayan, K. M., Boyle, J. P., Geiss, L. S., Saaddine, J. B., & Thompson, T. J. (2006). Impact of recent increase in incidence on future diabetes burden: U.S., 2005-2050. *Diabetes Care*, *29*(9), 2114–2116. https://doi.org/10.2337/dc06-1136.

Nuriel, T., Angulo, S. L., Khan, U., Ashok, A., Chen, Q., Figueroa, H. Y., et al. (2017). Neuronal hyperactivity due to loss of inhibitory tone in APOE4 mice lacking Alzheimer's disease-like pathology. *Nature Communications*, *8*(1), 1464. https://doi.org/10.1038/s41467-017-01444-0.

Ong, Q. R., Chan, E. S., Lim, M. L., Cole, G. M., & Wong, B. S. (2014). Reduced phosphorylation of brain insulin receptor substrate and Akt proteins in apolipoprotein-E4 targeted replacement mice. *Scientific Reports*, *4*, 3754. https://doi.org/10.1038/srep03754.

Oresic, M., Hyotylainen, T., Herukka, S. K., Sysi-Aho, M., Mattila, I., Seppanan-Laakso, T., et al. (2011). Metabolome in progression to Alzheimer's disease. *Translational Psychiatry*, *1*, e57. https://doi.org/10.1038/tp.2011.55.

Pallas, F., & Larson, D. F. (1996). Cerebral blood flow in the diabetic patient. *Perfusion*, *11*(5), 363–370. https://doi.org/10.1177/026765919601100502.

Palmer, A. M. (1999). The activity of the pentose phosphate pathway is increased in response to oxidative stress in Alzheimer's disease. *Journal of Neural Transmission (Vienna)*, *106*(3–4), 317–328. https://doi.org/10.1007/s007020050161.

Panov, A., Orynbayeva, Z., Vavilin, V., & Lyakhovich, V. (2014). Fatty acids in energy metabolism of the central nervous system. *BioMed Research International*, *2014*, 472459. https://doi.org/10.1155/2014/472459.

Peila, R., Rodriguez, B. L., Launer, L. J., & Honolulu-Asia Aging, S. (2002). Type 2 diabetes, APOE gene, and the risk for dementia and related pathologies: The Honolulu-Asia Aging Study. *Diabetes*, *51*(4), 1256–1262.

Pendse, A. A., Arbones-Mainar, J. M., Johnson, L. A., Altenburg, M. K., & Maeda, N. (2009). Apolipoprotein E knock-out and knock-in mice: Atherosclerosis, metabolic syndrome, and beyond. *Journal of Lipid Research*, *50*(Suppl), S178–S182. https://doi.org/10.1194/jlr.R800070-JLR200.

Perkins, M., Wolf, A. B., Chavira, B., Shonebarger, D., Meckel, J. P., Leung, L., et al. (2016). Altered energy metabolism pathways in the posterior cingulate in young adult apolipoprotein E varepsilon4 carriers. *Journal of Alzheimer's Disease*, *53*(1), 95–106. https://doi.org/10.3233/JAD-151205.

Phillips, M. C. (2014). Apolipoprotein E isoforms and lipoprotein metabolism. *IUBMB Life*, *66*(9), 616–623. https://doi.org/10.1002/iub.1314.

Potter, M., Newport, E., & Morten, K. J. (2016). The Warburg effect: 80 years on. *Biochemical Society Transactions*, *44*(5), 1499–1505. https://doi.org/10.1042/BST20160094.

Raber, J., Huang, Y., & Ashford, J. W. (2004). ApoE genotype accounts for the vast majority of AD risk and AD pathology. *Neurobiology of Aging*, *25*(5), 641–650. https://doi.org/10.1016/j.neurobiolaging.2003.12.023.

Ramassamy, C., Averill, D., Beffert, U., Bastianetto, S., Theroux, L., Lussier-Cacan, S., et al. (1999). Oxidative damage and protection by antioxidants in the frontal cortex of Alzheimer's disease is related to the apolipoprotein E genotype. *Free Radical Biology & Medicine*, *27*(5–6), 544–553.

Ramassamy, C., Averill, D., Beffert, U., Theroux, L., Lussier-Cacan, S., Cohn, J. S., et al. (2000). Oxidative insults are associated with apolipoprotein E genotype in Alzheimer's disease brain. *Neurobiology of Disease*, 7(1), 23–37. https://doi.org/10.1006/nbdi.1999.0273.

Randle, P. J., Garland, P. B., Hales, C. N., & Newsholme, E. A. (1963). The glucose fatty-acid cycle. Its role in insulin sensitivity and the metabolic disturbances of diabetes mellitus. *Lancet*, *1*(7285), 785–789.

Reiman, E. M., Caselli, R. J., Chen, K., Alexander, G. E., Bandy, D., & Frost, J. (2001). Declining brain activity in cognitively normal apolipoprotein E epsilon 4 heterozygotes: A foundation for using positron emission tomography to efficiently test treatments to prevent Alzheimer's disease. *Proceedings of the National Academy of Sciences of the United States of America*, *98*(6), 3334–3339. https://doi.org/10.1073/pnas.061509598.

Reiman, E. M., Caselli, R. J., Yun, L. S., Chen, K., Bandy, D., Minoshima, S., et al. (1996). Preclinical evidence of Alzheimer's disease in persons homozygous for the epsilon 4 allele for apolipoprotein E. *The New England Journal of Medicine*, *334*(12), 752–758. https://doi.org/10.1056/NEJM199603213341202.

Reiman, E. M., Chen, K., Alexander, G. E., Caselli, R. J., Bandy, D., Osborne, D., et al. (2004). Functional brain abnormalities in young adults at genetic risk for late-onset Alzheimer's dementia. *Proceedings of the National Academy of Sciences of the United States of America*, *101*(1), 284–289. https://doi.org/10.1073/pnas.2635903100.

Salameh, T. S., Bullock, K. M., Hujoel, I. A., Niehoff, M. L., Wolden-Hanson, T., Kim, J., et al. (2015). Central nervous system delivery of intranasal insulin: Mechanisms of uptake and effects on cognition. *Journal of Alzheimer's Disease*, *47*(3), 715–728. https://doi.org/10.3233/JAD-150307.

Schonfeld, P., & Reiser, G. (2013). Why does brain metabolism not favor burning of fatty acids to provide energy? Reflections on disadvantages of the use of free fatty acids as fuel for brain. *Journal of Cerebral Blood Flow and Metabolism*, *33*(10), 1493–1499. https://doi.org/10.1038/jcbfm.2013.128.

Shi, H. B., Tang, B., Liu, Y. W., Wang, X. F., & Chen, G. J. (2015). Alzheimer disease and cancer risk: A meta-analysis. *Journal of Cancer Research and Clinical Oncology*, *141*(3), 485–494. https://doi.org/10.1007/s00432-014-1773-5.

Shinomiya, S., Sasaki, J., Kiyohara, C., Tsuji, E., Inoue, H., Marugame, T., et al. (2001). Apolipoprotein E genotype, serum lipids, and colorectal adenomas in Japanese men. *Cancer Letters*, *164*(1), 33–40. https://doi.org/10.1016/s0304-3835(00)00724-2.

Small, G. W., Ercoli, L. M., Silverman, D. H., Huang, S. C., Komo, S., Bookheimer, S. Y., et al. (2000). Cerebral metabolic and cognitive decline in persons at genetic risk for Alzheimer's disease. *Proceedings of the National Academy of Sciences of the United States of America*, *97*(11), 6037–6042. https://doi.org/10.1073/pnas.090106797.

Small, G. W., Mazziotta, J. C., Collins, M. T., Baxter, L. R., Phelps, M. E., Mandelkern, M. A., et al. (1995). Apolipoprotein E type 4 allele and cerebral glucose metabolism in relatives at risk for familial Alzheimer disease. *JAMA*, *273*(12), 942–947.

Souza, D. R., Nakazone, M. A., Pinhel, M. A., Alvares, R. M., Monaco, A. C., Pinheiro, A., et al. (2009). Association between apolipoprotein E genotype, serum lipids, and colorectal cancer in Brazilian individuals. *Brazilian Journal of Medical and Biological Research*, *42*(5), 397–403. https://doi.org/10.1590/s0100-879x2009000500001.

Spitzer, J. J. (1973). CNS and fatty acid metabolism. *Physiologist*, *16*(1), 55–68.

Takahashi, S., Izawa, Y., & Suzuki, N. (2012). Astroglial pentose phosphate pathway rates in response to high-glucose environments. *ASN Neuro*, *4*(2), e00078. https://doi.org/10.1042/AN20120002.

Talbot, K., Wang, H. Y., Kazi, H., Han, L. Y., Bakshi, K. P., Stucky, A., et al. (2012). Demonstrated brain insulin resistance in Alzheimer's disease patients is associated with IGF-1 resistance, IRS-1 dysregulation, and cognitive decline. *The Journal of Clinical Investigation*, *122*(4), 1316–1338. https://doi.org/10.1172/jci59903.

Talmud, P. J., Stephens, J. W., Hawe, E., Demissie, S., Cupples, L. A., Hurel, S. J., et al. (2005). The significant increase in cardiovascular disease risk in APOEepsilon4 carriers is evident only in men who smoke: Potential relationship between reduced antioxidant status and ApoE4. *Annals of Human Genetics*, *69*(Pt. 6), 613–622. https://doi.org/10.1111/j.1529-8817.2005.00205.x.

Tamaoka, A., Miyatake, F., Matsuno, S., Ishii, K., Nagase, S., Sahara, N., et al. (2000). Apolipoprotein E allele-dependent antioxidant activity in brains with Alzheimer's disease. *Neurology*, *54*(12), 2319–2321.

Tejedor, M. T., Garcia-Sobreviela, M. P., Ledesma, M., & Arbones-Mainar, J. M. (2014). The apolipoprotein E polymorphism rs7412 associates with body fatness independently of plasma lipids in middle aged men. *PLoS One*, *9*(9), e108605. https://doi.org/10.1371/journal.pone.0108605.

Thambisetty, M., Beason-Held, L., An, Y., Kraut, M. A., & Resnick, S. M. (2010). APOE epsilon4 genotype and longitudinal changes in cerebral blood flow in normal aging. *Archives of Neurology*, *67*(1), 93–98. https://doi.org/10.1001/archneurol.2009.913.

Tian, Y., Wang, J., Ye, Y., Sun, L., Fan, Y., Wang, L., et al. (2014). Apolipoprotein E polymorphism and colorectal neoplasm: Results from a meta-analysis. *PLoS One*, *9*(7), e102477. https://doi.org/10.1371/journal.pone.0102477.

To, A. W., Ribe, E. M., Chuang, T. T., Schroeder, J. E., & Lovestone, S. (2011). The epsilon3 and epsilon4 alleles of human APOE differentially affect tau phosphorylation in hyperinsulinemic and pioglitazone treated mice. *PLoS One*, *6*(2), e16991. https://doi.org/10.1371/journal.pone.0016991.

Tramutola, A., Lanzillotta, C., Perluigi, M., & Butterfield, D. A. (2017). Oxidative stress, protein modification and Alzheimer disease. *Brain Research Bulletin*, *133*, 88–96. https://doi.org/10.1016/j.brainresbull.2016.06.005.

Vaughn, A. E., & Deshmukh, M. (2008). Glucose metabolism inhibits apoptosis in neurons and cancer cells by redox inactivation of cytochrome c. *Nature Cell Biology*, *10*(12), 1477–1483. https://doi.org/10.1038/ncb1807.

Walker, J. M., & Harrison, F. E. (2015). Shared neuropathological characteristics of obesity, Type 2 diabetes and Alzheimer's disease: Impacts on cognitive decline. *Nutrients*, 7(9), 7332–7357. https://doi.org/10.3390/nu7095341.

Wang, X., & Michaelis, E. K. (2010). Selective neuronal vulnerability to oxidative stress in the brain. *Frontiers in Aging Neuroscience*, *2*, 12. https://doi.org/10.3389/fnagi.2010.00012.

Watson, M. A., Gay, L., Stebbings, W. S., Speakman, C. T., Bingham, S. A., & Loktionov, A. (2003). Apolipoprotein E gene polymorphism and colorectal cancer: Gender-specific modulation of risk and prognosis. *Clinical Science (London, England)*, *104*(5), 537–545. https://doi.org/10.1042/cs20020329.

Williams, H. C., Farmer, B. C., Piron, M. A., Walsh, A. E., Bruntz, R., Gentry, M., et al. (2019). APOE alters glucose flux through central carbon pathways in astrocytes. *Neurobiology of Disease*, *136*, 104742.

Wilson, R. S., Bienias, J. L., Berry-Kravis, E., Evans, D. A., & Bennett, D. A. (2002). The apolipoprotein E epsilon 2 allele and decline in episodic memory. *Journal of Neurology, Neurosurgery, and Psychiatry*, *73*(6), 672–677.

Wilson, P. W., Schaefer, E. J., Larson, M. G., & Ordovas, J. M. (1996). Apolipoprotein E alleles and risk of coronary disease. A meta-analysis. *Arteriosclerosis, Thrombosis, and Vascular Biology*, *16*(10), 1250–1255.

Wolf, A. B., Caselli, R. J., Reiman, E. M., & Valla, J. (2013). APOE and neuroenergetics: An emerging paradigm in Alzheimer's disease. *Neurobiology of Aging*, *34*(4), 1007–1017. https://doi.org/10.1016/j.neurobiolaging.2012.10.011.

Wu, L., Zhang, X., & Zhao, L. (2018). Human ApoE isoforms differentially modulate brain glucose and ketone body metabolism: Implications for Alzheimer's disease risk reduction and early intervention. *The Journal of Neuroscience*, *38*(30), 6665–6681. https://doi.org/10.1523/jneurosci.2262-17.2018.

Xu, Q., Bernardo, A., Walker, D., Kanegawa, T., Mahley, R. W., & Huang, Y. (2006). Profile and regulation of apolipoprotein E (ApoE) expression in the CNS in mice with targeting of green fluorescent protein gene to the ApoE locus. *The Journal of Neuroscience*, *26*(19), 4985–4994. https://doi.org/10.1523/JNEUROSCI.5476-05.2006.

Yassine, H. N., Croteau, E., Rawat, V., Hibbeln, J. R., Rapoport, S. I., Cunnane, S. C., et al. (2017). DHA brain uptake and APOE4 status: A PET study with [1-(11)C]-DHA. *Alzheimer's Research & Therapy*, *9*(1), 23. https://doi.org/10.1186/s13195-017-0250-1.

Yassine, H. N., Rawat, V., Mack, W. J., Quinn, J. F., Yurko-Mauro, K., Bailey-Hall, E., et al. (2016). The effect of APOE genotype on the delivery of DHA to cerebrospinal fluid in Alzheimer's disease. *Alzheimer's Research & Therapy*, *8*, 25. https://doi.org/10.1186/s13195-016-0194-x.

Zechner, R., Moser, R., Newman, T. C., Fried, S. K., & Breslow, J. L. (1991). Apolipoprotein E gene expression in mouse 3T3-L1 adipocytes and human adipose tissue and its regulation by differentiation and lipid content. *The Journal of Biological Chemistry*, *266*(16), 10583–10588.

Zhao, Y., Li, J., Tang, Q., Gao, J., Chen, C., Jing, L., et al. (2014). Apolipoprotein E mimetic peptide protects against diffuse brain injury. *Neural Regeneration Research*, *9*(5), 463–473. https://doi.org/10.4103/1673-5374.130060.

Zhao, N., Liu, C. C., Van Ingelgom, A. J., Martens, Y. A., Linares, C., Knight, J. A., et al. (2017). Apolipoprotein E4 impairs neuronal insulin signaling by trapping insulin receptor in the endosomes. *Neuron*, *96*(1), 115–129.e115. https://doi.org/10.1016/j.neuron.2017.09.003.

Zhongyin, Z., Hesheng, L., Jun, L., & Jihong, C. (2006). Association of serum lipids and apolipoprotein E gene polymorphism with the risk of colorectal adenomas. *Saudi Medical Journal*, *27*(2), 161–164.

CHAPTER SIX

Status and future directions of clinical trials in Parkinson's disease

Grażyna Söderbom*
Klipspringer AB, Sävedalen, Sweden
*Corresponding author: e-mail address: grazyna@klipspringer.se

Contents

1. Parkinson's disease is a complex, multifactorial, clinical syndrome 155
2. Risk factors and pathology associated with Parkinson's disease 156
3. Current status of disease-modifying therapies and repurposing strategies 158
   3.1 Targeting mitochondria 158
   3.2 Targeting metabolic dysfunction, inflammation and repurposing strategies 165
4. Parkinson's disease has many unmet needs: Diagnosis, current clinical therapies, drug development and precision medicine 168
   4.1 The clinical diagnostic accuracy of Parkinson's disease can be improved 168
   4.2 Limitations and advances in clinical therapy 169
   4.3 Translating Parkinson's disease from lab to clinic 171
   4.4 The trouble with intervention trials 171
5. Biomarkers for diagnostics, drug development and precision medicine 172
   5.1 Advances in novel biomarkers 173
6. Future considerations and conclusion 174
Acknowledgments 177
References 177

Abstract

Novel therapies are needed to treat Parkinson's disease (PD) in which the clinical unmet need is pressing. Currently, no clinically available therapeutic strategy can either retard or reverse PD or repair its pathological consequences. L-DOPA (levodopa) is still the gold standard therapy for motor symptoms yet symptomatic therapies for both motor and non-motor symptoms are improving. Many on-going, intervention trials cover a broad range of targets, including cell replacement and gene therapy approaches, quality of life improving technologies, and disease-modifying strategies (e.g., controlling aberrant α-synuclein accumulation and regulating cellular/neuronal bioenergetics). Notably, the repurposing of glucagon-like peptide-1 analogues with potential disease-modifying effects based on metabolic pathology associated with PD has been promising. Nevertheless, there is a clear need for improved therapeutic and diagnostic options,

International Review of Neurobiology, Volume 154
ISSN 0074-7742
https://doi.org/10.1016/bs.irn.2020.02.009

disease progression tracking and patient stratification capabilities to deliver personalized treatment and optimize trial design. This review discusses some of the risk factors and consequent pathology associated with PD and particularly the metabolic aspects of PD, novel therapies targeting these pathologies (e.g., mitochondrial and lysosomal dysfunction, oxidative stress, and inflammation/neuroinflammation), including the repurposing of metabolic therapies, and unmet needs as potential drivers for future clinical trials and research in PD.

Abbreviations

AD	Alzheimer's disease
AKT	protein kinase B
ATP	adenosine triphosphate
BBB	blood-brain barrier
CNS	central nervous system
CoQ10	coenzyme Q, aka ubiquinone
CSF	cerebrospinal fluid
DaT-SPECT	dopamine transporter single-photon emission computed tomography imaging
DJ-1	gene encoding protein deglycase DJ-1 (*PARK 7*)
DLB	dementia with Lewy bodies
FDG-PET	2-deoxy-2[F^{18}]fluoro-D-glucose positron emission tomography
GBA1	gene encoding β-glucocerebrosidase
GCase	β-glucocerebrosidase
GLP-1	glucagon-like peptide-1
iPD	idiopathic/sporadic Parkinson's disease
JNK	c-Jun N-terminal kinase
L-DOPA	l-3,4-dihydroxyphenylalanine, levodopa
LRRK2	gene encoding leucine-rich repeat kinase 2 (LRRK2) protein (*PARK 8*)
MPC	mitochondrial pyruvate carrier
MPP^+	1-methyl-4-phenylpyridinium
MPTP	1-methyl-4-phenyl-1,2,3,6-tetrahydropyridine
NAC	*N*-acetylcysteine
NAD^+	oxidized nicotinamide adenine dinucleotide
NADH	reduced nicotinamide adenine dinucleotide
Nfl	neurofillament light chain
NIF	neuroinflammation
NLRP3	nucleotide-binding oligomerization domain (NOD)-, leucine-rich repeat-(LRR)-, and pyrin domain-containing protein 3
NR	nicotinamide riboside
NRPT	nicotinamide riboside and pterostilbene
NSAIDs	non-steroidal anti-inflammatory drugs
OxS	oxidative stress
PD	Parkinson's disease
PDD	Parkinson's disease with dementia

PD-GBA1	gene encoding β-glucocerebrosidase in Parkinson's disease
PD-GBA1	Parkinson's disease associated with the *GBA1* mutation
PINK1	gene encoding the mitochondrial kinase PTEN-induced kinase 1, PINK1 (*PARK 6*)
PINK1	mitochondrial serine/threonine-protein kinase encoded by *PINK1*
p-NF-κB	phospho-nuclear factor kappa-light-chain-enhancer of activated B cells
PPARγ	peroxisome proliferation-activated receptor-gamma
PRKN	gene encoding the E3 ubiquitin ligase Parkin (*PARK 2*)
SNCA	gene encoding α-synuclein (*PARK 1*)
SNpc	substantia nigra pars compacta
T2DM	type 2 diabetes mellitus
TZDs	thiazolidinediones
UDCA	ursodeoxycholic acid
α-Syn	α-synuclein

1. Parkinson's disease is a complex, multifactorial, clinical syndrome

Parkinson's disease (PD) is the most common neurodegenerative disorder after Alzheimer's disease (AD) and a "pandemic" currently affecting 6 million people worldwide that is predicted to impact 17 million by 2040 (Dorsey et al., 2018; GBD 2016 Parkinson's Disease Collaborators, 2018). Chronic and progressive, PD has been considered a movement disorder based on pathology associated with nigrostriatal pathway degeneration in the substantia nigra pars compacta (SNpc), where characteristic neuronal inclusions called Lewy Bodies containing insoluble α-synuclein (α-Syn) accumulate (Hornykiewicz, 2001; Spillantini et al., 1998; Toulorge et al., 2016).

More widespread pathology and association with non-motor symptoms (e.g., cognitive impairment, dementia, mood disorders and depression), however, indicates PD is a complex, etiologically multifactorial, heterogeneous "clinical syndrome" (Aarsland et al., 2017; Jenner et al., 2013). Many non-motor symptoms are associated with peripheral nervous system dysfunction during the prodromal stage when motor symptoms are unapparent. Central nervous system (CNS) α-Syn pathology may originate in the enteric nervous system or brainstem and entorhinal cortex and ascend retrogradely into the brain (Braak and Del Tredici, 2017) and evidence for propagation of misfolded α-Syn suggests that PD is a prion-like synucleinopathy (Visanji et al., 2013).

2. Risk factors and pathology associated with Parkinson's disease

Parkinson disease is associated with complex pathological interactions involving mitochondrial dysfunction and oxidative stress (OxS), endosomal-lysosomal dysfunction, aberrant protein pathology and endoplasmic reticulum stress, and immune response activation and neuroinflammation (NIF) (Billingsley et al., 2018; Booth et al., 2017; Klemann et al., 2017). Aging is associated with similar pathological processes and is the major risk factor for PD (Mattson and Arumugam, 2018; Reeve et al., 2014; Rodriguez et al., 2015). Thus, aging influences the onset, progression and severity of PD and may exacerbate underlying neurodegenerative pathology and genetic abnormalities that predispose many individuals to PD.

Genetic risk factors impact almost one fifth of the individuals affected by PD. Mutations in *SNCA* (*PARK 1*) encoding α-Syn are rare, associated with autosomal dominant PD, and lysosome dysfunction and aberrant chaperone-mediated autophagy. Mutations which are autosomal recessive and associated with early-onset PD (EOPD) include *PTEN-induced kinase* (*PINK1*; *PARK 6*) encoding the mitochondrial kinase PINK1 and linked to 1–15% EOPD cases depending on the population; *PRKN* (*PARK 2*) encoding the E3 ubiquitin ligase Parkin and accounting for 49% of familial EOPD and 15% of sporadic EOPD; and, *DJ-1* (*PARK 7*) encoding the antioxidant protein DJ-1, and associated with 1% of EOPD cases. The *PRKN*, *PINK1* and *DJ-1* mutations are linked to mitochondrial loss of function (Cookson, 2012; Karimi-Moghadam et al., 2018; Selvaraj and Piramanayagam, 2019). The mutation in *LRRK2* (*PARK 8*) encoding leucine-rich repeat kinase 2 (LRRK2) protein is associated with at least 4% of autosomal dominant, late onset PD, familial PD and 1% of idiopathic/sporadic PD (iPD) cases (Blauwendraat et al., 2018; Karimi-Moghadam et al., 2018; Shihabuddin et al., 2018), and consequences include disturbed mitochondrial fission/fusion dynamics (Niu et al., 2012), mitochondrial fission in microglia associated with pro-inflammatory responses (Ho et al., 2018) and the innate immune system (Lee et al., 2017).

Clearly, these mutations translate into pathological processes that cause mitochondrial dysfunction, i.e., impaired mitochondrial dynamics and bioenergetics (Chen et al., 2019a; Park et al., 2018). Evidence suggests that mitochondrial stress occurs early in disease, and even in prodromal PD. Glucose metabolism is dysregulated (Dunn et al., 2014) and peripheral blood cells exhibit mitochondrial dysfunction and increased glycolysis (Smith et al., 2018). Moreover, PD and type 2 diabetes mellitus (T2DM) are bi-directional risk factors (Söderbom and Zeng, 2020) indicating a metabolic component to PD. Esterline et al. (2018) have suggested that the lack of metabolic flexibility in cells due to nutritional overload in T2DM interferes with mitochondrial dynamics and lysosomal function, leading to accumulation of dysfunctional mitochondria and impaired neuronal bioenergetics and so should also be investigated in PD.

Mitochondrial stress can cause OxS, which triggers a vicious cycle of mitochondrial dysfunction and NIF early in PD. This exacerbates OxS and leads to aberrant accumulation of α-Syn that can impair complex I in mitochondria and positively feed the neurodegenerative loop (Bose and Beal, 2016; Devi et al., 2008; Ouchi et al., 2005; Zhou et al., 2011). In effect, mitochondrial dysfunction leads to neurodegeneration associated with OxS and NIF via complex, inter-related pathways. These pathways offer many potential targets for drug development whilst supporting mitochondrial dysfunction as central to PD pathology.

The brain is not immune-privileged even in healthy individuals and may be vulnerable to unfavorable peripheral conditions. Systemic inflammation associated with metabolic disorders, such as T2DM, can damage the blood-brain barrier (BBB) and increase CNS vulnerability to peripheral OxS, inflammatory mediators, lipotoxins, glucotoxins and fatty acids, which may impact PD (Söderbom and Zeng, 2020). Data from preclinical models of PD and individuals with PD suggest that peripheral-central inflammation cross-talk is important in PD etiology and progression (Fuzzati-Armentero et al., 2019), although no concrete data indicate that PD is directly associated with peripheral pathological changes. The systemic influence on PD is, however, evident. For example, systemic exposure to environmental toxins (e.g., pesticides like rotenone and paraquat) is associated with PD (Nandipati and Litvan, 2016). Moreover, systemic inflammation (Ferrari and Tarelli, 2011; Karpenko et al., 2018) and peripheral inflammatory triggers, such as those associated with the gut and its microbiome (Adams et al., 2019; Chen et al., 2019d; Villumsen et al., 2019) suggest that peripheral factors can impact the brain.

The systemic inflammatory factor in PD is also supported by studies with anti-inflammatory agents, such as immunosuppressant corticosteroids and inosine monophosphate dehydrogenase inhibitors, which appear to lower the risk of PD (Racette et al., 2018). Individuals with inflammatory bowel disease (IBD) have an increased incidence of PD yet early exposure to anti-inflammatory and anti-tumor necrosis factor therapy reduces this incidence (Peter et al., 2018). Interestingly, and contrary to individuals with IBD, newly diagnosed rheumatoid arthritis patients may be at lower risk of developing PD independent of treatment with anti-rheumatic drugs, which themselves reduce the risk further (Sung et al., 2016). This study supports an earlier study of Rugbjerg et al. (2009), and whilst the underlying mechanism for this inverse relationship needs further investigation, it may be related to long-term use of non-steroidal anti-inflammatory drugs (NSAIDs) that reduced the risk of PD via an anti-inflammatory effect (Mehta and Tanner, 2016), again implicating chronic inflammation in PD etiology. Interestingly, the nucleotide-binding oligomerization domain (NOD)-, leucine-rich repeat- (LRR)-, and pyrin domain-containing protein 3 (NLRP3) inflammasome activation inhibitor MCC950 significantly suppresses pro-inflammatory cytokine release in a mouse model of human ulcerative colitis (Perera et al., 2018), which may be relevant for PD.

Peripheral factors, therefore, play an important and possibly overlooked role in PD, and intervention trials investigating the impact of treating systemic inflammation as a consequence of metabolic disease or inflammatory conditions may provide novel neuroprotective therapies in PD.

3. Current status of disease-modifying therapies and repurposing strategies

3.1 Targeting mitochondria

The central role of mitochondria in PD and its consequences, supported by gene mutations and metabolic risk factors, has stimulated interest in modulating redox imbalance, apoptosis, OxS and NIF, and downstream lysosomal dysfunction as therapeutic targets in PD. Mitochondrial dysfunction as a therapeutic target in PD is reviewed elsewhere (Macdonald et al., 2018; Park et al., 2018) and the focus here will be on redox imbalance, apoptosis, OxS, NIF and lysosomal dysfunction.

3.1.1 Mitochondrial dysfunction and redox imbalance

Mitochondrial dysfunction occurs at the level of the NADH-CoQ reductase of complex I, which oxidizes NADH to NAD^+ in PD and PD with dementia (PDD) (Garcia-Esparcia et al., 2018; Gatt et al., 2016; Schapira et al., 1990). Aging is also associated mitochondrial dysfunction and redox imbalance (Lautrup et al., 2019; Tapias et al., 2019) so restoring mitochondrial function and redox balance may be beneficial in PD, particularly as modulating redox imbalance with the NAD^+ precursor nicotinamide riboside (NR) reduces mitochondrial defects, OxS, and neuronal dopaminergic damage in basal ganglia, and motor deficits in preclinical studies (Schöndorf et al., 2018; Shan et al., 2019).

Chronic supplementation with NR and other forms of NAD^+ supplements that are well-tolerated (e.g., pterostilbene or NRPT) stimulate NAD^+ metabolism, and increase and maintain NAD^+ levels in healthy adults (Dellinger et al., 2017; Martens et al., 2018) so are now being tested in PD. The impact of NR supplementation on mitochondrial function and slowing progression of newly diagnosed PD, for example, is being assessed in a double-blind, randomized, controlled trial (NOPARK; NCT03568968) and, since mitochondrial dysfunction is common to both PD and AD (Lautrup et al., 2019), a multi-center, double-blind, randomized, placebo-controlled, Phase 2 trial will investigate whether dietary metabolic cofactor supplementation improves mitochondrial function in AD and PD subjects (NCT04044131). The benefits of modulating NAD^+ levels in PD are reviewed comprehensively by Aman et al. (2018), and provide a promising lead as redox imbalance may be an upstream step of OxS and consequent NIF that may mitigate neurodegenerative pathology.

3.1.2 Mitochondrial dysfunction, apoptosis and inflammation

Ursodeoxycholic acid (UDCA) is a secondary bile acid with anti-apoptotic and anti-inflammatory effects indicated as a first-line treatment of primary biliary cirrhosis (Santiago et al., 2018). In preclinical PD studies, UDCA shows neuroprotective, anti-apoptotic and anti-inflammatory effects associated with improved motor performance (Abdelkader et al., 2016; Mortiboys et al., 2015). Its taurine conjugate form tauroursodeoxycholic acid has also shown neuroprotective effects (mediated via anti-oxidation and JNK activity modulation) and an AKT-dependent, pro-survival pathway (Castro-Caldas et al., 2012).

These studies all support investigation of UDCA in PD, and two are now underway, specifically a randomized, double-blind, placebo-controlled, proof-of-concept, Phase 2 study to determine its safety, tolerability and ability to retard disease progression in early PD (NCT03840005), and an open label Phase 1 study that will explore bioenergetic impairments associated with PD and determine the impact of repeated high-dose, oral administration of UDCA on aberrant mitochondrial function (NCT02967250).

3.1.3 Mitochondrial dysfunction, oxidative stress and neuroinflammation

Oxidative stress, NIF and inflammation are implicated in PD and are both cause and consequence of each other. Glial cells are activated by α-Syn, reactive oxygen species, mitochondrial DNA and neurodegeneration, and release a cocktail of pro-inflammatory cytokines and mediators that exacerbate damage in the brain (Chien et al., 2016; Codolo et al., 2013; Lee et al., 2010; Mogi et al., 1994; Zhou et al., 2016). The NLRP3 inflammasome is involved and may mediate environmental toxin exposure and consequent cellular stress associated with NIF and neurodegeneration in PD (Anderson et al., 2018).

The SNpc is vulnerable to OxS and NIF in PD due to its low antioxidant capability, high iron and glial cell levels and the vulnerability of dopaminergic neurons *per se* to OxS by nature of their structural properties (Ayton et al., 2014; Chien et al., 2016; Dölle et al., 2016; Giguère et al., 2019; Kastner et al., 1992; Sutachan et al., 2012). Consequent damage is clearly evident in post-mortem PD brain tissue (Toulorge et al., 2016). Thus, antioxidant intervention and anti-inflammatory strategies may be neuroprotective against this inflammatory milieu of OxS, NIF and neurodegenerative damage and have been evaluated in PD.

3.1.4 Antioxidant approaches in Parkinson's disease

The benefit of antioxidant vitamins in PD is unresolved. Early intervention studies showed that long-term treatment with Vitamin E had no effect on PD progression (Parkinson Study Group, 1993, 1998), although more recent studies suggest that vitamins may reduce the incidence of PD and improve clinical symptoms in affected individuals (reviewed by Zhao et al., 2019).

A wide range of other potential antioxidant approaches is also being tested. Idebenone (CV-2619), for example, is a synthetic derivative of CoQ10 first developed for AD that enhances electron transfer chain function and potentially has antioxidant effects (El-Hattab et al., 2017). Its impact on both motor and non-motor symptoms in early PD will be investigated in a multi-center, randomized, double-blind, placebo-controlled Phase 4 study (NCT03727295), which exemplifies a repurposing strategy and the speed of drug development.

N-acetylcysteine (NAC) increases cysteine availability and glutathione synthesis (El-Hattab et al., 2017), which may be important since nigral glutathione levels are decreased by 40% in PD (Sian et al., 1994). Since catecholamines and their metabolites may be autotoxic (Goldstein et al., 2014), ability of NAC to reduce brain dopamine oxidation as a disease-modifying effect is being investigated in a Phase 1 study in mild, levodopa-untreated PD (NCT03104725).

Urate is a plasma antioxidant and potentially neuroprotective in PD (Crotty et al., 2017). Its precursor inosine is safe, tolerable and effectively raises serum and cerebrospinal fluid (CSF) urate levels in early PD (Parkinson Study Group SURE-PD Investigators et al., 2014) so its impact on retarding clinical decline in early PD is being investigated in a multi-center, randomized, double-blind, placebo-controlled, Phase 3 trial (SURE-PD3; NCT02642393).

These studies are scheduled to end by 2021 and may provide novel antioxidant breakthroughs. Other potential approaches for clinical evaluation include flavonoids (e.g., resveratrol and curcumin) that show both antioxidant and anti-inflammatory effects and target α-Syn pathology (Hussain et al., 2018; Javed et al., 2019), and the ketone body D-β-hydroxybutyrate has been investigated as adjunctive nutritional therapy to replenish neuronal ATP stores and protect against OxS (Norwitz et al., 2019). To date, antioxidant strategies have shown limited success in clinical studies yet OxS is still a target based on its pathological association with mitochondrial dysfunction and inflammation in PD.

3.1.5 Non-steroidal anti-inflammatory drugs in Parkinson's disease

Whilst preclinical studies have demonstrated anti-inflammatory effects of NSAIDs associated with neuroprotection in PD (Swiątkiewicz et al., 2013; Teismann et al., 2003; Vijitruth et al., 2006), clinical evidence is mixed. Some studies do, however, suggest that NSAIDs (e.g., ibuprofen) are associated with a lower risk of developing PD (Gagne and Power, 2010; Gao et al., 2011; Rees et al., 2011; Wahner et al., 2007) and based on the pathology of PD, this is a promising and much investigated concept.

3.1.6 NAD^+ supplements as anti-inflammatory agents in Parkinson's disease

Niacin appears to induce macrophage polarization from the pro-inflammatory M1 to anti-inflammatory M2 phenotype via the anti-inflammatory niacin receptor GPR109A, the expression of which is significantly upregulated in blood, and the substantia nigra and its microglia in PD (Wakade et al., 2014, 2018). This is associated with a reduction in nuclear translocation of phospho-nuclear factor kappa-light-chain-enhancer of activated B cells (p-NF-κB) and subsequent expression of pro-inflammatory cytokines interleukin 1-β and interleukin-6 (Giri et al., 2019), suggesting that niacin may be a beneficial immunotherapy in PD.

Two double-blind, randomized trials are investigating the impact of over-the-counter Vitamin B3 supplements on inflammation, symptoms and disease severity in PD (NCT03462680 and NCT03808961), whilst the NAD-PARK study (NCT03816020) is a double-blind, randomized pilot trial of NAD-supplementation in drug naïve PD evaluating the impact of NR on the neurometabolic profile and motor symptoms as the primary and secondary objectives, respectively, in PD patients.

Interestingly, Ho et al. (2019) reported on CNM-Au8, a suspension of clean-surfaced, faceted nanocrystalline gold with unique nanocatalytic properties that converts NADH to NAD^+ to increase intracellular ATP levels, whilst independently lowering OxS via superoxide dismutase-like catalytic activity. In preclinical PD models, CNM-Au8 appeared to enhance cellular bioenergetics; it showed neuroprotective activity and improved functional behavioral tests associated with increasing total intracellular NAD^+ levels, reducing α-Syn aggregates, and preventing the accumulation of reactive oxygen species. The REPAIR-PD single-center, open label, Phase 2 pilot study will assess the safety, dynamics and impact of CNM-Au8 on the impaired neuronal redox state in PD patients diagnosed within 3 years of screening (NCT03815916).

3.1.7 NLRP3 inflammasome inhibitors as anti-inflammatory drugs in Parkinson's disease

Loss of function mutations in the aforementioned *PRKN* and *PINK1* genes are associated with mitochondrial quality control and innate immunity pathway regulation. Loss of Park 2 (Parkin) function is associated with over-activation of microglial cells in response to lipopolysaccharide, and loss of Park2/Pink1 function can exacerbate NLRP3 inflammasome activation by lipopolysaccharide-based stimuli (Mouton-Liger et al., 2018) so targeting the NLRP3 inflammasome in PD may provide beneficial anti-inflammatory effects.

The potent, highly specific, small molecule inhibitor of NLRP3 inflammasome activation, MCC950, protects mice against traumatic brain injury associated with reduced microglial activation, leukocyte recruitment, and pro-inflammatory cytokine production, preservation of BBB integrity and attenuated cell death (Ismael et al., 2018; Xu et al., 2018). In mouse microglial cells, MCC950 retards fibrillar α-Syn-mediated inflammasome activation and extracellular ASC adaptor protein release, and alleviates motor deficits, nigrostriatal dopaminergic degeneration and α-Syn accumulation (Gordon et al., 2018). It also protects human retinal endothelial cells from hyperglycemia-induced damage by reducing upregulated pro-inflammatory interleukin 1-β levels associated with hyperglycemia (Zhang et al., 2017). Whilst a preclinical trial on MCC950 in PD has been funded by the Michael J Fox Foundation (2016), no published results are available.

Other pharmacological inflammasome inhibitors are reviewed by Zahid et al. (2019), who also describe the ability of β-hydroxybutyrate to inhibit the NLRP3 inflammasome. Whilst the NLRP3 inflammasome is a relatively novel target in early stages of evaluation, it may lead to disease-modifying anti-inflammatory therapies that can impact both systemic and central inflammation associated with PD, particularly as it is also implicated in peripheral metabolic disorders and inflammatory conditions.

3.1.8 Leucine-rich repeat kinase 2 inhibitors as anti-inflammatory drugs in Parkinson's disease

The aforementioned mutation in *LRRK2* (*PARK 8*) is linked to impaired mitochondrial dynamics, bioenergetics, and endosomal-lysosomal dysfunction. All of these pathways are associated with NIF. Thus, evaluation of LRRK2 inhibitors may provide neuroprotective effects in PD.

The highly selective, brain-penetrant LRRK2 kinase inhibitor PF-360 improves endosomal-lysosomal function and prevents neurodegeneration in preclinical studies (Rocha et al., 2019) and LRRK2 inhibitors are potential disease-modifying therapies in PD by way of reducing inflammation and aberrant α-Syn accumulation and pathology (West, 2017). Two potent, selective and brain-penetrant, small molecule LRRK2 inhibitors DNL201 and DNL151 are currently being evaluated in multi-center, randomized, placebo-controlled, double-blind Phase 1b studies in PD (NCT03710707 and NCT04056689).

Antisense oligonucleotides can also inhibit the consequences of *LRRK2* gain of function mutations by targeting LRRK2 mRNA and reducing LRKK2 protein expression in the CNS with consequent reduced formation of brain α-Syn inclusions in mouse models of PD (Zhao et al., 2017). Thus, BIIB094 (aka ION859, IONIS-BIIB7$_{Rx}$) is being investigated in a Phase 1 study in PD subjects with/without verified *LRRK2* mutations, and individuals without any verified PD-related genetic mutations (NCT03976349).

Other LRRK2 inhibition strategies are still in preclinical testing (Daher et al., 2015; Ding et al., 2018; Greshock et al., 2016), but lysosome-targeted interventions should ease the pathological impact of α-Syn accumulation in the brain, improve mitochondrial dynamics and function, and ameliorate the OxS-NIF loop in PD.

3.1.9 Glucocerebrosidase as a therapeutic target in Parkinson's disease

Mutations in *GBA1* encoding the lysosomal hydrolase β-glucocerebrosidase (GCase) are the most common genetic risk factor for early-onset iPD, which occurs in at least 5% of patients and confers a 20- to 30-fold increased risk of PD (PD-GBA1) (Migdalska-Richards and Schapira, 2016). The *PD-GBA1* mutations are associated with decreased GCase levels and pathology identical to that of sporadic/iPD (Gegg and Schapira, 2018) and impaired autophagy-lysosomal function that enhances aberrant α-Syn aggregation and accumulation (Du et al., 2015; Murphy et al., 2014). This is important as aberrant α-Syn accumulation in PD exacerbates the OxS-NIF loop and can also directly disrupt mitochondrial function. Interestingly, mitochondrial dysfunction may influence loss of GCase activity (see Gegg and Schapira, 2018). Increasing the function of or restoring the consequences of normal GCase function may, therefore, ameliorate lysosomal dysfunction and impaired protein clearance that contributes to α-Syn pathology and NIF in PD.

Ambroxol is an over-the-counter drug indicated for cough. It also acts as a chaperone aiding mutant GCase molecule folding in the endoplasmic reticulum and thus trafficking to lysosomes (Bendikov-Bar et al., 2013), and increases GCase activity and reduces OxS markers in fibroblasts carrying *GBA* mutations in heterozygous *GBA* carriers with and without PD (McNeill et al., 2014). In the prospective, single-center, open label Phase 2a study of ambroxol in both *GBA*-positive and negative PD subjects as a disease-modifying strategy, ambroxol was shown to penetrate the CSF and modulate GCase (indicated by increased CSF levels of α-Syn interpreted as increased extracellular export of α-Syn from brain parenchyma) (NCT02941822; Mullin et al., 2020). Whilst only a small proof-of-concept trial with no placebo control, it is a good example of a genetic biomarker-driven, "repurposing" strategy in PD where the primary outcomes were met as a disease-modifying therapy for PD. Another study evaluating ambroxol is underway in mild-to-moderate PD (Phase 2; NCT02914366; Silveira et al., 2019).

3.2 Targeting metabolic dysfunction, inflammation and repurposing strategies

Metabolic dysfunction, both as intrinsic to PD and as an associated risk factor by way of metabolic disorders like T2DM, is an important therapeutic target. Recent intervention studies evaluating anti-diabetic therapies, such as glucagon-like peptide-1 (GLP-1) analogues, suggest this class of drugs has neuroprotective effects in PD (Athauda and Foltynie, 2016b; Hölscher, 2014). This supports the importance of T2DM as risk factor for PD, may potentially lead to disease-modifying therapies in PD, and is an example of a repurposing strategy in PD.

The GLP-1 analogues are agonists at the GLP-1 receptor, which not only induce insulin release and enhance insulin sensitivity to treat hyperglycemia, but can cross the BBB (e.g., exenatide; Daniele et al., 2015) and enhance mitochondrial function, neurogenesis, synaptic function, reduce inflammation and positively impact cognitive dysfunction associated with PD (Athauda and Foltynie, 2016b; Green et al., 2019; Hölscher, 2014). The neuroprotective mechanism of action of the GLP-1 analogues is yet to be established, but may result from direct actions on dopaminergic neurons (Li et al., 2009). Their effects may be anti-inflammatory (Athauda and Foltynie, 2016b) and could, therefore, also benefit PD patients by targeting peripheral metabolic dysfunction and consequent systemic inflammation; however, Ventorp et al. (2017) did not observe any anti-inflammatory effects with exendin-4 and suggest that reduced inflammation may be consequent to alleviating neuronal death.

3.2.1 Repurposing the GLP-1 analogues in Parkinson's disease

The encouraging results of the single-center, randomized, double-blind, placebo-controlled, Phase 2 trial in which exenatide significantly improved motor and cognitive symptoms in mild-to-moderate PD after 12 months of treatment (Athauda et al., 2017) present a novel opportunity as a disease-modifying therapy for PD. Its longitudinal effect as a symptomatic strategy for motor and cognitive function, as well as a potential disease-modifying strategy is being investigated in an open label, Phase 1 trial in exenatide-naïve PD patients (NCT03456687).

Other drugs in this class are also being evaluated. The impact of liraglutide on motor and non-motor symptoms and peripheral insulin resistance in iPD is being investigated in a single-center, randomized, double-blind, placebo-controlled Phase 2 trial (NCT02953665) and a

French, multi-center, randomized, double-blind, placebo-controlled, proof-of-concept Phase 2 trial is evaluating the impact of lixisenatide on motor and non-motor symptoms and disease modification in early PD (NCT03439943). The neuroprotective and anti-inflammatory properties of semaglutide in newly diagnosed iPD patients will also be investigated in the Phase 2, single-center, double-blind, placebo-controlled *GLP1R in Parkinson's Disease (GIPD)* study (NCT03659682). A disease-modifying breakthrough that could benefit a wide range of patients and target both motor and non-motor and symptoms may, therefore, occur in the near future.

Interestingly, both short-acting (e.g., lixisenatide) and long-acting (e.g., exenatide) GLP1 analogues are being investigated, each of which has a differential impact on peripheral glycemic control (Gentilella et al., 2019). In terms of a neuroprotective impact on the brain, the limiting parameter appears to be the ability of the GLP1 analogue to cross the BBB. The rate of BBB penetration is fundamental to the effectiveness of GLP1 analogues, where peptides lacking functional groups cross the BBB more effectively than peptides with attached fatty acids, whilst pegylated peptides do not readily enter the brain (Dr C Hölscher, personal communication). Studies have shown that exenatide, liraglutide and lixisenatide all cross the BBB and have beneficial effects on the brain in terms of neurogenesis and neuroprotection (Athauda and Foltynie, 2016b; Hunter and Hölscher, 2012). The novel dual GLP1/gastric inhibitory polypeptide (an incretin that stimulates insulin secretion) molecule DA-CH5, however, crosses the BBB more effectively than acetylated GLP1 analogues (e.g., liraglutide) and studies indicate that it may, therefore, be superior to liraglutide in terms of neuroprotective effects in a 1-methyl-4-phenyl-1,2,3,6-tetrahydropyridine (MPTP) mouse model of PD (Zhang et al., 2020) and in the intra-cerebroventricular streptozotocin-induced insulin desensitization rat model of AD (Li et al., 2020).

Brain penetration may be key to differentiating the relative efficacy of short- and long-acting GLP1 analogues and other novel molecules in the brain. Clinical studies with exenatide in PD have shown it to be well-tolerated with a similar frequency of adverse events to those of exenatide trials in T2DM, which must be considered when repurposing drugs for elderly populations with comorbidities. Ease of administration will also be important in this patient population and may favor long-acting, sustained release formulations; importantly, these formulations may also improve compliance and efficacy in individual with cognitive dysfunction (see Kim et al., 2017). Whilst it is still too early for clinical head-to-head trials in this arena, since the GLP1 analogues are generally well-tolerated in T2DM and in PD trials to date, they are a promising class of potential disease-modifying therapies for PD.

3.2.2 Other metabolic targets of Parkinson's disease

Another metabolic target associated with metabolic disorders is the peroxisome proliferation-activated receptor-gamma (PPARγ). This regulates mitochondrial function and turnover, energy metabolism, antioxidant defenses, redox balance and immune responses (Corona and Duchen, 2016). Thiazolidinediones (TZDs, aka glitazones) activate PPARγ yet inhibit the mitochondrial pyruvate carrier (MPC). Since the MPC may be main target through which TZDs modulate insulin-sensitivity, novel TZDs (e.g., MSCD-0160) have been developed (Corona and Duchen, 2016) with the aim of improving the side-effect profile in PD. First generation TZDs, such as pioglitazone and rosiglitazone, have been investigated in PD and despite promising neuroprotective effects in animal models, these have not yet translated into the clinic (Cardoso and Moreira, 2020). Other novel metabolic drugs have also been investigated as reviewed by Quansah et al. (2018).

Based on their preclinical work and recent analysis of two large databases in subjects treated with terazosin (an α1-adrenergic receptor antagonist used for hypertension and benign prostatic hyperplasia) and similar drugs, Cai et al. (2019) showed that terazosin may enhance phosphoglycerate kinase 1 activity to stimulate glycolysis and increase cellular ATP levels, thereby reducing PD progression and related complications. Thus, the safety and tolerability of terazosin in PD and whether it can reverse PD-associated energy deficits is being investigated in a single-center, randomized, double-blind, placebo-controlled, pilot Phase 2 trial, *Terazosin for Parkinson's Disease* (TZ-PD; NCT03905811).

In summary, mitochondrial dysfunction, OxS and NIF are intricately linked and provide many therapeutic targets in PD. Targeting OxS with antioxidant approaches alone may be too downstream of a more critical step to have significant clinical impact. Modulating upstream steps, such as redox imbalance and metabolic dysfunction may ameliorate OxS and NIF more effectively and modify PD progression. The GLP-1 analogues can target many of these pathways and are, to date, showing promise as disease-modifying therapies in PD.

4. Parkinson's disease has many unmet needs: Diagnosis, current clinical therapies, drug development and precision medicine

4.1 The clinical diagnostic accuracy of Parkinson's disease can be improved

No diagnostic biomarkers for PD exist in clinical practice. Early disease, characterized by peripheral, non-motor symptoms, is often misdiagnosed and untreated. Parkinson's disease is primarily diagnosed on clinical symptomatology related to late-stage motor symptoms, by which time 50–80% of nigral dopaminergic neurons has been lost (Hornykiewicz, 2001; Kordower et al., 2013). A diagnosis of PD can be confirmed by an individual's clinical response to dopaminergic medication (i.e., L-DOPA/levodopa) or using an ioflupane I^{123} injection plus single-photon emission computed tomography (SPECT) imaging (DaTscan™) that allows striatal dopamine transporter visualization. Unfortunately, not all individuals respond or respond only briefly to L-DOPA, particularly those diagnosed late in disease progression, whilst DaTscan™ does not differentiate between idiopathic PD and parkinsonism (Bajaj et al., 2013; Roussakis et al., 2013). Thus, the clinical diagnostic accuracy is low particularly in early disease and when the response to L-DOPA is unclear (Adler et al., 2014). The clinical diagnosis of PD, however, can be confirmed at post mortem.

As a clinical syndrome, PD is viewed as a complex, multifactorial yet single heterogeneous disease; it is diagnosed by unique pathophysiology into subtypes with Lewy bodies as the one common pathological denominator (Espay et al., 2017). Whilst this approach suits the development of dopamine replacement strategies, it fails in the hunt for disease-modifying therapies, particularly since it may not be possible to ascribe a single gene or pathway to a specific pathophenotype (Espay et al., 2017; Menche et al., 2017).

Good quality biomarkers are, therefore, essential to diagnose PD early, accurately and routinely, and fundamental to improving patient recruitment and stratification in trials for drug development, and delivering precision-based medicine, as will be discussed. Studies in AD have indicated that targeting neurodegenerative disorders early may be important. This is vital in the case of PD since repair techniques for the extensive neuronal loss in the dopaminergic system and beyond are not yet available.

4.2 Limitations and advances in clinical therapy

The clinical unmet need in PD is huge. Currently, no clinically available, disease-modifying strategies that retard, reverse or repair the consequences of underlying pathology exist. Therapeutic strategies are symptomatic. They focus on restoring and optimally regulating deficient dopaminergic signaling associated with the motor symptoms used to diagnose PD, managing therapy-associated side effects, and treating the non-motor symptoms that become increasingly apparent with disease progression.

4.2.1 Symptomatic therapy

L-DOPA, first developed to treat motor dysfunction in PD over 50 years ago, is still gold standard therapy. Whilst new formulations of L-DOPA and novel adjunct therapies are improving overall control of motor symptoms (Aquilonius and Nyholm, 2017; Lees et al., 2017; Schapira et al., 2017), the clinical need is significant, particularly as chronic L-DOPA therapy is associated with dyskinesia.

Similarly, non-motor symptoms like cognitive impairment and dementia can also be extremely problematic in PD and contribute to the unmet need. Dementia is associated with reduced brain glucose metabolism and atrophy, and 80% of PD patients develop it within 20 years (Aarsland et al., 2017; Chen et al., 2019b; Jia et al., 2019; Trošt et al., 2019). Cognitive impairment may even be exacerbated by deep brain stimulation that is used to treat the aforementioned L-DOPA side effects in some individuals (Herrington et al., 2016). The only available therapy for PDD is that used for AD (Chen et al., 2019c; Rektorova, 2019; van der Steen et al., 2019) with limited clinical impact (Briggs et al., 2016; Catania et al., 2019). Interestingly, based on the link between PD, T2DM and metabolic dysfunction (Söderbom and Zeng, 2020), and similar effects in AD (Cardoso and Moreira, 2020), intranasal insulin has been evaluated as a treatment for cognitive impairment in PD and demonstrated clinically relevant, functional improvement (NCT02064166; Novak et al., 2019). This may offer novel research directions and therapies for this debilitating aspect of PD.

Symptomatic therapy (reviewed by Carrarini et al., 2019; Oertel and Schulz, 2016) is limited in its scope, duration and impact on many aspects of PD. The multifactorial nature of PD and neurodegeneration that extends beyond the nigrostriatal dopamine system demands a multiple therapy approach targeting multiple pathways to increase therapeutic success (Lang and Obeso, 2004).

4.2.2 Neuro-regenerative cell and neuroprotective therapies

The consequent neurodegeneration of PD pathology has inspired neuro-regenerative cell and neuroprotective therapies. Cell therapy techniques are increasingly successful yet even healthy transplanted cells have shown α-Syn pathology at post mortem, suggesting that they too are subject to the toxic factor(s) affecting host cells (Kordower et al., 2008; Li et al., 2008). Recent evidence, however, shows that fetal dopamine neuron transplants are functional for up to 15–18 years in PD patients (Hallett et al., 2014) and genetic engineering strategies using CRISPR/Cas9n technology to delete endogenous *SNCA* in transplanted cells may ameliorate the potential α-Syn pathology (Brundin and Coetzee, 2019; Chen et al., 2019e).

These therapeutic strategies are reviewed elsewhere (Barker et al., 2017; Sundberg and Isacson, 2014; Whone et al., 2019), but the inability to halt or reverse disease progression means that individuals who have already reached the stage of motor symptoms are likely to require these strategies in combination with symptomatic and disease-modifying therapies.

4.2.3 A lack of disease-modifying therapy in Parkinson's disease

The discovery of genetic drivers of PD in specific populations has dramatically advanced our understanding of the pathogenic pathways underlying PD and provided novel targets for drug and biomarker development. Yet no breakthrough disease-modifying therapy exists in clinical practice and how these mutations translate into disease within the overall risk factor profile of PD patients that includes aging, environmental toxins, and peripheral metabolic factors needs more investigation.

Chen-Plotkin et al. (2018) reported that of more than 550 open studies in PD, only 10% were evaluating disease-modifying or neuroprotective strategies, of which only two had reached Phase 3. This is strongly influenced by the limitations of existing animal models of PD, and a lack of biomarkers to identify suitable individuals for trials, track PD progression and enable evaluation of long-term effects on PD in well-designed intervention trials.

4.3 Translating Parkinson's disease from lab to clinic

The serendipitous discovery of MPTP and its neurotoxic metabolite 1-methyl-4-phenylpyridinium (MPP^+) reinforced the upstream role of mitochondrial dysfunction in PD and enabled nigral degeneration akin to iPD to be reproduced in animal models (Gibb et al., 1989; Jenner et al., 1984).

Whilst useful for preclinical testing of symptomatic therapies targeting the motor symptoms of PD, such neurotoxic animal models are static—one-hit toxicity with MPTP may destroy the nigrostriatal pathway, but does not replicate the chronic, progressive or extensive nature of PD (Garvey et al., 1986; Gibb et al., 1989). This limits our understanding of the complex pathology of PD and hampers effective evaluation of disease-modifying therapies targeting disease progression.

Since the first gene mutation associated with PD (*SNCA*) was discovered (Nussbaum, 2017), many gene mutations that affect specific patient populations have been identified. These translate into complex molecular interactions, including mitochondrial dysfunction and OxS, endosomal-lysosomal dysfunction and NIF, and have led to the development of transgenic animal models of PD (Creed and Goldberg, 2018; Dawson et al., 2018) and biomarker development studies, such as the Rostock International Parkinson's Disease Study (ROPAD) on *LRRK2* (NCT03866603). This is important as more effective animal models based on pathology-driven biomarkers that can be tracked progressively along a PD continuum and which more accurately replicate the chronic and widespread and progressive pathology of PD are needed.

4.4 The trouble with intervention trials

Clinical trial design is sub-optimal, particularly when evaluating disease-modifying therapies. Patients are often recruited *de novo*, according to a specific disease stage, or as symptomatically-treated stable patients. An inability to accurately stratify patients means that recruited cohorts are often heterogeneous populations based on "patients with PD"—these comprise patients with early- and late-stage PD, and different comorbidities and/or severity of complications. Subsequently, "too many of the wrong type of PD patient" may mask a potentially positive signal and disqualify a drug that may have benefited many of the "right type of PD patient." Moreover, identifying patients with faster rates of disease progression and at earlier stages of the disease continuum would benefit disease-modifying trials (Chen-Plotkin et al., 2018).

Randomized, placebo-controlled clinical trials are ideal yet time consuming and expensive. Novel therapies are often compared to best available therapy, which is not ideal. Surrogate biomarkers of appropriate clinical endpoints are also crucial, and disease-modifying therapies may require novel endpoints reliant on validated biomarkers (Athauda and Foltynie, 2016a), which could be addressed by identifying and validating suitable biomarkers. Quantitative disease progression modeling may be a valuable advance here. Based on longitudinal data from clinical trials and observational studies, these models predict disease trajectory and correlate clinical stages with brain structures or biomarkers that change over time. Such empirical disease progression models can be used in drug development and in clinical trial design and simulation, and are particularly useful for disease-modifying studies, as reviewed by Venuto et al. (2016).

5. Biomarkers for diagnostics, drug development and precision medicine

Clearly, there is a need for biomarkers for early diagnosis, drug development and precision medicine in PD. Biomarkers must be accurate, reliable, and routinely available for routine clinical diagnosis. They must also be interpretable, feasible, cost effective and reproducible across patient cohorts (Holland, 2016).

They are fundamental for improving animal models of PD; accurately and routinely identifying patients at different stages of PD; tracking PD progression along its neurodegenerative continuum; improving patient identification and stratification for trial design; generating validated biomarkers as trial endpoints associated with clinical outcomes and optimizing trial design; targeting therapies to patients most likely to benefit based on their specific genetic and molecular signatures; and, ideally identifying and targeting PD early.

Biomarker development will also enhance systems biology strategies, which themselves will enable diagnosis and classification of PD subtypes to improve patient stratification for trials and targeting of therapies to the right patient with the right biomarker-driven pathology. This already occurs in cancer therapy (Korsunsky et al., 2014; Werner et al., 2014), is being considered for RCTs in general (Ayers and Day, 2015), and has begun in AD (Hampel et al., 2018) and potentially in PD (Menche et al., 2017; Sardi et al., 2018).

Nevertheless, the complexity of PD and its distinctive pathological phenotype in each individual means that a combination of biomarkers will likely be needed to accurately describe a patient's pathological signature and drive personalized treatment and therapy optimization (Chahine and Stern, 2017; Chen-Plotkin and Zetterberg, 2018; Emamzadeh and Surguchov, 2018; Redenšek et al., 2018).

5.1 Advances in novel biomarkers

Many biomarker studies are driving this paradigm shift in PD and many targets exist. For routine clinical diagnostics, clearly the less invasive and expensive the biomarker the better, and those based on biofluids (e.g., CSF, blood or saliva) would be ideal.

In terms of diagnosis, individuals with PD have been identified based on significantly decreased saliva levels of the micro-RNAs, miR-153 and miR-223, which are associated with α-Syn production and down-regulated in transgenic PD models (Cressatti et al., 2020). In addition, CSF and serum neurofillament light chain (Nfl), indicative of axonal damage in PD, show promise in distinguishing between PD patients and non-diseased controls, particularly when CSF-Nfl is combined with other, α-Syn-based parameters of pathology, and may also track disease progression (Oosterveld et al., 2020).

Determination of early- vs. advanced-stage PD may be distinguished by a blood-based analysis of protein autoantibodies combined with proteomics technology that identified mitochondrial recycling factor (MRRF) and ribosomal protein S18 (RPS18) as distinct, high confidence biomarkers associated with protein translation and mitochondrial function (Wu et al., 2020). Rate of disease progression, meanwhile, may be detected using biomarkers of cellular senescence (telomere length, p16 and p21) and a five-cytokine pro-inflammatory profile from blood-derived leucocytes and serum, respectively; this predicts a change in cognitive function in *de novo* PD, with p16 levels in particular predicting faster rates of motor and cognitive decline over 36 months of follow up (Martin-Ruiz et al., 2020).

Although neuroimaging techniques are clearly less versatile for primary healthcare diagnostics, they are valuable in clinical trials for accurate patient segmentation and monitoring therapeutic impact and disease progression. Abbasi et al. (2019) have used Diffusion Imaging Analysis to investigate changes in brain structural properties with disease progression based on clinical outcomes to differentiate clinical subtypes of *de novo* PD and monitor its progression and found that brain connectivity disruption was associated with decline in cognitive function. Huber et al. (2020) used DaT-SPECT and FDG-PET imaging data from Dementia with Lewy Body (DLB) patients to show that brain metabolism is altered depending on dopaminergic degeneration in specific regions of the basal ganglia and limbic systems and could be used as a biomarker for DLB and early dopaminergic loss.

These biomarkers, based on molecular markers and pathology central to PD, represent novel methods of diagnosing PD at different stages and tracking its progression, all of which are vital for disease-modifying therapy development. They may also allow re-alignment of therapy in patients with disease progression until methods to retard or halt PD progression are discovered. Importantly, a disease-modifying therapy may work in early disease, but not in late disease (Athauda and Foltynie, 2016a) so each step along the PD continuum must be specifically evaluated. Biomarkers offer early diagnostic capabilities, including those targeting cognitive dysfunction, which also needs early diagnosis for effective management as has been suggested in AD intervention studies.

Gwinn et al. (2017) review the main biomarker programs in PD, and non-interventional studies focusing specifically on biomarker development include the *Systems Medicine of Mitochondrial and Biochemical Parkinson's Disease and Other Related Movement Disorders* (*SysMedPD*) study geared to develop patient stratification strategies associated with PD-related biochemical pathway dysfunction (NCT03421899); an observational, prospective inflammatory biomarker profiling study designed to identify biomarkers that correlate with a clinical diagnosis of PD and enable patient stratification by severity (NCT03633513); a study evaluating α-Syn, LRRK2 and Parkin individually or in combination as peripheral biomarkers for PD-GBA diagnosis and progression (NCT03811496); and biomarker development for cognitive decline in early PD (NCT04101968). These concerted efforts should revolutionize diagnostics, drug development, and intervention trials in PD.

6. Future considerations and conclusion

The near-future issue in PD research is whether it will be re-defined from a single, heterogeneous disease into unique diseases based on specific clinical, pathological, genetic and molecular/biological signatures and associated biomarkers representing different clinical manifestations of disease. Advances in omics research, novel, pathology-based biomarkers development, and systems biology may drive this revolution (Chen-Plotkin and Zetterberg, 2018; Espay and Lang, 2018). Notably, breast cancer is seen as a disease continuum driven by biomarker diagnosis and disease staging. This may be a valid strategy for PD; such a paradigm shift would allow PD to be intercepted and treated earlier rather than later, and precision medicine (targeting the right therapy to the right person at the right time) may become a reality.

This may also clarify which pathological trigger(s) should be targeted for treatment at what stage of disease in which patient, and how early can/should (prodromal) PD be targeted to prevent irreversible damage must also be defined. Furthermore, it would drive development of animal models with biomarker-driven pathology to improve preclinical to clinical translation (Espay and Lang, 2018).

Biomarkers are fundamental for accurate and timely diagnosis of PD, which is important since recent AD intervention trials targeting aberrant protein pathology have suggested that accurately targeting therapy at the right dose early in disease may be crucial to retarding its progression. Pathology-driven biomarkers will improve diagnostics and may involve more accessible methods of visualizing the brain, such as early ocular changes that reflect brain pathology in PD (Arrigo et al., 2017; Guo et al., 2018; Jiménez et al., 2014). They will also revolutionize drug development; biomarkers would optimize patient stratification for clinical trials that would be smaller and target specific stages of disease. This would lead to cheaper and shorter trials, as in AD (Veitch et al., 2019). Quantitative modeling techniques would also improve trial design, particularly for disease-modifying drugs. Importantly, biomarkers might enable re-evaluation of certain therapies previously disqualified because of a poorly-defined patient population in smaller, better targeted patient cohorts with the right biomarker(s).

Aberrant α-Syn accumulation targeting strategies have spearheaded PD research over the last few decades and have focused on its removal/control (see Brundin et al., 2017; Dehay et al., 2015). This area of research has many interesting aspects and future implications for PD. On-going, proof-of-concept trials are investigating the impact of (i) reducing α-Syn levels via inhibiting *SNCA* transcription/translation; (ii) reducing/preventing formation of α-Syn aggregates; (iii) enhancing lysosomal or proteosomal enzyme activity to promote intracellular α-Syn clearance; and (iv) neutralizing/clearing extracellular α-Syn and/or inhibiting α-Syn cell-to-cell transmission (Merchant et al., 2019). Whilst it has been suggested that α-Syn may be neuroprotective (De Franceschi et al., 2017; Hashimoto et al., 2002;

Labrie and Brundin, 2017), its aberrant accumulation clearly contributes to the OxS-NIF-mitochondrial dysfunction pathology of PD. This makes it a valid therapeutic target, and the aforementioned biomarker-driven, proof-of-concept study using ambroxol in PD-GBA1 indicates the disease-modifying benefits of α-Syn removal in PD (Mullin et al., 2020). Moreover, a multipronged therapeutic strategy of tackling α-Syn and disease-modifying approaches may be required for PD (Takamatsu et al., 2017).

Currently, strategies that repurpose existing therapies indicated for non-PD diseases (e.g., GLP-1 agonists targeting the metabolic aspects of PD or ambroxol that targets the consequences of PD-associated genetic mutations) are promising and may more rapidly provide breakthrough disease-modifying treatments. Importantly, the repurposing of non-PD drugs for PD and other NDDs like AD can save significant financial costs and time in the drug development process. Many hurdles must be overcome, including regulatory affairs and intellectual property laws, and a major challenge is how to stimulate research into sometimes off-patent drugs with incentivizing programs. This is also an issue in other therapy areas [e.g., the cancer realm, as discussed by Hernandez et al. (2017)] and may be aided by governmental granting agencies and philanthropic organizations that provide financial support and incentives to overcome financial disincentives in the development of novel treatment strategies in an area of high unmet clinical need (see Grooendorst et al., 2011).

The impact of peripheral factors and disease, including metabolic disorders and systemic inflammation, also offers opportunities for disease modification strategies, particularly the many anti-diabetic options available. Regenerative/repair/cell replacement strategies will also be vital in the near future, and likely to be deployed alongside disease-modifying and symptomatic strategies. Yet restoring and optimally regulating dopaminergic signaling and managing associated side effects will remain the mainstay of PD therapy in the absence of disease-modifying therapy.

Parkinson's disease severely impacts the quality of life of both patient and carer. A holistic strategy with a multidisciplinary and personalized approach is important (Cugusi et al., 2014; Kalampokini et al., 2019; Nag and Jelinek, 2019; Scherbaum et al., 2020; Wu et al., 2017), and *mens sane in corpore sano* may apply based on increasing evidence from lifestyle intervention strategies in AD (Mattson, 2020; Sindi et al., 2015), particularly if PD is a systemic problem. Nevertheless, the advances into many avenues of basic and therapeutic research suggest a radical overhaul to our approach to PD is afoot and a hopeful future.

Acknowledgments

The author would like to thank Dr Bai-Yun Zeng and Dr Eva-Corina Caragounis for reviewing this manuscript, and Dr Christian Hölscher for information on the relative effectiveness of long- vs. short-acting GLP1 agonists in the brain.

References

Aarsland, D., Creese, B., Politis, M., Chaudhuri, K. R., Ffytche, D. H., Weintraub, D., et al. (2017). Cognitive decline in Parkinson disease. *Nature Reviews Neurology*, *13*(4), 217–231.

Abbasi, N., Fereshtehnejad, S. M., Zeighami, Y., Larcher, K. M., Postuma, R. B., & Dagher, A. (2019). Predicting severity and prognosis in Parkinson's disease from brain microstructure and connectivity. *NeuroImage: Clinical*, *25*, 102111.

Abdelkader, N. F., Safar, M. M., & Salem, H. A. (2016). Ursodeoxycholic acid ameliorates apoptotic cascade in the rotenone model of Parkinson's disease: Modulation of mitochondrial perturbations. *Molecular Neurobiology*, *53*(2), 810–817.

Adams, B., Nunes, J. M., Page, M. J., Roberts, T., Carr, J., Nell, T. A., et al. (2019). Parkinson's disease: A systemic inflammatory disease accompanied by bacterial inflammagens. *Frontiers in Aging Neuroscience*, *11*, 210.

Adler, C. H., Beach, T. G., Hentz, J. G., Shill, H. A., Caviness, J. N., Driver-Dunckley, E., et al. (2014). Low clinical diagnostic accuracy of early vs advanced Parkinson disease: Clinicopathologic study. *Neurology*, *83*(5), 406–412.

Aman, Y., Qiu, Y., Tao, J., & Fang, E. F. (2018). Therapeutic potential of boosting NAD+ in aging and age-related diseases. *Translational Medicine of Aging*, *1*(2), 30–37.

Anderson, F. L., Coffey, M. M., Berwin, B. L., & Havrda, M. C. (2018). Inflammasomes: An emerging mechanism translating environmental toxicant exposure into neuroinflammation in Parkinson's disease. *Toxicological Sciences*, *166*(1), 3–15.

Aquilonius, S. M., & Nyholm, D. (2017). Development of new levodopa treatment strategies in Parkinson's disease-from bedside to bench to bedside. *Upsala Journal of Medical Sciences*, *122*(2), 71–77.

Arrigo, A., Calamuneri, A., Milardi, D., Mormina, E., Rania, L., Postorino, E., et al. (2017). Visual system involvement in patients with newly diagnosed Parkinson disease. *Radiology*, *285*(3), 885–895.

Athauda, D., & Foltynie, T. (2016a). Challenges in detecting disease modification in Parkinson's disease clinical trials. *Parkinsonism & Related Disorders*, *32*, 1–11.

Athauda, D., & Foltynie, T. (2016b). The glucagon-like peptide 1 (GLP) receptor as a therapeutic target in Parkinson's disease: Mechanisms of action. *Drug Discovery Today*, *21*(5), 802–818.

Athauda, D., Maclagan, K., Skene, S. S., Bajwa-Joseph, M., Letchford, D., Chowdhury, K., et al. (2017). Exenatide once weekly versus placebo in Parkinson's disease: A randomised, double-blind, placebo-controlled trial. *Lancet*, *390*(10103), 1664–1675.

Ayers, D., & Day, P. J. (2015). Systems medicine: The application of systems biology approaches for modern medical research and drug development. *Molecular Biology International*, *2015*, 698169.

Ayton, S., Lei, P., Adlard, P. A., Volitakis, I., Cherny, R. A., Bush, A. I., et al. (2014). Iron accumulation confers neurotoxicity to a vulnerable population of nigral neurons: Implications for Parkinson's disease. *Molecular Neurodegeneration*, *9*, 27.

Bajaj, N., Hauser, R. A., & Grachev, I. D. (2013). Clinical utility of dopamine transporter single photon emission CT (DaT-SPECT) with (123I) ioflupane in diagnosis of parkinsonian syndromes. *Journal of Neurology, Neurosurgery, and Psychiatry*, *84*(11), 1288–1295.

Barker, R. A., Parmar, M., Studer, L., & Takahashi, J. (2017). Human trials of stem cell-derived dopamine neurons for Parkinson's disease: Dawn of a new era. *Cell Stem Cell*, *21*(5), 569–573.

Bendikov-Bar, I., Maor, G., Filocamo, M., & Horowitz, M. (2013). Ambroxol as a pharmacological chaperone for mutant glucocerebrosidase. *Blood Cells, Molecules & Diseases*, *50*(2), 141–145.

Billingsley, K. J., Bandres-Ciga, S., Saez-Atienzar, S., & Singleton, A. B. (2018). Genetic risk factors in Parkinson's disease. *Cell and Tissue Research*, *373*(1), 9–20.

Blauwendraat, C., Reed, X., Kia, D. A., Gan-Or, Z., Lesage, S., Pihlstrøm, L., et al. (2018). Frequency of loss of function variants in LRRK2 in Parkinson disease. *JAMA Neurology*, *75*(11), 1416–1422.

Booth, H. D. E., Hirst, W. D., & Wade-Martins, R. (2017). The role of astrocyte dysfunction in Parkinson's disease pathogenesis. *Trends in Neurosciences*, *40*(6), 358–370.

Bose, A., & Beal, M. F. (2016). Mitochondrial dysfunction in Parkinson's disease. *Journal of Neurochemistry*, *139*(Suppl. 1), 216–231.

Braak, H., & Del Tredici, K. (2017). Neuropathological staging of brain pathology in sporadic Parkinson's disease: Separating the wheat from the chaff. *Journal of Parkinson's Disease*, 7(s1), S71–S85.

Briggs, R., Kennelly, S. P., & O'Neill, D. (2016). Drug treatments in Alzheimer's disease. *Clinical Medicine (London, England)*, *16*(3), 247–253.

Brundin, P., & Coetzee, G. A. (2019). Genetically engineered stem cell-derived neurons can be rendered resistant to alpha-synuclein aggregate pathology. *The European Journal of Neuroscience*, *49*(3), 316–319.

Brundin, P., Dave, K. D., & Kordower, J. H. (2017). Therapeutic approaches to target alpha-synuclein pathology. *Experimental Neurology*, *298*(Pt. B), 225–235.

Cai, R., Zhang, Y., Simmering, J. E., Schultz, J. L., Li, Y., Fernandez-Carasa, I., et al. (2019). Enhancing glycolysis attenuates Parkinson's disease progression in models and clinical databases. *The Journal of Clinical Investigation*, *129*(10), 4539–4549.

Cardoso, S., & Moreira, P. I. (2020). Antidiabetic drugs for Alzheimer´s and Parkinson´s diseases: Repurposing insulin, metformin and thiazolidinediones. *International Review of Neurobiology*, *155*. (Part 2 Chapter 2).

Carrarini, C., Russo, M., Dono, F., Di Pietro, M., Rispoli, M. G., Di Stefano, V., et al. (2019). A stage-based approach to therapy in Parkinson's disease. *Biomolecules*, *9*(8), pii: E388.

Castro-Caldas, M., Carvalho, A. N., Rodrigues, E., Henderson, C. J., Wolf, C. R., Rodrigues, C. M., et al. (2012). Tauroursodeoxycholic acid prevents MPTP-induced dopaminergic cell death in a mouse model of Parkinson's disease. *Molecular Neurobiology*, *46*(2), 475–486.

Catania, M., Giaccone, G., Salmona, M., Tagliavini, F., & Di Fede, G. (2019). Dreaming of a new world where Alzheimer's is a treatable disorder. *Frontiers in Aging Neuroscience*, *11*, 317.

Chahine, L. M., & Stern, M. B. (2017). Parkinson's disease biomarkers: Where are we and where do we go next? *Movement Disorders Clinical Practice*, *4*(6), 796–805.

Chen, C., Turnbull, D. M., & Reeve, A. K. (2019a). Mitochondrial dysfunction in Parkinson's disease-cause or consequence? *Biology (Basel)*, *8*(2), pii: E38.

Chen, F., Wu, T., Luo, Y., Li, Z., Guan, Q., Meng, X., et al. (2019b). Amnestic mild cognitive impairment in Parkinson's disease: White matter structural changes and mechanisms. *PLoS One*, *14*(12), e0226175.

Chen, J. H., Hong, C. T., Wu, D., Chi, W. C., Yen, C. F., Liao, H. F., et al. (2019c). Dementia-related functional disability in moderate to advanced Parkinson's disease: Assessment using the world health organization disability assessment schedule 2.0. *International Journal of Environmental Research and Public Health*, *16*(12), pii: E2230.

Chen, Q. Q., Haikal, C., Li, W., & Li, J. Y. (2019d). Gut inflammation in association with pathogenesis of Parkinson's disease. *Frontiers in Molecular Neuroscience*, *12*, 218.

Chen, Y., Dolt, K. S., Kriek, M., Baker, T., Downey, P., Drummond, N. J., et al. (2019e). Engineering synucleinopathy-resistant human dopaminergic neurons by CRISPR-mediated deletion of the SNCA gene. *The European Journal of Neuroscience*, *49*(4), 510–524.

Chen-Plotkin, A. S., Albin, R., Alcalay, R., Babcock, D., Bajaj, V., Bowman, D., et al. (2018). Finding useful biomarkers for Parkinson's disease. *Science Translational Medicine*, *10*(454), pii: eaam6003.

Chen-Plotkin, A. S., & Zetterberg, H. (2018). Updating our definitions of Parkinson's disease for a molecular age. *Journal of Parkinson's Disease*, *8*(s1), S53–S57.

Chien, C. H., Lee, M. J., Liou, H. C., Liou, H. H., & Fu, W. M. (2016). Microglia-derived cytokines/chemokines are involved in the enhancement of LPS-induced loss of nigrostriatal dopaminergic neurons in DJ-1 knockout mice. *PLoS One*, *11*(3), e0151569.

Codolo, G., Plotegher, N., Pozzobon, T., Brucale, M., Tessari, I., Bubacco, L., et al. (2013). Triggering of inflammasome by aggregated α-synuclein, an inflammatory response in synucleinopathies. *PLoS One*, *8*(1), e55375.

Cookson, M. R. (2012). Parkinsonism due to mutations in PINK1, parkin, and DJ-1 and oxidative stress and mitochondrial pathways. *Cold Spring Harbor Perspectives in Medicine*, *2*(9), a009415.

Corona, J. C., & Duchen, M. R. (2016). PPARγ as a therapeutic target to rescue mitochondrial function in neurological disease. *Free Radical Biology & Medicine*, *100*, 153–163.

Creed, R. B., & Goldberg, M. S. (2018). New developments in genetic rat models of Parkinson's disease. *Movement Disorders*, *33*(5), 717–729.

Cressatti, M., Juwara, L., Galindez, J. M., Velly, A. M., Nkurunziza, E. S., Marier, S., et al. (2020). Salivary microR-153 and microR-223 Levels as potential diagnostic biomarkers of idiopathic Parkinson's disease. *Movement Disorders*, *35*(3), 468–477.

Crotty, G. F., Ascherio, A., & Schwarzschild, M. A. (2017). Targeting urate to reduce oxidative stress in Parkinson disease. *Experimental Neurology*, *298*(Pt. B), 210–224.

Cugusi, L., Solla, P., Zedda, F., Loi, M., Serpe, R., Cannas, A., et al. (2014). Effects of an adapted physical activity program on motor and non-motor functions and quality of life in patients with Parkinson's disease. *Neurorehabilitation*, *35*(4), 789–794.

Daher, J. P., Abdelmotilib, H. A., Hu, X., Volpicelli-Daley, L. A., Moehle, M. S., Fraser, K. B., et al. (2015). Leucine-rich repeat kinase 2 (LRRK2) pharmacological inhibition abates α-synuclein gene-induced neurodegeneration. *The Journal of Biological Chemistry*, *290*(32), 19433–19444.

Daniele, G., Iozzo, P., Molina-Carrion, M., Lancaster, J., Ciociaro, D., Cersosimo, E., et al. (2015). Exenatide regulates cerebral glucose metabolism in brain areas associated with glucose homeostasis and reward system. *Diabetes*, *64*(10), 3406–3412.

Dawson, T. M., Golde, T. E., & Lagier-Tourenne, C. (2018). Animal models of neurodegenerative diseases. *Nature Neuroscience*, *21*(10), 1370–1379.

De Franceschi, G., Fecchio, C., Sharon, R., Schapira, A. H. V., Proukakis, C., Bellotti, V., et al. (2017). α-Synuclein structural features inhibit harmful polyunsaturated fatty acid oxidation, suggesting roles in neuroprotection. *The Journal of Biological Chemistry*, *292*(17), 6927–6937.

Dehay, B., Bourdenx, M., Gorry, P., Przedborski, S., Vila, M., Hunot, S., et al. (2015). Targeting α-synuclein for treatment of Parkinson's disease: Mechanistic and therapeutic considerations. *Lancet Neurology*, *14*(8), 855–866.

Dellinger, R. W., Santos, S. R., Morris, M., Evans, M., Alminana, D., Guarente, L., et al. (2017). Repeat dose NRPT (nicotinamide riboside and pterostilbene) increases NAD+ levels in humans safely and sustainably: A randomized, double-blind, placebo-controlled study. *NPJ Aging and Mechanisms of Disease*, *3*, 17.

Devi, L., Raghavendran, V., Prabhu, B. M., Avadhani, N. G., & Anandatheerthavarada, H. K. (2008). Mitochondrial import and accumulation of alpha-synuclein impair complex I in human dopaminergic neuronal cultures and Parkinson disease brain. *The Journal of Biological Chemistry*, *283*(14), 9089–9100.

Ding, X., Stasi, L. P., Ho, M. H., Zhao, B., Wang, H., Long, K., et al. (2018). Discovery of 4-ethoxy-7H-pyrrolo[2,3-d]pyrimidin-2-amines as potent, selective and orally bioavailable LRRK2 inhibitors. *Bioorganic & Medicinal Chemistry Letters*, *28*(9), 1615–1620.

Dölle, C., Flønes, I., Nido, G. S., Miletic, H., Osuagwu, N., Kristoffersen, S., et al. (2016). Defective mitochondrial DNA homeostasis in the substantia nigra in Parkinson disease. *Nature Communications*, 7, 13548.

Dorsey, E. R., Sherer, T., Okun, M. S., & Bloem, B. R. (2018). The emerging evidence of the Parkinson pandemic. *Journal of Parkinson's Disease*, *8*(s1), S3–S8.

Du, T. T., Wang, L., Duan, C. L., Lu, L. L., Zhang, J. L., Gao, G., et al. (2015). GBA deficiency promotes SNCA/α-synuclein accumulation through autophagic inhibition by inactivated PPP2A. *Autophagy*, *11*(10), 1803–1820.

Dunn, L., Allen, G. F., Mamais, A., Ling, H., Li, A., Duberley, K. E., et al. (2014). Dysregulation of glucose metabolism is an early event in sporadic Parkinson's disease. *Neurobiology of Aging*, *35*(5), 1111–1115.

El-Hattab, A. W., Zarante, A. M., Almannai, M., & Scaglia, F. (2017). Therapies for mitochondrial diseases and current clinical trials. *Molecular Genetics and Metabolism*, *122*(3), 1–9.

Emamzadeh, F. N., & Surguchov, A. (2018). Parkinson's disease: Biomarkers, treatment, and risk factors. *Frontiers in Neuroscience*, *12*, 612.

Espay, A. J., & Lang, A. E. (2018). Parkinson diseases in the 2020s and beyond: Replacing clinico-pathologic convergence with systems biology divergence. *Journal of Parkinson's Disease*, *8*(s1), S59–S64.

Espay, A. J., Schwarzschild, M. A., Tanner, C. M., Fernandez, H. H., Simon, D. K., Leverenz, J. B., et al. (2017). Biomarker-driven phenotyping in Parkinson's disease: A translational missing link in disease-modifying clinical trials. *Movement Disorders*, *32*(3), 319–324.

Esterline, R. L., Vaag, A., Oscarsson, J., & Vora, J. (2018). Mechanisms in endocrinology: SGLT2 inhibitors: Clinical benefits by restoration of normal diurnal metabolism? *European Journal of Endocrinology*, *178*(4), R113–R125.

Ferrari, C. C., & Tarelli, R. (2011). Parkinson's disease and systemic inflammation. *Parkinson's Disease*, *2011*, 436813.

Fuzzati-Armentero, M. T., Cerri, S., & Blandini, F. (2019). Peripheral-central neuroimmune crosstalk in Parkinson's disease: What do patients and animal models tell us? *Frontiers in Neurology*, *10*, 232.

Gagne, J. J., & Power, M. C. (2010). Anti-inflammatory drugs and risk of Parkinson disease: A meta-analysis. *Neurology*, *74*(12), 995–1002.

Gao, X., Chen, H., Schwarzschild, M. A., & Ascherio, A. (2011). Use of ibuprofen and risk of Parkinson disease. *Neurology*, *76*(10), 863–869.

Garcia-Esparcia, P., Koneti, A., Rodríguez-Oroz, M. C., Gago, B., Del Rio, J. A., & Ferrer, I. (2018). Mitochondrial activity in the frontal cortex area 8 and angular gyrus in Parkinson's disease and Parkinson's disease with dementia. *Brain Pathology*, *28*(1), 43–57.

Garvey, J., Petersen, M., Waters, C. M., Rose, S. P., Hunt, S., Briggs, R., et al. (1986). Administration of MPTP to the common marmoset does not alter cortical cholinergic function. *Movement Disorders*, *1*(2), 129–134.

Gatt, A. P., Duncan, O. F., Attems, J., Francis, P. T., Ballard, C. G., & Bateman, J. M. (2016). Dementia in Parkinson's disease is associated with enhanced mitochondrial complex I deficiency. *Movement Disorders*, *31*(3), 352–359.

GBD 2016 Parkinson's Disease Collaborators. (2018). Global, regional, and national burden of Parkinson's disease, 1990-2016: A systematic analysis for the Global Burden of Disease Study 2016. *Lancet Neurology*, *17*(11), 939–953.
Gegg, M. E., & Schapira, A. H. V. (2018). The role of glucocerebrosidase in Parkinson disease pathogenesis. *The FEBS Journal*, *285*(19), 3591–3603.
Gentilella, R., Pechtner, V., Corcos, A., & Consoli, A. (2019). Glucagon-like peptide-1 receptor agonists in type 2 diabetes treatment: Are they all the same? *Diabetes/ Metabolism Research and Reviews*, *35*(1), e3070.
Gibb, W. R. G., Terruli, M., Less, A. J., Jenner, P., & Marsden, C. D. (1989). The evolution and distribution of morphological changes in the nervous system of the common marmoset following the acute administration of 1-methyl-4-phenyl-1,2,3,6-tetra-hydropyridine. *Movement Disorders*, *4*(1), 53–74.
Giguère, N., Delignat-Lavaud, B., Herborg, F., Voisin, A., Li, Y., Jacquemet, V., et al. (2019). Increased vulnerability of nigral dopamine neurons after expansion of their axonal arborization size through D2 dopamine receptor conditional knockout. *PLoS Genetics*, *15*(8), e1008352.
Giri, B., Belanger, K., Seamon, M., Bradley, E., Purohit, S., Chong, R., et al. (2019). Niacin ameliorates neuro-inflammation in Parkinson's disease via GPR109A. *International Journal of Molecular Sciences*, *20*(18), pii: E4559.
Goldstein, D. S., Kopin, I. J., & Sharabi, Y. (2014). Catecholamine autotoxicity. Implications for pharmacology and therapeutics of Parkinson disease and related disorders. *Pharmacology & Therapeutics*, *144*(3), 268–282.
Gordon, R., Albornoz, E. A., Christie, D. C., Langley, M. R., Kumar, V., Mantovani, S., et al. (2018). Inflammasome inhibition prevents α-synuclein pathology and dopaminergic neurodegeneration in mice. *Science Translational Medicine*, *10*(465), pii: eaah4066.
Green, H., Tsitsi, P., Markaki, I., Aarsland, D., & Svenningsson, P. (2019). Novel treatment opportunities against cognitive impairment in Parkinson's disease with an emphasis on diabetes-related pathways. *CNS Drugs*, *33*(2), 143–160.
Greshock, T. J., Sanders, J. M., Drolet, R. E., Rajapakse, H. A., Chang, R. K., Kim, B., et al. (2016). Potent, selective and orally bioavailable leucine-rich repeat kinase 2 (LRRK2) inhibitors. *Bioorganic & Medicinal Chemistry Letters*, *26*(11), 2631–2635.
Grootendorst, P., Hollis, A., Levine, D. K., Pogge, T., & Edwards, A. M. (2011). New approaches to rewarding pharmaceutical innovation. *CMAJ*, *183*(6), 681–685.
Guo, L., Normando, E. M., Shah, P. A., De Groef, L., & Cordeiro, M. F. (2018). Oculo-visual abnormalities in Parkinson's disease: Possible value as biomarkers. *Movement Disorders*, *33*(9), 1390–1406.
Gwinn, K., David, K. K., Swanson-Fischer, C., Albin, R., Hillaire-Clarke, C. S., Sieber, B. A., et al. (2017). Parkinson's disease biomarkers: Perspective from the NINDS Parkinson's disease biomarkers program. *Biomarkers in Medicine*, *11*(6), 451–473.
Hallett, P. J., Cooper, O., Sadi, D., Robertson, H., Mendez, I., & Isacson, O. (2014). Long-term health of dopaminergic neuron transplants in Parkinson's disease patients. *Cell Reports*, *7*(6), 1755–1761.
Hampel, H., Toschi, N., Babiloni, C., Baldacci, F., Black, K. L., Bokde, A. L. W., et al. (2018). Revolution of Alzheimer precision neurology. Passageway of systems biology and neurophysiology. *Journal of Alzheimer's Disease*, *64*(s1), S47–S105.
Hashimoto, M., Hsu, L. J., Rockenstein, E., Takenouchi, T., Mallory, M., & Masliah, E. (2002). alpha-Synuclein protects against oxidative stress via inactivation of the c-Jun N-terminal kinase stress-signaling pathway in neuronal cells. *The Journal of Biological Chemistry*, *277*(13), 11465–11472.
Hernandez, J. J., Pryszlak, M., Smith, L., Yanchus, C., Kurji, N., Shahani, V. M., et al. (2017). Giving drugs a second chance: Overcoming regulatory and financial hurdles in repurposing approved drugs as cancer therapeutics. *Frontiers in Oncology*, *7*, 273.

Herrington, T. M., Cheng, J. J., & Eskandar, E. N. (2016). Mechanisms of deep brain stimulation. *Journal of Neurophysiology*, *115*(1), 19–38.

Ho, K. S., Hotchkin, M. T., & Mortenson, M. G. (2019). Gold nanocatalysis as a novel therapeutic for neuroprotection in Parkinson's disease. In *Program no. 743.04. 2019 neuroscience meeting planner*. Chicago, IL: Society for Neuroscience, Online.

Ho, D. H., Je, A. R., Lee, H., Son, I., Kweon, H. S., Kim, H. G., et al. (2018). LRRK2 kinase activity induces mitochondrial fission in microglia via Drp1 and modulates neuroinflammation. *Experimental Neurobiology*, *27*(3), 171–180.

Holland, R. L. (2016). What makes a good biomarker? *Advances in Precision Medicine*, *1*(1), 4–11.

Hölscher, C. (2014). Drugs developed for treatment of diabetes show protective effects in Alzheimer's and Parkinson's diseases. *Sheng Li Xue Bao*, *66*(5), 497–510.

Hornykiewicz, O. (2001). Chemical neuroanatomy of the basal ganglia—Normal and in Parkinson's disease. *Journal of Chemical Neuroanatomy*, *22*(1–2), 3–12.

Huber, M., Beyer, L., Prix, C., Schönecker, S., Palleis, C., Rauchmann, B. S., et al. (2020). Metabolic correlates of dopaminergic loss in dementia with lewy bodies. *Movement Disorders*, *35*(4), 595–605.

Hunter, K., & Hölscher, C. (2012). Drugs developed to treat diabetes, liraglutide and lixisenatide, cross the blood brain barrier and enhance neurogenesis. *BMC Neuroscience*, *13*, 33.

Hussain, G., Zhang, L., Rasul, A., Anwar, H., Sohail, M. U., Razzaq, A., et al. (2018). Role of plant-derived flavonoids and their mechanism in attenuation of Alzheimer's and Parkinson's diseases: An update of recent data. *Molecules*, *23*(4), pii: E814.

Ismael, S., Nasoohi, S., & Ishrat, T. (2018). MCC950, the selective inhibitor of nucleotide oligomerization domain-like receptor protein-3 inflammasome, protects mice against traumatic brain injury. *Journal of Neurotrauma*, *35*(11), 1294–1303.

Javed, H., Nagoor Meeran, M. F., Azimullah, S., Adem, A., Sadek, B., et al. (2019). Plant extracts and phytochemicals targeting α-synuclein aggregation in Parkinson's disease models. *Frontiers in Pharmacology*, *9*, 1555.

Jenner, P., Morris, H. R., Robbins, T. W., Goedert, M., Hardy, J., Ben-Shlomo, Y., et al. (2013). Parkinson's disease—The debate on the clinical phenomenology, aetiology, pathology and pathogenesis. *Journal of Parkinson's Disease*, *3*(1), 1–11.

Jenner, P. J., Rupniak, N. M. J., Rose, S., Kelly, E., Kilpatrick, G., Lees, A., et al. (1984). 1-Methyl-4-phenyl-1,2,3,6-tetrahydropyridine-induced parkinsonism in the common marmoset. *Neuroscience Letters*, *50*, 85–90.

Jia, X., Wang, Z., Yang, T., Li, Y., Gao, S., Wu, G., et al. (2019). Entorhinal cortex atrophy in early, drug-naive Parkinson's disease with mild cognitive impairment. *Aging and Disease*, *10*(6), 1221–1232.

Jiménez, B., Ascaso, F. J., Cristóbal, J. A., & López del Val, J. (2014). Development of a prediction formula of Parkinson disease severity by optical coherence tomography. *Movement Disorders*, *29*(1), 68–74.

Kalampokini, S., Becker, A., Fassbender, K., Lyros, E., & Unger, M. M. (2019). Nonpharmacological modulation of chronic inflammation in Parkinson's disease: Role of diet interventions. *Parkinson's Disease*, *2019*, 7535472.

Karimi-Moghadam, A., Charsouei, S., Bell, B., & Jabalameli, M. R. (2018). Parkinson disease from mendelian forms to genetic susceptibility: New molecular insights into the neurodegeneration process. *Cellular and Molecular Neurobiology*, *38*(6), 1153–1178.

Karpenko, M. N., Vasilishina, A. A., Gromova, E. A., Muruzheva, Z. M., Miliukhina, I. V., & Bernadotte, A. (2018). Interleukin-1β, interleukin-1 receptor antagonist, interleukin-6, interleukin-10, and tumor necrosis factor-α levels in CSF and serum in relation to the clinical diversity of Parkinson's disease. *Cellular Immunology*, *327*, 77–82.

Kastner, A., Hirsch, E. C., Lejeune, O., Javoy-Agid, F., Rascol, O., & Agid, Y. (1992). Is the vulnerability of neurons in the substantia nigra of patients with Parkinson's disease related to their neuromelanin content? *Journal of Neurochemistry*, *59*(3), 1080–1089.

Kim, D. S., Choi, H. I., Wang, Y., Luo, Y., Hoffer, B. J., & Greig, N. H. (2017). A new treatment strategy for Parkinson's disease through the gut-brain axis: The glucagon-like peptide-1 receptor pathway. *Cell Transplantation*, *26*(9), 1560–1571.

Klemann, C. J. H. M., Martens, G. J. M., Sharma, M., Martens, M. B., Isacson, O., Gasser, T., et al. (2017). Integrated molecular landscape of Parkinson's disease. *NPJ Parkinson's Disease*, *3*, 14.

Kordower, J. H., Chu, Y., Hauser, R. A., Freeman, T. B., & Olanow, C. W. (2008). Lewy body-like pathology in long-term embryonic nigral transplants in Parkinson's disease. *Nature Medicine*, *14*(5), 504–506.

Kordower, J. H., Olanow, C. W., Dodiya, H. B., Chu, Y., Beach, T. G., Adler, C. H., et al. (2013). Disease duration and the integrity of the nigrostriatal system in Parkinson's disease. *Brain*, *136*(Pt. 8), 2419–2431.

Korsunsky, I., McGovern, K., LaGatta, T., Olde Loohuis, L., Grosso-Applewhite, T., Griffeth, N., et al. (2014). Systems biology of cancer: A challenging expedition for clinical and quantitative biologists. *Frontiers in Bioengineering and Biotechnology*, *2*, 27.

Labrie, V., & Brundin, P. (2017). Alpha-Synuclein to the rescue: Immune cell recruitment by alpha-synuclein during gastrointestinal infection. *Journal of Innate Immunity*, *9*(5), 437–440.

Lang, A. E., & Obeso, J. A. (2004). Challenges in Parkinson's disease: Restoration of the nigrostriatal dopamine system is not enough. *Lancet Neurology*, *3*(5), 309–316.

Lautrup, S., Sinclair, D. A., Mattson, M. P., & Fang, E. F. (2019). NAD+ in brain aging and neurodegenerative disorders. *Cell Metabolism*, *30*(4), 630–655.

Lee, H., James, W. S., & Cowley, S. A. (2017). LRRK2 in peripheral and central nervous system innate immunity: Its link to Parkinson's disease. *Biochemical Society Transactions*, *45*(1), 131–139.

Lee, H. J., Suk, J. E., Patrick, C., Bae, E. J., Cho, J. H., Rho, S., et al. (2010). Direct transfer of a-synuclein from neuron to astroglia causes inflammatory responses in synucleinopathies. *The Journal of Biological Chemistry*, *285*, 9262–9272.

Lees, A. J., Ferreira, J., Rascol, O., Poewe, W., Rocha, J. F., McCrory, M., et al. (2017). Opicapone as adjunct to levodopa therapy in patients with Parkinson disease and motor fluctuations: A randomized clinical trial. *JAMA Neurology*, *74*(2), 197–206.

Li, J. Y., Englund, E., Holton, J. L., Soulet, D., Hagell, P., Lees, A. J., et al. (2008). Lewy bodies in grafted neurons in subjects with Parkinson's disease suggest host-to-graft disease propagation. *Nature Medicine*, *14*(5), 501–503.

Li, C., Liu, W., Li, X., Zhang, Z., Qi, H., Liu, S., et al. (2020). The novel GLP-1/GIP analogue DA5-CH reduces tau phosphorylation and normalizes theta rhythm in the icv. STZ rat model of AD. *Brain and Behavior*, *10*(3), e01505.

Li, Y., Perry, T., Kindy, M. S., Harvey, B. K., Tweedie, D., Holloway, H. W., et al. (2009). GLP-1 receptor stimulation preserves primary cortical and dopaminergic neurons in cellular and rodent models of stroke and Parkinsonism. *Proceedings of the National Academy of Sciences of the United States of America*, *106*(4), 1285–1290.

Macdonald, R., Barnes, K., Hastings, C., & Mortiboys, H. (2018). Mitochondrial abnormalities in Parkinson's disease and Alzheimer's disease: Can mitochondria be targeted therapeutically? *Biochemical Society Transactions*, *46*(4), 891–909.

Martens, C. R., Denman, B. A., Mazzo, M. R., Armstrong, M. L., Reisdorph, N., McQueen, M. B., et al. (2018). Chronic nicotinamide riboside supplementation is well-tolerated and elevates NAD+ in healthy middle-aged and older adults. *Nature Communications*, *9*(1), 1286.

Martin-Ruiz, C., Williams-Gray, C. H., Yarnall, A. J., Boucher, J. J., Lawson, R. A., Wijeyekoon, R. S., et al. (2020). Senescence and inflammatory markers for predicting clinical progression in Parkinson's disease: The ICICLE-PD study. *Journal of Parkinson's Disease*, *10*(1), 193–206.

Mattson, M. P. (2020). Involvement of GABAergic interneuron dysfunction and neuronal network hyperexcitability in Alzheimer's disease: Amelioration by metabolic switching. *International Review of Neurobiology*, *154*, 191–206.

Mattson, M. P., & Arumugam, T. V. (2018). Hallmarks of brain aging: Adaptive and pathological modification by metabolic states. *Cell Metabolism*, *27*(6), 1176–1199.

McNeill, A., Magalhaes, J., Shen, C., Chau, K. Y., Hughes, D., Mehta, A., et al. (2014). Ambroxol improves lysosomal biochemistry in glucocerebrosidase mutation-linked Parkinson disease cells. *Brain*, *137*(Pt 5), 1481–1495.

Mehta, S. H., & Tanner, C. M. (2016). Role of neuroinflammation in Parkinson disease: The enigma continues. *Mayo Clinic Proceedings*, *91*(10), 1328–1330.

Menche, J., Guney, E., Sharma, A., Branigan, P. J., Loza, M. J., Baribaud, F., et al. (2017). Integrating personalized gene expression profiles into predictive disease-associated gene pools. *NPJ Systems Biology and Applications*, *3*, 10.

Merchant, K. M., Cedarbaum, J. M., Brundin, P., Dave, K. D., Eberling, J., Espay, A. J., et al. (2019). A proposed roadmap for Parkinson's disease proof of concept clinical trials investigating compounds targeting alpha-synuclein. *Journal of Parkinson's Disease*, *9*(1), 31–61.

Michael J Fox Foundation. (2016). https://www.michaeljfox.org/grant/blocking-inflammasome-induced-neuroinflammation-parkinsons-disease-potent-orally-available?grant_id=1521.

Migdalska-Richards, A., & Schapira, A. H. (2016). The relationship between glucocerebrosidase mutations and Parkinson disease. *Journal of Neurochemistry*, *139*(Suppl. 1), 77–90.

Mogi, M., Harada, M., Kondo, T., Riederer, P., Inagaki, H., Minami, M., et al. (1994). Interleukin-1 beta, interleukin-6, epidermal growth factor and transforming growth factor-alpha are elevated in the brain from parkinsonian patients. *Neuroscience Letters*, *180*(2), 147–150.

Mortiboys, H., Furmston, R., Bronstad, G., Aasly, J., Elliott, C., & Bandmann, O. (2015). UDCA exerts beneficial effect on mitochondrial dysfunction in LRRK2(G2019S) carriers and in vivo. *Neurology*, *85*(10), 846–852.

Mouton-Liger, F., Rosazza, T., Sepulveda-Diaz, J., Ieang, A., Hassoun, S. M., Claire, E., et al. (2018). Parkin deficiency modulates NLRP3 inflammasome activation by attenuating an A20-dependent negative feedback loop. *Glia*, *66*(8), 1736–1751.

Mullin, S., Smith, L., Lee, K., D'Souza, G., Woodgate, P., Elflein, J., et al. (2020). Ambroxol for the treatment of patients with Parkinson disease with and without glucocerebrosidase gene mutations: A nonrandomized, noncontrolled trial. *JAMA Neurology*, 77(4), 427–434.

Murphy, K. E., Gysbers, A. M., Abbott, S. K., Tayebi, N., Kim, W. S., Sidransky, E., et al. (2014). Reduced glucocerebrosidase is associated with increased α-synuclein in sporadic Parkinson's disease. *Brain*, *137*(Pt 3), 834–848.

Nag, N., & Jelinek, G. A. (2019). More research is needed on lifestyle behaviors that influence progression of Parkinson's disease. *Frontiers in Neurology*, *10*, 452.

Nandipati, S., & Litvan, I. (2016). Environmental Exposures and Parkinson's Disease. *International Journal of Environmental Research and Public Health*, *13*(9), pii: E881.

Niu, J., Yu, M., Wang, C., & Xu, Z. (2012). Leucine-rich repeat kinase 2 disturbs mitochondrial dynamics via Dynamin-like protein. *Journal of Neurochemistry*, *122*(3), 650–658.

Norwitz, N. G., Hu, M. T., & Clarke, K. (2019). The mechanisms by which the ketone body D-β-hydroxybutyrate may improve the multiple cellular pathologies of Parkinson's disease. *Frontiers in Nutrition*, *6*, 63.

Novak, P., Pimentel Maldonado, D. A., & Novak, V. (2019). Safety and preliminary efficacy of intranasal insulin for cognitive impairment in Parkinson disease and multiple system atrophy: A double-blinded placebo-controlled pilot study. *PLoS One*, *14*(4), e0214364.

Nussbaum, R. L. (2017). The identification of alpha-synuclein as the first Parkinson disease gene. *Journal of Parkinson's Disease*, 7(s1), S43–S49.

Oertel, W., & Schulz, J. B. (2016). Current and experimental treatments of Parkinson disease: A guide for neuroscientists. *Journal of Neurochemistry*, *139*(Suppl. 1), 325–337.

Oosterveld, L. P., Verberk, I. M. W., Majbour, N. K., El-Agnaf, O. M., Weinstein, H. C., Berendse, H. W., et al. (2020). CSF or serum neurofilament light added to α-synuclein panel discriminates Parkinson's from controls. *Movement Disorders*, *35*(2), 288–295.

Ouchi, Y., Yoshikawa, E., Sekine, Y., Futatsubashi, M., Kanno, T., Ogusu, T., et al. (2005). Microglial activation and dopamine terminal loss in early Parkinson's disease. *Annals of Neurology*, *57*(2), 168–175.

Park, J. S., Davis, R. L., & Sue, C. M. (2018). Mitochondrial dysfunction in Parkinson's disease: New mechanistic insights and therapeutic perspectives. *Current Neurology and Neuroscience Reports*, *18*(5), 21.

Parkinson Study Group. (1993). Effects of tocopherol and deprenyl on the progression of disability in early Parkinson's disease. *The New England Journal of Medicine*, *328*(3), 176–183.

Parkinson Study Group. (1998). Mortality in DATATOP: A multicenter trial in early Parkinson's disease. *Annals of Neurology*, *43*(3), 318–325.

Parkinson Study Group SURE-PD Investigators, Schwarzschild, M. A., Ascherio, A., Beal, M. F., Cudkowicz, M. E., Curhan, G. C., et al. (2014). Inosine to increase serum and cerebrospinal fluid urate in Parkinson disease: A randomized clinical trial. *JAMA Neurology*, *71*(2), 141–150.

Perera, A. P., Fernando, R., Shinde, T., Gundamaraju, R., Southam, B., Sohal, S. S., et al. (2018). MCC950, a specific small molecule inhibitor of NLRP3 inflammasome attenuates colonic inflammation in spontaneous colitis mice. *Scientific Reports*, *8*(1), 8618.

Peter, I., Dubinsky, M., Bressman, S., Park, A., Lu, C., Chen, N., et al. (2018). Anti-tumor necrosis factor therapy and incidence of Parkinson disease among patients with inflammatory bowel disease. *JAMA Neurology*, *75*(8), 939–946.

Quansah, E., Peelaerts, W., Langston, J. W., Simon, D. K., Colca, J., & Brundin, P. (2018). Targeting energy metabolism via the mitochondrial pyruvate carrier as a novel approach to attenuate neurodegeneration. *Molecular Neurodegeneration*, *13*(1), 28.

Racette, B. A., Gross, A., Vouri, S. M., Camacho-Soto, A., Willis, A. W., & Searles, N. S. (2018). Immunosuppressants and risk of Parkinson disease. *Annals of Clinical Translational Neurology*, *5*(7), 870–875.

Redenšek, S., Dolžan, V., & Kunej, T. (2018). From genomics to omics landscapes of Parkinson's disease: Revealing the molecular mechanisms. *OMICS*, *22*(1), 1–16.

Rees, K., Stowe, R., Patel, S., Ives, N., Breen, K., Clarke, C. E., et al. (2011). Non-steroidal anti-inflammatory drugs as disease-modifying agents for Parkinson's disease: Evidence from observational studies. *Cochrane Database of Systematic Reviews*, *11*CD008454.

Reeve, A., Simcox, E., & Turnbull, D. (2014). Ageing and Parkinson's disease: Why is advancing age the biggest risk factor? *Ageing Research Reviews*, *14*, 19–30.

Rektorova, I. (2019). Current treatment of behavioral and cognitive symptoms of Parkinson's disease. *Parkinsonism & Related Disorders*, *59*, 65–73.

Rocha, E. M., De Miranda, B. R., Castro, S., Drolet, R., Hatcher, N. G., Yao, L., et al. (2019). LRRK2 inhibition prevents endolysosomal deficits seen in human Parkinson's disease. *Neurobiology of Disease*, *134*, 104626.

Rodriguez, M., Rodriguez-Sabate, C., Morales, I., Sanchez, A., & Sabate, M. (2015). Parkinson's disease as a result of aging. *Aging Cell*, *14*(3), 293–308.

Roussakis, A. A., Piccini, P., & Politis, M. (2013). Clinical utility of DaTscan™ (123I-Ioflupane Injection) in the diagnosis of Parkinsonian Syndromes. *Degenerative Neurological and Neuromuscular Disease*, *3*, 33–39.

Rugbjerg, K., Friis, S., Ritz, B., Schernhammer, E. S., Korbo, L., & Olsen, J. H. (2009). Autoimmune disease and risk for Parkinson disease: A population-based case-control study. *Neurology*, *73*(18), 1462–1468.

Santiago, P., Scheinberg, A. R., & Levy, C. (2018). Cholestatic liver diseases: New targets, new therapies. *Therapeutic Advances in Gastroenterology*, *11*, 1756284818787400.

Sardi, S. P., Cedarbaum, J. M., & Brundin, P. (2018). Targeted therapies for Parkinson's disease: From genetics to the clinic. *Movement Disorders*, *33*(5), 684–696.

Schapira, A. H., Cooper, J. M., Dexter, D., Clark, J. B., Jenner, P., & Marsden, C. D. (1990). Mitochondrial complex I deficiency in Parkinson's disease. *Journal of Neurochemistry*, *54*(3), 823–827.

Schapira, A. H., Fox, S. H., Hauser, R. A., Jankovic, J., Jost, W. H., Kenney, C., et al. (2017). Assessment of safety and efficacy of safinamide as a levodopa adjunct in patients with Parkinson disease and motor fluctuations: A randomized clinical trial. *JAMA Neurology*, *74*(2), 216–224.

Scherbaum, R., Hartelt, E., Kinkel, M., Gold, R., Muhlack, S., & Tönges, L. (2020). Parkinson's disease multimodal complex treatment improves motor symptoms, depression and quality of life. *Journal of Neurology*, *267*(4), 954–965.

Schöndorf, D. C., Ivanyuk, D., Baden, P., Sanchez-Martinez, A., De Cicco, S., Yu, C., et al. (2018). The NAD+ precursor nicotinamide riboside rescues mitochondrial defects and neuronal loss in iPSC and fly models of Parkinson's disease. *Cell Reports*, *23*(10), 2976–2988.

Selvaraj, S., & Piramanayagam, S. (2019). Impact of gene mutation in the development of Parkinson's disease. *Genes & Diseases*, *6*(2), 120–128.

Shan, C., Gong, Y. L., Zhuang, Q. Q., Hou, Y. F., Wang, S. M., Zhu, Q., et al. (2019). Protective effects of β- nicotinamide adenine dinucleotide against motor deficits and dopaminergic neuronal damage in a mouse model of Parkinson's disease. *Progress in Neuro-Psychopharmacology & Biological Psychiatry*, *94*, 109670.

Shihabuddin, L. S., Brundin, P., Greenamyre, J. T., Stephenson, D., & Sardi, S. P. (2018). New frontiers in Parkinson's disease: From genetics to the clinic. *The Journal of Neuroscience*, *38*(44), 9375–9382.

Sian, J., Dexter, D. T., Lees, A. J., Daniel, S., Agid, Y., Javoy-Agid, F., et al. (1994). Alterations in glutathione levels in Parkinson's disease and other neurodegenerative disorders affecting basal ganglia. *Annals of Neurology*, *36*(3), 348–355.

Silveira, C. R. A., MacKinley, J., Coleman, K., Li, Z., Finger, E., Bartha, R., et al. (2019). Ambroxol as a novel disease-modifying treatment for Parkinson's disease dementia: Protocol for a single-centre, randomized, double-blind, placebo-controlled trial. *BMC Neurology*, *19*(1), 20.

Sindi, S., Mangialasche, F., & Kivipelto, M. (2015). Advances in the prevention of Alzheimer's disease. *F1000Prime Reports*, 7, 50.

Smith, A. M., Depp, C., Ryan, B. J., Johnston, G. I., Alegre-Abarrategui, J., Evetts, S., et al. (2018). Mitochondrial dysfunction and increased glycolysis in prodromal and early Parkinson's blood cells. *Movement Disorders*, *33*(10), 1580–1590.

Söderbom, G., & Zeng, B. Y. (2020). The NLRP3 inflammasome as a bridge between neuro-inflammation in metabolic and neurodegenerative diseases. *International Review of Neurobiology*, *154*, 345–392.

Spillantini, M. G., Crowther, R. A., Jakes, R., Hasegawa, M., & Goedert, M. (1998). alpha-Synuclein in filamentous inclusions of Lewy bodies from Parkinson's disease and dementia with lewy bodies. *Proceedings of the National Academy of Sciences of the United States of America*, *95*(11), 6469–6473.

Sundberg, M., & Isacson, O. (2014). Advances in stem-cell—Generated transplantation therapy for Parkinson's disease. *Expert Opinion on Biological Therapy*, *14*(4), 437–453.

Sung, Y. F., Liu, F. C., Lin, C. C., Lee, J. T., Yang, F. C., Chou, Y. C., et al. (2016). Reduced risk of Parkinson disease in patients with rheumatoid arthritis: A nationwide population-based study. *Mayo Clinic Proceedings*, *91*(10), 1346–1353.

Sutachan, J. J., Casas, Z., Albarracin, S. L., Stab, B. R., 2nd, Samudio, I., Gonzalez, J., et al. (2012). Cellular and molecular mechanisms of antioxidants in Parkinson's disease. *Nutritional Neuroscience*, *15*(3), 120–126.

Świątkiewicz, M., Zaremba, M., Joniec, I., Członkowski, A., & Kurkowska-Jastrzębska, I. (2013). Potential neuroprotective effect of ibuprofen, insights from the mice model of Parkinson's disease. *Pharmacological Reports*, *65*(5), 1227–1236.

Takamatsu, Y., Ho, G., Koike, W., Sugama, S., Takenouchi, T., Waragai, M., et al. (2017). Combined immunotherapy with "anti-insulin resistance" therapy as a novel therapeutic strategy against neurodegenerative diseases. *NPJ Parkinson's Disease*, *3*, 4.

Tapias, V., McCoy, J. L., & Greenamyre, J. T. (2019). Phenothiazine normalizes the NADH/NAD+ ratio, maintains mitochondrial integrity and protects the nigrostriatal dopamine system in a chronic rotenone model of Parkinson's disease. *Redox Biology*, *24*, 101164.

Teismann, P., Tieu, K., Choi, D. K., Wu, D. C., Naini, A., Hunot, S., et al. (2003). Cyclooxygenase-2 is instrumental in Parkinson's disease neurodegeneration. *Proceedings of the National Academy of Sciences of the United States of America*, *100*(9), 5473–5478.

Toulorge, D., Schapira, A. H., & Hajj, R. (2016). Molecular changes in the postmortem parkinsonian brain. *Journal of Neurochemistry*, *139*(Suppl. 1), 27–58.

Trošt, M., Perovnik, M., & Pirtošek, Z. (2019). Correlations of neuropsychological and metabolic brain changes in Parkinson's disease and other α-Synucleinopathies. *Frontiers in Neurology*, *10*, 1204.

van der Steen, J. T., Lennaerts, H., Hommel, D., Augustijn, B., Groot, M., Hasselaar, J., et al. (2019). Dementia and Parkinson's disease: Similar and divergent challenges in providing palliative care. *Frontiers in Neurology*, *10*, 54.

Veitch, D. P., Weiner, M. W., Aisen, P. S., Beckett, L. A., Cairns, N. J., Green, R. C., et al. (2019). Understanding disease progression and improving Alzheimer's disease clinical trials: Recent highlights from the Alzheimer's disease neuroimaging Initiative. *Alzheimer's & Dementia*, *15*(1), 106–152.

Ventorp, F., Bay-Richter, C., Nagendra, A. S., Janelidze, S., Matsson, V. S., Lipton, J., et al. (2017). Exendin-4 treatment improves LPS-induced depressive-like behavior without affecting pro-inflammatory cytokines. *Journal of Parkinson's Disease*, 7(2), 263–273.

Venuto, C. S., Potter, N. B., Dorsey, E. R., & Kieburtz, K. (2016). A review of disease progression models of Parkinson's disease and applications in clinical trials. *Movement Disorders*, *31*(7), 947–956.

Vijitruth, R., Liu, M., Choi, D. Y., Nguyen, X. V., Hunter, R. L., & Bing, G. (2006). Cyclooxygenase-2 mediates microglial activation and secondary dopaminergic cell death in the mouse MPTP model of Parkinson's disease. *Journal of Neuroinflammation*, *3*, 6.

Villumsen, M., Aznar, S., Pakkenberg, B., Jess, T., & Brudek, T. (2019). Inflammatory bowel disease increases the risk of Parkinson's disease: A Danish nationwide cohort study 1977-2014. *Gut*, *68*(1), 18–24.

Visanji, N. P., Brooks, P. L., Hazrati, L. N., & Lang, A. E. (2013). The prion hypothesis in Parkinson's disease: Braak to the future. *Acta Neuropathologica Communications*, *1*, 2.

Wahner, A. D., Bronstein, J. M., Bordelon, Y. M., & Ritz, B. (2007). Nonsteroidal anti-inflammatory drugs may protect against Parkinson disease. *Neurology*, *69*(19), 1836–1842.

Wakade, C., Chong, R., Bradley, E., Thomas, B., & Morgan, J. (2014). Upregulation of GPR109A in Parkinson's disease. *PLoS One*, *9*(10), e109818.

Wakade, C., Giri, B., Malik, A., Khodadadi, H., Morgan, J. C., Chong, R. K., et al. (2018). Niacin modulates macrophage polarization in Parkinson's disease. *Journal of Neuroimmunology*, *320*, 76–79.

Werner, H. M., Mills, G. B., & Ram, P. T. (2014). Cancer systems biology: A peek into the future of patient care? *Nature Reviews Clinical Oncology*, *11*(3), 167–176.

West, A. B. (2017). Achieving neuroprotection with LRRK2 kinase inhibitors in Parkinson disease. *Experimental Neurology*, *298*(Pt. B), 236–245.

Whone, A. L., Boca, M., Luz, M., Woolley, M., Mooney, L., Dharia, S., et al. (2019). Extended treatment with glial cell line-derived neurotrophic factor in Parkinson's disease. *Journal of Parkinson's Disease*, *9*(2), 301–313.

Wu, P. L., Lee, M., & Huang, T. T. (2017). Effectiveness of physical activity on patients with depression and Parkinson's disease: A systematic review. *PLoS One*, *12*(7), e0181515.

Wu, Y., Yao, Q., Jiang, G. X., Wang, G., & Cheng, Q. (2020). Identification of distinct blood-based biomarkers in early stage of Parkinson's disease. *Neurological Sciences*, *41*(4), 893–901.

Xu, X., Yin, D., Ren, H., Gao, W., Li, F., Sun, D., et al. (2018). Selective NLRP3 inflammasome inhibitor reduces neuroinflammation and improves long-term neurological outcomes in a murine model of traumatic brain injury. *Neurobiology of Disease*, *117*, 15–27.

Zahid, A., Li, B., Kombe, A. J. K., Jin, T., & Tao, J. (2019). Pharmacological Inhibitors of the NLRP3 Inflammasome. *Frontiers in Immunology*, *10*, 2538.

Zhang, Y., Lv, X., Hu, Z., Ye, X., Zheng, X., Ding, Y., et al. (2017). Protection of MCC950 against high-glucose-induced human retinal endothelial cell dysfunction. *Cell Death & Disease*, *8*(7), e2941.

Zhang, L., Zhang, L., Li, Y., Li, L., Melchiorsen, J. U., Rosenkilde, M., et al. (2020). The novel dual GLP-1/GIP receptor agonist DA-CH5 is superior to single GLP-1 receptor agonists in the MPTP model of Parkinson's disease. *Journal of Parkinson's Disease*, *10*(2), 523–542.

Zhao, H. T., John, N., Delic, V., Ikeda-Lee, K., Kim, A., Weihofen, A., et al. (2017). LRRK2 antisense oligonucleotides ameliorate α-synuclein inclusion formation in a Parkinson's disease mouse model. *Molecular Therapy—Nucleic Acids*, *8*, 508–519.

Zhao, X., Zhang, M., Li, C., Jiang, X., Su, Y., & Zhang, Y. (2019). Benefits of vitamins in the treatment of Parkinson's disease. *Oxidative Medicine and Cellular Longevity*, *2019*, 9426867.

Zhou, Y., Lu, M., Du, R. H., Qiao, C., Jiang, C. Y., Zhang, K. Z., et al. (2016). MicroRNA-7 targets Nod-like receptor protein 3 inflammasome to modulate neuroinflammation in the pathogenesis of Parkinson's disease. *Molecular Neurodegeneration*, *11*, 28.

Zhou, R., Yazdi, A. S., Menu, P., & Tschopp, J. (2011). A role for mitochondria in NLRP3 inflammasome activation. *Nature*, *469*(7329), 221–225.

SECTION 2

Metabolic approaches to neurodegenerative disease

CHAPTER SEVEN

Involvement of GABAergic interneuron dysfunction and neuronal network hyperexcitability in Alzheimer's disease: Amelioration by metabolic switching

Mark P. Mattson*
Department of Neuroscience, Johns Hopkins University School of Medicine, Baltimore, MD, United States
*Corresponding author: e-mail address: mmattso2@jhmi.edu

Contents

1. Neuronal network hyperexcitability in Alzheimer's disease 192
2. Their high metabolic demand renders GABAergic interneurons vulnerable to Alzheimer's disease 196
3. Intermittent metabolic challenges, ketones and NAD^+ enhance the plasticity and resilience of neuronal networks 197
4. Conclusions 201
References 201
Further reading 205

Abstract

Evidence accumulated over the past three decades suggests that the neurons that degenerate in Alzheimer's disease (AD) suffer metabolic compromise, hyperexcitability and consequent calcium ion (Ca^{2+}) overload-mediated dysfunction and death. Contributing to the neuronal network hyperexcitability in AD is the degeneration of inhibitory GABAergic interneurons, apparently as a result of their unusually high firing frequency and metabolic demand. Aging, genetic factors, and unhealthy lifestyles including lack of exercise and overconsumption of calorie-rich foods may compromise the ability of neurons to sustain mitochondrial function and to remove damaged molecules. Here I briefly review evidence supporting a role for early GABAergic neuron degeneration and consequent neuronal network hyperexcitability in AD. I then highlight data suggesting pivotal roles for ketones, NAD^+ and the mitochondrial protein deacetylase sirtuin 3 (SIRT3) in protecting against hyperexcitability in AD.

International Review of Neurobiology, Volume 154
ISSN 0074-7742
https://doi.org/10.1016/bs.irn.2020.01.006

Abbreviations

Aβ	amyloid beta-peptide
AD	Alzheimer's disease
APP	beta-amyloid precursor protein
BDNF	brain-derived neurotrophic factor
BHB	beta-hydroxybutyrate
CR	calretinin
EEG	electroencephalogram
FGF2	fibroblast growth factor 2
GABA	gamma amino-butyric acid
GLUT3	glucose transporter 3
MCI	mild cognitive impairment
NAD	nicotinamide adenine dinucleotide
NLRP3	NLR family pryrin domain containing 3
NR	nicotinamide riboside
PET	positron emission tomography
PS1	presenilin 1
PV	parvalbumin
SIRT3	sirtuin 3
SOD2	superoxide dismutase 2

1. Neuronal network hyperexcitability in Alzheimer's disease

The core neuronal circuitry throughout the brain consists of excitatory glutamatergic neurons and inhibitory GABAergic neurons (Fig. 1). Glutamatergic neurons are typically large and pyramidal shaped and have long axons that project within and between brain regions. GABAergic interneurons are much smaller and have axons that synapse on multiple glutamatergic neurons within their immediate vicinity. In the brain regions prominently affected in AD such as the hippocampus, entorhinal cortex and frontal cortex the large glutamatergic neurons often exhibit Tau neurofibrillary tangles as they degenerate. Because the large glutamatergic neurons are more conspicuous than the small GABAergic neurons, the fact that GABAergic neurons degenerate early in AD was not recognized for many decades.

People with AD are much more likely to experience epileptic seizures than are age-matched neurologically normal people (Vossel et al., 2017). Moreover, a retrospective study showed that the prevalence of epilepsy prior to cognitive decline was 17 times higher than age-matched controls (DiFrancesco et al., 2017). What are the underlying alterations that

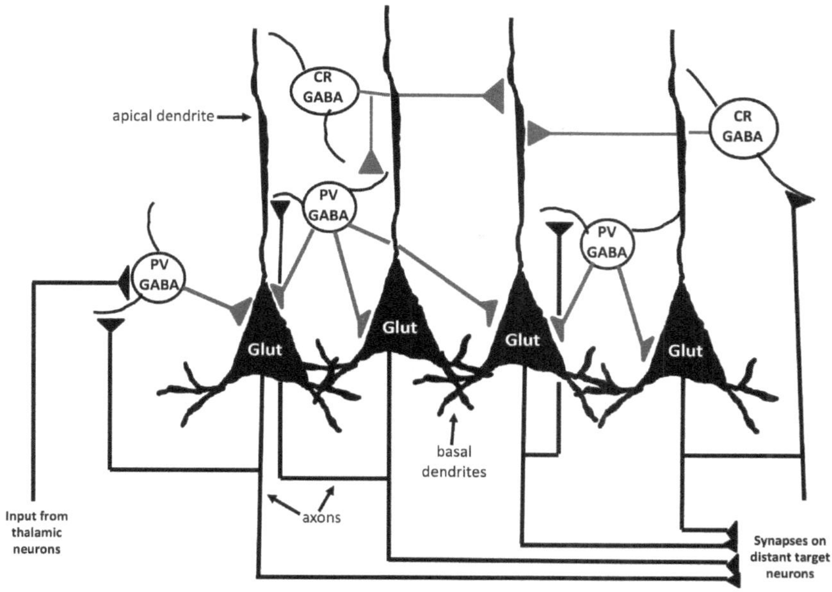

Fig. 1 Core neuronal circuitry of the brain consists of excitatory neurons that deploy the neurotransmitter glutamate (Glut) and inhibitory neurons that deploy GABA. Glutamatergic neurons are relatively large with axons that project long distances within and between brain regions. GABAergic interneurons synapse upon glutamatergic neurons within their immediate vicinity. GABAergic neurons that express the calcium-binding proteins parvalbumin (PV) or calretinin (CR) have a very high firing frequency and energy demand which may make them particularly vulnerable to aging and age-related neurodegenerative disorders, including Alzheimer's disease.

result in neuronal network hyperexcitability in AD? Age is the major risk factor for AD and several age-related alterations that occur in the brain during aging increase the vulnerability of neurons to excitotoxicity (Fig. 2). These 'hallmarks of brain aging' have been reviewed in detail elsewhere and include oxidative damage, mitochondrial dysfunction, impaired autophagy, glial cell activation and associated inflammation, impaired adaptive cellular stress responses and dysregulated neuronal Ca^{2+} homeostasis (Mattson and Arumugum, 2018). For example, exposure of neurons to oxidative stress or subtoxic doses of mitochondrial toxins increases their vulnerability to being damaged and killed by glutamate (Keller et al., 1997; Nicholls, 2008). Likewise, impaired neurotrophic signaling or DNA repair render neurons vulnerable to excitotoxicity (Mattson, 2008; Yang et al., 2010).

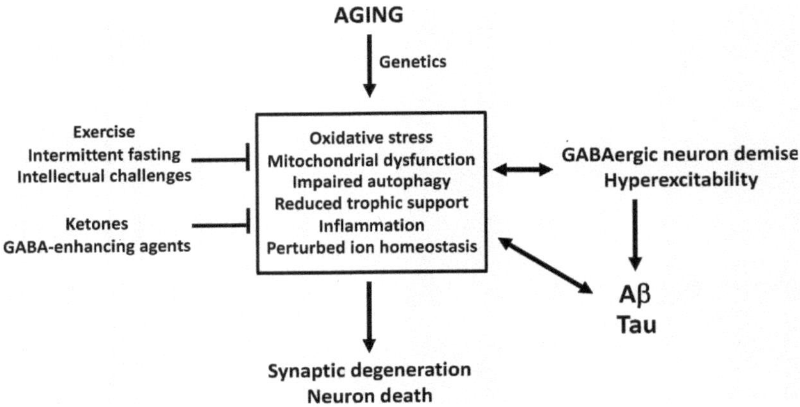

Fig. 2 Working model of cellular and molecular mechanisms upstream and downstream of neuronal network excitability in Alzheimer's disease pathogenesis. As the brain ages cells have propensity to accumulate oxidatively damaged molecules, and the abilities of their mitochondria, autophagic and neurotrophic systems to function well declines. Microglia and astrocytes contribute to local brain tissue inflammation. All of these age-related changes likely contribute to dysregulation of neuronal ion homeostasis. Because of their high metabolic demand, GABAergic interneurons are particularly vulnerable to aging and their demise results in loss of control of glutamatergic neuron activity and neuronal network hyperexcitabilty in AD. Age-related factors and hyperexcitablity promote Aβ and Tau pathologies and, conversely, Aβ and Tau pathologies exacerbate hyperexcitablity and neuronal degeneration. Increasing evidence suggests that the bioenergetic challenges of exercise and intermittent fasting, as well as intellectual challenges can counteract the age-related cellular changes that render neuronal networks vulnerable to hyperexcitability and AD. Supplements that elevate ketones (e.g., ketone ester, medium chain triglycerides) or drugs that activate GABA receptors, or otherwise prevent aberrant neuronal network activity, are potential treatments for the prevention and treatment of AD.

The question then becomes if and how age-related alterations act upstream and/or downstream of the disease-defining neuropathological 'lesions' of AD, namely amyloid β-peptide (Aβ) plaques and Tau tangles. Direct exposure of cultured primary human cortical neurons to aggregating Aβ greatly increases their vulnerability to glutamate-induced Ca^{2+} overload and excitotoxicity (Mattson et al., 1992). As it aggregates on the plasma membrane of neurons, Aβ causes oxidative stress and lipid peroxidation which, in turn, impairs the function of ion-motive ATPases (Na^{2+} and Ca^{2+} 'pumps'). As a result the neuronal membrane potential remains depolarized for an abnormally long time period after an action potential and intracellular Ca^{2+} levels become excessively elevated. In addition, via a similar mechanism involving membrane lipid peroxidation and the generation of the aldehyde

4-hydroxynonenal, aggregating Aβ impairs the function of the neuronal glucose transporter GLUT3 (Mark et al., 1997). The latter effect of Aβ likely contributes to the well-established impairment of neuronal glucose utilization in patients with mild cognitive impairment (MCI) and AD. However, positron emission tomography (PET) studies have shown that brain cells of AD patients are able to utilize ketones (Castellano et al., 2015) and there is, therefore, considerable interest in applying ketogenic diets and intermittent fasting to the prevention and treatment of AD (Fortier et al., 2019).

Mutations in the gene encoding the integral membrane protein presenilin 1 (PS1) are responsible for the majority of cases of dominantly inherited early-onset AD. Presenilin 1 cleaves the β-amyloid precursor protein (APP) to liberate Aβ, and PS1 mutations tend to increase the production of Aβ. When he was a postdoc in my laboratory, Qing Guo discovered that several PS1 mutations perturb neuronal Ca^{2+} handling and can render neurons vulnerable to seizures and excitotoxicity (Guo et al., 1999). This effect of PS1 mutations is apparently independent of Aβ and instead involves an abnormal accumulation of Ca^{2+} in the endoplasmic reticulum (Bezprozvanny and Mattson, 2008).

In my opinion, Tau pathology is downstream of metabolic, oxidative and inflammatory stress, impaired autophagy and proteasome function, and an excitatory imbalance in AD. This conclusion is bolstered by studies of animal models showing that these age-related cellular stressors are sufficient to induce or exacerbate Tau pathology. A very recent example comes from a study demonstrating that NLRP3 inflammasome activation can cause Tau pathology in a mouse model (Ising et al., 2019). Thirty years ago, I reported that overactivation of glutamate receptors elicits tangle-like changes in Tau in cultured hippocampal neurons by a mechanism involving cellular Ca^{2+} overload (Mattson, 1990). My colleague Robert Sapolsky and I then showed that experimentally induced seizures can cause similar Tau pathology in hippocampal neurons of adult rats (Elliott et al., 1993). We then found that several anticonvulsant drugs used to treat epilepsy can prevent Aβ-induced Ca^{2+} dysregulation and Tau pathology in cultured neurons (Mark et al., 1995). Subsequent findings from several different laboratories have demonstrated the ability of a range of drugs that reduce neuronal excitability (e.g., diazoxide and valproic acid) to ameliorate cognitive deficits and Aβ and Tau pathologies in mouse AD models (Liu et al., 2010; Qing et al., 2008). However, anticonvulsants can themselves impair cognition because they not only suppress neuronal hyperexcitability, but also reduce the Ca^{2+} signal that mediates the synaptic plasticity that underlies learning and memory.

2. Their high metabolic demand renders GABAergic interneurons vulnerable to Alzheimer's disease

Neurofibrillary tangles are most prominent in large glutamatergic neurons of the cerebral cortex and hippocampus. These neurons were, therefore, thought to the most vulnerable and degenerate early in MCI and AD. However, recent findings point to a less obvious type of neuron—small GABAergic interneurons—as being among the first to degenerate in AD. Why might this be so? Clues come from studies of the biochemistry and electrophysiology of interneurons. Different types of GABAergic interneurons are distinguished by the expression of high levels of Ca^{2+}-binding proteins; these include those with parvalbumin (PV), calbindin and calretinin (CR). My laboratory and others have shown that such Ca^{2+}-binding proteins can buffer Ca^{2+} loads sufficiently to protect neurons against excitotoxicity (Figueredo-Cardenas et al., 1998; Mattson et al., 1991). Parvalbumin GABAergic interneurons have a remarkably high firing frequency (Kawaguchi and Kubota, 1993). They therefore have a correspondingly high metabolic rate because the Na^+ and Ca^{2+} pumps require large amounts of ATP to enable them to restore the ion gradients following each action potential. Consistent with their high energy expenditure, PV interneurons possess relatively large numbers of mitochondria (Kann, 2016). Parvalbumin neurons would, therefore, be expected to suffer relatively more stress than other neurons.

Indeed, studies of post-mortem brain tissue of AD patients and AD mouse models are consistent with the notion that PV neurons become dysfunctional and degenerate early in the disease process (Cheng et al., 2020; Villette and Dutar, 2017). Here I highlight a few examples of this evidence. The entorhinal and perirhinal cortices are brain regions affected early in the disease process in AD. These brain regions 'funnel' information from visual and auditory cortices into the hippocampus and play critical roles in spatial pattern recognition, including spoken and written language. A recent study showed that neurons expressing PV or somatostatin are highly vulnerable to degeneration in AD (Sanchez-Mejias et al., 2019). Using an APP mutant 'knockin' mouse model, Petrache et al. (2019) found that there was an early hyperexcitability of neurons in the lateral entorhinal cortex. They further found that treatment with GABAA receptor agonists normalized neuronal excitability in the lateral entorhinal cortex. A study of old 3xTgAD mice

(mice expressing mutant human APP, PS1 and Tau) revealed highly significant selective loss of PV and CR interneurons in region CA1 of the hippocampus (Zallo et al., 2018). By recording electrical activity in individual PV interneurons and pyramidal neurons in hippocampal slices from APP/PS1 mutant mice and wild type mice, Hijazi et al. (2019) showed that PV neurons exhibit hyperexcitability associated with Aβ accumulation and cognitive impairment. They expressed a drug-activated inhibitory receptor in PV neurons and showed that this normalized hippocampal neuronal network activity and ameliorated cognitive deficits.

Fast-spiking GABAergic neurons play a key role in establishment and maintenance of electroencephalogram (EEG) gamma oscillations which are involved in memory formation in cortical hippocampal neuronal networks (Kann, 2016). Reductions of gamma oscillations have been documented in AD patients (Gleichmann et al., 2011). In studies of a mouse model of AD, Iaccarino et al. (2016) found that behaviorally driven gamma oscillations were reduced before the onset of Aβ plaque accumulation and cognitive impairment. Using optogenetic technology they selectively stimulated PV interneurons at gamma frequency and found that this reduced Aβ levels. They then subjected the AD mice to a gamma frequency light-flickering treatment, which also reduced Aβ accumulation. These findings suggest the possibility that restoring gamma oscillations might have a therapeutic benefit in AD patients.

3. Intermittent metabolic challenges, ketones and NAD^+ enhance the plasticity and resilience of neuronal networks

We developed a working model for optimizing brain health and resistance to neurodegenerative disorders which requires switching back-and-forth between a metabolic challenge state and a recovery state (de Cabo and Mattson, 2019; Mattson et al., 2018). The model is based upon accumulated knowledge of how brain cells respond to intermittent exercise and intermittent fasting. During the challenge period (e.g., running for 1 h or fasting for 16 h), neurons switch from using primarily glucose to ketones as their main energy source. They downregulate overall protein synthesis and upregulate autophagy, DNA repair and antioxidant pathways. Then, during the recovery period (resting, eating, sleeping), the neurons upregulate protein synthesis enabling their axons and dendrites to grow and new synapses to form. Over extended periods of time (weeks and months) neuronal networks optimize their functionality (e.g., improved cognition and mood) and become resistant to the kinds of stress that occur in brain aging and neurodegenerative disorders.

Based upon considerable research performed in animal studies and human clinical trials, intermittent fasting (IF) has become a popular approach for people to lose weight and reduce risk factors for cardiovascular disease, diabetes and cancers (Anton et al., 2018; de Cabo and Mattson, 2019; Hutchison et al., 2019; Longo and Mattson, 2014). During the past three decades, studies in my laboratory have addressed the questions of if and how IF affects neuroplasticity, brain aging, and neuronal vulnerability to injury and neurodegenerative disorders (see Mattson, 2012; Mattson et al., 2018 for reviews). In our first study of the effects of IF on the brain, we showed that rats maintained on an alternate day fasting regimen were more resistant to seizures induced by the glutamate receptor agonist kainic acid (Bruce-Keller et al., 1999). Since then we have shown that IF can bolster neuronal stress resistance and neuroplasticity (synapse formation and maintenance), and neurogenesis (Cheng et al., 2016; Lee et al., 2002). Intermittent fasting protects neurons in rats and mice against degeneration caused by mitochondrial toxins and improves functional outcomes in rat and mouse models of Parkinson's and Huntington's diseases (Duan et al., 2003; Duan and Mattson, 1999) and stroke (Arumugam et al., 2010; Yu and Mattson, 1999). Interestingly, inhibition of glucose metabolism by administration of 2-deoxyglucose was also neuroprotective in animal models of epilepsy, Parkinson's disease and stroke (Duan and Mattson, 1999; Lee et al., 1999; Yu and Mattson, 1999). These early studies of the mechanism underlying the neuroprotective effects of IF and 2-deoxyglucose suggested roles for upregulation of the expression of protein chaperones (HSP70 and GRP78), brain-derived neurotrophic factor (BDNF) and fibroblast growth factor 2 (FGF2).

In our first study of IF in an AD animal model, we found that daily fasting (time-restricted feeding) and alternate day fasting ameliorated cognitive deficits in 3xTgAD transgenic mice (Halagappa et al., 2007). More recently, we found that IF prevents seizures and improves cognitive function in an APP mutant knockin mouse model of AD (Liu et al., 2019). In the latter study we also performed electrophysiological recordings from hippocampal pyramidal neurons and found that IF restores hippocampal synaptic plasticity in a SIRT3-dependent manner. These findings provide evidence that a mitochondrial protein deacetylase is critical for synaptic plasticity and cognition. The protein or proteins that mediate these effects of SIRT3 are unknown. However, we did provide evidence that mitochondrial superoxide dismutase 2 (SOD2) is deacetylated by SIRT3 (Cheng et al., 2016) and that GABAergic synaptic activity is reduced in neurons from SOD2+/− mice (Liu et al., 2019). Deacetylation of SOD2 increases its enzymatic activity, but it is unclear exactly how a reduction of superoxide levels might enhance GABAergic synaptic activity.

Exercise is a metabolic challenge that is beneficial for the brain. Regular exercise enhances cognition and improves mood (see Dauwan et al., 2019; Mattson, 2012; for review). Adlard et al. (2005) showed that voluntary wheel running reduces brain Aβ load and improves spatial learning and memory in a mouse model of AD. Subsequent studies have confirmed beneficial effects of exercise in mouse AD models and have further provided evidence that exercise reduces neuroinflammation and stimulates neurogenesis (Ryan and Kelly, 2016). Moreover, running wheel exercise protects cerebral cortical and hippocampal neurons against excitotoxicity by a mechanism involving suppression of glutamate induced Ca^{2+} overload (Cheng et al., 2016). Reduced PV expression and neuronal network hyperexcitability in APP mutant mice are reversed when the mice are maintained an enriched environment that include voluntary running wheel exercise (Cattaud et al., 2018). The ability of exercise to constrain neuronal network excitability within normal limits requires the mitochondrial deacetylase SIRT3 suggesting a critical role for mitochondria (Cheng et al., 2016) and possibly a bolstering of GABAergic neurons (Cheng et al., 2020).

The fasted state is defined by liver glycogen stores being depleted, fatty acids being released from adipocytes and converted to the ketones β-hydroxybutyrate (BHB) and acetoacetate in the liver (Mattson et al., 2018). It has been clearly established that ketogenic diets can suppress epileptic seizures in humans (Poff et al., 2019). Dietary supplementation with the energy substrates BHB and pyruvate can ameliorate hyperactivity of hippocampal neurons in an APP/PS1 mutant mouse AD model (Zilberter et al., 2013). We found that dietary supplementation with a BHB ketone ester protected against seizures and associated death in an APP/PS1 double mutant mouse AD model (Cheng et al., 2020). We crossed APP/PS1 double mutant AD mice with SIRT3 knockout mice and found that AD mice with SIRT3 haploinsufficiency (heterozygotes) developed epileptiform EEG activity within their first 2 months of age and die of seizures within 5 months

(Cheng et al., 2020). The cerebral neuronal network hyperactivity in AD mice with reduced SIRT3 levels was associated with the loss of PV and CR GABAergic interneurons. Dietary supplementation with a ketone ester stimulated SIRT3 expression and prevented the cerebral hyperexcitability and early death in the AD mice (Cheng et al., 2020). Dietary supplementation with the BHB ester ameliorated cognitive deficits and lessened Aβ and Tau pathologies in the 3xTgAD mouse model of AD (Kashiwaya et al., 2013). The ketone ester protected PV and CR GABAergic interneurons against degeneration in the APP/PS1 double mutant mouse AD model demonstrating the ability of BHB to protect these metabolically challenged interneurons (Cheng et al., 2020).

As elaborated upon in reviews based on work from the laboratories of Dimitrios Kapogiannis (Kapogiannis and Avgerinos, 2020) and Kieran Clarke (Norwitz et al., 2020), there is considerable evidence accumulating that supports a potential therapeutic benefit of ketone-elevating treatments in AD and other neurodegenerative disorders. This notion is supported not only by studies of animal models, but also by recent human studies. For example, Steve Cunnane used an acetoacetate radiotracer and PET imaging to show that when healthy adult humans consume a ketogenic diet there is a six-fold increase in the utilization of the ketone by brain cells while their glucose utilization decreases by 20% (Courchesne-Loyer et al., 2017). The latter study also demonstrated a direct correlation between plasma ketones and brain ketone utilization. Using the same PET methods, the authors showed that whereas the ability of brain cells to use glucose is compromised in patients with AD, their ability to use ketones is not compromised (Castellano et al., 2015). In a more recent study, MCI patients that were treated with medium chain triglycerides (ketone precursors) exhibited improved executive function, episodic memory and processing speed during a six-month period (Fortier et al., 2019). Perhaps even more exciting is the therapeutic potential for the BHB ester which elevates circulating BHB to levels that occur with extended fasting periods even while the person is consuming regular meals (Solo-Mota et al., 2019). The BHB ester improves running endurance and learning and memory in rats (Murray et al., 2016) and can boost endurance in well-trained humans (Cox et al., 2016).

Finally, another approach that would be expected to protect GABAergic interneurons and prevent or reverse neuronal hyperexcitability in AD is to boost cellular NAD^+ levels. Treatment with nicotinamide for 8 months improved spatial learning and memory and lessened Aβ and Tau pathologies in 3xTgAD mice (Liu et al., 2013). These beneficial effects of nicotinamide were associated with improved mitochondrial function and enhanced autophagy.

Treatment of APP mutant AD mice with nicotinamide riboside (NR) for 3 months increased brain NAD^+ levels, ameliorated cognitive deficits and restored hippocampal synaptic plasticity (Gong et al., 2013). More recently we found that NR treatment protects hippocampal neurons against dysfunction and degeneration in a novel DNA repair-deficient mouse model of AD (Hou et al., 2018). Similarly, Xie et al. (2019) found that NR treatment improved cognition and reduced microglial activation in APP/PS1 mutant AD mice. Given its safety and potential to boost neuronal NAD^+ levels, NR should be tested in clinical trials in AD patients.

4. Conclusions

Abundant evidence from human and animal studies demonstrates that hyperexcitability of cortical neuronal networks occurs early in AD. Animal and cell culture studies have shown that hyperexcitability can increase Aβ production and cause Tau pathology, and that aggregating Aβ can render neurons vulnerable to hyperexcitability. Because of their very high firing frequency and consequent high energy demand, PV and CR GABAergic interneurons may be unusually vulnerable to metabolic, ionic and oxidative stress during aging. These GABAergic neurons degenerate early in AD and animal models of AD which would be expected to escalate neuronal network hyperexcitability and likely explains, at least in part, the increased seizure incidence in AD patients. The physiological bioenergetic challenges of intermittent fasting and exercise can enhance GABAergic tone and can protect against hyperexcitability and neuronal degeneration in animal models of AD. Boosting ketone levels with a ketone ester and boosting neuronal NAD^+ levels with NR are promising interventions for protecting GABAergic neurons and so preventing neuronal network hyperexcitability in AD.

References

Adlard, P. A., Perreau, V. M., Pop, V., & Cotman, C. W. (2005). Voluntary exercise decreases amyloid load in a transgenic model of Alzheimer's disease. *The Journal of Neuroscience*, *25*, 4217–4221.

Anton, S. D., Moehl, K., Donahoo, W. T., Marosi, K., Lee, S. A., Mainous, A. G., 3rd, et al. (2018). Flipping the metabolic switch: Understanding and applying the health benefits of fasting. *Obesity (Silver Spring), 26*, 254–268.

Arumugam, T. V., Phillips, T. M., Cheng, A., Morrell, C. H., Mattson, M. P., & Wan, R. (2010). Age and energy intake interact to modify cell stress pathways and stroke outcome. *Annals of Neurology, 67*(1), 41–52.

Bezprozvanny, I., & Mattson, M. P. (2008). Neuronal calcium mishandling and the pathogenesis of Alzheimer's disease. *Trends in Neurosciences, 31*, 454–463.

Bruce-Keller, A. J., Umberger, G., McFall, R., & Mattson, M. P. (1999). Food restriction reduces brain damage and improves behavioral outcome following excitotoxic and metabolic insults. *Annals of Neurology, 45*, 8–15.

Castellano, C. A., Nugent, S., Paquet, N., Tremblay, S., Bocti, C., Lacombe, G., et al. (2015). Lower brain 18F-fluorodeoxyglucose uptake but normal 11C-acetoacetate metabolism in mild Alzheimer's disease dementia. *Journal of Alzheimer's Disease, 43*, 1343–1353.

Cattaud, V., Bezzina, C., Rey, C. C., Lejards, C., Dahan, L., & Verret, L. (2018). Early disruption of parvalbumin expression and perineuronal nets in the hippocampus of the Tg2576 mouse model of Alzheimer's disease can be rescued by enriched environment. *Neurobiology of Aging, 72*, 147–158.

Cheng, A., Wang, J., Ghena, N., Zhao, Q., Perone, I., King, M. T., et al. (2020). SIRT3 Haploinsufficiency aggravates loss of GABAergic Interneurons and neuronal network hyperexcitability in an Alzheimer's disease model. *Journal of Neuroscience, 40*, 694–709.

Cheng, A., Yang, Y., Zhou, Y., Maharana, C., Lu, D., Peng, W., et al. (2016). Mitochondrial SIRT3 mediates adaptive responses of neurons to exercise and metabolic and excitatory challenges. *Cell Metabolism, 23*, 128–142.

Courchesne-Loyer, A., Croteau, E., Castellano, C. A., St-Pierre, V., Hennebelle, M., & Cunnane, S. C. (2017). Inverse relationship between brain glucose and ketone metabolism in adults during short-term moderate dietary ketosis: A dual tracer quantitative positron emission tomography study. *Journal of Cerebral Blood Flow and Metabolism, 37*, 2485–2493.

Cox, P. J., Kirk, T., Ashmore, T., Willerton, K., Evans, R., Smith, A., et al. (2016). Nutritional ketosis alters fuel preference and thereby endurance performance in athletes. *Cell Metabolism, 24*, 256–268.

Dauwan, M., Begemann, M. J. H., Slot, M. I. E., Lee, E. H. M., Scheltens, P., & Sommer, I. E. C. (2019). Physical exercise improves quality of life, depressive symptoms, and cognition across chronic brain disorders: A transdiagnostic systematic review and meta-analysis of randomized controlled trials. *Journal of Neurology*. https://doi.org/10.1007/s00415-019-09493-9, 2019 Aug 14. (Epub ahead of print).

de Cabo, R., & Mattson, M. P. (2019). Effects of intermittent fasting on health, aging, and disease. *The New England Journal of Medicine, 381*, 2541–2551.

DiFrancesco, J. C., Tremolizzo, L., Polonia, V., Giussani, G., Bianchi, E., & Franchi, C. (2017). Adult-onset epilepsy in presymptomatic Alzheimer's disease: A retrospective study. *Journal of Alzheimer's Disease, 60*, 1267–1274.

Duan, W., Guo, Z., Jiang, H., Ware, M., Li, X. J., & Mattson, M. P. (2003). Dietary restriction normalizes glucose metabolism and BDNF levels, slows disease progression, and increases survival in huntingtin mutant mice. *Proceedings of the National Academy of Sciences of the United States of America, 100*, 2911–2916.

Duan, W., & Mattson, M. P. (1999). Dietary restriction and 2-deoxyglucose administration improve behavioral outcome and reduce degeneration of dopaminergic neurons in models of Parkinson's disease. *Journal of Neuroscience Research, 57*, 195–206.

Elliott, E. M., Mattson, M. P., Vanderklish, P., Lynch, G., Chang, I., & Sapolsky, R. M. (1993). Corticosterone exacerbates kainate-induced alterations in hippocampal tau immunoreactivity and spectrin proteolysis in vivo. *Journal of Neurochemistry*, *61*, 57–67.

Figueredo-Cardenas, G., Harris, C. L., Anderson, K. D., & Reiner, A. (1998). Relative resistance of striatal neurons containing calbindin or parvalbumin to quinolinic acid-mediated excitotoxicity compared to other striatal neuron types. *Experimental Neurology*, *149*, 356–372.

Fortier, M., Castellano, C. A., Croteau, E., Langlois, F., Bocti, C., St-Pierre, V., et al. (2019). A ketogenic drink improves brain energy and some measures of cognition in mild cognitive impairment. *Alzheimer's & Dementia*, *15*, 625–634.

Gleichmann, M., Chow, V. W., & Mattson, M. P. (2011). Homeostatic disinhibition in the aging brain and Alzheimer's disease. *Journal of Alzheimer's Disease*, *24*, 15–24.

Gong, B., Pan, Y., Vempati, P., Zhao, W., Knable, L., Ho, L., et al. (2013). Nicotinamide riboside restores cognition through an upregulation of proliferator-activated receptor-γ coactivator 1α regulated β-secretase 1 degradation and mitochondrial gene expression in Alzheimer's mouse models. *Neurobiology of Aging*, *34*, 1581–1588.

Guo, Q., Fu, W., Sopher, B. L., Miller, M. W., Ware, C. B., Martin, G. M., et al. (1999). Increased vulnerability of hippocampal neurons to excitotoxic necrosis in presenilin-1 mutant knock-in mice. *Nature Medicine*, *5*, 101–106.

Halagappa, V. K., Guo, Z., Pearson, M., Matsuoka, Y., Cutler, R. G., Laferla, F. M., et al. (2007). Intermittent fasting and caloric restriction ameliorate age-related behavioral deficits in the triple-transgenic mouse model of Alzheimer's disease. *Neurobiology of Disease*, *26*, 212–220.

Hijazi, S., et al. (2019). Early restoration of parvalbumin interneuron activity prevents memory loss and network hyperexcitability in a mouse model of Alzheimer's disease. *Molecular Psychiatry*. https://doi.org/10.1038/s41380-019-0483-4, 2019 Aug 20. (Epub ahead of print).

Hou, Y., Lautrup, S., Cordonnier, S., Wang, Y., Croteau, D. L., Zavala, E., et al. (2018). NAD_+ supplementation normalizes key Alzheimer's features and DNA damage responses in a new AD mouse model with introduced DNA repair deficiency. *Proceedings of the National Academy of Sciences of the United States of America*, *115*, E1876–E1885.

Hutchison, A. T., Regmi, P., Manoogian, E. N. C., Fleischer, J. G., Wittert, G. A., Panda, S., et al. (2019). Time-restricted feeding improves glucose tolerance in men at risk for type 2 diabetes: A randomized crossover trial. *Obesity (Silver Spring)*, *27*, 724–732.

Iaccarino, H. F., et al. (2016). Gamma frequency entrainment attenuates amyloid load and modifies microglia. *Nature*, *540*, 230–235.

Ising, C., et al. (2019). NLRP3 inflammasome activation drives tau pathology. *Nature*, *575*, 669–673.

Kann, O. (2016). The interneuron energy hypothesis: Implications for brain disease. *Neurobiology of Disease*, *90*, 75–85.

Kapogiannis, D., & Avgerinos, K. I. (2020). Brain glucose and ketone utilization in brain ageing and neurodegenerative disorders. In *Metabolic and bioenergetic drivers of neurodegenerative disease: neurodegenerative disease research and commonalities with metabolic diseases. Section 1: Overview of neurodegenerative research and commonalities with metabolic diseases* (Vol. 154, Chapter 3).

Kashiwaya, Y., Bergman, C., Lee, J. H., Wan, R., King, M. T., Mughal, M. R., et al. (2013). A ketone ester diet exhibits anxiolytic and cognition-sparing properties, and lessens amyloid and tau pathologies in a mouse model of Alzheimer's disease. *Neurobiology of Aging*, *34*(6), 1530–1539.

Kawaguchi, Y., & Kubota, Y. (1993). Correlation of physiological subgroupings of non-pyramidal cells with parvalbumin- and calbindinD28k-immunoreactive neurons in layer V of rat frontal cortex. *Journal of Neurophysiology*, *70*, 387–396.

Keller, J. N., Mark, R. J., Bruce, A. J., Blanc, E., Rothstein, J. D., Uchida, K., et al. (1997). 4-Hydroxynonenal, an aldehydic product of membrane lipid peroxidation, impairs glutamate transport and mitochondrial function in synaptosomes. *Neuroscience*, *80*, 685–696.

Lee, J., Bruce-Keller, A. J., Kruman, Y., Chan, S. L., & Mattson, M. P. (1999). 2-Deoxy-D-glucose protects hippocampal neurons against excitotoxic and oxidative injury: Evidence for the involvement of stress proteins. *Journal of Neuroscience Research*, *57*, 48–61.

Lee, J., Seroogy, K. B., & Mattson, M. P. (2002). Dietary restriction enhances neurotrophin expression and neurogenesis in the hippocampus of adult mice. *Journal of Neurochemistry*, *80*, 539–547.

Liu, D., Pitta, M., Jiang, H., Lee, J. H., Zhang, G., Chen, X., et al. (2013). Nicotinamide forestalls pathology and cognitive decline in Alzheimer mice: Evidence for improved neuronal bioenergetics and autophagy procession. *Neurobiology of Aging*, *34*, 1564–1580.

Liu, D., Pitta, M., Lee, J. H., Ray, B., Lahiri, D. K., Furukawa, K., et al. (2010). The KATP channel activator diazoxide ameliorates amyloid-β and tau pathologies and improves memory in the 3xTgAD mouse model of Alzheimer's disease. *Journal of Alzheimer's Disease*, *22*, 443–457.

Liu, Y., et al. (2019). SIRT3 mediates hippocampal synaptic adaptations to intermittent fasting and ameliorates deficits in APP mutant mice. *Nature Communications*, *10*(1), 1886.

Longo, V. D., & Mattson, M. P. (2014). Fasting: Molecular mechanisms and clinical applications. *Cell Metabolism*, *19*, 181–192.

Mark, R. J., Ashford, J. W., Goodman, Y., & Mattson, M. P. (1995). Anticonvulsants attenuate amyloid beta-peptide neurotoxicity, Ca2+ deregulation, and cytoskeletal pathology. *Neurobiology of Aging*, *16*, 187–198.

Mark, R. J., Pang, Z., Geddes, J. W., Uchida, K., & Mattson, M. P. (1997). Amyloid beta-peptide impairs glucose transport in hippocampal and cortical neurons: Involvement of membrane lipid peroxidation. *The Journal of Neuroscience*, *17*, 1046–1054.

Mattson, M. P. (1990). Antigenic changes similar to those seen in neurofibrillary tangles are elicited by glutamate and Ca2+ influx in cultured hippocampal neurons. *Neuron*, *4*, 105–117.

Mattson, M. P. (2008). Glutamate and neurotrophic factors in neuronal plasticity and disease. *Annals of the New York Academy of Sciences*, *1144*, 97–112.

Mattson, M. P. (2012). Energy intake and exercise as determinants of brain health and vulnerability to injury and disease. *Cell Metabolism*, *16*, 706–722.

Mattson, M. P., & Arumugam, T. (2018). Hallmarks of brain aging: Adaptive and pathological modification by metabolic states. *Cell Metabolism*, *27*, 1176–1199.

Mattson, M. P., Cheng, B., Davis, D., Bryant, K., Lieberburg, I., & Rydel, R. E. (1992). beta-amyloid peptides destabilize calcium homeostasis and render human cortical neurons vulnerable to excitotoxicity. *The Journal of Neuroscience*, *12*, 376–388.

Mattson, M. P., Rychlik, B., Chu, C., & Christakos, S. (1991). Evidence for calcium-reducing and excito-protective roles for the calcium-binding protein calbindin-D28k in cultured hippocampal neurons. *Neuron*, *6*, 41–51.

Murray, A. J., Knight, N. S., Cole, M. A., Cochlin, L. E., Carter, E., & Tchabanenko, K. (2016). Novel ketone diet enhances physical and cognitive performance. *The FASEB Journal*, *30*, 4021–4032.

Nicholls, D. G. (2008). Oxidative stress and energy crises in neuronal dysfunction. *Annals of the New York Academy of Sciences*, *1147*, 53–60.

Norwitz, N. G., Jaramillo, J. G., Clarke, K., & Soto, A. (2020). Ketotherapeutics for neurodegenerative disorders. In *Metabolic and bioenergetic drivers of neurodegenerative disease: Treating neurodegenerative disorders as metabolic diseases* (Vol. 155, Chapter 6).

Petrache, A. L., Rajulawalla, A., Shi, A., Wetzel, A., Saito, T., Saido, T. C., et al. (2019). Aberrant excitatory-inhibitory synaptic mechanisms in entorhinal cortex microcircuits during the pathogenesis of Alzheimer's disease. *Cerebral Cortex, 29*, 1834–1850.

Poff, A. M., Rho, J. M., & D'Agostino, D. P. (2019). Ketone Administration for Seizure Disorders: History and rationale for ketone esters and metabolic alternatives. *Frontiers in Neuroscience, 13*, 1041.

Qing, H., He, G., Ly, P. T., Fox, C. J., Staufenbiel, M., Cai, F., et al. (2008). Valproic acid inhibits Abeta production, neuritic plaque formation, and behavioral deficits in Alzheimer's disease mouse models. *The Journal of Experimental Medicine, 205*, 2781–2789.

Ryan, S. M., & Kelly, Á. M. (2016). Exercise as a pro-cognitive, pro-neurogenic and anti-inflammatory intervention in transgenic mouse models of Alzheimer's disease. *Ageing Research Reviews, 27*, 77–92.

Sanchez-Mejias, E., et al. (2019). Distinct disease-sensitive GABAergic neurons in the perirhinal cortex of Alzheimer's mice and patients. *Brain Pathology, 109*. https://doi.org/10.1111/bpa.12785, 2019 Sep 6. (Epub ahead of print).

Soto-Mota, A., Vansant, H., Evans, R. D., & Clarke, K. (2019). Safety and tolerability of sustained exogenous ketosis using ketone monoester drinks for 28 days in healthy adults. *Regulatory Toxicology and Pharmacology, 109*, 104506.

Villette, V., & Dutar, P. (2017). GABAergic microcircuits in Alzheimer's disease models. *Current Alzheimer Research, 14*, 30–39.

Vossel, K. A., Tartaglia, M. C., Nygaard, H. B., Zeman, A. Z., & Miller, B. L. (2017). Epileptic activity in Alzheimer's disease: Causes and clinical relevance. *Lancet Neurology, 16*, 311–322.

Xie, X., Gao, Y., Zeng, M., Wang, Y., Wei, T. F., Lu, Y. B., et al. (2019). Nicotinamide ribose ameliorates cognitive impairment of aged and Alzheimer's disease model mice. *Metabolic Brain Disease, 34*, 353–366.

Yang, J. L., Tadokoro, T., Keijzers, G., Mattson, M. P., & Bohr, V. A. (2010). Neurons efficiently repair glutamate-induced oxidative DNA damage by a process involving CREB-mediated up-regulation of apurinic endonuclease 1. *The Journal of Biological Chemistry, 285*, 28191–28199.

Yu, Z. F., & Mattson, M. P. (1999). Dietary restriction and 2-deoxyglucose administration reduce focal ischemic brain damage and improve behavioral outcome: Evidence for a preconditioning mechanism. *Journal of Neuroscience Research, 57*, 830–839.

Zallo, F., Gardenal, E., Verkhratsky, A., & Rodríguez, J. J. (2018). Loss of calretinin and parvalbumin positive interneurones in the hippocampal CA1 of aged Alzheimer's disease mice. *Neuroscience Letters, 681*, 19–25.

Zilberter, M., et al. (2013). Dietary energy substrates reverse early neuronal hyperactivity in a mouse model of Alzheimer's disease. *Journal of Neurochemistry, 125*, 157–171.

Further reading

Cheng, A., Wan, R., Yang, J. L., Kamimura, N., Son, T. G., Ouyang, X., et al. (2012). Involvement of PGC-1α in the formation and maintenance of neuronal dendritic spines. *Nature Communications, 3*, 1250.

Nencioni, A., Caffa, I., Cortellino, S., & Longo, V. D. (2018). Fasting and cancer: Molecular mechanisms and clinical application. *Nature Reviews Cancer, 18*, 707–719.

CHAPTER EIGHT

The mitochondrial hypothesis: Dysfunction, bioenergetic defects, and the metabolic link to Alzheimer's disease

Russell H. Swerdlow*
University of Kansas Alzheimer's Disease Center, University of Kansas Medical Center, Kansas City, KS, United States
*Corresponding author: e-mail address: rswerdlow@kumc.edu

Contents

1. Introduction 209
2. Normal mitochondrial and metabolism function 210
   2.1 General mitochondrial structure and organization 210
   2.2 Mitochondrial functional aspects 212
   2.3 Other relevant metabolism features 214
3. Alzheimer's disease-relevant alterations in mitochondrial and metabolism function 215
   3.1 General mitochondrial structure and organization 215
   3.2 Mitochondrial functional aspects 216
   3.3 Other relevant metabolism features 217
4. Metabolic endophenotypes 218
5. Relationship to proteostasis 219
6. Genetic contributions 220
   6.1 Nuclear genes 221
   6.2 Mitochondrial genes 222
7. A mitochondrial cascade hypothesis 224
8. Conclusions and implications 226
Acknowledgments 228
Conflict of interests 228
References 228

Abstract

Alzheimer's disease (AD) features mitochondrial dysfunction and altered metabolism. Other pathologies could drive these changes, or alternatively these changes could drive other pathologies. In considering this question, it is worth noting that perturbed AD patient mitochondrial and metabolism dysfunction extend beyond the brain and to some extent define a systemic phenotype. It is difficult to attribute this systemic

International Review of Neurobiology, Volume 154
ISSN 0074-7742
https://doi.org/10.1016/bs.irn.2020.01.008

phenotype to brain beta-amyloid or tau proteins. Conversely, mitochondria increasingly appear to play a critical role in cell proteostasis, which suggests that mitochondrial dysfunction may promote protein aggregation. Mitochondrial and metabolism-related characteristics also define AD endophenotypes in cognitively normal middle-aged individuals, which suggests that mitochondrial and metabolism-related AD characteristics precede clinical decline. Genetic analyses increasingly implicate mitochondria and metabolism-relevant genes in AD risk. Collectively these factors suggest that mitochondria are more relevant to the causes of AD than its consequences, and support the view that a mitochondrial cascade features prominently in AD. This chapter reviews the case for mitochondrial and metabolism dysfunction in AD and the challenges of proving that a primary mitochondrial cascade is pertinent to the disease.

Abbreviations

Aβ	beta-amyloid
AD	Alzheimer's disease
APP	amyloid precursor protein
ATP synthase	complex V
CNS	central nervous system
coenzyme Q:cytochrome C oxidoreductase or cytochrome *bc*1 complex	complex III
cytochrome oxidase	complex IV
DM	diabetes mellitus
Drp1	dynamin-related protein 1
ER	endoplasmic reticulum
ETC	electron transport chain
FDG PET	fluorodeoxyglucose positron emission tomography
Fis1	mitochondrial fission 1
GWAS	Genome wide association
MAGIC	mitochondria as guardians in the cytosol
MAMs	mitochondrial-associated membranes
MCI	mild cognitive impairment
Mfn1	mitofusin 1
Mfn2	mitofusin 2
MICOS	mitochondrial contact site and cristae organizing system
MRI	magnetic resonance imaging
mtDNA	mitochondrial DNA
NADH:ubiquinone oxidoreductase	complex I
NT2	Ntera2
OPA1	optic atrophy 1
PCR	polymerase chain reaction
PS1	presenilin 1
PS2	presenilin 2

ROS	reactive oxygen species
RNS	reactive nitrogen species
rRNA	ribosomal RNA
succinate dehydrogenase or succinate: ubiquinone oxidoreductase	complex II
TDP43	transactivation response DNA-binding protein 43
TFAM	transcription factor A of the mitochondria
TOMM40	translocator of the outer mitochondrial membrane 40
UCPs	uncoupling proteins
WES	whole exon sequencing
WGS	whole genome sequencing

1. Introduction

Dementia commonly occurs in aged individuals, as do accumulations within the brain of extracellular beta-amyloid (Aβ) protein plaques and intraneuronal, neurofibrillary tau protein tangles (Swerdlow, 2007a). The classic definition of Alzheimer's disease (AD) emphasizes the concomitant presence of all three features (dementia, plaques, and tangles), although recently proposed criteria replace this triad with qualified gradations along an "Alzheimer's continuum" (Jack et al., 2018). For example, the presence of amyloid plaques without neurofibrillary tangles defines a state called Alzheimer's pathologic change, and concomitant plaque and tangle histology is simply AD. Beyond this, an individual's clinical status serves mostly to qualify the AD stage. Individuals with plaques but not tangles can have preclinical Alzheimer's pathological change, Alzheimer's pathological change with mild cognitive impairment (MCI), or Alzheimer's pathological change with dementia. Depending on their cognitive state, individuals with both plaques and tangles can have preclinical AD, AD with MCI/prodromal AD, or AD with dementia.

The plaques and tangles may cause cognitive decline or represent byproducts of other processes that cause cognitive decline. One popular hypothesis proposes the appearance of toxic Aβ species initiates the pathological events that ultimately result in other histology changes, neurodegeneration, and clinical decline (Hardy and Selkoe, 2002). This "amyloid cascade hypothesis" does not specifically address why toxic Aβ species first appear in the brain, although it is presumed that the initial introduction may involve a stochastic misfolding of a nontoxic Aβ peptide into a toxic Aβ peptide, or that Aβ may trigger a stochastic misfolding of tau protein. Misfolded Aβ or tau molecules might then convert other Aβ or tau molecules to the misfolded pattern; newly misfolded Aβ or tau templates promote further Aβ or tau misfolding; and the disease spreads throughout the brain in a prion-like fashion.

While stochastic protein misfolding is a recognized phenomenon, the biological background against which the misfolding event occurs is not stochastic. This background includes factors, such as protein posttranslational modifications, chaperones and chaperonins, energy levels, levels of gene expression, mRNA splicing, rates of protein translation, rates of protein removal, changes in amino acid chirality, and the specific mechanisms that mediate protein removal. The cell environment, therefore, influences protein levels, processing and folding. A deeper consideration of why protein homeostasis changes in AD could provide clues into the overall disease process. This applies whether an altered protein drives disease or simply marks its presence.

Metabolism refers to the chemical processes that play out within a cell, organ, or set of organs in order to maintain viability. Metabolism includes anabolism, which refers to the synthesis of molecules, and catabolism, which refers to their breakdown. Anabolic processes frequently consume energy, while catabolic processes release energy. In addition to playing critical roles in both anabolic and catabolic metabolism at the cellular level, mitochondria also contribute much of the energy a cell needs to build and maintain infrastructure, and in general to resist entropy. Viewed from the perspective of those who study mitochondria, a substantial proportion of cell function is geared toward monitoring energy levels and mitochondrial status, and for responding to what is monitored. It is difficult to separate metabolism and mitochondria from protein homeostasis, and on a technical level protein synthesis and removal represents a type of metabolism. This review considers AD clinical characteristics and protein hallmarks from the perspective of mitochondrial and general metabolism, and presents the case for a primary mitochondrial/metabolism etiology in AD.

2. Normal mitochondrial and metabolism function

2.1 General mitochondrial structure and organization

Mitochondria are rather large cell components whose lengths typically exceed 500 nM. Their size, though, varies dramatically because individual mitochondrial units can fuse to create larger units. Each mitochondrion is bounded by two membranes, an outer and an inner membrane, which differ in terms of phospholipid and protein content. The phospholipid components of both membranes are transferred from the endoplasmic reticulum (ER), in regions of close physical apposition termed mitochondrial-associated membranes (MAMs) (Tatsuta et al., 2014). Extensive modification of imported phospholipids occurs, for example, phosphatidylserine transferred from the ER to the inner mitochondrial membrane is converted to phosphoethanolamine, and within the inner mitochondrial membrane phosphatidic acid transferred from the ER can undergo a multistep conversion to cardiolipin.

The phospholipid composition within the inner membrane causes it to form invaginated envelopes called cristae. The space that defines the compartment formed by the outer and inner membranes is called the intermembrane space, and contains two definable subcompartments, an intermembrane space lined on one side by the inner membrane and the other by the outer membrane, and an intermembrane space that is mostly bound by the inner membrane cristae invaginations. A protein complex called the mitochondrial contact site and cristae organizing system (MICOS) helps maintain the inner membrane structure and inner-outer membrane structural relationships (Kozjak-Pavlovic, 2017).

Proteins residing within the outer mitochondrial membrane and intermembrane space are exclusively expressed by nuclear genes, translated within the cytoplasm, and moved to their destination. Nuclear encoded, mitochondrial-targeted proteins frequently contain N-terminal targeting signals composed of positively charged amino acids that track to negatively charged mitochondria. Most proteins in the inner mitochondrial membrane arise from nuclear genes, although some inner mitochondrial membrane proteins are expressed by genes located within mitochondria and translated by ribosomes located within mitochondria.

The mitochondrial DNA (mtDNA) genes comprise plasmids that sit within the matrix, the interior compartment bounded by the inner membrane. Mitochondrial DNA plasmids do not associate with histones, as nuclear DNA does, although the mtDNA does associate with other proteins, such as the transcription factor A of the mitochondria (TFAM) (Scarpulla, 2008). Mitochondrial DNA plasmids are packaged into condensed structures called nucleoids, which tether to the mitochondrial inner membrane. The matrix contains over 1000 proteins which support a variety of pathways and functions, including the Kreb's cycle, enzymes involved in the urea cycle, mtDNA replication and transcription, fatty acid oxidation, and fatty acid synthesis.

2.2 Mitochondrial functional aspects

Mitochondria are a major site of cell oxygen consumption. Oxygen is consumed during the process of oxidative phosphorylation, in which high energy electrons are passed down an electron transport chain (ETC) that contains four holoenzyme complexes. These complexes are designated complex I (NADH:ubiquinone oxidoreductase), complex II (succinate dehydrogenase or succinate: ubiquinone oxidoreductase), complex III (coenzyme Q:cytochrome C oxidoreductase or cytochrome bc_1 complex) and complex IV (cytochrome oxidase). Energy released from electrons passing through the chain is used to pump protons from the matrix, across the inner membrane, and into the intermembrane space. This creates a proton gradient, with a negatively charged matrix separated from a positively charged intermembrane space by the inner membrane. The proton separation creates a mitochondrial membrane potential and a protonmotive force, $\Delta\psi$, which defines how avidly protons are driven to re-enter the matrix. Protons can return to the matrix through another holoenzyme, complex V (ATP synthase), which uses the energy of the proton flux it facilitates to phosphorylate ADP to ATP. The respiratory chain consists of the ETC plus the ATP synthase.

Normally, energy-depleted electrons are delivered to molecular oxygen by the cytochrome oxidase, which consumes molecular oxygen and generates water. Under various circumstances this electron transfer may terminate prior to its completion. As the membrane potential hyperpolarizes the amount of energy required to pump protons from the matrix increases, which can hinder electrons from completing their ETC journey. Electrons released from the ETC react with molecular oxygen and nitrogen to generate reactive oxygen species (ROS) and reactive nitrogen species (RNS). Conversely, depolarization across the inner membrane may induce an increase in electron passage through the ETC as the mitochondrion attempts to regenerate its membrane potential. Depolarization can arise due to proton leak through the inner membrane and back into the matrix, independent of the ATP synthase or physiologically through channels formed by uncoupling proteins (UCPs). A return of protons to the matrix through an inner membrane leak or through UCPs "uncouples" electron transport from ATP generation, redirecting energy captured from electron transport to heat production. Structural ETC perturbations can also reduce the integrity of the chain and release electrons. In this scenario, ROS and RNS production occur within the context of a relative mitochondrial membrane potential depolarization.

Mitochondria can undergo fission into smaller structures or fuse together (Chan, 2006). These are enzyme mediated processes. Critical fission enzymes include dynamin-related protein 1 (Drp1) and mitochondrial fission 1 (Fis1). Optic atrophy 1 (OPA1), mitofusin 1 (Mfn1), and Mfn2 mediate fusion. Mitochondrial fission can help create smaller mitochondria that can travel easier within cells or segregate defective mitochondrial components through a disposal process called mitophagy (Youle and Narendra, 2011). Mitochondrial fusion produces organelles that are more respiration-efficient and allows for the functional assistance of missing or defective components through a process called complementation.

The respiratory chain holoenzymes are multimeric proteins that collectively contain over 80 subunits. The nucleus encodes all complex II subunits, while the other complexes contain a mix of nuclear DNA and mtDNA-encoded subunits. To be exact, the approximately 16.5 kB mtDNA includes seven genes that transcribe complex I subunits, one gene that transcribes a complex III subunit, three genes that transcribe complex IV subunits, and two genes that transcribe complex V subunits. In addition to these 13 structural genes, the mtDNA also has 24 synthetic genes that generate infrastructure needed to translate the structural gene transcripts. Two of these synthetic genes encode ribosomal RNA (rRNA) subunits that create mitochondrial ribosomes, and 22 synthetic genes create the mitochondrial tRNAs. The mtDNA genes lack introns; there is very little noncoding DNA aside from a stretch called the control region, which contains the promoters for RNA synthesis, as well as sequence that initiates mtDNA replication. The mtDNA physically associates with a protein called TFAM, which contributes to mtDNA replication and transcription (Scarpulla, 2008).

On several parameters mtDNA fundamentally differs from nuclear DNA. Cells can contain thousands of copies. Mitochondrial DNA plasmids replicate as the mitochondria in which they reside undergo cycles of elimination and division, so mtDNA constantly regenerates. One consequence of ongoing replication is that mistakes occur which leads to different nucleotides at specific nucleotide positions. Such variation is called heteroplasmy. Heteroplasmy can manifest as highly variable ratios of wild type to mutant sequence, from very low to very high percentages. The percentage of mutation required to confer a biochemical consequence depends on the functional impact of the variant. If the variant has little impact, very high or even homoplasmic levels of variation are necessary to alter cell integrity. Variants with strong effects can affect cell function at lower levels of heteroplasmy. The variant's functional impact defines the "threshold" at which the heteroplasmy becomes relevant. Heteroplasmic variants can arise somatically, and when a heteroplasmic cell divides the variant code can asymmetrically segregate between the daughter cells. This produces lineages with enriched or diluted variants. Further distinguishing it from nuclear DNA, mtDNA inheritance is maternal. Ova contain many copies, and under physiological conditions, the sperm's contribution is absent.

2.3 Other relevant metabolism features

Mitochondria participate pervasively in carbon flux. This obviously includes catabolic fluxes, which dismantle carbon skeletons to directly or indirectly produce energy. Examples include fatty acid β-oxidation and parts of the Krebs cycle. It also includes anabolic fluxes, which enlarge carbon skeletons and frequently consume energy. These carbon fluxes connect to myriad metabolic pathways that exist outside of mitochondria.

Cells import glucose via glucose transporters. After undergoing phosphorylation, glucose carbon can enter glycolysis, and proceed to pyruvate that enters mitochondria and undergoes further catabolism. Carbon skeletons can also leave the glycolysis pathway at points at which glycolysis intermediates are in equilibrium with certain amino acids, and phosphorylated glucose additionally enters the pentose phosphate shunt where it generates ribose sugars required for nucleic acid synthesis. Glucose carbon can also terminate as lactate, which in turn is often removed by the cell that produces it. Even in the presence of adequate oxygen, therefore, not all glucose carbon is used to support oxidative phosphorylation with a consequent production of water and CO_2. Some refer to the processes that produce glucose-derived, non-CO_2 carbon products in the presence of adequate oxygen as aerobic glycolysis (Vaishnavi et al., 2010).

It is perhaps cliché to point out, but many metabolic fluxes feature carbon skeleton manipulations. These manipulations are designed to produce energy in some cases, or to build carbon infrastructure through energy-consuming pathways. Catabolic and anabolic pathways occur within and outside of mitochondria, but even in cases in which fluxes occur outside of mitochondria, the mitochondria play a role. For example, the redirection of lactate carbon back to glucose, through gluconeogenesis, requires some transit through mitochondria. Even when glycolysis-generated pyruvate enters mitochondria and is converted to acetyl CoA, that acetyl

CoA carbon can exit the mitochondria following its entry into the Krebs cycle, where it forms citrate. Citrate that has exited the mitochondria can undergo a reaction in the cytosol to generate cytosolic acetyl CoA, which is then used in lipogenesis. This represents a feature of Warburg metabolism, in which growing cells convert glucose carbon into other macromolecules (Vander Heiden et al., 2009). In general, mitochondria play a key role in cell carbon laundering.

3. Alzheimer's disease-relevant alterations in mitochondrial and metabolism function

3.1 General mitochondrial structure and organization

Mitochondria in the brains of AD subjects differ physically from those in the brains of non-AD subjects. On average they are smaller (Baloyannis, 2006), which coincides with a shift in the mitochondrial fission-fusion balance toward fission (Manczak et al., 2011; Wang et al., 2009). Despite this, AD brains show more frequent enlargement of individual mitochondria with cristae disruption (Baloyannis, 2006).

While there is a clear reduction in the number of normal appearing AD brain mitochondria, how mitochondrial mass changes in the disease is less straightforward. Autophagosomes reportedly contain increased amounts of mitochondrial material, which could suggest increased mitochondrial turnover (Hirai et al., 2001). On the other hand, mRNA and protein levels of the peroxisome proliferator-activated receptor gamma coactivator 1-alpha (PGC-1α), a transcriptional coactivator that drives mitochondrial biogenesis, are reduced (Qin et al., 2009; Sheng et al., 2012). The overall data seem to support a picture of enhanced mitochondrial elimination, with a concurrent reduction in replacement by new mitochondria.

A similar situation pertains to mtDNA copy number. Most studies that quantify mtDNA use polymerase chain reaction (PCR)-based methods, and there is a clear reduction in PCR-amplifiable brain mtDNA (Brown et al., 2001). However, one study that used an immunolabeling approach to assess AD brain mtDNA reported a substantial increase in mtDNA within autophagosomes (Hirai et al., 2001). In this case, mtDNA is presumably simply along for the ride as cells remove their bad mitochondria, consistent with the broader finding that AD brains contain increased amounts of mitochondrial debris.

3.2 Mitochondrial functional aspects

Direct assessments of mitochondrial respiration require fresh mitochondria. Freeze-thawed mitochondria are less appropriate as this disrupts membranes and uncouples mitochondrial oxygen consumption from ATP production. Despite this challenge, studies report AD brain mitochondrial respiration differs from mitochondria from control brain tissue (Sims et al., 1987).

Freeze-thawed mitochondria are suitable for determining enzyme Vmax activities, and AD brain mitochondria show a variety of Vmax activity reductions. This includes lower activities for two thiamine and lipoic acid-dependent matrix enzymes, pyruvate dehydrogenase complex and α-ketoglutarate dehydrogenase complex (Gibson et al., 1988). Within the respiratory chain, complex IV Vmax activities are consistently lower in AD than control brain tissue (Parker et al., 1994; Swerdlow, 2012). Complex I activity reductions are also occasionally reported, and some terminal Krebs cycle enzymes activities are possibly increased (Gibson et al., 1998; Swerdlow, 2012).

Interestingly and importantly, some of these AD biochemical phenotypes are not brain limited. Fibroblast cultures prepared from AD patients, including sporadic AD patients without identifiable amyloid precursor protein (APP) or presenilin gene mutations, show similar Vmax activity reductions (Curti et al., 1997; Gibson et al., 1988). In fact, the AD complex IV Vmax activity reduction was first identified in platelet mitochondria (Parker et al., 1990), which prompted investigators to measure its activity in brain mitochondria. In that case, the reduced systemic activity predicted a low brain activity (Swerdlow, 2012).

Mitochondria can produce free radicals, and AD mitochondria show increased free radical modifications. This includes increased markers of mtDNA oxidative damage (Lovell and Markesbery, 2007; Mecocci et al., 1994). Mitochondrial DNA shows additional damage that manifests in the form of increased mtDNA deletions (Corral-Debrinski et al., 1994). Increased mitochondrial fission and reduced mitochondrial motility are evident, and these changes too are not brain-limited (Wang et al., 2008a, 2008b).

The reasons that underlie these biochemical and molecular perturbations warrant consideration. These biochemical and molecular phenotypes could generally arise due to a general failure of the organelle, or alternatively any of these specific phenotypes could cause a general organelle failure.

Complicating this point is the fact that Aβ protein appears to accumulate within AD brain mitochondria (Caspersen et al., 2005; Lustbader et al., 2004). It is unclear whether Aβ accumulates within AD brain mitochondria and damages them or if Aβ accumulates within AD brain mitochondria because they are damaged. Amyloid precursor protein itself targets mitochondria, and increasing mitochondrial APP or exposing them to mutant APP alters mitochondrial function (Anandatheerthavarada et al., 2003; Anandatheerthavarada and Devi, 2007; Devi et al., 2006). Presenilin proteins also associate with mitochondria (Hansson et al., 2004), either through direct contact or indirectly through their presence in mitochondria-associated cell structures (Area-Gomez and Schon, 2016). Clearly links exist between mitochondria and proteins implicated in AD through genetic or histologic studies. A better understanding of these links could potentially advance our understanding of this disease.

3.3 Other relevant metabolism features

Glucose utilization is altered in AD brains. This is evident on fluorodeoxyglucose positron emission tomography (FDG PET) studies (Mosconi et al., 2009). FDG PET is accomplished by injecting an individual with radiolabelled FDG. FDG accesses the brain through the circulation, with more metabolically active brain regions taking up greater amounts of FDG. The FDG undergoes phosphorylation, which traps it within cells. The deoxyglucose modification blocks further passage through glycolysis and imaging the brain via PET reveals which brain areas took up the most glucose and by extension were most metabolically active. In AD, certain relatively posterior brain regions, including the posterior temporal, parietal, posterior cingulate and precuneal cortices, show diminished glucose uptake.

Direct molecular surveys of AD autopsy brains show reductions in insulin-initiated signaling pathway activity. This pattern suggests a likely state of brain insulin resistance, leading some to classify AD as a form of diabetes mellitus (DM) (Steen et al., 2005).

Even though AD brain insulin resistance can occur in the absence of obvious systemic insulin resistance, epidemiological studies suggest a connection between central nervous system (CNS) and peripheral insulin resistance (Morris et al., 2014). The meaning of this association is unclear. It could indicate DM or peripheral insulin resistance contributes to AD. Alternatively, this association could reflect a common underlying physiologic state. To this point, a systemic mitochondrial lesion in AD could account for the concomitant presence of CNS and systemic insulin resistance phenotypes (Morris et al., 2014).

Genome wide association (GWAS) studies implicate lipid homeostasis in the evolution of AD. The APOE gene, which encodes the cholesterol/lipid transporting protein apolipoprotein E, contains amino acid-altering genetic variants. These variants constitute the best studied and most powerful determinant of sporadic AD risk (Corder et al., 1993). One copy of the APOE4 variant increases the AD odds ratio about 3.5 times over baseline risk, and the APOE2 variant decreases AD risk. The AD research field remains uncertain over why APOE variants so profoundly influence AD odds ratios, but within the context of the GWAS studies it certainly is possible that the impact of APOE derives from differences in how the various apolipoprotein E forms influence lipid homeostasis. While the aggregate of these genetic data do not directly implicate a role for mitochondria, because mitochondria and lipid metabolism are connected, it seems reasonable to postulate genetic data implicating lipid homeostasis indirectly implicate a role for mitochondria in AD.

4. Metabolic endophenotypes

Investigators increasingly rely on biomarkers to reveal the secrets of AD. Biomarkers can demonstrate on-going physiologic, molecular, or subtle clinical events are underway before clearly noticeable clinical changes arise in individuals. Biomarkers can inform the earliest definable presence of the disease, mark the order in which different events occur, indicate the rate at which alterations evolve, or make mechanistic inferences. Biomarker-defined states, in the absence of a full phenotype, are endophenotypes.

Mosconi et al. (2007) reported an FDG PET-defined phenotype in which cognitively intact, middle-aged individuals with an AD-affected mother were more likely to show AD-typical brain regions with reduced glucose utilization than cognitively intact, middle-aged individuals with an AD-affected father. This FDG PET-defined endophenotype, therefore, follows a pattern of maternal inheritance that is potentially consistent with mtDNA inheritance. The same investigators used amyloid PET imaging to show the brains of middle-aged individuals with an AD-affected mother were more likely to contain fibrillar amyloid than brains from AD-affected fathers (Mosconi et al., 2010a, 2010b). Spinal fluid from the children of AD mothers are more likely to show AD-consistent Aβ and oxidative stress characteristics than spinal fluid from the children of AD fathers (Mosconi et al., 2010a, 2010b). Brain perfusion, demonstrated through a magnetic resonance imaging (MRI)-based technique called arterial spin labeling, shows a similar pattern (Okonkwo et al., 2012), and volumetric MRI studies shows brain regions compromised in AD show greater atrophy in the children of AD mothers than in the children of AD fathers (Honea et al., 2011). Platelet mitochondria cytochrome oxidase activities are lower in blood samples from the children of AD mothers than they are in blood samples from the children of AD fathers (Mosconi et al., 2011). Maternal inheritance of AD endophenotypes, therefore, is noted across a range of studies and suggests a possible contributory role for mtDNA.

In an analysis of Framingham Longevity Study participants, memory test scores in cognitively intact middle-aged individuals were lowest in APOE4 carriers who also had an AD-affected mother (Debette et al., 2009). These scores were lower than those from APOE4 carriers who also had an AD-affected father.

5. Relationship to proteostasis

Proteins aggregate in the AD brain. This includes Aβ plaques, tau tangles, and cytosolic inclusions consisting of the transactivation response DNA-binding protein 43 (TDP43). One potential explanation is that one or more of these aggregating proteins are cytotoxic. Another is that proteostasis is perturbed in the dysfunctional or degenerating AD brain, which leads to protein aggregation.

On one hand, proteins implicated in AD associate with mitochondria. For example, APP and Aβ are found in or on mitochondria (Anandatheerthavarada et al., 2003; Lustbader et al., 2004), and driving these proteins to mitochondria interferes with their function. Transactivation response DNA-binding protein 43 and even tau protein as well reportedly localize to mitochondria (Choi et al., 2014; Gao et al., 2019). This suggests that altered proteostasis may drive AD brain mitochondrial dysfunction. One limitation of this hypothesis is that while it can reasonably account for brain mitochondrial dysfunction, it is challenging to see how this might lead to altered mitochondrial function in nonbrain tissues.

Alternatively, mitochondrial dysfunction could alter the cell environment in ways that promote protein aggregation (Weidling and Swerdlow, 2019). Compromised mitochondrial function could cause bioenergetic or oxidative stress, which in turn could change levels of protein expression, alter posttranslational protein modifications or cause protein oxidative damage. Protein chaperone or chaperonin levels could change as a result, as could levels or activities of secretase enzymes.

Another possibility, increasingly supported by experimental data, is that mitochondria may directly participate in proteostasis by acting as de facto protein disposal units. This basic principle was initially demonstrated in yeast, which direct peptides with aggregation potential into mitochondria. This process, called "mitochondria as guardians in the cytosol" (MAGIC), raises the aggregation threshold of the implicated protein (Ruan et al., 2017). Similarly, knock-out of the protein PINK1, which helps mediate at least one mitophagy pathway, increases amyloidosis in the brains of transgenic mice (Du et al., 2017). Another study reported promoting mitophagy with the compound urolithin A reduced brain amyloidosis in an AD transgenic mouse model (Sorrentino et al., 2017).

The relationship between brain aging and protein aggregation is important to consider. Those who develop AD protein aggregation can live for decades before those aggregations clinically manifest. It is unclear whether a brain switches from an aggregation-free state to an aggregation-containing state because it loses its capacity to prevent protein aggregations, or because its proteins gain a unique capacity to aggregate. In the latter case, there is evidence to support the view that protein misfolding occurs as a stochastic event. As years go by, the chance of one these misfolding events occurring increases, and once it does the misfolded peptide templates the misfolding of other peptides and seeds an aggregation. Even if this is correct, though, changes in mitochondrial function would alter the background against which a stochastic protein misfolding event occurs. In this scenario, while mitochondrial dysfunction might not directly drive a specific protein misfolding event, it could still lower the threshold at which that protein misfolding event occurs.

6. Genetic contributions

Only very rarely, probably quite well-below frequencies of 1% of cases, does AD present within a Mendelian context. Gene mutations implicated with early onset, autosomal dominant familial AD reside in the genes for APP, presenilin 1 (PS1), and presenilin 2 (PS2). Cases of AD due to mutations in these genes typically show a relatively young age of onset, with symptoms manifesting in the fifth or sixth decade (Swerdlow, 2007a). There are infrequent exceptions to this, but in general if an AD-affected individual does not present before the end of their sixth decade, and does not have a parent who was also demented before the end of their sixth decade, genetic testing is not likely to reveal an autosomal dominant mutation in the APP, PS1, or PS2 genes.

By far most cases present sporadically or at least pseudo-sporadically. The pseudo-sporadic qualification clarifies that genes do influence the risk of sporadic disease, but not to the extent at which Mendelian inheritance is obvious. This complicates how we think about AD. Advancing age constitutes the greatest sporadic AD risk factor, and so how long an individual survives profoundly impacts whether they die with or without AD. The best way to avoid developing AD, perhaps, is to simply not survive into advanced old age.

Sporadic cases can nevertheless also present in middle age. In fact, the number of early onset sporadic cases outnumbers the autosomal dominant cases. Among the sporadic cases, an earlier age of onset suggests a greater genetic contribution. Early onset, sporadic cases frequently carry APOE4 alleles, which supports the view that possessing an APOE4 allele lowers the age of onset in those who do develop AD (Blacker and Tanzi, 1998). It turns out that given long enough survival, the proportion of those with the potential to develop AD could exceed the proportion of the population which does not have the potential to develop AD (Corrada et al., 2008).

6.1 Nuclear genes

Sporadic AD GWAS now implicate over 30 nuclear loci that imply a reduced or increased AD risk (Kunkle et al., 2019). GWAS most effectively reveal common variant associations. Methods better equipped to detect rare variants, including whole exon sequencing (WES) and whole genome sequencing (WGS), are increasingly utilized.

With GWAS, locus association is a first step. The next includes determining the responsible gene within the locus. While locus identification is bias-free, pinpointing the appropriate gene is not. Next, it is important to consider why the candidate gene may influence AD risk. This also introduces bias. Within the context of these challenges, genetic studies indicate a role for microglial, lipid, and endocytosis biology in AD (Karch and Goate, 2015).

At least one gene study implicates mitochondria. This study found rare variants in the PTCD1 gene influence AD risk (Fleck et al., 2019). The PTCD1 gene encodes a protein that localizes to mitochondria and likely affects the translation of mtDNA-derived mRNAs. A preliminary study also implicates a rare variant in the TAMM41 gene, which encodes a mitochondria-localized protein required for cardiolipin synthesis (Zhang et al., 2019).

It is possible that other AD-relevant genes affect mitochondria. Some of these genes produce proteins that localize to mitochondria (Pa et al., 2019). Other proteins indirectly influence, or are influenced by, mitochondrial function. In general, it is hard to dissociate lipid homeostasis from mitochondrial function.

One GWAS based study broadly considered nuclear genes whose products localize to mitochondria. These genes naturally fall within loci with previously determined AD odds ratios. By assigning the appropriate locus-associated odds ratio to each nuclear-encoded mitochondrial gene, the investigators created mitochondria-relevant polygenic risk scores (Andrews et al., 2019). Subjects with the highest mitochondrial polygenic risk scores were more likely to have AD. Essentially, the mitochondrial polygenic risk score tracked AD risk.

APOE represents a more specific case in point. APOE4-derived apolipoprotein E folds differently than APOE2 or APOE3-derived apolipoprotein E, which impacts the enzymatic processing of apolipoprotein E. APOE4-derived apolipoprotein E undergoes a cleavage event that opens a mitochondrial targeting signal. The mitochondria-targeting peptide accesses mitochondria and alters organelle function in a way that lowers complex IV activity (Chang et al., 2005; Chen et al., 2011).

The translocator of the outer mitochondrial membrane 40 (TOMM40) gene sits adjacent to the APOE gene. TOMM40 variants also associate with AD risk (Roses, 2010), but some of these variants are in linkage disequilibrium with APOE4. It remains unclear whether TOMM40 independently contributes to AD risk, although several studies report this is potentially the case (Watts et al., 2019).

The three proteins implicated in familial AD also show mitochondrial ties. APP localizes to mitochondria (Anandatheerthavarada et al., 2003). Presenilin 1 and 2 are seen either at mitochondria or in structures that interact with mitochondria, such as mitochondrial-associated membranes (MAMs) (Area-Gomez and Schon, 2016; Hansson et al., 2004).

6.2 Mitochondrial genes

Data to date suggest that mtDNA contributes to AD, but the nature of this contribution remains unresolved (Swerdlow, 2012). In this respect, the complexities of mtDNA render unique challenges. This includes issues related to heteroplasmy and heteroplasmy detection, its propensity for somatic mutation, and a high degree of interindividual variability in which linked variants often occur. Interactions between mtDNA and nuclear DNA could also complicate relevant analyses (Andrews et al., 2019).

Substantial indirect data suggest an mtDNA role. One epidemiologic study found AD-diagnosed subjects were more likely to have a demented mother than father (Edland et al., 1996). This is consistent with a maternal inheritance bias. The fact that women live longer than men, paternal health records are not as comprehensive as maternal health records, and women may have a higher AD risk can confound this type of study. Multiple endophenotype studies also indirectly implicate mtDNA inheritance, as the middle-aged children of AD mothers show higher rates of AD endophenotypes than those with AD fathers (Swerdlow et al., 2014).

Other support comes from studies of cytoplasmic hybrids (cybrids), which were initially deployed to test whether mtDNA contributes to reduced AD complex IV activity (Swerdlow et al., 1997). To perform these cybrid studies, a cell line was depleted of its endogenous mtDNA to create what is called a ρ0 cell line (Swerdlow, 2007b). Platelets from AD and control subjects were isolated and mixed with ρ0 cells in the presence of a membrane-permeabilizing detergent to permit cytosol mixing to occur. This creates cybrid cells that contain the nucleus of the ρ0 cell and mtDNA from the platelet donor. Because the nuclear background remains consistent between cybrid cell lines created from different platelet donors, functional differences between cybrid cell lines most likely reflect differences in mtDNA. In these experiments, complex IV activity in cybrid cell lines containing mtDNA from AD subjects on average was lower than the activity in lines containing mtDNA from control subjects (Swerdlow et al., 1997). This outcome indicates mtDNA to at least some extent contributes to lower AD complex IV activity.

Further analyses of cybrid cell lines generated from AD and non-AD subjects find further differences. Alzheimer's disease cybrid lines show evidence of increased oxidative stress, a reduced mitochondrial membrane potential, impaired mitochondrial calcium uptake, a reduced peak respiration rate, an overall smaller mitochondrial size, positive modification of a mitochondrial fission protein, a lower glycolysis flux, less ATP, activation of the NFκB transcription factor, increased apoptosis activity, and increased amounts of Aβ protein (Swerdlow et al., 2017). While data like these do not establish a causal role for mtDNA in AD, they do indicate mtDNA can influence several recognized AD pathologies, and may cause other functional changes.

Genetic association studies could establish a more traditional connection between mtDNA and AD, reveal specific responsible mtDNA features or variants, and address uncertainty over cause versus consequence relationships. The literature includes notable attempts, but studies are neither as extensive nor comprehensive as those involving nuclear DNA, and there are no definitive conclusions (Ridge and Kauwe, 2018). Common variants do not present in isolation, but rather within the context of other common variants. This makes it difficult to quantify the impact of a single variant. Rare variants may confer more clear-cut associations, but studies of rare variants require the complete sequencing of large numbers of cases and controls. The datasets needed for such analysis are only now being generated (Zhang et al., 2019). Interactions between mtDNA and nuclear DNA variants represent another potential confounding factor.

Some studies focus on mtDNA haplogroups, sets of mtDNA variants that present in conjunction with each other. During prehistoric human migrations mutations arose in different populations and were fixed in those populations (Torroni et al., 1996). While haplogroups do not permit the interrogation of individual variants, they do to some extent simplify comparisons of some common variant combinations. One limitation of haplogroup analyses is that they can potentially combine mtDNA variants that increase AD risk with variants that decrease AD risk. Haplogroup studies do report effects on AD risk, but results are inconsistent between studies and preclude confident conclusions (Ridge and Kauwe, 2018).

The role of mtDNA in mediating AD pathology or influencing AD risk resides in a state of equipoise. Resolving this conundrum may ultimately require the development of mtDNA polygenic risk scores. This remains a challenging endeavor since common variants tend to travel in groups, which complicates assigning odds ratios to individual variants. Hopefully this problem will prove tractable.

7. A mitochondrial cascade hypothesis

Mitochondrial changes in AD could represent irrelevant artifacts of other disease pathologies, mediate their effects, or trigger them. The essential role of the mitochondria in so many aspects of cell function makes the first possibility less likely. Mitochondrial abnormalities also occur in nondegenerating tissues, and many abnormalities occur in the absence of other obvious pathologies (Swerdlow, 2012).

The AD field is increasingly open to the possibility that mitochondria are important in AD. Much of this enthusiasm derives from experiments in which other AD pathologies are used to perturb mitochondria in ways that recapitulate AD-specific mitochondrial dysfunction and drive other abnormalities. For example, delivering APP, Aβ, or APOE4-derived peptides to mitochondria can impair the respiratory chain or shift the mitochondrial fission-fusion balance (Anandatheerthavarada and Devi, 2007; Chen et al., 2011; Du et al., 2008; Wang et al., 2008a, 2008b). In this capacity, mitochondria could help elucidate other AD hypotheses.

The amyloid cascade hypothesis proposes that Aβ interferes with cell function and survival. Exactly how Aβ damages cells is an open question. Some postulate mitochondria in some way mediate Aβ toxicity. In one study, Aβ exposure was lethal to the Ntera2 (NT2) neuronal cell line but not its ρ0 derivative (Cardoso et al., 2001). This implies the respiratory chain mediates Aβ toxicity and is consistent with data that show Aβ interferes with respiratory chain function (Casley et al., 2002). Beta amyloid also appears to bind other mitochondrial components, including an alcohol dehydrogenase enzyme and cyclophilin D (Du et al., 2008; Lustbader et al., 2004). Studies such as these suggest a potential important role for mitochondria in AD, within the context of the amyloid cascade hypothesis.

It is more difficult to reconcile the amyloid cascade hypothesis with the systemic AD mitochondrial phenotype (Morris et al., 2014). It remains possible, albeit unlikely, that systemically generated Aβ causes that phenotype, but if this does occur it is still important to consider what alters systemic Aβ homeostasis. Certainly, cells carefully manage APP and Aβ production and elimination, and mitochondria may contribute to that regulation.

Cybrid data are also relevant to this point (Swerdlow, 2012; Swerdlow et al., 2017). The cybrid approach specifically reveals the functional effects of specific mtDNA sequences, and even extra-mitochondrial effects presumably reflect mtDNA-derived changes in mitochondrial function. Generating cybrid cell lines relies on cytosolic mixing, which could cotransfer Aβ residing in either the platelet mitochondria or its free cytosol, but any transferred component that cannot replicate itself degrades and dilutes over the course of cell cycling and division. If transferred toxic Aβ was to account for the cybrid data, it would need to manifest prion-like properties.

While Aβ may indeed cause or amplify mitochondrial dysfunction, at least in the brains of AD subjects, it is important to consider the possibility that at least some mitochondrial changes arise independently of Aβ. Several conceptual points support this view.

Mitochondrial function changes with advancing age (Boveris and Navarro, 2008; Navarro and Boveris, 2007), which could potentially tie sporadic AD to its greatest single risk factor, advancing age (Swerdlow, 2007a). Pseudosporadic epidemiology that occurs in conjunction with maternal inheritance bias, in both the fully-fledged disease state and with endophenotypes, is consistent with mitochondrial genetics. The ability to replicate several AD-associated pathologies in cultured cells following mtDNA transfer from AD patients argues that mtDNA contributes to those pathologies and is not simply a consequence of those pathologies (Swerdlow, 2012; Swerdlow et al., 2017).

Various investigators initially envisioned a primary role for mitochondria in AD (Beal, 1995; Gibson et al., 1998; Parker et al., 1990; Wallace, 1992). Some further proposed this role derived ultimately from mtDNA, with Parker (1990) postulating mtDNA inheritance and Wallace (1992) postulating mtDNA somatic mutation as the critical factors. Whether a consequence of inherited mtDNA variation, or somatic mtDNA mutation, or both, it is important to tie these putative mtDNA effects to nuclear genes, and at this point data support this overall concept (Andrews et al., 2019; Pa et al., 2019; Swerdlow et al., 2017). It is also important to tie mitochondrial function to protein homeostasis and aggregation in ways that allow mitochondria to drive those protein changes to some extent, even if those protein changes cycle back to alter the mitochondria (Weidling and Swerdlow, 2019). Ordering the molecular changes of AD in this way defines a mitochondrial cascade hypothesis, which uses mitochondrial dysfunction and bioenergetic defects to establish a potentially causal metabolic link to AD (Swerdlow, 2018; Swerdlow et al., 2010, 2014; Swerdlow and Khan, 2004, 2009).

8. Conclusions and implications

Alzheimer's disease features altered mitochondrial function, structure, maintenance, and dynamics (Swerdlow, 2012). These features are seen not only within the brain, the organ associated with clinical AD symptoms and signs, but also outside the brain. Their localization to nondegenerating and non-neuronal sites suggests they are not exclusive by-products of neurodegeneration. Outside the brain, the scope of the systemic biochemical phenotype defined by these mitochondrial pathologies exceeds currently reported systemic Aβ or tau pathology, which suggests that systemic Aβ or tau pathology should not entirely account for systemic mitochondrial pathology.

Mitochondrial dysfunction appears to drive at least some non-mitochondrial AD pathologies, including an increase in Aβ (Gabuzda et al., 1994; Khan et al., 2000; Onyango et al., 2010). Cybrid studies indicate AD mitochondrial dysfunction at least to some degree tracks with mtDNA (Swerdlow et al., 2017). Mitochondria also feature prominently in theories of aging, which could help explain potential ties between sporadic AD and its greatest risk factor, advancing age (Swerdlow, 2007a). Age-related mitochondrial changes could first initiate a period of compensated brain aging, which evolves into a state of uncompensated brain aging. Considerations such as these support the argument that mitochondria independently contribute to AD risk and progression, and support speculation of a pathological mitochondrial cascade (Swerdlow, 2018; Swerdlow et al., 2010, 2014, 2017; Swerdlow and Khan, 2004, 2009).

Advancing a mitochondrial cascade hypothesis at this point requires advances on three fronts. First, firm associations with specific genetic variants will prove critical. This could include associations with nuclear genes that encode mitochondria-localized proteins, nuclear genes that encode proteins that do not localize to mitochondria but influence mitochondrial function, or genes encoded on the mtDNA. Substantial indirect data, including cybrid, epidemiologic, and endophenotype data are consistent with this possibility (Swerdlow et al., 2017) and recent genetic association studies implicate genes whose products localize to or influence mitochondria (Pa et al., 2019; Swerdlow et al., 2017). Additional studies should further explore these phenomena.

Second, a mitochondrial cascade hypothesis should explain altered proteostasis and protein aggregation. Decades of research provide a sophisticated picture of how cells handle AD-aggregating proteins, especially Aβ and tau, but also α-synuclein and TDP43. Despite this, it is difficult to conclude that we comprehensively understand the factors that initiate altered Aβ and tau homeostasis. It is important to determine if factors other than stochastic misfolding events are relevant, and if so whether mitochondria contribute to those factors. Data to date leave no doubt that mitochondrial function and bioenergetics influence protein homeostasis (Weidling and Swerdlow, 2019), and prior research indicates links between mitochondrial function, APP, Aβ, and tau (Anandatheerthavarada and Devi, 2007;

Choi et al., 2014; Hoglinger et al., 2005; Lustbader et al., 2004). A more complete picture of how and why mitochondria could influence APP, Aβ, tau, α-synuclein, and TDP43 homeostasis in AD is worth pursuing.

Third, the AD field will need to show therapies that target mitochondria and bioenergetics can favorably influence the disease. If mitochondria are truly critical in AD, then manipulations that reverse mitochondrial dysfunction should benefit those with the disease. This will prove challenging as specific mitochondrial derangements may require very specific manipulations, it remains unclear what specific manipulations are needed most, executing the desired manipulations could prove difficult, and targeting the necessary manipulations to where they are needed may present problems. Ultimately, improving the lives of those afflicted remains the ultimate test from a conceptual perspective and the goal from a practical perspective. Currently, though, the aggregate of the evidence argues it is reasonable to target mitochondria and bioenergetics for the treatment of AD (Swerdlow, 2011, 2014).

Acknowledgments

R.H.S. is supported by the National Institute on Aging (P30 AG035982, R01 AG060733, R01 AG061194) and the Department of Defense (AZ 170111).

Conflict of interests

The author has no conflict of interests with any organizations or entity mentioned here.

References

Anandatheerthavarada, H. K., Biswas, G., Robin, M. A., & Avadhani, N. G. (2003). Mitochondrial targeting and a novel transmembrane arrest of Alzheimer's amyloid precursor protein impairs mitochondrial function in neuronal cells. *The Journal of Cell Biology*, *161*(1), 41–54.

Anandatheerthavarada, H. K., & Devi, L. (2007). Amyloid precursor protein and mitochondrial dysfunction in Alzheimer's disease. *The Neuroscientist*, *13*(6), 626–638.

Andrews, S. J., Fulton-Howard, B., Patterson, C., McFall, G. P., Gross, A., Michaelis, E. K., et al. (2019). Mitonuclear interactions influence Alzheimer's disease risk. *Neurobiology of Aging*. https://doi.org/10.1016/j.neurobiolaging.2019.09.007, [Epub ahead of print].

Area-Gomez, E., & Schon, E. A. (2016). Mitochondria-associated ER membranes and Alzheimer disease. *Current Opinion in Genetics & Development*, *38*, 90–96.

Baloyannis, S. J. (2006). Mitochondrial alterations in Alzheimer's disease. *Journal of Alzheimer's Disease*, *9*(2), 119–126.

Beal, M. F. (1995). Aging, energy, and oxidative stress in neurodegenerative diseases. *Annals of Neurology*, *38*(3), 357–366.

Blacker, D., & Tanzi, R. E. (1998). The genetics of Alzheimer disease: Current status and future prospects. *Archives of Neurology, 55*(3), 294–296.
Boveris, A., & Navarro, A. (2008). Brain mitochondrial dysfunction in aging. *IUBMB Life, 60*(5), 308–314.
Brown, A. M., Sheu, R. K., Mohs, R., Haroutunian, V., & Blass, J. P. (2001). Correlation of the clinical severity of Alzheimer's disease with an aberration in mitochondrial DNA (mtDNA). *Journal of Molecular Neuroscience, 16*(1), 41–48.
Cardoso, S. M., Santos, S., Swerdlow, R. H., & Oliveira, C. R. (2001). Functional mitochondria are required for amyloid beta-mediated neurotoxicity. *The FASEB Journal, 15*(8), 1439–1441.
Casley, C. S., Canevari, L., Land, J. M., Clark, J. B., & Sharpe, M. A. (2002). Beta-amyloid inhibits integrated mitochondrial respiration and key enzyme activities. *Journal of Neurochemistry, 80*(1), 91–100.
Caspersen, C., Wang, N., Yao, J., Sosunov, A., Chen, X., Lustbader, J. W., et al. (2005). Mitochondrial Abeta: A potential focal point for neuronal metabolic dysfunction in Alzheimer's disease. *The FASEB Journal, 19*(14), 2040–2041.
Chan, D. C. (2006). Mitochondria: Dynamic organelles in disease, aging, and development. *Cell, 125*(7), 1241–1252.
Chang, S., ran Ma, T., Miranda, R. D., Balestra, M. E., Mahley, R. W., & Huang, Y. (2005). Lipid- and receptor-binding regions of apolipoprotein E4 fragments act in concert to cause mitochondrial dysfunction and neurotoxicity. *Proceedings of the National Academy of Sciences of the United States of America, 102*(51), 18694–18699.
Chen, H. K., Ji, Z. S., Dodson, S. E., Miranda, R. D., Rosenblum, C. I., Reynolds, I. J., et al. (2011). Apolipoprotein E4 domain interaction mediates detrimental effects on mitochondria and is a potential therapeutic target for Alzheimer disease. *The Journal of Biological Chemistry, 286*(7), 5215–5221.
Choi, J., Chandrasekaran, K., Demarest, T. G., Kristian, T., Xu, S., Vijaykumar, K., et al. (2014). Brain diabetic neurodegeneration segregates with low intrinsic aerobic capacity. *Annals of Clinical Translational Neurology, 1*(8), 589–604.
Corder, E. H., Saunders, A. M., Strittmatter, W. J., Schmechel, D. E., Gaskell, P. C., Small, G. W., et al. (1993). Gene dose of apolipoprotein E type 4 allele and the risk of Alzheimer's disease in late onset families. *Science, 261*(5123), 921–923.
Corrada, M. M., Brookmeyer, R., Berlau, D., Paganini-Hill, A., & Kawas, C. H. (2008). Prevalence of dementia after age 90: Results from the 90+ study. *Neurology, 71*(5), 337–343.
Corral-Debrinski, M., Horton, T., Lott, M. T., Shoffner, J. M., McKee, A. C., Beal, M. F., et al. (1994). Marked changes in mitochondrial DNA deletion levels in Alzheimer brains. *Genomics, 23*(2), 471–476.
Curti, D., Rognoni, F., Gasparini, L., Cattaneo, A., Paolillo, M., Racchi, M., et al. (1997). Oxidative metabolism in cultured fibroblasts derived from sporadic Alzheimer's disease (AD) patients. *Neuroscience Letters, 236*(1), 13–16.
Debette, S., Wolf, P. A., Beiser, A., Au, R., Himali, J. J., Pikula, A., et al. (2009). Association of parental dementia with cognitive and brain MRI measures in middle-aged adults. *Neurology, 73*(24), 2071–2078.
Devi, L., Prabhu, B. M., Galati, D. F., Avadhani, N. G., & Anandatheerthavarada, H. K. (2006). Accumulation of amyloid precursor protein in the mitochondrial import channels of human Alzheimer's disease brain is associated with mitochondrial dysfunction. *The Journal of Neuroscience, 26*(35), 9057–9068.
Du, H., Guo, L., Fang, F., Chen, D., Sosunov, A. A., McKhann, G. M., et al. (2008). Cyclophilin D deficiency attenuates mitochondrial and neuronal perturbation and ameliorates learning and memory in Alzheimer's disease. *Nature Medicine, 14*(10), 1097–1105.

Du, F., Yu, Q., Yan, S., Hu, G., Lue, L. F., Walker, D. G., et al. (2017). PINK1 signalling rescues amyloid pathology and mitochondrial dysfunction in Alzheimer's disease. *Brain*, *140*(12), 3233–3251.

Edland, S. D., Silverman, J. M., Peskind, E. R., Tsuang, D., Wijsman, E., & Morris, J. C. (1996). Increased risk of dementia in mothers of Alzheimer's disease cases: Evidence for maternal inheritance. *Neurology*, *47*(1), 254–256.

Fleck, D., Phu, L., Verschueren, E., Hinkle, T., Reichelt, M., Bhangale, T., et al. (2019). PTCD1 is required for mitochondrial oxidative-phosphorylation: Possible genetic association with Alzheimer's disease. *The Journal of Neuroscience*, *39*(24), 4636–4656.

Gabuzda, D., Busciglio, J., Chen, L. B., Matsudaira, P., & Yankner, B. A. (1994). Inhibition of energy metabolism alters the processing of amyloid precursor protein and induces a potentially amyloidogenic derivative. *The Journal of Biological Chemistry*, *269*(18), 13623–13628.

Gao, J., Wang, L., Yan, T., Perry, G., & Wang, X. (2019). TDP-43 proteinopathy and mitochondrial abnormalities in neurodegeneration. *Molecular and Cellular Neurosciences*, *100*, 103396.

Gibson, G. E., Sheu, K. F., & Blass, J. P. (1998). Abnormalities of mitochondrial enzymes in Alzheimer disease. *Journal of Neural Transmission*, *105*(8–9), 855–870.

Gibson, G. E., Sheu, K. F., Blass, J. P., Baker, A., Carlson, K. C., Harding, B., et al. (1988). Reduced activities of thiamine-dependent enzymes in the brains and peripheral tissues of patients with Alzheimer's disease. *Archives of Neurology*, *45*(8), 836–840.

Hansson, C. A., Frykman, S., Farmery, M. R., Tjernberg, L. O., Nilsberth, C., Pursglove, S. E., et al. (2004). Nicastrin, presenilin, APH-1, and PEN-2 form active gamma-secretase complexes in mitochondria. *The Journal of Biological Chemistry*, *279*(49), 51654–51660.

Hardy, J., & Selkoe, D. J. (2002). The amyloid hypothesis of Alzheimer's disease: Progress and problems on the road to therapeutics. *Science*, *297*(5580), 353–356.

Hirai, K., Aliev, G., Nunomura, A., Fujioka, H., Russell, R. L., Atwood, C. S., et al. (2001). Mitochondrial abnormalities in Alzheimer's disease. *The Journal of Neuroscience*, *21*(9), 3017–3023.

Hoglinger, G. U., Lannuzel, A., Khondiker, M. E., Michel, P. P., Duyckaerts, C., Feger, J., et al. (2005). The mitochondrial complex I inhibitor rotenone triggers a cerebral tauopathy. *Journal of Neurochemistry*, *95*(4), 930–939.

Honea, R. A., Swerdlow, R. H., Vidoni, E., & Burns, J. M. (2011). Progressive regional atrophy in normaladults with a maternal history of Alzheimer disease. *Neurology*, *76*, 822–829.

Jack, C. R., Jr., Bennett, D. A., Blennow, K., Carrillo, M. C., Dunn, B., Haeberlein, S. B., et al. (2018). NIA-AA research framework: Toward a biological definition of Alzheimer's disease. *Alzheimer's Dementia*, *14*(4), 535–562.

Karch, C. M., & Goate, A. M. (2015). Alzheimer's disease risk genes and mechanisms of disease pathogenesis. *Biological Psychiatry*, 77(1), 43–51.

Khan, S. M., Cassarino, D. S., Abramova, N. N., Keeney, P. M., Borland, M. K., Trimmer, P. A., et al. (2000). Alzheimer's disease cybrids replicate beta-amyloid abnormalities through cell death pathways. *Annals of Neurology*, *48*(2), 148–155.

Kozjak-Pavlovic, V. (2017). The MICOS complex of human mitochondria. *Cell and Tissue Research*, *367*(1), 83–93.

Kunkle, B. W., Grenier-Boley, B., Sims, R., Bis, J. C., Damotte, V., Naj, A. C., et al. (2019). Genetic meta-analysis of diagnosed Alzheimer's disease identifies new risk loci and implicates Abeta, tau, immunity and lipid processing. *Nature Genetics*, *51*(3), 414–430.

Lovell, M. A., & Markesbery, W. R. (2007). Oxidative DNA damage in mild cognitive impairment and late-stage Alzheimer's disease. *Nucleic Acids Research*, *35*(22), 7497–7504.

Lustbader, J. W., Cirilli, M., Lin, C., Xu, H. W., Takuma, K., Wang, N., et al. (2004). ABAD directly links Abeta to mitochondrial toxicity in Alzheimer's disease. *Science*, *304*(5669), 448–452.

Manczak, M., Calkins, M. J., & Reddy, P. H. (2011). Impaired mitochondrial dynamics and abnormal interaction of amyloid beta with mitochondrial protein Drp1 in neurons from patients with Alzheimer's disease: Implications for neuronal damage. *Human Molecular Genetics*, *20*(13), 2495–2509.

Mecocci, P., MacGarvey, U., & Beal, M. F. (1994). Oxidative damage to mitochondrial DNA is increased in Alzheimer's disease. *Annals of Neurology*, *36*(5), 747–751.

Morris, J. K., Honea, R. A., Vidoni, E. D., Swerdlow, R. H., & Burns, J. M. (2014). Is Alzheimer's disease a systemic disease? *Biochimica et Biophysica Acta*, *1842*(9), 1340–1349.

Mosconi, L., Brys, M., Switalski, R., Mistur, R., Glodzik, L., Pirraglia, E., et al. (2007). Maternal family history of Alzheimer's disease predisposes to reduced brain glucose metabolism. *Proceedings of the National Academy of Sciences of the United States of America*, *104*(48), 19067–19072.

Mosconi, L., de Leon, M., Murray, J., E, L., Lu, J., Javier, E., et al. (2011). Reduced mitochondria cytochrome oxidase activity in adult children of mothers with Alzheimer's disease. *Journal of Alzheimer's Disease*, *27*(3), 483–490.

Mosconi, L., Glodzik, L., Mistur, R., McHugh, P., Rich, K. E., Javier, E., et al. (2010a). Oxidative stress and amyloid-beta pathology in normal individuals with a maternal history of Alzheimer's. *Biological Psychiatry*, *68*(10), 913–921.

Mosconi, L., Mistur, R., Switalski, R., Tsui, W. H., Glodzik, L., Li, Y., et al. (2009). FDG-PET changes in brain glucose metabolism from normal cognition to pathologically verified Alzheimer's disease. *European Journal of Nuclear Medicine and Molecular Imaging*, *36*(5), 811–822.

Mosconi, L., Rinne, J. O., Tsui, W. H., Berti, V., Li, Y., Wang, H., et al. (2010b). Increased fibrillar amyloid-{beta} burden in normal individuals with a family history of late-onset Alzheimer's. *Proceedings of the National Academy of Sciences of the United States of America*, *107*(13), 5949–5954.

Navarro, A., & Boveris, A. (2007). The mitochondrial energy transduction system and the aging process. *American Journal of Physiology Cell Physiology*, *292*(2), C670–C686.

Okonkwo, O. C., Xu, G., Oh, J. M., Dowling, N. M., Carlsson, C. M., Gallagher, C. L., et al. (2012). Cerebral blood flow is diminished in asymptomatic middle-aged adults with maternal history of Alzheimer's disease. *Cerebral Cortex*, *24*, 978–988.

Onyango, I. G., Ahn, J. Y., Tuttle, J. B., Bennett, J. P., Jr., & Swerdlow, R. H. (2010). Nerve growth factor attenuates oxidant-induced beta-amyloid neurotoxicity in sporadic Alzheimer's disease cybrids. *Journal of Neurochemistry*, *114*(6), 1605–1618.

Pa, J., Andrews, S. J., & Swerdlow, R. H. (2019). Mitochondria and Alzheimer's: Is PTCD1 the smoking gun? *Trends in Neurosciences*, *42*(11), 759–762.

Parker, W. D. (1990). Sporadic neurologic disease and the electron transport chain: A hypothesis. In R. M. Pascuzzi (Ed.), *Proceedings of the 1989 Scientfic Meeting of the American Society for Neurological Investigation: New Developments in Neuromuscular Disease*, Bloomington, Indiana: Indiana University Printing Services.

Parker, W. D., Jr., Filley, C. M., & Parks, J. K. (1990). Cytochrome oxidase deficiency in Alzheimer's disease. *Neurology*, *40*(8), 1302–1303.

Parker, W. D., Jr., Parks, J., Filley, C. M., & Kleinschmidt-DeMasters, B. K. (1994). Electron transport chain defects in Alzheimer's disease brain. *Neurology*, *44*(6), 1090–1096.

Qin, W., Haroutunian, V., Katsel, P., Cardozo, C. P., Ho, L., Buxbaum, J. D., et al. (2009). PGC-1alpha expression decreases in the Alzheimer disease brain as a function of dementia. *Archives of Neurology*, *66*(3), 352–361.

Ridge, P. G., & Kauwe, J. S. K. (2018). Mitochondria and Alzheimer's disease: The role of mitochondrial genetic variation. *Current Genetic Medicine Reports*, *6*(1), 1–10.
Roses, A. D. (2010). An inherited variable poly-T repeat genotype in TOMM40 in Alzheimer disease. *Archives of Neurology*, *67*(5), 536–541.
Ruan, L., Zhou, C., Jin, E., Kucharavy, A., Zhang, Y., Wen, Z., et al. (2017). Cytosolic proteostasis through importing of misfolded proteins into mitochondria. *Nature*, *543*(7645), 443–446.
Scarpulla, R. C. (2008). Transcriptional paradigms in mammalian mitochondrial biogenesis and function. *Physiological Reviews*, *88*(2), 611–638.
Sheng, B., Wang, X., Su, B., Lee, H. G., Casadesus, G., Perry, G., et al. (2012). Impaired mitochondrial biogenesis contributes to mitochondrial dysfunction in Alzheimer's disease. *Journal of Neurochemistry*, *120*(3), 419–429.
Sims, N. R., Finegan, J. M., Blass, J. P., Bowen, D. M., & Neary, D. (1987). Mitochondrial function in brain tissue in primary degenerative dementia. *Brain Research*, *436*(1), 30–38.
Sorrentino, V., Romani, M., Mouchiroud, L., Beck, J. S., Zhang, H., D'Amico, D., et al. (2017). Enhancing mitochondrial proteostasis reduces amyloid-beta proteotoxicity. *Nature*, *552*(7684), 187–193.
Steen, E., Terry, B. M., Rivera, E. J., Cannon, J. L., Neely, T. R., Tavares, R., et al. (2005). Impaired insulin and insulin-like growth factor expression and signaling mechanisms in Alzheimer's disease—Is this type 3 diabetes? *Journal of Alzheimer's Disease*, 7(1), 63–80.
Swerdlow, R. H. (2007a). Is aging part of Alzheimer's disease, or is Alzheimer's disease part of aging? *Neurobiology of Aging*, *28*(10), 1465–1480.
Swerdlow, R. H. (2007b). Mitochondria in cybrids containing mtDNA from persons with mitochondriopathies. *Journal of Neuroscience Research*, *85*(15), 3416–3428.
Swerdlow, R. H. (2011). Role and treatment of mitochondrial DNA-related mitochondrial dysfunction in sporadic neurodegenerative diseases. *Current Pharmaceutical Design*, 17(31), 3356–3373.
Swerdlow, R. H. (2012). Mitochondria and cell bioenergetics: Increasingly recognized components and a possible etiologic cause of Alzheimer's disease. *Antioxidants & Redox Signaling*, *16*(12), 1434–1455.
Swerdlow, R. H. (2014). Bioenergetic medicine. *British Journal of Pharmacology*, *171*(8), 1854–1869.
Swerdlow, R. H. (2018). Mitochondria and mitochondrial cascades in Alzheimer's disease. *Journal of Alzheimer's Disease*, *62*(3), 1403–1416.
Swerdlow, R. H., Burns, J. M., & Khan, S. M. (2010). The Alzheimer's disease mitochondrial cascade hypothesis. *Journal of Alzheimer's Disease*, *20*(Suppl. 2), S265–S279.
Swerdlow, R. H., Burns, J. M., & Khan, S. M. (2014). The Alzheimer's disease mitochondrial cascade hypothesis: Progress and perspectives. *Biochimica et Biophysica Acta*, *1842*(8), 1219–1231.
Swerdlow, R. H., & Khan, S. M. (2004). A "mitochondrial cascade hypothesis" for sporadic Alzheimer's disease. *Medical Hypotheses*, *63*(1), 8–20.
Swerdlow, R. H., & Khan, S. M. (2009). The Alzheimer's disease mitochondrial cascade hypothesis: An update. *Experimental Neurology*, *218*(2), 308–315.
Swerdlow, R. H., Koppel, S., Weidling, I., Hayley, C., Ji, Y., & Wilkins, H. M. (2017). Mitochondria, cybrids, aging, and Alzheimer's disease. *Progress in Molecular Biology and Translational Science*, *146*, 259–302.
Swerdlow, R. H., Parks, J. K., Cassarino, D. S., Maguire, D. J., Maguire, R. S., Bennett, J. P., Jr., et al. (1997). Cybrids in Alzheimer's disease: A cellular model of the disease? *Neurology*, *49*(4), 918–925.
Tatsuta, T., Scharwey, M., & Langer, T. (2014). Mitochondrial lipid trafficking. *Trends in Cell Biology*, *24*(1), 44–52.

Torroni, A., Huoponen, K., Francalacci, P., Petrozzi, M., Morelli, L., Scozzari, R., et al. (1996). Classification of European mtDNAs from an analysis of three European populations. *Genetics*, *144*(4), 1835–1850.
Vaishnavi, S. N., Vlassenko, A. G., Rundle, M. M., Snyder, A. Z., Mintun, M. A., & Raichle, M. E. (2010). Regional aerobic glycolysis in the human brain. *Proceedings of the National Academy of Sciences of the United States of America*, *107*(41), 17757–17762.
Vander Heiden, M. G., Cantley, L. C., & Thompson, C. B. (2009). Understanding the Warburg effect: The metabolic requirements of cell proliferation. *Science*, *324*(5930), 1029–1033.
Wallace, D. C. (1992). Mitochondrial genetics: A paradigm for aging and degenerative diseases? *Science*, *256*(5057), 628–632.
Wang, X., Su, B., Fujioka, H., & Zhu, X. (2008a). Dynamin-like protein 1 reduction underlies mitochondrial morphology and distribution abnormalities in fibroblasts from sporadic Alzheimer's disease patients. *The American Journal of Pathology*, *173*(2), 470–482.
Wang, X., Su, B., Lee, H. G., Li, X., Perry, G., Smith, M. A., et al. (2009). Impaired balance of mitochondrial fission and fusion in Alzheimer's disease. *The Journal of Neuroscience*, *29*(28), 9090–9103.
Wang, X., Su, B., Siedlak, S. L., Moreira, P. I., Fujioka, H., Wang, Y., et al. (2008b). Amyloid-beta overproduction causes abnormal mitochondrial dynamics via differential modulation of mitochondrial fission/fusion proteins. *Proceedings of the National Academy of Sciences of the United States of America*, *105*(49), 19318–19323.
Watts, A., Wilkins, H. M., Michaelis, E., & Swerdlow, R. H. (2019). TOMM40 '523 associations with baseline and longitudinal cognition in APOE varepsilon3 homozygotes. *Journal of Alzheimer's Disease*, *70*(4), 1059–1068.
Weidling, I., & Swerdlow, R. H. (2019). Mitochondrial dysfunction and stress responses in Alzheimer's disease. *Biology (Basel)*, *8*(2), 39.
Youle, R. J., & Narendra, D. P. (2011). Mechanisms of mitophagy. *Nature Reviews. Molecular Cell Biology*, *12*(1), 9–14.
Zhang, X., Farrell, J., Zhu, C., Mayeux, R., Haines, J. L., Pericak-Vance, M. A., et al. (2019). Mitochondrial variants and haplogroups associated with late-onset Alzheimer's disease identified by whole exome sequencing. *Alzheimer's and Dementia*, *15*, P565, AAIC.

CHAPTER NINE

MAM and C99, key players in the pathogenesis of Alzheimer's disease

Marta Pera[a,b,*], Jorge Montesinos[a,c], Delfina Larrea[a], Rishi R. Agrawal[d], Kevin R. Velasco[a], Irina G. Stavrovskaya[a], Taekyung D. Yun[a], Estela Area-Gomez[a,c,d,*]

[a]Department of Neurology, Columbia University Irving Medical Center, New York, NY, United States
[b]Basic Sciences Department, Faculty of Medicine and Health Sciences, Universitat Internacional de Catalunya, Sant Cugat del Vallés, Barcelona, Spain
[c]Taub Institute for Research on Alzheimer's Disease and the Aging Brain, Columbia University Irving Medical Center, New York, NY, United States
[d]Institute of Human Nutrition, Columbia University Irving Medical Center, New York, NY, United States
*Corresponding authors: e-mail address: martaperamunoz@gmail.com; eag2118@columbia.edu

Contents

1. Introduction 237
2. C99 in Alzheimer's disease animal models and patients 240
   2.1 C99 in Alzheimer's disease animal models 241
   2.2 C99 in familial and sporadic Alzheimer's disease patients 243
3. Mitochondria-associated endoplasmic reticulum membranes in Alzheimer's disease 244
   3.1 The MAM and γ-secretase 247
   3.2 In Alzheimer's disease, C99 accumulates at the MAM 249
4. Lipid dyshomeostasis in Alzheimer's disease is MAM-related 250
   4.1 Cholesterol 251
   4.2 Sphingolipids 253
   4.3 Phospholipids 254
5. Mitochondrial alterations in Alzheimer's disease and MAM 255
   5.1 Mitochondrial dysfunction in Alzheimer's disease: Biochemical and dynamic alterations 256
   5.2 C99 accumulation at the MAM impacts mitochondrial function 258
6. Concluding remarks 260
References 262

Abstract

Inter-organelle communication is a rapidly-expanding field that has transformed our understanding of cell biology and pathology. Organelle-organelle contact sites can generate transient functional domains that act as enzymatic hubs involved in the

International Review of Neurobiology, Volume 154
ISSN 0074-7742
https://doi.org/10.1016/bs.irn.2020.03.016

regulation of cellular metabolism and intracellular signaling. One of these hubs is located in areas of the endoplasmic reticulum (ER) connected to mitochondria, called mitochondria-associated ER membranes (MAM). These MAM are transient lipid rafts intimately involved in cholesterol and phospholipid metabolism, calcium homeostasis, and mitochondrial function and dynamics. In addition, γ-secretase-mediated proteolysis of the amyloid precursor protein 99-aa C-terminal fragment (C99) to form amyloid β also occurs at the MAM. Our most recent data indicates that in Alzheimer's disease, increases in uncleaved C99 levels at the MAM provoke the upregulation of MAM-resident functions, resulting in the loss of lipid homeostasis, and mitochondrial dysfunction. Here, we discuss the relevance of these findings in the field, and the contribution of C99 and MAM dysfunction to Alzheimer's disease neuropathology.

List of abbreviations

Aβ	amyloid beta peptide
ABCA7	ATP-binding cassette transporter A
ACAT1	Acyl-CoA:colesterol acyl-transferase 1
AD	Alzheimer's disease
AICD	amyloid precursor protein intracelular domain
APOE	apolipoprotein E
APP	amyloid precusor protein
APP mouse	trangenic mice overexpressing human APP Swedish (K670N/M671L) and APP751 isoform under the control of the murine Thy1 promoter
APPE693Q mouse	APP Dutch mutation expressing mouse
APP(OSK)-Tg mouse	the APP(OSK)-Tg model expresses APP harboring the Osaka (E693Δ) mutation
BACE1	β-secretase
BBB	blood-brain barrier
B6C3-Tg mouse	APP/PS1 mouse model (APPswe,PSEN1dE9) 85Dbo/J
CEs	cholesteryl esters
CLU	clusterin
CSF	cerebrospinal fluid
C83	amyloid precursor protein 83 aa C-terminal fragment
C99	amyloid precursor protein 99 aa C-terminal fragment
DHA	docosahexaenoic acid
DRP1	dynamin-related protein 1
ER	endoplasmic reticulum
FACL4	acyl-CoA synthetase 4
FAD	familial Alzheimer's disease
GWAS	genome-wide association studies
hBACE1 and hBACE1/hAPP Tg mice	transgenic mice expressing human BACE1 under the Thy1 promoter in combination with human APP
HDL	high-density lipoprotein
HMGCoAR	hydroxymethylglutaryl-CoA reductase
iEM	immunogold electron microscopy

IP_3	inositol triphosphate
J20 mouse	mice expressing a mutant form of the human APP bearing both the Swedish (K670N/M671L) and the Indiana (V717F) mutations (APPSwInd) under the control of the PDGFB promoter [B6. Cg-Tg(PDGFB-APPSwInd)]
LDL	low-density lipoprotein
MAM	mitochondria-associated endoplasmic reticulum membranes
MAPT	microtubule-associated protein tau
MEFs	mouse embryonic fibroblasts
OXPHOS	oxidative phosphorylation
PICALM	phosphatidylinositol-binding clathrine assembly protein
PS cDKO mouse	forebrain-specific conditional double knock-out mice lacking both PS1 and PS2
PS-DKO	presenilin double knock-out
PSS	phosphatidyl synthase
PS1	presenilin 1
PS1 mouse	mice expressing human presenilin with mutation M146 L or M146 V via the PDGF-β promoter
PS1-KI^{M146V} mouse	mouse that expresses the PS1 M146V targeted allele at normal physiological levels
PS2	presenilin 2
PtdCho	phosphatidylcholine
PtdEtn	phosphatidylethanolamine
PtdSer	phosphatidylserine
PUFA	polyunsaturated fatty acid
SAD	sporadic Alzheimer disease
SORL1	sortilin-related receptor 1
STARD1	steroidogenic acute regulatory protein 1
TEM	transmission electron microscopy
TgCRND8 mouse	mouse model which encodes human amyloid-protein precursor (APP) bearing the Swedish (KM670/671NL) and Indiana (V717F) APP mutations
Tg2576 mouse	transgenic mice overexpressing the human APP695 isoform with the Swedish mutation (K670N/M671L) via the hamster prion promoter.
TREM2	triggering receptor expressed on myeloid cells 2
VDAC2	voltage-dependent anion channel 2
3xTg mouse	triple-transgenic model (APPswe, PS1M106V, TauP301L)

1. Introduction

The first case of Alzheimer's disease (AD) was described by Alois Alzheimer more than a century ago. As yet, the cause of the disease remains unknown and no treatment has been found to prevent or ameliorate AD, which remains the most common cause of dementia. Increasing longevity in western societies has had an enormous impact on the incidence of aging-associated diseases like AD. An estimated 5.8 million Americans of all ages were living with AD in 2019 (AD Facts and Figure Report, 2019, AD Association). The economic and social burdens of this devastating disease are huge; as the western population ages, AD is becoming a more common cause of death, and it is the only top-ten cause of death that cannot be prevented, cured or even slowed (AD Facts and Figure Report, 2019, AD Association). The main hurdle to finding a cure is the lack of knowledge about the mechanism that causes AD. Despite ongoing efforts to understand the events leading to the disease and to find a treatment, there remain more questions than answers 100 years after the first case was described.

Most AD cases are sporadic (SAD), characterized by the inheritance of genetically determined risk factors or polymorphisms that do not single-handedly cause AD but rather confer increased probability of developing the disease. Different isoforms of apolipoprotein E (APOE), a component of the lipoproteins that transport cholesterol and lipids throughout the body, are the most common and validated of these risk factors (Holtzman et al., 2012). Carriers of one or two copies of the ε4 variant of APOE (ApoE4) have an ~4-fold or ~12-fold increased dose-dependent risk for developing AD compared to individuals carrying ApoE3, the most common isotype, and the rarest variant, ApoE2, which is believed to be protective against AD (Farrer et al., 1997).

On the other hand, ~1–5% of AD cases are familial (FAD), characterized by mutations in amyloid precursor protein (APP), a membrane protein of yet unknown function, and presenilin-1 (PS1) or presenilin-2 (PS2), the catalytic subunits of the γ-secretase complex. This complex is responsible for the cleavage of several proteins in the cell, including APP. In the non-amyloidogenic pathway, APP is first cleaved by α-secretase in non-lipid raft membrane domains, to produce the 83-aa long C-terminal fragment (C83). This C83 peptide is then cleaved by the γ-secretase complex to produce p3 and the APP intracellular domain (AICD). Less frequently, APP can also be cleaved via the amyloidogenic pathway, which occurs in lipid raft membrane domains, where APP is first cleaved by β-secretase (BACE1) to produce the 99-aa long C-terminal fragment (C99). As in the case of C83, C99 is subsequently cleaved by the γ-secretase complex to produce amyloid β (Aβ) along with AICD (Fig. 1).

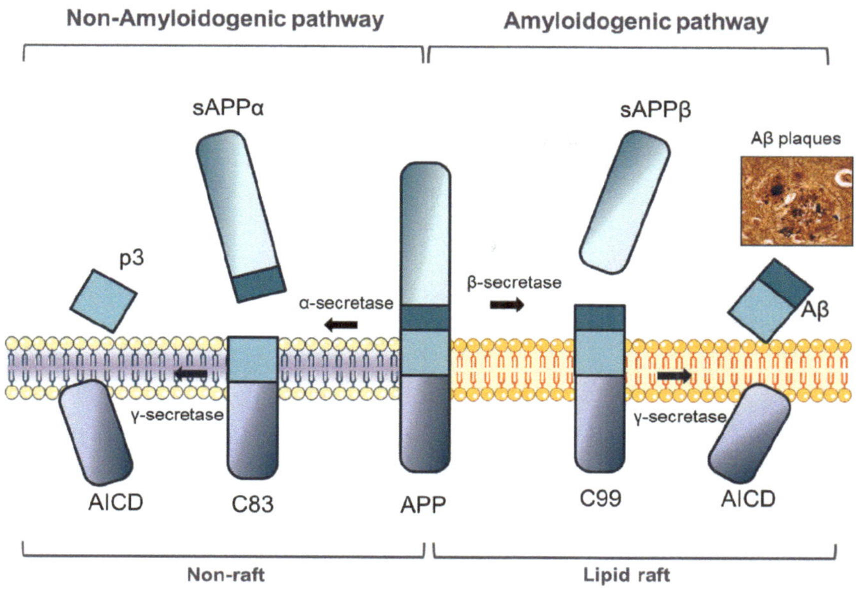

Fig. 1 Amyloid precusor protein processing

The main neuropathological hallmarks of AD are Aβ plaques and neurofibrillary tau tangles, with Aβ deposition receiving more attention by far. The "amyloid cascade hypothesis" was proposed by Hardy and Higgins (1992) and suggests that Aβ deposition triggers all other pathogenic events of Alzheimer's disease. Billions of dollars have been invested into drugs that reduce γ-secretase activity and Aβ accumulation but, up to now, all such clinical trials have failed, from γ-secretase inhibitors to anti-Aβ vaccines that completely remove Aβ from the brain (Anderson et al., 2017; Mehta et al., 2018). Other approaches have also been tested without success, including BACE1 inhibitors, tau aggregation inhibitors and microglia activation inhibitors. Current drug treatments are merely palliative, such as acetylcholinesterase inhibitors (rivastigmine, donepezil and galantamine) and NMDA receptor antagonists (memantine), which alleviate some of the disease symptoms but do not halt disease progression.

Since 2003, over 200 compounds have reached Phase 2 clinical evaluations but no new drugs have been approved for AD treatment (Cummings et al., 2014; Godyń et al., 2016). Other challenges facing drug development in AD is the lack of robust biological biomarkers that allow for an early diagnosis. The "Aβ-centric vision" of the field has focused AD biomarker research on Aβ, but detectable alterations in Aβ happen later in the disease progression than other AD features. Hence, it is important to expand the field of focus beyond Aβ and investigate novel candidate biomarkers that enable early disease detection. An early diagnosis, before the appearance of plaques and tangles, is necessary to treat the disease before the presence of symptoms and decelerate the neurodegenerative process, increasing the quality of life of patients and caregivers.

Indeed, many different alterations that happen during early stages of the disease cannot be explained by the appearance of Aβ plaques later in the disease. Among these other symptoms and features are the altered metabolism of fatty acids and phospholipids (Fraser et al., 2010; Pettegrew et al., 2001), elevated levels of circulating cholesterol (Stefani and Liguri, 2009), accumulation of lipid droplets in cells (Gómez-Ramos and Morán, 2007; Pani et al., 2009a, 2009b), aberrant calcium regulation (Bezprozvanny, 2008), reduced levels of brain glucose (Mosconi, 2005), and mitochondrial dysfunction (Gillardon et al., 2007; Wang et al., 2009a, 2009b). Thus, these "other" features have been either ignored entirely or relegated to downstream events secondary to plaques and tangles. Therefore, to solve this riddle and get to the root of the matter, it is fundamental to put together all the pieces of the complex AD puzzle.

Due to the considerable number of publications showing that Aβ plaque appearance is not an early event in AD pathogenesis, combined with the failure of Aβ-targeted clinical trials, the amyloid cascade hypothesis has been called into question. The deleterious effects of Aβ are undeniable, but the role of Aβ in initial phases is still debated. The field is moving towards overcoming these obstacles and seeking different perspectives, including inflammation (Ringheim and Szczepanik, 2006; Salminen et al., 2009; Unterberger et al., 2006), hypoxic stress (Sun et al., 2006; Zhang and Le, 2010) and mitochondrial dysfunction and oxidative stress (Castellani et al., 2002; Wang et al., 2009a, 2009b). The main objective of this review is to explore new horizons and introduce a novel AD perspective displacing the focus of attention from Aβ to C99, and mitochondria-associated endoplasmic reticulum membranes (MAM).

2. C99 in Alzheimer's disease animal models and patients

The amyloidogenic pathway of APP processing has received massive attention from scientists and pharmaceutical companies because the main goal of the AD field since the amyloid cascade hypothesis was proposed two decades ago, has been to reduce Aβ production to avoid the formation of senile plaques. The most attractive elements of the pathway have been γ-secretase and Aβ, but all the components of this pathway, including BACE1 and C99, have been extensively studied as well. Numerous animal models have been developed as important tools to understand the disease and moreover many studies have been performed in AD patients to better understand disease progression in humans. There is a growing body of evidence suggesting a critical role of C99 in AD neuropathology using mouse models. The possibility that C99 contributes to AD etiology independently of Aβ is supported by C99-induced MAM dysfunction (Pera et al., 2017), endosome and lysosome dysfunction (Lauritzen et al., 2012, 2019), neuroinflammation (Kaur et al., 2017) and electrophysiological alterations (Bourgeois et al., 2018; Mondragón-Rodríguez et al., 2018) in various cellular and mouse models at early stages. Since at least some of these alterations are also documented in human pathology, one may speculate on a possible role of C99 accumulation in humans as well.

2.1 C99 in Alzheimer's disease animal models

Rodents (mice and rats) and various non-human primates (Crawley, 1999; Sarter et al., 1992a, 1992b) are the species typically used to model cognition, dementia, and AD in drug discovery. Nearly 170 transgenic, knock-in and knock-out models of AD have been developed to date (ALZ FORUM Research Models Database https://www.alzforum.org/research-models) and principally focus on human gene mutations in APP, PS1, microtubule-associated protein tau (MAPT), triggering receptor expressed on myeloid cells 2 (TREM2), and APOE, as well as transfection of BACE1, alone or in combination with another gene. In 2015, Foley et al. (2015) extensively reviewed the relationship between Aβ levels and cognitive function in five AD mouse models, i.e., Tg2576 (Westerman et al., 2002), APP23 (Van Dam et al., 2005), PS1 (Lok et al., 2013), 3xTg (Sterniczuk et al., 2010), and APP(OSK)-Tg (Umeda et al., 2012), and no correlation between quantified Aβ levels and cognitive function was found, once again contradicting the amyloid hypothesis. Conversely, mounting data indicate a crucial role of C99 in AD animal models.

First, studies analyzing BACE1 activity and C99 accumulation in AD models date from the early 2000s (Cai et al., 2010; Fukumoto et al., 2004; Rockenstein et al., 2005). These works show a clear increase of BACE1 activity and C99 accumulation in different AD models: the

Tg2576 mouse (Fukumoto et al., 2004), hBACE1 and hBACE1/hAPP tg mice (Rockenstein et al., 2005) and the rhesus monkey (*Macaca mulatta*) (Cai et al., 2010; Fukumoto et al., 2004). Interestingly, BACE1 activity and C99 accumulation in the brain increase with aging (Fukumoto et al., 2004) and changes are more pronounced in brain areas like the temporal and frontal cortex that are deeply affected by AD (Cai et al., 2010; Fukumoto et al., 2004; Rockenstein et al., 2005). Furthermore, work performed in the 3xTgAD (APPswe, PS1M106V, TauP301L) mouse show a particularly strong age-dependent intraneuronal accumulation of C99 (Lauritzen et al., 2012). Importantly, in the 3xTgAD mouse, C99 accumulation appears at very early stages, before Aβ levels are detectable, in AD-related areas like the hippocampus and cortex. Intraneuronal C99 accumulation has been recently corroborated by different groups in several mouse models, including the APPE693Q (Dutch mutation) expressing mouse (Kaur et al., 2017), the TgCRND8 (Swedish and Indiana mutations mouse) (Cavanagh et al., 2013) and the J20 mouse (Swedish and Indiana mutations) (Mondragón-Rodríguez et al., 2018). Moreover, partial BACE1 reduction has been shown to rescue synaptic pathology and recent and remote memory deficits in different AD mouse models (Kimura et al., 2010; McConlogue et al., 2007), underlining the relevance of C99 in AD pathology.

In addition, analysis of subcellular fractions isolated from WT and PS1-KI^{M146V} mouse (Guo et al., 1999) brain also showed an increase in the levels of C99 in the mutant samples, and especially in the MAM fractions, compared to that in controls (Fig. 2), while the relative concentration of AICD was not changed significantly (Pera et al., 2017). The PS1-KI^{M146V} mouse expresses the PS1 M146V targeted allele at normal physiological levels. This specific accumulation of C99 at MAM suggests that the C99-MAM interaction is potentially an unexplored piece in the AD pathogenesis puzzle.

Interestingly, C99 alterations are also characteristic of SAD models, as treatment with exogenous ApoE increases APP C-terminal fragment (APP-CTF) accumulation in different cell lines, indicating that ApoE modulates γ-secretase cleavage (Irizarry et al., 2004). These data indicate that alterations in C99 are a common element of both sporadic and familial AD models.

The animal model that better supports the important role of C99 in AD pathogenesis is the forebrain-specific, conditional double knock-out mice lacking both PS1 and PS2 (PS cDKO) generated by Jia Shen's lab in 2004 (Saura et al., 2004). The PS cDKO mice show reduced levels of Aβ and accumulation of APP-CTFs in the cortex (Beglopoulos et al., 2004). By 6 months of age, these mice failed to learn by the Morris water maze and contextual fear

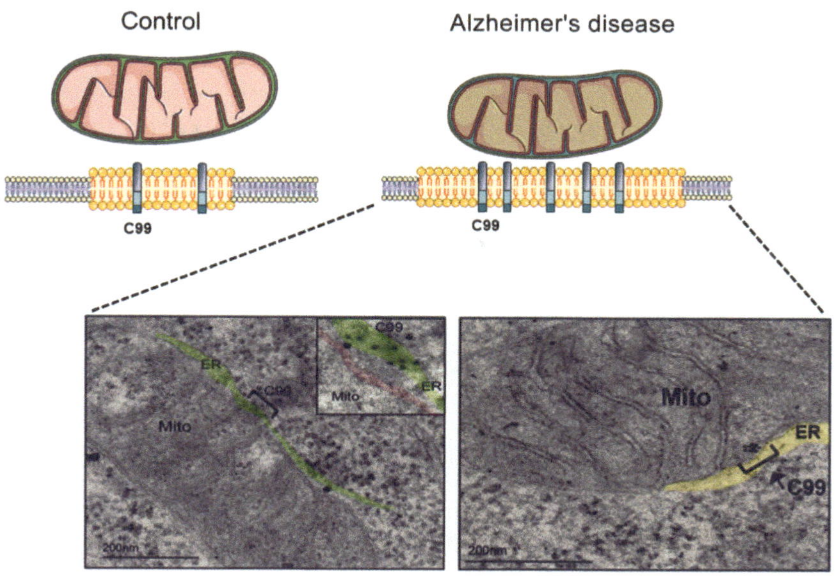

Fig. 2 Accumulation of C99 at MAM in cell modes of AD. Representative immunoelectron microscopy image of cells KO for both Presenilin-1 and -2, and incubated with antibodies against APP-CTF conjugated with immunogold particles. (Left panel) The ER is colored in green and outer mitochondria membrane colored in pink; (Right) ER membranes are colored in yellow. *From Pera, M., Larrea, D., Guardia-Laguarta, C., Montesinos, J., Velasco, K. R., Agrawal, R. R., et al. (2017). Increased localization of APP-C99 in mitochondria-associated ER membranes causes mitochondrial dysfunction in Alzheimer disease.* The EMBO Journal, *e201796797. https://doi.org/10.15252/embj.201796797.*

conditioning tasks and also exhibited deficits in open field and rotarod tests (Saura et al., 2004). The lack of presenilins also resulted in many other impaired physiological functions including neurotransmitter release and calcium release. These data, supported by other studies demonstrating toxicity linked to the absence of or low γ-secretase activity (Doody et al., 2013; Mitani et al., 2012; Tamayev and Dadamio, 2012), strongly argue against γ-secretase inhibition as a therapeutic strategy for the treatment of AD. These results define essential roles and molecular targets of presenilins in synaptic plasticity, learning and memory, and neuronal survival in the adult cerebral cortex. Thus, C99 appears to be a common biochemical denominator in AD mouse models and could be an etiological trigger of AD pathology.

2.2 C99 in familial and sporadic Alzheimer's disease patients

Due to the complexity of AD pathogenesis, relevant data in humans is essential. Extensive work has been done to elucidate the role of BACE1 and C99 in human brain samples. However, the use of human samples has important limitations because brain samples are scarce, as many investigators are interested in the same areas, and because the tissues are obtained post mortem and so represent the end stage of the disease rather than earlier stages that can be therapeutically targeted. Nevertheless, some papers have indeed reported significantly elevated C99 levels in human AD brain tissue compared to control brains in both FAD and SAD (Fukumoto et al., 2002; Holsiger, 2002; Pera et al., 2013; Yang et al., 2003) indicating that C99 accumulation also occurs in human pathology and can be observed at late stages of the disease.

As in animal models, BACE1 activity increases with aging in humans (Fukumoto et al., 2004) and AD-related brain regions like the temporal cortex show more pronounced alterations in BACE1 activity (Fukumoto et al., 2002). Levels of C99 are also increased in FAD human fibroblasts compared to controls (Pera et al., 2017). Determining whether C99 accumulation also happens to be more prominent at the MAM in human brain tissues would be extremely useful to support the C99/MAM hypothesis. Unfortunately, due to the difficulties in obtaining fresh human brain tissue, these assays have not been done yet. Nevertheless, the mechanism leading to the increased levels of BACE1 protein and C99 in human AD brains is currently under investigation as it may constitute a common determining factor in the pathogenesis of both sporadic and familial AD.

3. Mitochondria-associated endoplasmic reticulum membranes in Alzheimer's disease

Mitochondria-associated endoplasmic reticulum membranes (MAM) are a domain of the ER connected physically and biochemically to the mitochondria and are involved in a number of key metabolic functions (Hayashi et al., 2009), including the synthesis and transfer of phospholipids between the ER and mitochondria (Vance, 2003), cholesterol metabolism (Rusiñol et al., 1994), and calcium homeostasis (Csordás et al., 2010). The MAM is an extremely dynamic compartment that can change and adapt depending on cellular requirements. Based on its lipid composition, the MAM is an intracellular lipid raft, a specialized domain rich in cholesterol and sphingolipids that has a liquid-ordered structure with unique biophysical characteristics. MAM recruits and orients various signaling proteins needed to regulate the cross-talk between the ER and the mitochondria. Among the proteins enriched in or localized to the MAM are those involved in lipid metabolism [e.g., phosphatidylserine synthase, PSS (Stone and Vance, 2000;

Vance et al., 1997)]; cholesterol metabolism [e.g., acyl-CoA:cholesterol acyl-transferase 1, ACAT1 (Rusiñol et al., 1994) and, steroidogenic acute regulatory protein 1, STARD1 (Prasad et al., 2015)]; lipid transfer between the ER and mitochondria [e.g., fatty acid transfer protein 4, FACL4 (Jia et al., 2007)]; calcium homeostasis [e.g., inositol triphosphate, IP_3, receptors and sigma1 receptors (Mendes et al., 2005; Szabadkai et al., 2006)]; and the maintenance of mitochondrial function [e.g., voltage-dependent anion channel 2, VDAC2 (Prasad et al., 2015)] and morphology [e.g., dynamin-related protein 1, DRP1 (Friedman et al., 2011)]. In addition, other proteins stabilize and regulate the interaction of ER and mitochondria, including mitofusin 2 (De Brito and Scorrano, 2008) and phosphofurin acidic cluster sorting protein 2 (Simmen et al., 2005) (Fig. 3A).

MAM formation is the consequence of close apposition between the ER and the mitochondria, separated by approximately by 10–40 nm (Csordás et al., 2006). The small distance between the two compartments creates some limitations in properly visualizing contact sites. The MAM can be visualized using different microscopy approaches, with confocal microscopy and transmission electron microscopy (TEM) being the most broadly used until now. Confocal microscopes are readily available, but this technique has two limitations: the resolution limit is far larger than the size of most contact sites and fixation for immunofluorescence may introduce artifacts (Scorrano et al., 2019). Transmission electron microscopy is an atomic-resolution microscopy system that provides high-resolution ultrastructure information on contact sites within the cellular context (Csordás et al., 2006). Hence, TEM is often considered the "gold standard" for the study of the fine architecture of contact sites and to reveal their morphological and functional diversity. Imaging approaches are very informative about the physical characteristics of contacts-sites, but insufficient to determine MAM functionality, so they should be accompanied by biochemical assays as the MAM also contain well-described measurable activities that can be used as indicators of the status of ER-mitochondria communication. Functional assays include the analysis of phosphatidylserine (PtdSer) and phosphatidylethanolamine (PtdEtn) synthesis (Schumacher et al., 2002), cholesterol esterification by ACAT1 (gene *SOAT1*) (Chang et al., 2009), and calcium transport and homeostasis between the ER and the mitochondria (Patergnani et al., 2011). This review is focused on lipid-related MAM activities.

In biology, as in architecture, "form follows function." The lipid composition of the MAM is crucial for the proper recruitment, folding and

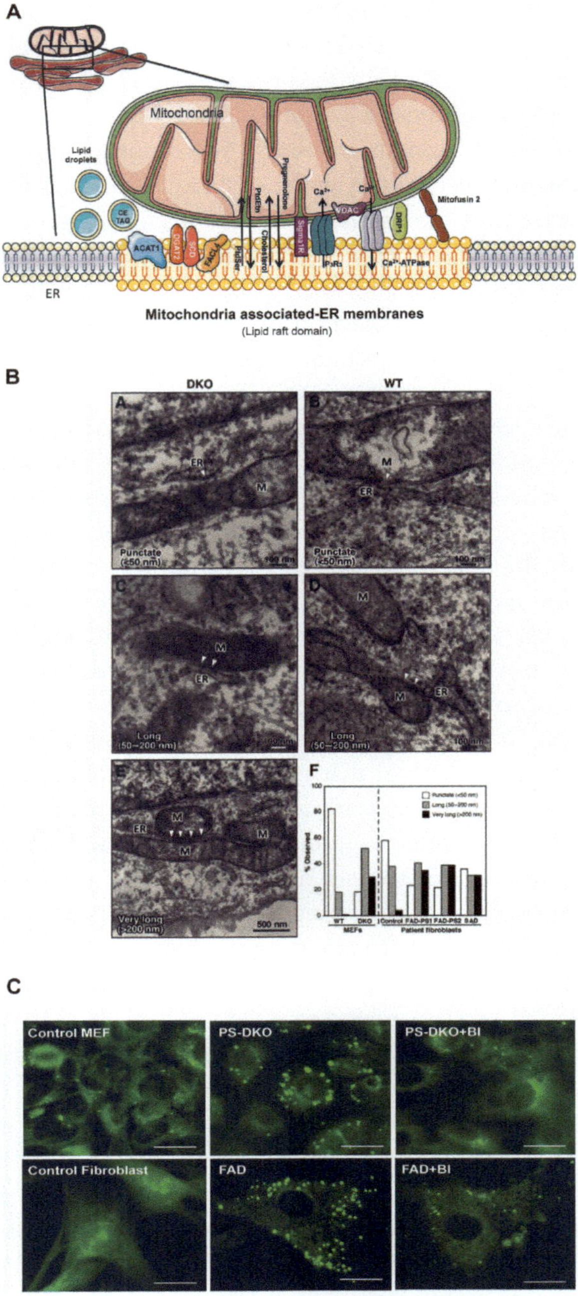

Fig. 3 The MAM is upregulated in cell modes of AD. (A) Schematic representation of MAM-resident proteins. (B) Representative electron microscopy images showing more

function of the proteins that localize to this domain. The fine-tuning of this functional compartment is essential to maintain homeostasis of the processes orchestrated here, and indeed alterations in MAM have been described in different neuropathological conditions, such as Parkinson disease (Guardia-Laguarta et al., 2015), Charcot-Marie-Tooth (Larrea et al., 2019), amyotrophic lateral sclerosis (De vos et al., 2012; Watanabe et al., 2016) and AD (Area-Gomez et al., 2012; Pera et al., 2017).

3.1 The MAM and γ-secretase

Early symptoms of AD include alterations in lipid metabolism (Fraser et al., 2010; Gómez-Ramos and Morán, 2007; Pani et al., 2009a, 2009b; Pettegrew et al., 2001), calcium homeostasis (Bezprozvanny, 2008) and mitochondrial dysfunction (Gillardon et al., 2007; Wang et al., 2009a, 2009b). All of these activities converge at the same cellular location: MAM. The conclusion that many important functions perturbed in AD are MAM-related led to the hypothesis that MAM might play a critical role in AD. Based on that supposition, different research groups found that PS1 and PS2 (Area-Gomez et al., 2009; Newman et al., 2014) and γ-secretase activity itself (Area-Gomez et al., 2012; Schreiner et al., 2015) are enriched predominantly at the MAM. In addition, the finding that MAM is an intracellular lipid raft (Area-Gomez et al., 2009; Hayashi and Fujimoto, 2010) supports the view that the lipid rafts in which PS1 and γ-secretase activity are known to reside (Vetrivel et al., 2004) are located not only at the plasma membrane (Area-Gomez et al., 2009; Marambaud et al., 2002). but also intracellularly at the MAM (Area-Gomez et al., 2009; Hayashi and Fujimoto, 2010).

Given the enrichment of the presenilins and γ-secretase activity at MAM, the possibility that MAM function might be altered in AD has also been explored. First observations using confocal microscopy and TEM showed more abundant and longer ER-mitochondria connections in

abundant and longer ER-mitochondria connections in AD cellular models and fibroblasts from AD patients compared to controls. (C) Confocal image of PS-DKO MEFs and FAD fibroblasts stained for neutral lipids (LipidTox. Life Technologies) show significantly more CEs than corresponding controls. *From Area-Gomez, E., Del Carmen Lara Castillo, M., Tambini, M. D., Guardia-Laguarta, C., De Groof, A. J. C., Madra, M., et al. (2012). Upregulated function of mitochondria-associated ER membranes in Alzheimer disease.* EMBO Journal, 31*(21), 4106–4123. https://doi.org/10.1038/emboj.2012.202.*

cellular AD models and fibroblasts from AD patients (Area-Gomez et al., 2012) (Fig. 3B). The observation of increased physical apposition between the two organelles was validated by the study of the measurable MAM functions: phospholipid synthesis and transfer and cholesterol esterification.

In the broader view, phospholipids can be synthesized *de novo* via the "Kennedy" pathway (Voelker, 2000) or at the MAM via base-exchange (Flis and Daum, 2013). Briefly, the Kennedy pathway involves the synthesis of PtdEtn and Phosphatidylcholine (PtdCho) from ethanolamine and choline precursors via CDP-Ethanolamine and CDP-Choline intermediates. On the other hand, in the MAM, PtdSer translocates from the ER to mitochondria, where it is converted to PtdEtn; PtdEtn then travels back to the ER to be converted into PtdCho (Voelker, 2005). The synthesis and transport of PtdSer and PtdEtn are increased significantly in PS-DKO mouse embryonic fibroblasts (MEFs) and FAD patient fibroblasts (Area-Gomez et al., 2012), indicating upregulated ER-mitochondria cross-talk in FAD.

Another measurable MAM-localized activity that is clearly altered in AD is the intracellular conversion by ACAT1 of free cholesterol to cholesteryl esters (CEs) that are then stored in cytoplasmic lipid droplets (Puglielli et al., 2001). The PS-DKO MEFs and FAD fibroblasts contain significantly more CEs than corresponding controls (Area-Gomez et al., 2012) (Fig. 3C). Notably, the increase in CEs in these cells is paralleled by a concomitant increase in cytoplasmic neutral lipid droplets (containing CEs) (Area-Gomez et al., 2012). Patients with AD have elevated circulating cholesterol (Stefani et al., 2009), increased lipid droplet formation (Gómez-Ramos and Morán, 2007; Pani et al., 2009a, 2009b) and ACAT1 activity appears to be required for the production of Aβ (Puglielli et al., 2004, 2001). These previously unexplained observations can now be explained by a pathological increase in ER-mitochondria apposition.

Remarkably, these findings on altered MAM-mediated phospholipid and cholesterol homeostasis and on increased ER-mitochondria apposition are not limited to PS-DKO MEFs and FAD fibroblasts, as the same results were obtained in SAD fibroblasts and cellular models where different effects of ApoE4 versus ApoE3 on MAM function have been evaluated (Area-Gomez et al., 2012; Tambini et al., 2016). Moreover, changes in the proteome of the MAM have been detected in the B6C3 Tg (APPswe,PSEN1dE9) 85Dbo/J(APP/PS1) mouse at 3 months of age (Völgyi et al., 2018). This mouse is a double transgenic, amyloid overexpressing model of AD (Jankowsky et al., 2001, 2004). At 3 months of age, these mice perform without detectable memory deficit in behavioral tests (Reiserer et al., 2007);

however, changes in mitochondrial proteome had been described at the same age (Volgyi et al., 2017), corroborating the idea that MAM dysregulation is an early event in AD pathogenesis.

The fact that lipid dyshomeostasis and altered MAM apposition could be observed in different cellular and mouse FAD and SAD models at early stages implies that upregulated ER-mitochondria communication and dysregulated MAM function are likely to be a general property of cells from essentially all AD patients, and that altered MAM behavior is a pathogenic event upstream of plaque and tangle formation. Taken together, these findings support the view that the functional cause of AD is increased ER-mitochondria communication and upregulated MAM function.

3.2 In Alzheimer's disease, C99 accumulates at the MAM

Increased ER-mitochondria communication and upregulated MAM function are key players in AD pathogenesis. However, what is the biochemical cause of ER-mitochondria hyperconnectivity and its relationship to aberrant APP processing? Recent reports elucidate these upstream events leading to perturbed MAM structure and function.

Amyloid precursor protein processing occurs in lipid raft domains (Cordy et al., 2006). While most of these domains are found in the plasma membrane, intracellular lipid rafts have also been described (Browman, 2006). One of these intracellular lipid rafts is, as we have already reported, MAM. Presenilins and γ-secretase activity have been shown to localize at the MAM (Area-Gomez et al., 2009; Newman et al., 2014; Schreiner et al., 2015) along with their substrate, C99. Unprocessed C99 can be localized at MAM and its concentration at MAM is increased in cell (MEFs) and animal models of AD and in cells from AD patients. To assess the subcellular localization of C99, diverse experimental approaches have been used, including different types of microscopy and subcellular fractionation.

Confocal microscopy analysis revealed that C99 is mainly present in endosomes and in ER membranes (Das et al., 2016; Pera et al., 2017). In addition to those sites, C99 also localizes to regions where both ER and mitochondria are present. This suggests that, like the presenilins (Area-Gomez et al., 2009), C99 can be localized to areas of the ER apposed to the mitochondria and this is consistent with the fact that γ-secretase activity is present in this compartment (Area-Gomez et al., 2009; Schreiner et al., 2015). Localization of unprocessed C99 has also been verified by immunogold electron microscopy (iEM) of PS-DKO cells using antibodies against

C-terminal regions of APP. In agreement with the confocal analyses, iEM images indicated that, when uncleaved, C99 can be localized in MAM regions of the ER.

The MAM can be isolated by differential centrifugation due to its lower density compared to the bulk ER (Simons and Vaz, 2004). Therefore, the subcellular localization of C99 can be analyzed in the MAM enriched fractions. Subcellular fraction analysis supported the imaging data: C99 can localize at the MAM where it is rapidly processed by γ-secretase in standard conditions while, in AD models and patients, unprocessed C99 accumulates at the MAM fraction upregulating MAM functionality and ER-mitochondrial apposition (Pera et al., 2017). Inhibition of BACE1, which abrogates C99 production, reverses MAM hyperconnectivity in different AD models and FAD human fibroblasts, reinforcing the idea that C99 plays a primary role in AD pathogenesis (Pera et al., 2017). Altogether, these results show that the increase and the retention of uncleaved C99 at the MAM induce both a physical and a functional enhancement of ER-mitochondria connections.

4. Lipid dyshomeostasis in Alzheimer's disease is MAM-related

Lipids comprise a large number of chemically distinct molecules arising from combinations of fatty acids with various backbone structures. The regulation of lipid metabolism is particularly critical in the brain (Borroni et al., 2016; Petrov et al., 2016) as this organ has a high concentration of lipids and also because neurons are so dependent on lipid homeostasis for synapsis and action potentials. The crucial role of lipids in tissue physiology and cell signaling makes it unsurprising that lipid dysregulation has been described numerous times during early stages of neurodegenerative diseases, such as Parkinson's, Niemann Pick, Huntington's and Alzheimer's diseases (Allinquant et al., 2014). Lipid homeostasis alterations in AD include different lipid species like sterol lipids, phospholipids, sphingolipids and fatty acids.

Identification of the gene encoding the E4 variant of ApoE4 (*APOE ε4* allele) as a significant risk factor for late-onset AD provided evidence for a role of cholesterol in the pathogenesis of AD (Puglielli et al., 2003) as ApoE is one of the major apolipoproteins in plasma and the principal cholesterol carrier protein in the brain. Other AD risk factors like obesity and dyslipidemia also point to a deregulation of lipid metabolism in AD. Meta-analyses conclude that obesity results in a 42% increase in the risk for dementia in adults (Beydoun et al., 2008) and elevated cholesterol levels have been proposed as risk factors for the development of AD (Meng et al., 2014; Pappolla et al., 2003). The processing of APP is also strongly influenced by the lipid environment, indicating a link between lipid metabolism and APP cleavage (Mett et al., 2015). Furthermore, large-scale collaborative GWAS and the International Genomics of Alzheimer's Project significantly advanced the knowledge regarding the genetic underpinnings of SAD, identifying at least 20 additional genetic risk loci (Harold et al., 2009; Hollingworth et al., 2011; Lambert, 2013; Lambert et al., 2009; Naj et al., 2011; Seshadri et al., 2010). None of these risk loci confer as great a risk for AD as *APOE* $\varepsilon 4$ does but some are related to lipid metabolism, namely *ATP-binding cassette transporter A* (*ABCA7*; locus19p13.3), *clusterin* (*CLU*; locus 8p21.1), *phosphatidylinositol-binding clathrin assembly protein* (*PICALM*; locus 11q14.2) and *sortilin-related receptor 1* (*SORL1*; locus 11q24.1).

Hence, strong evidence suggests an important role for lipid metabolism in AD; however, the cause of these lipid metabolic alterations is still a matter of debate. The following section section reviews the main lipid classes that are affected in AD and reveals their relationship with C99 and MAM.

4.1 Cholesterol

The brain is the most cholesterol rich organ, containing about 25% of whole body cholesterol (Björkhem et al., 2004). Most of this cholesterol is in myelin (Segatto et al., 2013), but it is also present in glial cells and neuronal membranes in the unesterified (free) form. In the brain, most of the cholesterol is synthesized *de novo* because peripheral cholesterol cannot enter due to the low permeability of the blood-brain barrier (BBB) (Papassotiropoulos et al., 2000). In mature neurons, cholesterol is synthesized and supplied by glial cells, primarily by astrocytes (Poirier et al., 1993).

In AD patients and models, alterations in cholesterol homeostasis have been broadly investigated. Several epidemiological studies indicate that high levels of plasma cholesterol are an AD risk factor. Specifically, high serum cholesterol levels in midlife are associated with a higher risk of developing AD (Meng et al., 2014; Pappolla et al., 2003). Additionally, enhanced levels of low-density lipoprotein (LDL) cholesterol and reduced levels of high-density lipoprotein (HDL) cholesterol in serum correlate with cerebral amyloid deposition in humans (Reed et al., 2014). In human, post-mortem AD brains, cholesterol was found to be elevated and highly enriched in amyloid plaques (Cutler et al., 2004; Panchal et al., 2010). Often, during early AD stages, patients have higher levels of 24(S)-hydroxycholesterol (24-OH cholesterol) in peripheral circulation and cerebrospinal fluid (CSF) than healthy individuals (Khandelwal et al., 2011; Kölsch et al., 2004; Lütjohann et al., 2000) because free cholesterol can also be converted to 24-OH cholesterol (Björkhem, 2006; Björkhem et al., 2009; Shobab et al., 2005), which can then be transported into the peripheral circulation and tissues through the BBB.

The large amount of evidence that demonstrates the importance of cholesterol dyshomeostasis in AD patients points to statins as a potential therapy for AD. Statins inhibit hydroxymethylglutaryl-CoA reductase (HMGCoAR), the enzyme that catalyzes the rate-limiting reaction of *de novo* cholesterol synthesis (Di Paolo and Kim, 2011; Martins et al., 2009). The effectiveness of statins was analyzed in observational and randomized trials, leading to inconsistent conclusions. Statin intake is associated with a reduced incidence of AD or dementia in general in most, but not all, of these studies (Arvanitakis et al., 2008; Haag et al., 2009; Rea et al., 2005; Wang et al., 2018; Wolozin et al., 2000). Of note, the reduction of serum cholesterol levels by the intake of statins in midlife might have a preventive effect toward the development of AD (Kivipelto et al., 2002; Pappolla et al., 2003; Shinohara et al., 2014). In contrast, most clinical trials failed to observe any benefit of statins in individuals already suffering from AD (Feldman et al., 2010; Mcguinness et al., 2016; Sano et al., 2011), indicating that cholesterol-lowering drugs have a preventive rather than a therapeutic potential in AD. All these data obtained in humans indicate remarkable alterations in cholesterol homeostasis and enhanced cholesterol turnover in AD patients.

Cholesterol is enriched in lipid rafts, the membrane microdomains where amyloidogenic processing of APP takes place. Therefore, it is not surprising that changes in cholesterol metabolism disturb lipid raft structure and fluidity, affecting APP processing and vice-versa. Cellular cholesterol increases lead to increased lipid rafts, consequently enhancing amyloidogenic APP processing (Hao et al., 2001; Hicks et al., 2012; Simons et al., 1998). In several animal models, hypercholesterolemia has been correlated with enhanced Aβ levels (Maulik et al., 2013; Refolo et al., 2000; Sparks et al., 1994). Accumulating evidence also indicates that the balance between free cholesterol and CEs is essential for regulating amyloidosis in AD pathogenesis. The release of Aβ increases with CE levels in cultured cells (Puglielli et al., 2001) and Aβ release and CE levels are significantly reduced after inhibiting ACAT1 activity (Hutter-Paier et al., 2004). Moreover, ACAT inhibition reduces aberrant phosphorylated tau accumulation (Van Der Kant et al., 2019).

The supposition that cholesterol plays a central role in AD pathogenesis is also consistent with the identification of allelic variants in other cholesterol metabolism-related genes as risk factors for AD (Banerjee and Mukherjee, 2018), including *ABCA7* (Dong et al., 2017), which promotes the cellular efflux of cholesterol and phospholipids (Burns and Rebeck, 2010).

In AD, C99 abnormally accumulates at the MAM, provoking an increase in MAM apposition and function. Specifically, it has an enormous impact on ACAT1-mediated esterification of free cholesterol, which is subsequently stored in lipid droplets (Fig. 3C). Alzheimer disease models and SAD and FAD patient fibroblasts show a clear increase in CEs and lipid droplets (Area-Gomez et al., 2012), an increase that is interestingly reversed when C99 production is abrogated (Pera et al., 2017). This experimental evidence underlines the key role of the C99-MAM relationship in the regulation of cholesterol metabolism, indicating that alterations in this relationship can potentially be the cause of cholesterol dyshomeostasis in AD.

4.2 Sphingolipids

Sphingolipids are highly enriched in the nervous system, where they are critical constituents of plasma membranes and important for proper brain development and function. Sphingolipids are not merely structural elements, but are also recognized as regulators of cellular events by their ability to form lipid rafts. Thus, perturbations of sphingolipid metabolism can lead to rearrangements in cellular membranes that have been linked to the development of various neurological diseases. The first evidence for a role of sphingolipids in neurodegeneration came from the observation of lysosomal storage diseases, inherited disorders characterized by the lysosomal accumulation of different sphingolipids.

The role of sphingolipids in AD has been studied for the last two decades. Several reports, including post-mortem studies, have shown that ceramide levels are elevated in cortical regions of the AD brain (Cutler et al., 2004; Han et al., 2002; He et al., 2010) and CSF (Satoi et al., 2005). Whitematter ceramides are increased three-fold in the AD temporal cortex and cerebellum during the early stage of the disease (Han et al., 2002). In late stages, ceramides remain elevated about two-fold compared to age-matched controls. *In vitro* studies indicated that ceramide influences Aβ production by stabilizing BACE1 and promoting the amyloidogenic pathway of APP processing (Patil et al., 2007; Puglielli et al., 2003). In contrast to ceramide, sphingomyelin is reduced in AD models (Pera et al., 2017) and has been demonstrated to reduce Aβ levels (Grimm et al., 2005). Recent studies showed that the levels of acid sphingomyelinase, the hydrolysis of sphingomyelins, and the generation of ceramide are positively correlated with hyperphosphorylated tau and Aβ levels in the AD brain (He et al., 2010).

Total ganglioside levels in the AD brain are significantly reduced, mainly in the temporal and frontal cortices and the nucleus basalis of Meynert (Kracun et al., 1992; Svennerholm et al., 1994). In detail, in AD brains, complex gangliosides tend to decrease while simple gangliosides are elevated. The sulfatide levels are depleted up to 92% in gray matter and 58% in whitematter of all examined brain regions in the early stages of AD (Han et al., 2002). Moreover, sulfatide levels are higher in AD patients who are APOE ε4 allele carriers compared to non-APOE-ε4E allele carriers (Han et al., 2003).

The MAM plays a key role in the regulation of lipid homeostasis, including sphingolipid metabolism. Recent data indicate that brain tissues from AD models and AD patient-derived cells display significant increases in sphingomyelinase-mediated sphingomyelin hydrolysis within MAM domains (Pera et al., 2017). This increased sphingomyelinase activity reduced sphingomyelin content and elevated ceramide (the product of sphingomyelin hydrolysis) at ER-mitochondria contact sites and in mitochondrial membranes (Pera et al., 2017). Intriguingly, C99 production inhibition results in restored levels of sphingolipids in AD models (Pera et al., 2017), indicating a causative role of C99 accumulation at the MAM in sphingolipid dysregulation.

4.3 Phospholipids

Phospholipids account for the majority of structural lipids in eukaryotic membranes. They form the major structural unit, the phospholipid bilayer. Each of the subclasses plays individual roles, with varying structural characteristics conferring functional variation. In the brain, the majority of choline used for neurotransmission is stored in the membrane as PtdCho (Blusztajn et al., 1987). As such, it serves as a vital reservoir for essential brain function. Phosphatidylethanolamine is a minor component and, for the most part, is found on the inner leaflet of the plasma membrane (Fadeel and Xue, 2009). The PtdEtn fatty acid chain is enriched in the polyunsaturated fatty acid (PUFA), arachidonic acid, and in the brain arachidonic acid is an essential precursor to a number of important neuromodulatory molecules, such as the prostaglandins and anandamides. As a negatively charged phospholipid, PtdSer is thought to act as an electrostatic mediator for a number of proteins

(Maksymiw et al., 1987) and also as a buffer for essential bioactive fatty acids like docosahexaenoic acid (DHA) (Guo et al., 2007; Van Meer et al., 2008). Docosahexaenoic acid is a PUFA that accounts for 40% of all esterified fatty acids in neuronal plasma membrane phospholipids and is essential for brain development and function (Lauritzen et al., 2001; Muskiet et al., 2006). It is therefore hypothesized that PtdSer in membranes acts as a release/storage facility for DHA. In the serum/plasma of AD patients, as well as in certain post-mortem AD brains regions, DHA content is reduced (Conquer et al., 2000; Söderberg et al., 1991; Tully et al., 2003).

Cardiolipin, a metabolite of phosphatidylglycerol, is a major constituent of mitochondrial membranes (De Kroon et al., 1997; Morita and Terada, 2015), accounting for as much as 15% of all lipids in the mitochondria. Cardiolipin maintains the membrane potential of the inner mitochondrial membrane (IMM), while also supporting proteins involved in mitochondrial respiration (Jiang et al., 2000). Recent reports showed a drop in cardiolipin content in AD brain samples (Monteiro-Cardoso et al., 2015)

Nuclear magnetic resonance studies of Folch (lipid extraction protocol) extracts of autopsy material revealed significant reductions in AD brain levels of PtdEtn and phosphatidylinositol, and elevations in sphingomyelin and the plasmalogen derivative of PtdEtn (Pettegrew et al., 2001). These correlated with neuropathological hallmarks of the diseases and measures of cognitive decline. PtdSer and PtdEtn transport and synthesis are increased in AD cellular models and FAD and SAD patient fibroblasts (Area-Gomez et al., 2012) and, as mentioned above, the enzymes responsible for phospholipid transport and synthesis reside at the MAM and can be affected by the hyperconnectivity of this compartment.

Notably, the main lipid alterations described in AD patients and models converge at the same point: MAM, a hot spot for regulation of lipid metabolism. Consequently, C99 accumulation at the MAM can be a hitherto unknown cause of lipid dyshomeostasis in AD.

5. Mitochondrial alterations in Alzheimer's disease and MAM

In the last decade, many reports have demonstrated mitochondrial impairment in AD. The connection between mitochondria and AD emerged from reports of mitochondrial morphological alterations in post-mortem brain sections (Saraiva et al., 1985; Sumpter et al., 1986) and metabolic alterations in fibroblasts from patients, such as reduced glucose and deficits in calcium homeostasis (Peterson and Goldman, 1986).

This structural and functional evidence prompted the investigators to study the causes and the relevance of mitochondrial dysfunction in AD. The origin of AD mitochondrial changes is still debated, but there are two viewpoints that are prevalent. The first claims that Aβ accumulation induces mitochondrial dysfunction, a theory that does not challenge the amyloid cascade hypothesis and is the most accepted mitochondrial theory in the AD field (Manczak et al., 2006, 2016). However, this perspective has many weaknesses, as most of the experimental evidence was obtained using non-physiological concentrations of Aβ *in vitro* (Casley et al., 2002) and because mitochondrial alterations happen before the deposition of plaques and tangles in the progression of the disease (Balietti et al., 2013; Wang et al., 2014; Yao et al., 2009).

The second hypothesis is the "mitochondrial cascade hypothesis", which proposes that mitochondrial dysfunction sparks all the other pathologic characteristics of the disease (Chen et al., 2006; Coskun, 2012; Demetrius and Driver, 2015; Moreira et al., 2010; Reddy, 2009; Spuch et al., 2012; Swerdlow, 2011; Swerdlow et al., 2014). Nonetheless, while this possibility is intriguing, an abundance of publications (Atamna and Frey, 2007; Cai and Tammineni, 2016; DuBoff et al., 2013; Kivipelto and Solomon, 2006; Santos et al., 2010; Sims et al., 1987; Stokin, 2005; Su et al., 2010; Young-Collier et al., 2012) seems to indicate that the impairment of mitochondrial function in AD is a consequence or a 'side effect' of the biochemical and morphological changes seen in AD (Cabezas-Opazo et al., 2015; Mancuso et al., 2006; Pera et al., 2017). Herein we focus on a third hypothesis that highlights the role of C99 at the MAM as a mechanism of mitochondrial dysfunction in AD.

5.1 Mitochondrial dysfunction in Alzheimer's disease: Biochemical and dynamic alterations

Initial reports of AD brain electron microscopy images revealed altered mitochondrial infrastructure (Johnson, 1970; Wisniewski et al., 1970). In the early 1980s, fluorodeoxyglucose positron emission tomography studies showed less glucose utilization in brains from AD patients than in brains from control subjects (de Leon et al., 1983; Ferris et al., 1980; Foster et al., 1983; Friedland, 1983), which provoked interest in a potential metabolic component of this disease (Beal, 1995; Blass and Zemcov, 1984; Hoyer, 1993; Swerdlow et al., 1994). Bioenergetic deficits have been extensively reported in AD patients and AD models. Specifically, AD patients showed decreases in enzymes of the mitochondrial tricarboxylic acid cycle (Gibson et al., 1988, 2012, Gibson and Huang, 2004, Sorbi et al., 1983) and alterations in pivotal metabolic enzymes, such as the pyruvate dehydrogenase complex. Additionally, reductions in respiratory chain activity and ATP production have been reported in AD models and patients (Atamna and Frey, 2007; Bosetti et al., 2002; Ohta and Ohsawa, 2006; Redjems-Bennani et al., 1998; Sims et al., 1987; Wilkins and Swerdlow, 2017; Young-Collier et al., 2012). The efficiency of ATP production can also be modulated by the association of individual respiratory complexes into supercomplexes (Acin-Perez and Enriquez, 2014) that are reduced in AD cells and animal models (Pera et al., 2017). The reduced efficiency of mitochondrial respiration could explain the high levels of reactive oxygen (ROS) and nitrogen (RNS) species found in AD cells (Du et al., 2010; Hirai et al., 2001; Manczak et al., 2006; Nunomura et al., 2006; Ohta and Ohsawa, 2006) and the consequent damage to DNA, protein and lipid structure (Manoharan et al., 2016).

Mitochondria are extremely dynamic organelles and alterations in mitochondrial fission and/or fusion machinery could have a dramatic effect on mitochondrial function, energy and redox homeostasis. In neurons, mitochondrial motility is essential for providing ATP to synapses, to promote axonal growth, for calcium buffering, and for ensuring mitochondrial repair and degradation (Sheng, 2017). Indeed, altered mitochondrial dynamics are well documented in AD models and patients with a bias toward increased mitochondrial fragmentation (Han et al., 2017; Manczak et al., 2011; Nakamura et al., 2010; Wang et al., 2008a, 2008b, 2009a, 2009b) and an impairment of mitochondrial axonal transport that precedes the accumulation of toxic protein aggregates, and is linked to disturbed axonal integrity and synaptic function (Cai and Tammineni, 2016; Calkins, 2011; Riemer and Kins, 2013; Stokin, 2005; Wang et al., 2010). Moreover, while mitochondria show a uniform distribution in control cells, they are predominantly perinuclear in AD cells (Bonda et al., 2010; Wang et al., 2008a, 2008b, 2009a, 2009b). This observation was supported using human brain samples, specifically in the hippocampal CA1 region of an autoptic brain from an AD patient with a pathogenic PS1 mutation (A434C) (Area-Gomez et al., 2018). Mitochondria were uniformly distributed in the cytosol of control hippocampus, as expected. In contrast, a 'ring' of mitochondria exists around the nucleus of the AD patient neurons, and there is depletion of the organelles in the distal region of hippocampal cell bodies. Thus, perinuclear mitochondria in AD patient cells *in vitro* likely reflect what is occurring clinically *in vivo* (Fig. 4).

Taken together, it is clear that there are numerous structural and functional alterations in mitochondria in both familial and sporadic AD. However, even though many of these mitochondrial phenotypes are evident before the appearance of plaques (Balietti et al., 2013; Wang et al., 2014; Yao et al., 2009),

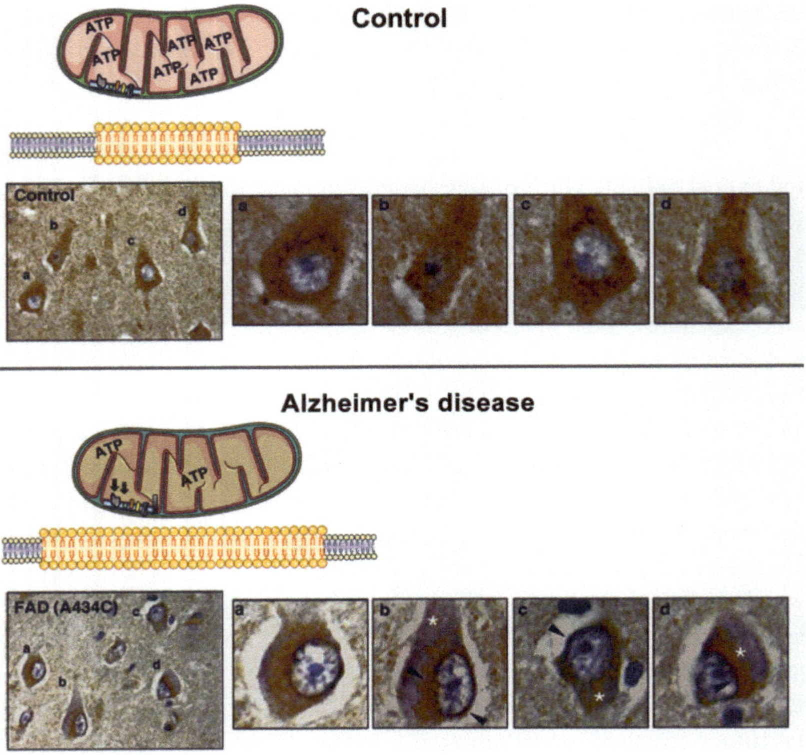

Fig. 4 Increased perinuclear mitochondria appears in brains tissues from AD patients. Mitochondrial staining in control and AD tissues shows significant depletion of these organelles in the distal region of hippocampal cell bodies (FAD brain carrying the *PS1*^{A434C} mutation).

mitochondria are at the center of a complex network of metabolic pathways that can maintain cellular homeostasis compensating reductions in oxidative phosphorylation (OXPHOS). Therefore, correlation between an OXPHOS defect and pathology is far from straightforward and mitochondrial dysfunction is unlikely to be found to be the underlying cause of the disease. Thus, mitochondrial dysfunction and altered mitodynamics, while occurring early, are probably downstream consequences of other specific primary events in AD progression, and are not a fundamental cause of pathogenesis.

5.2 C99 accumulation at the MAM impacts mitochondrial function

Mitochondrial dysfunction is an early event in AD progression; however, which upstream events are the cause is still a matter of debate. Recent publications support the hypothesis that alterations in the C99-MAM interaction can be one of these upstream events. In AD, as already discussed in previous sections, presenilins, γ-secretase activity itself and C99 are enriched at the MAM domains (Area-Gomez et al., 2009; Newman et al., 2014; Pera et al., 2017; Schreiner et al., 2015). Notably, activities localized in the MAM are altered in both familial and sporadic AD (Area-Gomez et al., 2012; Hedskog et al., 2013; Pera et al., 2017). The MAM acts as a lipid homeostasis hub since the cross-talk between ER and mitochondria lipid pathways converge there. Thus, disturbances in MAM-associated activities can impact mitochondrial biology by several mechanisms.

The MAM is involved in the regulation of lipid homeostasis, including phospholipid synthesis and transport and sphingolipid metabolism. In AD, sphingomyelinase activity is increased, reducing sphingomyelin content and elevating ceramide at ER-mitochondria contact sites and in mitochondrial membranes (Pera et al., 2017). Pharmacological reduction of ceramide levels in both AD models and cells from AD patients rescued mitochondrial respiratory deficits and restored supercomplex assembly (Pera et al., 2017), underscoring the harmful effect of ceramides on OXPHOS regulation (Kogot-Levin and Saada, 2014; Yu et al., 2007). Elevations in ceramide could alter mitochondrial membrane properties, impeding the conformation or assembly of respiratory complexes in the inner mitochondrial membrane (Pera et al., 2017) and negatively impacting respiratory rate (Lapuente-Brun et al., 2013). In addition, upon exceeding a threshold, ceramide can form channels in the outer mitochondrial membrane large enough to allow soluble proteins to translocate into the cytosol (Colombini, 2017). We also note that increases in the local concentration of ceramide at the MAM cause mitochondria to bind tubulin via the outer membrane protein porin VDAC (Kong et al., 2018), which regulates mitochondrial transport, subcellular distribution, and division (Ausman et al., 2018).

Interestingly, ceramide levels, OXPHOS and supercomplex assembly are restored in AD cells after BACE1 inhibitor treatment (Pera et al., 2017), which dramatically reduces C99 production, indicating that C99 accumulation is an event upstream of ceramide increase. These results illustrate how bioenergetic defects are logical consequences of alterations in the regulation of sphingolipid turnover at the MAM domains (Kogot-Levin and Saada, 2014).

Moreover, as we have already described, one of the best-known activities localized to the MAM regions is phospholipid synthesis and regulation (Vance, 2014). Alterations in MAM can affect the phospholipid composition of mitochondrial membranes, including the signature mitochondrial lipid, cardiolipin (Flis and Daum, 2013; van Vliet et al., 2014). Cardiolipin interacts with and regulates a number of mitochondrial proteins, including

OXPHOS complexes (Schlame et al., 2000). In fact, this lipid is required for the proper assembly and functioning of all membrane-bound respiratory complexes and ATP synthase (Acehan et al., 2011). Cardiolipin is also involved in the formation and stabilization of mitochondrial supercomplexes (Pfeiffer et al., 2003). Unsurprisingly, alterations in the concentration and/or transacylation of cardiolipin are associated with mitochondrial OXPHOS defects (Gohil et al., 2004). Cardiolipin could also influence bioenergetics indirectly, including regulation of the orientation of the ADP/ATP carrier in the IMM (Claypool et al., 2008) via modulation of mitochondrial creatine kinase activity (Schlattner et al., 2009) or even via regulation of Bax oligomeric pores on the mitochondrial outer membrane (Lai et al., 2019).

Finally, given the detrimental effect of ceramide on mitochondrial supercomplex assembly and respiratory chain activity (Zigdon et al., 2013), we conclude that its accumulation is likely to be a primary cause of mitochondrial dysfunction in AD. Increased levels of unprocessed MAM-localized C99 are a driver of mitochondrial dysfunction in AD, mediated by the loss of sphingolipid homeostasis at ER-mitochondria connections.

6. Concluding remarks

Alzheimer disease, the enigma that neuroscientists aim to untangle, is a compendium of seemingly disconnected pathological features. Altered phospholipid metabolism, elevated levels of cholesterol, accumulation of lipid droplets, aberrant calcium regulation, mitochondrial dysfunction and Aβ plaques appear as completely disparate events, but converge at a shared locus of regulation: MAM. This functional compartment is altered in FAD and SAD models and patients, and a growing body of evidence is building up the "MAM hypothesis" that supports MAM dysregulation as a trigger of many of the AD pathological features. The "MAM hypothesis," which has been comprehensively presented in this review, offers a completely new conceptual AD framework and paves the way for the discovery of new biomarkers and therapies. The toxic role of Aβ plaques in late stages of the disease is undeniable; however, it is crucial to identify upstream events that are causative in early stages to tackle AD-associated neurodegeneration as soon as possible. In AD, C99 accumulates at the MAM, impacting its apposition and function. This is the starting event that triggers lipid and calcium dyshomeostasis, mitochondrial dysfunction and Aβ accumulation. According to this theory, BACE1 partial inhibition, re-establishment of MAM apposition and function or sphingomyelinase modulation, could be potential interesting therapeutic targets to be explored (Fig. 5).

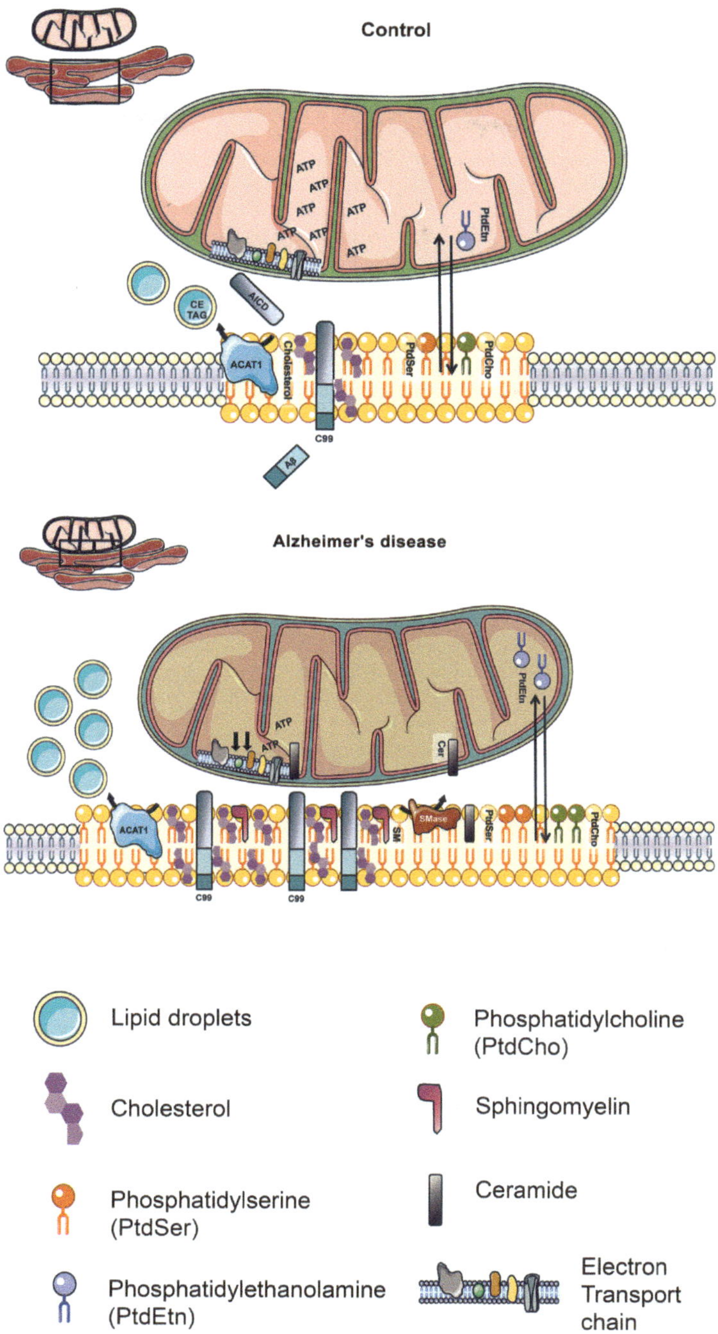

Fig. 5 Schematic representation of the "MAM hypothesis."

References

Acehan, D., Malhotra, A., Xu, Y., Ren, M., Stokes, D. L., & Schlame, M. (2011). Cardiolipin affects the supramolecular organization of ATP synthase in mitochondria. *Biophysical Journal, 100*(9), 2184–2192. https://doi.org/10.1016/j.bpj.2011.03.031.

Acin-Perez, R., & Enriquez, J. A. (2014). The function of the respiratory supercomplexes: The plasticity model. *Biochimica et Biophysica Acta—Bioenergetics, 1837*(4), 444–450. https://doi.org/10.1016/j.bbabio.2013.12.009.

Allinquant, B., Clamagirand, C., & Potier, M. C. (2014). Role of cholesterol metabolism in the pathogenesis of Alzheimer's disease. *Current Opinion in Clinical Nutrition and Metabolic Care, 17*(4), 319–323. https://doi.org/10.1097/MCO.0000000000000069.

Anderson, R. M., Hadjichrysanthou, C., Evans, S., & Wong, M. M. (2017). Why do so many clinical trials of therapies for Alzheimer's disease fail? *The Lancet, 390*(10110), 2327–2329. https://doi.org/10.1016/S0140-6736(17)32399-1.

Area-Gomez, E., De Groof, A. J. C., Boldogh, I., Bird, T. D., Gibson, G. E., Koehler, C. M., et al. (2009). Presenilins are enriched in endoplasmic reticulum membranes associated with mitochondria. *American Journal of Pathology, 175*(5), 1810–1816. https://doi.org/10.2353/ajpath.2009.090219.

Area-Gomez, E., De Groof, A., Bonilla, E., Montesinos, J., Tanji, K., Boldogh, I., et al. (2018). A key role for MAM in mediating mitochondrial dysfunction in Alzheimer disease. *Cell Death and Disease, 9*(3), 335. https://doi.org/10.1038/s41419-017-0215-0.

Area-Gomez, E., Del Carmen Lara Castillo, M., Tambini, M. D., Guardia-Laguarta, C., De Groof, A. J. C., Madra, M., et al. (2012). Upregulated function of mitochondria-associated ER membranes in Alzheimer disease. *EMBO Journal, 31*(21), 4106–4123. https://doi.org/10.1038/emboj.2012.202.

Arvanitakis, Z., Schneider, J. A., Wilson, R. S., Bienias, J. L., Kelly, J. F., Evans, D. A., et al. (2008). Statins, incident Alzheimer disease, change in cognitive function, and neuropathology. *Neurology, 70*(19 Pt 2), 1795–1802. https://doi.org/10.1212/01.wnl.0000288181.00826.63.

Atamna, H., & Frey, W. H. (2007). Mechanisms of mitochondrial dysfunction and energy deficiency in Alzheimer's disease. *Mitochondrion, 7*(5), 297–310. https://doi.org/10.1016/j.mito.2007.06.001.

Ausman, J., Abbade, J., Ermini, L., Farrell, A., Tagliaferro, A., Post, M., et al. (2018). Ceramide-induced BOK promotes mitochondrial fission in preeclampsia. *Cell Death & Disease, 9*(3). https://doi.org/10.1038/s41419-018-0360-0.

Balietti, M., Giorgetti, B., Casoli, T., Solazzi, M., Tamagnini, F., Burattini, C., et al. (2013). Early selective vulnerability of synapses and synaptic mitochondria in the hippocampal CA1 region of the Tg2576 mouse model of Alzheimer's disease. *Journal of Alzheimer's Disease, 34*(4), 887–896. https://doi.org/10.3233/JAD-121711.

Banerjee, S., & Mukherjee, S. (2018). Cholesterol: A key in the pathogenesis of Alzheimer's disease. *ChemMedChem, 13*(17), 1742–1743. https://doi.org/10.1002/cmdc.201800452.

Beal, M. F. (1995). Aging, energy, and oxidative stress in neurodegenerative diseases. *Annals of Neurology, 38*(3), 357–366. https://doi.org/10.1002/ana.410380304.

Beglopoulos, V., Sun, X., Saura, C. A., Lemere, C. A., Kim, R. D., & Shen, J. (2004). Reduced β-amyloid production and increased inflammatory responses in presenilin conditional knock-out mice. *Journal of Biological Chemistry, 279*(45), 46907–46914. https://doi.org/10.1074/jbc.M409544200.

Beydoun, M. A., Beydoun, H. A., & Wang, Y. (2008). Obesity and central obesity as risk factors for incident dementia and its subtypes: A systematic review and meta-analysis. *Obesity Reviews, 9*(3), 204–218. https://doi.org/10.1111/j.1467-789X.2008.00473.x.

Bezprozvanny, I. (2008). Neuronal calcium mishandling and the pathogenesis of Alzheimer's disease. *Trends in Neuroscience, 31*(9), 454–463.

Björkhem, I. (2006). Crossing the barrier: Oxysterols as cholesterol transporters and metabolic modulators in the brain. *Journal of Internal Medicine*, *260*(6), 493–508. https://doi.org/10.1111/j.1365-2796.2006.01725.x.

Björkhem, I., Cedazo-Minguez, A., Leoni, V., & Meaney, S. (2009). Oxysterols and neurodegenerative diseases. *Molecular Aspects of Medicine*, *30*(3), 171–179. https://doi.org/10.1016/j.mam.2009.02.001.

Björkhem, I., Meaney, S., & Fogelman, A. M. (2004). Brain cholesterol: Long secret life behind a barrier. *Arteriosclerosis, Thrombosis, and Vascular Biology*, *24*(5), 806–815. https://doi.org/10.1161/01.ATV.0000120374.59826.1b.

Blass, J. P., & Zemcov, A. (1984). Alzheimer's disease—A metabolic systems degeneration? *Neurochemical Pathology*, *2*(2), 103–114. https://doi.org/10.1007/BF02834249.

Blusztajn, J. K., Liscovitch, M., Mauron, C., Richardson, U. I., & Wurtman, R. J. (1987). Phosphatidylcholine as a precursor of choline for acetylcholine synthesis. *Journal of Neural Transmission. Supplementum*, *24*, 247–259.

Bonda, D. J., Wang, X., Perry, G., Smith, M. A., & Zhu, X. (2010). Mitochondrial dynamics in alzheimers disease: Opportunities for future treatment strategies. *Drugs and Aging*, *27*(3), 181–192. https://doi.org/10.2165/11532140-000000000-00000.

Borroni, M. V., Vallés, A. S., & Barrantes, F. J. (2016). The lipid habitats of neurotransmitter receptors in brain. *Biochimica et Biophysica Acta—Biomembranes*, *1858*(11), 2662–2670. https://doi.org/10.1016/j.bbamem.2016.07.005.

Bosetti, F., Brizzi, F., Barogi, S., Mancuso, M., Siciliano, G., Tendi, E. A., et al. (2002). Cytochrome c oxidase and mitochondrial F1F0-ATPase (ATP synthase) activities in platelets and brain from patients with Alzheimer's disease. *Neurobiology of Aging*, *23*(3), 371–376. https://doi.org/10.1016/S0197-4580(01)00314-1.

Bourgeois, A., Lauritzen, I., Lorivel, T., Bauer, C., Checler, F., & Pardossi-Piquard, R. (2018). Intraneuronal accumulation of C99 contributes to synaptic alterations, apathy-like behavior, and spatial learning deficits in 3 × TgAD and 2 × TgAD mice. *Neurobiology of Aging*, *71*, 21–31. https://doi.org/10.1016/j.neurobiolaging.2018.06.038.

Browman, D. T. (2006). Erlin-1 and erlin-2 are novel members of the prohibitin family of proteins that define lipid-raft-like domains of the ER. *Journal of Cell Science*, *119*(15), 3149–3160. https://doi.org/10.1242/jcs.03060.

Burns, M. P., & Rebeck, G. W. (2010). Intracellular cholesterol homeostasis and amyloid precursor protein processing. *Biochimica et Biophysica Acta—Molecular and Cell Biology of Lipids*, *1801*(8), 853–859. https://doi.org/10.1016/j.bbalip.2010.03.004.

Cabezas-Opazo, F. A., Vergara-Pulgar, K., Pérez, M. J., Jara, C., Osorio-Fuentealba, C., & Quintanilla, R. A. (2015). Mitochondrial dysfunction contributes to the pathogenesis of Alzheimer's disease. *Oxidative Medicine and Cellular Longevity*, *2015*(1). https://doi.org/10.1155/2015/509654.

Cai, Q., & Tammineni, P. (2016). Alterations in mitochondrial quality control in alzheimer's disease. *Frontiers in Cellular Neuroscience*, *10*(FEB)), 1–17. https://doi.org/10.3389/fncel.2016.00024.

Cai, Y., Xiong, K., Zhang, X. M., Cai, H., Luo, X. G., Feng, J. C., et al. (2010). B-Secretase-1 elevation in aged monkey and alzheimer's disease human cerebral cortex occurs around the vasculature in partnership with multisystem axon terminal pathogenesis and β-amyloid accumulation. *European Journal of Neuroscience*, *32*(7), 1223–1238. https://doi.org/10.1111/j.1460-9568.2010.07376.x.

Calkins, M. J. (2011). Amyloid. *Beta impairs Mitochondrial anterograde transport and degenerates synapses in Alzheimer's Disease neurons*, *1812*(4), 507–513. https://doi.org/10.1016/j.cortex.2009.08.003.Predictive.

Casley, C. S., Canevari, L., Land, J. M., Clark, J. B., & Sharpe, M. A. (2002). β-Amyloid inhibits integrated mitochondrial respiration and key enzyme activities. *Journal of Neurochemistry*, *80*(1), 91–100. https://doi.org/10.1046/j.0022-3042.2001.00681.x.

Castellani, R., Hirai, K., Aliev, G., Drew, K. L., Nunomura, A., Takeda, A., et al. (2002). Role of mitochondrial dysfunction in Alzheimer's disease. *Journal of Neuroscience Research, 70*(3), 357–360. https://doi.org/10.1002/jnr.10389.

Cavanagh, C., Colby-Milley, J., Bouvier, D., Farso, M., Chabot, J. G., Quirion, R., et al. (2013). βCTF-correlated burst of hippocampal TNFα occurs at a very early, pre-plaque stage in the TgCRND8 mouse model of Alzheimer's disease. *Journal of Alzheimer's Disease, 36*(2), 233–238. https://doi.org/10.3233/JAD-122131.

Chang, T. Y., Li, B. L., Chang, C. C. Y., & Urano, Y. (2009). Acyl-coenzyme A:cholesterol acyltransferases. *American Journal of Physiology - Endocrinology and Metabolism, 297*(1), 1–9. https://doi.org/10.1152/ajpendo.90926.2008.

Chen, X., Stern, D., & Yan, S. D. (2006). *Mitochondrial Dysfunction and Alzheimer ' s Disease, 2*, 515–520.

Claypool, S. M., Oktay, Y., Boontheung, P., Loo, J. A., & Koehler, C. M. (2008). Cardiolipin defines the interactome of the major ADP/ATP carrier protein of the mitochondrial inner membrane. *Journal of Cell Biology, 182*(5), 937–950. https://doi.org/10.1083/jcb.200801152.

Colombini, M. (2017). Ceramide channels and mitochondrial outer membrane permeability. *Journal of Bioenergetics and Biomembranes, 49*(1), 57–64. https://doi.org/10.1007/s10863-016-9646-z.

Conquer, J. A., Tierney, M. C., Zecevic, J., Bettger, W. J., & Fisher, R. H. (2000). Fatty acid analysis of blood plasma of patients with Alzheimer's disease, other types of dementia, and cognitive impairment. *Lipids, 35*(12), 1305–1312. https://doi.org/10.1007/s11745-000-0646-3.

Cordy, J. M., Hooper, N. M., & Turner, A. J. (2006, January). The involvement of lipid rafts in Alzheimer's disease (review). *Molecular Membrane Biology*. https://doi.org/10.1080/09687860500496417.

Coskun, P. (2012). NIH Public Access. *Biochimica et Biophysica Acta - Biomembranes, 1820*(5), 553–564. https://doi.org/10.1016/j.cortex.2009.08.003.Predictive.

Crawley, J. N. (1999). Behavioral phenotyping of transgenic and knockout mice: Experimental design and evaluation of general health, sensory functions, motor abilities, and specific behavioral tests. *Brain Research, 835*(1), 18–26. https://doi.org/10.1016/S0006-8993(98)01258-X.

Csordás, G., Renken, C., Várnai, P., Walter, L., Weaver, D., Buttle, K. F., et al. (2006). Structural and functional features and significance of the physical linkage between ER and mitochondria. *Journal of Cell Biology, 174*(7), 915–921. https://doi.org/10.1083/jcb.200604016.

Csordás, G., Várnai, P., Golenár, T., Roy, S., Purkins, G., Schneider, T. G., et al. (2010). Imaging Interorganelle contacts and local calcium dynamics at the ER-mitochondrial interface. *Molecular Cell, 39*(1), 121–132. https://doi.org/10.1016/j.molcel.2010.06.029.

Cummings, J. L., Morstorf, T., & Zhong, K. (2014). Alzheimer's disease drug-development pipeline: Few candidates, frequent failures. *Alzheimer's Research & Therapy, 6*(4), 1–7. https://doi.org/10.1186/alzrt269.

Cutler, R. G., Kelly, J., Storie, K., Pedersen, W. A., Tammara, A., Hatanpaa, K., et al. (2004). Involvement of oxidative stress-induced abnormalities in ceramide and cholesterol metabolism in brain aging and Alzheimer's disease. *Proceedings of the National Academy of Sciences of the United States of America, 101*(7), 2070–2075. https://doi.org/10.1073/pnas.0305799101.

Das, U., Wang, L., Ganguly, A., Saikia, J. M., Wagner, S. L., Koo, E. H., et al. (2016). Visualization of APP and BACE-1 approximation in neurons: New insights into the amyloidogenic pathway. *Nature Neuroscience, 19*(1), 55–64. https://doi.org/10.1038/nn.4188.Visualization.

De Brito, O. M., & Scorrano, L. (2008). Mitofusin 2 tethers endoplasmic reticulum to mitochondria. *Nature*, *456*(7222), 605–610. https://doi.org/10.1038/nature07534.
De Kroon, A. I. P. M., Dolis, D., Mayer, A., Lill, R., & De Kruijff, B. (1997). Phospholipid composition of highly purified mitochondrial outer membranes of rat liver and Neurospora crassa. Is cardiolipin present in the mitochondrial outer membrane? *Biochimica et Biophysica Acta—Biomembranes*, *1325*(1), 108–116. https://doi.org/10.1016/S0005-2736(96)00240-4.
de Leon, M. J., Ferris, S. H., George, A. E., Christman, D. R., Fowler, J. S., Gentes, C., et al. (1983). Positron emission tomographic studies of aging and Alzheimer disease. *AJNR. American Journal of Neuroradiology*, *4*(3), 568–571.
De Vos, K. J., Mórotz, G. M., Stoica, R., Tudor, E. L., Lau, K. F., Ackerley, S., et al. (2012). VAPB interacts with the mitochondrial protein PTPIP51 to regulate calcium homeostasis. *Human Molecular Genetics*, *21*(6), 1299–1311. https://doi.org/10.1093/hmg/ddr559.
Demetrius, L. A., & Driver, J. A. (2015). Preventing Alzheimer's disease by means of natural selection. *Journal of the Royal Society, Interface/the Royal Society*, *12*(102), 20140919. https://doi.org/10.1098/rsif.2014.0919.
Di Paolo, G., & Kim, T. W. (2011). Linking lipids to Alzheimer's disease: Cholesterol and beyond. *Nature Reviews Neuroscience*, *12*(5), 284–296. https://doi.org/10.1038/nrn3012.
Dong, H. K., Gim, J. A., Yeo, S. H., & Kim, H. S. (2017). Integrated late onset Alzheimer's disease (LOAD) susceptibility genes: Cholesterol metabolism and trafficking perspectives. *Gene*, *597*, 10–16. https://doi.org/10.1016/j.gene.2016.10.022.
Doody, R. S., Raman, R., Farlow, M., Iwatsubo, T., Vellas, B., Joffe, S., et al. (2013). A phase 3 trial of semagacestat for treatment of Alzheimer's disease. *New England Journal of Medicine*, *369*(4), 341–350. https://doi.org/10.1056/NEJMoa1210951.
Du, H., Guo, L., Yan, S., Sosunov, A. A., McKhann, G. M., & Yan, S. S. (2010). Early deficits in synaptic mitochondria in an Alzheimer's disease mouse model. *Proceedings of the National Academy of Sciences of the United States of America*, *107*(43), 18670–18675. https://doi.org/10.1073/pnas.1006586107.
DuBoff, B., Feany, M., & Götz, J. (2013). Why size matters—Balancing mitochondrial dynamics in Alzheimer's disease. *Trends in Neurosciences*, *36*(6), 325–335. https://doi.org/10.1016/j.tins.2013.03.002.
Fadeel, B., & Xue, D. (2009). The ins and outs of phospholipid asymmetry in the plasma membrane: Roles in health and disease. *Critical Reviews in Biochemistry and Molecular Biology*, *44*(5), 264–277. https://doi.org/10.1080/10409230903193307.
Farrer, L. A., Cupples, L. A., Haines, J. L., Hyman, B., Kukull, W. A., Mayeux, R., et al. (1997). Effects of age, sex, and ethnicity on the association between apolipoprotein E genotype and Alzheimer disease: A meta-analysis. *Journal of the American Medical Association*, *278*(16), 1349–1356.
Feldman, H. H., Doody, R. S., Kivipelto, M., Sparks, D. L., Waters, D. D., Jones, R. W., et al. (2010). Randomized controlled trial of atorvastatin in mild to moderate Alzheimer disease: LEADe. *Neurology*, *74*(12), 956–964. https://doi.org/10.1212/WNL.0b013e3181d6476a.
Ferris, S. H., deLeon, M. J., Wolf, A. P., Farkas, T., Christman, D. R., Reisberg, B., et al. (1980). Positron emission tomography in the study of aging and senile dementia. *Neurobiology of Aging*, *1*(2), 127–131.
Flis, V. V., & Daum, G. (2013). Lipid transport between the endoplasmic reticulum and mitochondria. *Cold Spring Harbor Perspectives in Biology*, *5*(6), pii: a013235. https://doi.org/10.1101/cshperspect.a013235.
Foley, A. M., Ammar, Z. M., Lee, R. H., & Mitchell, C. S. (2015). Systematic review of the relationship between amyloid β levels and measures of transgenic mouse cognitive deficit in Alzheimer's disease. *Journal of Alzheimer's Disease*, *44*(3), 787–795. https://doi.org/10.3233/JAD-142208.

Foster, N. L., Chase, T. N., Fedio, P., Patronas, N. J., Brooks, R. A., & Di Chiro, G. (1983). Alzheimer's disease: Focal cortical changes shown by positron emission tomography. *Neurology*, *33*(8), 961–965. https://doi.org/10.1212/wnl.33.8.961.

Fraser, T., Tayler, H., & Love, S. (2010). Fatty acid composition of frontal, temporal and parietal neocortex in the normal human brain and in Alzheimer's disease. *Neurochemical Research*, *35*(3), 503–513. https://doi.org/10.1007/s11064-009-0087-5.

Friedland, R. P. (1983). Regional cerebral metabolic alterations in dementia of the Alzheimer type: Positron emission tomography with [18F]fluorodeoxyglucose. *Journal of Computer Assisted Tomography*, 7(4), 590–598.

Friedman, J. R., Lackner, L. L., West, M., DiBenedetto, J. R., Nunnari, J., & Voeltz, G. K. (2011). ER tubules mark sites of mitochondrial division. *Science*, *334*(6054), 358–362. https://doi.org/10.1126/science.1207385.

Fukumoto, H., Cheung, B. S., Hyman, B. T., & Irizarry, M. C. (2002). Secretase protein and activity are increased in the neocortex in Alzheimer disease. *Archives of Neurology*, *59*(9), 1381–1389. Retrieved from http://archneur.jamanetwork.com/.

Fukumoto, H., Rosene, D. L., Moss, M. B., Raju, S., Hyman, B. T., & Irizarry, M. C. (2004). β-Secretase activity increases with aging in human, monkey, and mouse brain. *American Journal of Pathology*, *164*(2), 719–725. https://doi.org/10.1016/S0002-9440(10)63159-8.

Gibson, G. E., Chen, H. L., Xu, H., Qiu, L., Xu, Z., Denton, T. T., et al. (2012). Deficits in the mitochondrial enzyme α-ketoglutarate dehydrogenase lead to Alzheimer's disease-like calcium dysregulation. *Neurobiology of Aging*, 33(6), 1121.e13–1121.e24. https://doi.org/10.1016/j.neurobiolaging.2011.11.003.

Gibson, G. E., & Huang, H. M. (2004). Mitochondrial enzymes and endoplasmic reticulum calcium stores as targets of oxidative stress in neurodegenerative diseases. *Journal of Bioenergetics and Biomembranes*, *36*(4 SPEC.ISS), 335–340. https://doi.org/10.1023/B:JOBB.0000041764.45552.f3.

Gibson, G. E., Sheu, K. R., Blass, J. P., Baker, A., Carlson, K. C., Harding, B., et al. (1988). Reduced activities of thiamine-dependent enzymes in the brains and peripheral tissues of patients with Alzheimer's disease. *Archives of Neurology*, *45*(8), 836–840. https://doi.org/10.1001/archneur.1988.00520320022009.

Gillardon, F., Rist, W., Kussmaul, L., Vogel, J., Berg, M., Danzer, K., et al. (2007). Proteomic and functional alterations in brain mitochondria from Tg2576 mice occur before amyloid plaque deposition. *Proteomics*, 7(4), 605–616. https://doi.org/10.1002/pmic.200600728.

Godyń, J., Jończyk, J., Panek, D., & Malawska, B. (2016). Therapeutic strategies for Alzheimer's disease in clinical trials. *Pharmacological Reports*, *68*(1), 127–138. https://doi.org/10.1016/j.pharep.2015.07.006.

Gohil, V. M., Hayes, P., Matsuyama, S., Schägger, H., Schlame, M., & Greenberg, M. L. (2004). Cardiolipin biosynthesis and mitochondrial respiratory chain function are interdependent. *Journal of Biological Chemistry*, *279*(41), 42612–42618. https://doi.org/10.1074/jbc.M402545200.

Gómez-Ramos, P., & Morán, M. A. (2007). Ultrastructural localization of intraneuronal Aβ-peptide in Alzheimer disease brains. *Journal of Alzheimer's Disease*, *11*(1), 53–59. https://doi.org/10.3233/JAD-2007-11109.

Grimm, M. O. W., Grimm, H. S., Pätzold, A. J., Zinser, E. G., Halonen, R., Duering, M., et al. (2005). Regulation of cholesterol and sphingomyelin metabolism by amyloid β and presenilin. *Nature Cell Biology*, 7(11), 1118–1123. https://doi.org/10.1038/ncb1313.

Guardia-Laguarta, C., Area-Gomez, E., Schon, E. A., & Przedborski, S. (2015). Novel subcellular localization for Î±-synuclein: Possible functional consequences. *Frontiers in Neuroanatomy*, *9*(February), 1–7. https://doi.org/10.3389/fnana.2015.00017.

Guo, Q., Fu, W., Sopher, B. L., Miller, M. W., Ware, C. B., Martin, G. M., et al. (1999). Increased vulnerability of hippocampal neurons to excitotoxic necrosis in presenilin-1 mutant knock-in mice. *Nature Medicine*, *5*(1), 101–106. https://doi.org/10.1038/4789.

Guo, M., Stockert, L., Akbar, M., & Kim, H. Y. (2007). Neuronal specific increase of phosphatidylserine by docosahexaenoic acid. *Journal of Molecular Neuroscience*, *33*(1), 67–73. https://doi.org/10.1007/s12031-007-0046-z.

Haag, M. D. M., Hofman, A., Koudstaal, P. J., Stricker, B. H. C., & Breteler, M. M. B. (2009). Statins are associated with a reduced risk of Alzheimer disease regardless of lipophilicity. The Rotterdam study. *Journal of Neurology, Neurosurgery and Psychiatry*, *80*(1), 13–17. https://doi.org/10.1136/jnnp.2008.150433.

Han, X., Cheng, H., Fryer, J. D., Fagan, A. M., & Holtzman, D. M. (2003). Novel role for apolipoprotein E in the central nervous system: Modulation of sulfatide content. *Journal of Biological Chemistry*, *278*(10), 8043–8051. https://doi.org/10.1074/jbc.M212340200.

Han, X., Holtzman, D. M., McKeel, D. W., Kelley, J., & Morris, J. C. (2002). Substantial sulfatide deficiency and ceramide elevation in very early Alzheimer's disease: Potential role in disease pathogenesis. *Journal of Neurochemistry*, *82*(4), 809–818. https://doi.org/10.1046/j.1471-4159.2002.00997.x.

Han, X. J., Hu, Y. Y., Yang, Z. J., Jiang, L. P., Shi, S. L., Li, Y. R., et al. (2017). Amyloid β-42 induces neuronal apoptosis by targeting mitochondria. *Molecular Medicine Reports*, *16*(4), 4521–4528. https://doi.org/10.3892/mmr.2017.7203.

Hao, M., Mukherjee, S., & Maxfield, F. R. (2001). Cholesterol depletion induces large scale domain segregation in living cell membranes. *Proceedings of the National Academy of Sciences of the United States of America*, *98*(23), 13072–13077. https://doi.org/10.1073/pnas.231377398.

Hardy, J. A., & Higgins, G. A. (1992). Alzheimer's disease: The amyloid cascade hypothesis. *Science*, *256*, 184–185.

Harold, D., Abraham, R., Hollingworth, P., Sims, R., Hamshere, M., Pahwa, J. S., et al. (2009). Genome-wide association study identifies variants at CLU and PICALM associated with Alzheimer's disease, and shows evidence for additional susceptibility genes. *Nature Genetics*, *41*(10), 1088–1093. https://doi.org/10.1038/ng.440.Genome-wide.

Hayashi, T., & Fujimoto, M. (2010). Detergent-resistant microdomains determine the localization of σ-1 receptors to the endoplasmic reticulum-mitochondria junction. *Molecular Pharmacology*, 77(4), 517–528. https://doi.org/10.1124/mol.109.062539.

Hayashi, T., Rizzuto, R., Hajnoczky, G., & Su, T. P. (2009). MAM: More than just a housekeeper. *Trends in Cell Biology*, *19*(2), 81–88. https://doi.org/10.1016/j.tcb.2008.12.002.

He, X., Huang, Y., Li, B., Gong, C. X., & Schuchman, E. H. (2010). Deregulation of sphingolipid metabolism in Alzheimer's disease. *Neurobiology of Aging*, *31*(3), 398–408. https://doi.org/10.1016/j.neurobiolaging.2008.05.010.

Hedskog, L., Pinho, C. M., Filadi, R., Ronnback, A., Hertwig, L., Wiehager, B., et al. (2013). Modulation of the endoplasmic reticulum-mitochondria interface in Alzheimer's disease and related models. *Proceedings of the National Academy of Sciences of the United States of America*, *110*(19), 7916–7921. https://doi.org/10.1073/pnas.1300677110.

Hicks, D. A., Nalivaeva, N. N., & Turner, A. J. (2012). Lipid rafts and Alzheimer's disease: Protein-lipid interactions and perturbation of signaling. *Frontiers in Physiology*, *3*, 189. https://doi.org/10.3389/fphys.2012.00189.

Hirai, K., Aliev, G., Nunomura, A., Fujioka, H., Russell, R. L., Atwood, C. S., et al. (2001). Mitochondrial abnormalities in Alzheimer's disease. *Journal of Neuroscience*, *21*(9), 3017–3023. https://doi.org/10.1017/s1431927611001875.

Hollingworth, P., Harold, D., Sims, R., Gerrish, A., Lambert, C., Carrasquillo, M. M., et al. (2011). UKPMC funders group author manuscript and CD2AP are associated with Alzheimer' s disease. *Nature Genetics*, *43*(5), 429–435. https://doi.org/10.1038/ng.803.Common.

Holsiger, D. (2002). Increased expression of the amyloid precursor B-secretase in Alzheimer's Disease. *Annals of Neurology*, *51*(6), 760–763. https://doi.org/10.1002/ana.10168.

Holtzman, D. M., Herz, J., & Bu, G. (2012). Apolipoprotein E and apolipoprotein E receptors: Normal biology and roles in Alzheimer disease. *Cold Spring Harbor Perspectives in Medicine*, *2*(3), a006312. https://doi.org/10.1101/cshperspect.a006312.

Hoyer, S. (1993). Brain Oxidative Energy and Related Metabolism, Neuronal Stress, and Alzheimer's Disease: A Speculative Synthesis. *Journal of Geriatric Psychiatry and Neurology*, *6*(1), 3–13. https://doi.org/10.1177/002383099300600101.

Hutter-Paier, B., Hofmeister, A., Windisch, M., Huttunen, H. J., Puglielli, L., Kim, D. Y., et al. (2004). The ACAT inhibitor CP-113,818 markedly reduces amyloid pathology in a mouse model of Alzheimer's disease. *Neuron*, *44*(2), 227–238. https://doi.org/10.1016/j.neuron.2004.08.043.

Irizarry, M. C., Deng, A., Lleo, A., Berezovska, O., Von Arnim, C. A. F., Martin-Rehrmann, M., et al. (2004). Apolipoprotein E modulates γ-secretase cleavage of the amyloid precursor protein. *Journal of Neurochemistry*, *90*(5), 1132–1143. https://doi.org/10.1111/j.1471-4159.2004.02581.x.

Jankowsky, J. L., Fadale, D. J., Anderson, J., Xu, G. M., Gonzales, V., Jenkins, N. A., et al. (2004). Mutant presenilins specifically elevate the levels of the 42 residue beta-amyloid peptide in vivo: Evidence for augmentation of a 42-specific gamma secretase. *Human Molecular Genetics*, *13*(2), 159–170. https://doi.org/10.1093/hmg/ddh019.

Jankowsky, J. L., Slunt, H. H., Ratovitski, T., Jenkins, N. A., Copeland, N. G., & Borchelt, D. R. (2001). Co-expression of multiple transgenes in mouse CNS: A comparison of strategies. *Biomolecular Engineering*, *17*(6), 157–165. https://doi.org/10.1016/s1389-0344(01)00067-3.

Jia, Z., Moulson, C. L., Pei, Z., Miner, J. H., & Watkins, P. A. (2007). Fatty acid transport protein 4 is the principal very long chain fatty acyl-CoA synthetase in skin fibroblasts. *Journal of Biological Chemistry*, *282*(28), 20573–20583. https://doi.org/10.1074/jbc.M700568200.

Jiang, F., Ryan, M. T., Schlame, M., Zhao, M., Gu, Z., Klingenberg, M., et al. (2000). Absence of cardiolipin in the crd1 null mutant results in decreased mitochondrial membrane potential and reduced mitochondrial function. *Journal of Biological Chemistry*, *275*(29), 22387–22394. https://doi.org/10.1074/jbc.M909868199.

Johnson, A. (1970). Nucleoside phosphatase activities associated with the tangles and plaques of Alzheimer's disease: A histochemical study of natural and experimental neurofibrillary tangles. Journal of Neuropathology and Experimental Neurology, 29, 463–478. Retrieved from http://www.ghbook.ir/index.php?name=ذوین های ر سانه و ف رهنگ&option=com_dbook&task=readonline&book_id=13650&page=73&chkhashk=ED9C9491B4&Itemid=218&lang=fa&tmpl=component.

Kaur, G., et al. (2017). 乳鼠心肌提取 HHS Public Access. *Molecular Psychiatry*, *22*(7), 981–989. https://doi.org/10.1038/mp.2016.189.

Khandelwal, P. J., Herman, A. M., & Moussa, C. E. H. (2011). Inflammation in the early stages of neurodegenerative pathology. *Journal of Neuroimmunology*, *238*(1–2), 1–11. https://doi.org/10.1016/j.jneuroim.2011.07.002.

Kimura, R., Devi, L., & Ohno, M. (2010). Partial reduction of BACE1 improves synaptic plasticity, recent and remote memories in Alzheimer's disease transgenic mice. *Journal of Neurochemistry*, *113*(1), 248–261. https://doi.org/10.1111/j.1471-4159.2010.06608.x.

Kivipelto, M., Helkala, E. L., Laakso, M. P., Hänninen, T., Hallikainen, M., Alhainen, K., et al. (2002). Apolipoprotein E ε4 allele, elevated midlife total cholesterol level, and high midlife systolic blood pressure are independent risk factors for late-life Alzheimer disease. *Annals of Internal Medicine*, *137*(3), 149–155. https://doi.org/10.7326/0003-4819-137-3-200208060-00006.

Kivipelto, M., & Solomon, A. (2006). Cholesterol as a risk factor for Alzheimer's disease - Epidemiological evidence. *Acta Neurologica Scandinavica*, *114*(Suppl. 185), 50–57. https://doi.org/10.1111/j.1600-0404.2006.00685.x.

Kogot-Levin, A., & Saada, A. (2014). Ceramide and the mitochondrial respiratory chain. *Biochimie*, *100*(1), 88–94. https://doi.org/10.1016/j.biochi.2013.07.027.

Kölsch, H., Heun, R., Kerksiek, A., Bergmann, K. V., Maier, W., & Lütjohann, D. (2004). Altered levels of plasma 24S- and 27-hydroxycholesterol in demented patients. *Neuroscience Letters*, *368*(3), 303–308. https://doi.org/10.1016/j.neulet.2004.07.031.

Kong, J.-N., Zhu, Z., Itokazu, Y., Wang, G., Dinkins, M. B., Zhong, L., et al. (2018). Novel function of ceramide for regulation of mitochondrial ATP release in astrocytes. *Journal of Lipid Research*, *59*(3), 488–506. https://doi.org/10.1194/jlr.M081877.

Kracun, I., Kalanj, S., Talan-Hranilovic, J., & Cosovic, C. (1992). Cortical distribution of gangliosides in Alzheimer's disease. *Neurochemistry International*, *20*(3), 433–438. https://doi.org/10.1016/0197-0186(92)90058-Y.

Lai, Y. C., Li, C. C., Sung, T. C., Chang, C. W., Lan, Y. J., & Chiang, Y. W. (2019). The role of cardiolipin in promoting the membrane pore-forming activity of BAX oligomers. *Biochimica et Biophysica Acta—Biomembranes*, *1861*(1), 268–280. https://doi.org/10.1016/j.bbamem.2018.06.014.

Lambert, J. C. (2013). Meta-analysis of 74,046 individuals identifies 11 new susceptibility loci for Alzheimer's disease. *Nature Genetics*, *45*(12), 1452–1458.

Lambert, J. C., Heath, S., Even, G., Campion, D., Sleegers, K., Hiltunen, M., et al. (2009). Genome-wide association study identifies variants at CLU and CR1 associated with Alzheimer's disease. *Nature Genetics*, *41*(10), 1094–1099. https://doi.org/10.1038/ng.439.

Lapuente-Brun, E., Moreno-Loshuertos, R., Aciń-Pérez, R., Latorre-Pellicer, A., Colaś, C., Balsa, E., et al. (2013). Supercomplex assembly determines electron flux in the mitochondrial electron transport chain. *Science*, *340*(6140), 1567–1570. https://doi.org/10.1126/science.1230381.

Larrea, D., Pera, M., Gonnelli, A., Quintana-Cabrera, R., Akman, H. O., Guardia-Laguarta, C., et al. (2019). MFN2 mutations in Charcot-Marie-Tooth disease alter mitochondria-associated ER membrane function but do not impair bioenergetics. *Human Molecular Genetics*, *28*(11), 1782–1800. https://doi.org/10.1093/hmg/ddz008.

Lauritzen, L., Hansen, H. S., Jorgensen, M. H., & Michaelsen, K. F. (2001). The essentiality of long chain n-3 fatty acids in relation to development and function of the brain and retina. *Progress in Lipid Research*, *40*(1–2), 1–94. https://doi.org/10.1016/S0163-7827(00)00017-5.

Lauritzen, I., Pardossi-Piquard, R., Bauer, C., Brigham, E., Abraham, J.-D., Ranaldi, S., et al. (2012). The -secretase-derived C-terminal fragment of APP, C99, but not A, is a key contributor to early intraneuronal lesions in triple-transgenic mouse hippocampus. *Journal of Neuroscience*, *32*(46), 16243–16255. https://doi.org/10.1523/jneurosci.2775-12.2012.

Lauritzen, I., Pardossi-Piquard, R., Bourgeois, A., Bécot, A., & Checler, F. (2019). Does intraneuronal accumulation of carboxyl-terminal fragments of the amyloid precursor protein trigger early neurotoxicity in Alzheimer's disease? *Current Alzheimer Research*, *16*(5), 453–457. https://doi.org/10.2174/1567205016666190325092841.

Lok, K., Zhao, H., Shen, H., Wang, Z., Gao, X., Zhao, W., et al. (2013). Characterization of the APP/PS1 mouse model of Alzheimer's disease in senescence accelerated background. *Neuroscience Letters*, 557(PB), 84–89. https://doi.org/10.1016/j.neulet.2013.10.051.

Lütjohann, D., Papassotiropoulos, A., Björkhem, I., Locatelli, S., Bagli, M., Oehring, R. D., et al. (2000). Plasma 24S-hydroxycholesterol (cerebrosterol) is increased in Alzheimer and vascular demented patients. *Journal of Lipid Research*, *41*(2), 195–198.

Maksymiw, R., Sui, S., Gaub, H., & Sackmann, E. (1987). Electrostatic coupling of spectrin dimers to phosphatidylserine containing lipid lamellae. *Biochemistry*, *26*(11), 2983–2990. https://doi.org/10.1021/bi00385a005.

Mancuso, M., Siciliano, G., Filosto, M., & Murri, L. (2006). Mitochondrial dysfunction and Alzheimer's disease: New developments. *Journal of Alzheimer's Disease*, *9*(2), 111–117. https://doi.org/10.3233/JAD-2006-9203.

Manczak, M., Anekonda, T. S., Henson, E., Park, B. S., Quinn, J., & Reddy, P. H. (2006). Mitochondria are a direct site of Aβ accumulation in Alzheimer's disease neurons: Implications for free radical generation and oxidative damage in disease progression. *Human Molecular Genetics*, *15*(9), 1437–1449. https://doi.org/10.1093/hmg/ddl066.

Manczak, M., Calkins, M. J., & Reddy, P. H. (2011). Impaired mitochondrial dynamics and abnormal interaction of amyloid beta with mitochondrial protein Drp1 in neurons from patients with Alzheimer's disease: Implications for neuronal damage. *Human Molecular Genetics*, *20*(13), 2495–2509. https://doi.org/10.1093/hmg/ddr139.

Manczak, M., Kandimalla, R., Fry, D., Sesaki, H., & Reddy, P. H. (2016). Protective effects of reduced dynamin-related protein 1 against amyloid beta-induced mitochondrial dysfunction and synaptic damage in Alzheimer's disease. *Human Molecular Genetics*, *25*(23), 5148–5166. https://doi.org/10.1093/hmg/ddw330.

Manoharan, S., Guillemin, G. J., Abiramasundari, R. S., Essa, M. M., Akbar, M., & Akbar, M. D. (2016). The role of reactive oxygen species in the pathogenesis of Alzheimer's disease, Parkinson's Disease, and Huntington's disease: A mini review. *Oxidative Medicine and Cellular Longevity*, *2016,* 8590578. https://doi.org/10.1155/2016/8590578.

Marambaud, P., Shioi, J., Serban, G., Georgakopoulos, A., Sarner, S., Nagy, V., et al. (2002). A presenilin-1/γ-secretase cleavage releases the E-cadherin intracellular domain and regulates disassembly of adherens junctions. *EMBO Journal*, *21*(8), 1948–1956. https://doi.org/10.1093/emboj/21.8.1948.

Martins, I. J., Berger, T., Sharman, M. J., Verdile, G., Fuller, S. J., & Martins, R. N. (2009). Cholesterol metabolism and transport in the pathogenesis of Alzheimer's disease. *Journal of Neurochemistry*. https://doi.org/10.1111/j.1471-4159.2009.06408.x.

Maulik, M., Westaway, D., Jhamandas, J. H., & Kar, S. (2013). Role of cholesterol in APP metabolism and its significance in Alzheimer's disease pathogenesis. *Molecular Neurobiology*, *47*(1), 37–63. Humana Press Inc. https://doi.org/10.1007/s12035-012-8337-y.

McConlogue, L., Buttini, M., Anderson, J. P., Brigham, E. F., Chen, K. S., Freedman, S. B., et al. (2007). Partial reduction of BACE1 has dramatic effects on Alzheimer plaque and synaptic pathology in APP transgenic mice. *Journal of Biological Chemistry*, *282*(36), 26326–26334. https://doi.org/10.1074/jbc.M611687200.

Mcguinness, B., Hare, O. J., Craig, D., Bullock, R., Malouf, R., & Passmore, P. (2016). Statins for the treatment of dementia (review) summary of findings for the main comparison. *Cochrane Review*, *8*. https://doi.org/10.1002/14651858.CD003160.pub3.www.cochranelibrary.com, https://www.cochranelibrary.com/cdsr/doi/10.1002/14651858.CD003160.pub3/full.

Mehta, D., Jackson, R., Paul, G., Shi, J., Sabbagh, M., Network, V. H., et al. (2018). Why do trials for. *Alzheimer's disease drugs keep failing?* *26*(6), 735–739. https://doi.org/10.1080/13543784.2017.1323868.Why.

Mendes, C. C. P., Gomes, D. A., Thompson, M., Souto, N. C., Goes, T. S., Goes, A. M., et al. (2005). The type III inositol 1,4,5-trisphosphate receptor preferentially transmits apoptotic Ca2+ signals into mitochondria. *Journal of Biological Chemistry*, *280*(49), 40892–40900. https://doi.org/10.1074/jbc.M506623200.

Meng, X. F., Yu, J. T., Wang, H. F., Tan, M. S., Wang, C., Tan, C. C., et al. (2014). Midlife vascular risk factors and the risk of alzheimer's disease: A systematic review and meta-analysis. *Journal of Alzheimer's Disease*, *42*(4), 1295–1310. https://doi.org/10.3233/JAD-140954.

Mett, J., Hartmann, T., & Grimm, M. O. W. (2015). The effects of glycerophospholipids and fatty acids on APP processing: Implications for alzheimer's disease. In *Handbook of lipids in human function. Fatty acids*. Elsevier Inc. https://doi.org/10.1016/B978-1-63067-036-8.00013-5.

Mitani, Y., Yarimizu, J., Saita, K., Uchino, H., Akashiba, H., Shitaka, Y., et al. (2012). Differential effects between γ-Secretase inhibitors and modulators on cognitive function in amyloid precursor protein-transgenic and nontransgenic mice. *Journal of Neuroscience*, *32*(6), 2037–2050. https://doi.org/10.1523/JNEUROSCI.4264-11.2012.

Mondragón-Rodríguez, S., Gu, N., Manseau, F., & Williams, S. (2018). Alzheimer's transgenic model is characterized by very early brain network alterations and β-CTF fragment accumulation: Reversal by β-secretase inhibition. *Frontiers in Cellular Neuroscience*, *12*(May), 1–17. https://doi.org/10.3389/fncel.2018.00121.

Monteiro-Cardoso, V., Oliveira, M. M., Melo, T., Domingues, M. R. M., Moreira, P. I., Ferreiro, E., et al. (2015). Cardiolipin profile changes are associated to the early synaptic mitochondrial dysfunction in Alzheimer's disease. *Journal of Alzheimer's Disease*, *43*, 1375–1392. https://doi.org/10.3233/JAD-141002.

Moreira, P. I., Carvalho, C., Zhu, X., Smith, M. A., & Perry, G. (2010). Mitochondrial dysfunction is a trigger of Alzheimer's disease pathophysiology. *Biochimica et Biophysica Acta—Molecular Basis of Disease*, *1802*(1), 2–10. https://doi.org/10.1016/j.bbadis.2009.10.006.

Morita, S. Y., & Terada, T. (2015). Enzymatic measurement of phosphatidylglycerol and cardiolipin in cultured cells and mitochondria. *Scientific Reports*, *5*(October 2014), 1–15. https://doi.org/10.1038/srep11737.

Mosconi, L. (2005). Brain glucose metabolism in the early and specific diagnosis of Alzheimer's disease: FDG-PET studies in MCI and AD. *European Journal of Nuclear Medicine and Molecular Imaging*, *32*(4), 486–510. https://doi.org/10.1007/s00259-005-1762-7.

Muskiet, F. A. J., van Goor, S. A., Kuipers, R. S., Velzing-Aarts, F. V., Smit, E. N., Bouwstra, H., et al. (2006). Long-chain polyunsaturated fatty acids in maternal and infant nutrition. *Prostaglandins, Leukotrienes, and Essential Fatty Acids*, *75*(3), 135–144. https://doi.org/10.1016/j.plefa.2006.05.010.

Naj, A. C., Jun, G., Beecham, G. W., Wang, L., Narayan, B., Buros, J., et al. (2011). NIH. *Public Access EPHA1 are associated with late-onset Alzheimer's disease*, *43*(5), 436–441. https://doi.org/10.1038/ng.801.Common.

Nakamura, T., Cieplak, P., Cho, D. H., Godzik, A., & Lipton, S. A. (2010). S-Nitrosylation of Drp1 links excessive mitochondrial fission to neuronal injury in neurodegeneration. *Mitochondrion*, *10*(5), 573–578. https://doi.org/10.1016/j.mito.2010.04.007.

Newman, M., Wilson, L., Verdile, G., Lim, A., Khan, I., Moussavi Nik, S. H., et al. (2014). Differential, dominant activation and inhibition of notch signalling and APP cleavage by truncations of PSEN1 in human disease deficits. *Human Molecular Genetics*, *23*(3), 602–617. https://doi.org/10.1093/hmg/ddt448.

Nunomura, A., Castellani, R. J., Zhu, X., Moreira, P. I., Perry, G., & Smith, M. A. (2006). Involvement of oxidative stress in Alzheimer disease. *Journal of Neuropathology and Experimental Neurology*, *65*(7), 631–641. https://doi.org/10.1097/01.jnen.0000228136.58062.bf.

Ohta, S., & Ohsawa, I. (2006). Dysfunction of mitochondria and oxidative stress in the pathogenesis of Alzheimer's disease: On defects in the cytochrome c oxidase complex and aldehyde detoxification. *Journal of Alzheimer's Disease*, *9*(2), 155–166. https://doi.org/10.3233/JAD-2006-9208.

Panchal, M., Loeper, J., Cossec, J. C., Perruchini, C., Lazar, A., Pompon, D., et al. (2010). Enrichment of cholesterol in microdissected Alzheimer's disease senile plaques as assessed by mass spectrometry. *Journal of Lipid Research*, *51*(3), 598–605. https://doi.org/10.1194/jlr.M001859.

Pani, A., Dessì, S., Diaz, G., La Colla, P., Abete, C., Mulas, C., et al. (2009a). Altered cholesterol ester cycle in skin fibroblasts from patients with Alzheimer's disease. *Journal of Alzheimer's Disease*, *18*(4), 829–841. https://doi.org/10.3233/JAD-2009-1193.

Pani, A., Mandas, A., Diaz, G., Abete, C., Cocco, P. L., Angius, F., et al. (2009b). Accumulation of neutral lipids in peripheral blood mononuclear cells as a distinctive trait of Alzheimer patients and asymptomatic subjects at risk of disease. *BMC Medicine*, 7, 66. https://doi.org/10.1186/1741-7015-7-66.

Papassotiropoulos, A., Lütjohann, D., Bagli, M., Locatelli, S., Jessen, F., Rao, M. L., et al. (2000). Plasma 24S-hydroxycholesterol: A peripheral indicator of neuronal degeneration and potential state marker for Alzheimer's disease. *NeuroReport*, *11*(9), 1959–1961. https://doi.org/10.1097/00001756-200006260-00030.

Pappolla, M. A., Bryant-Thomas, T. K., Herbert, D., Pacheco, J., Fabra Garcia, M., Manjon, M., et al. (2003). Mild hypercholesterolemia is an early risk factor for the development of Alzheimer amyloid pathology. *Neurology*, *61*(2), 199–205. https://doi.org/10.1212/01.WNL.0000070182.02537.84.

Patergnani, S., Suski, J. M., Agnoletto, C., Bononi, A., Bonora, M., De Marchi, E., et al. (2011). Calcium signaling around mitochondria associated membranes (MAMs). *Cell Communication and Signaling: CCS*, *9*, 19. https://doi.org/10.1186/1478-811X-9-19.

Patil, S., Melrose, J., & Chan, C. (2007). Involvement of astroglial ceramide in palmitic acid-induced Alzheimer-like changes in primary neurons. *European Journal of Neuroscience*, *26*(8), 2131–2141. https://doi.org/10.1111/j.1460-9568.2007.05797.x.

Pera, M., Alcolea, D., Sánchez-Valle, R., Guardia-Laguarta, C., Colom-Cadena, M., Badiola, N., et al. (2013). Distinct patterns of APP processing in the CNS in autosomal-dominant and sporadic Alzheimer disease. *Acta Neuropathologica*, *125*(2), 201–213. https://doi.org/10.1007/s00401-012-1062-9.

Pera, M., Larrea, D., Guardia-Laguarta, C., Montesinos, J., Velasco, K. R., Agrawal, R. R., et al. (2017). Increased localization of APP-C99 in mitochondria-associated ER membranes causes mitochondrial dysfunction in Alzheimer disease. *The EMBO Journal*, *36*, 3356–3371. https://doi.org/10.15252/embj.201796797.

Peterson, C., & Goldman, J. E. (1986). Alterations in calcium content and biochemical processes in cultured skin fibroblasts from aged and Alzheimer donors. *Proceedings of the National Academy of Sciences of the United States of America*, *83*(8), 2758–2762. https://doi.org/10.1073/pnas.83.8.2758.

Petrov, A. M., Kasimov, M. R., & Zefirov, A. L. (2016). Brain cholesterol metabolism and its defects: Linkage to neurodegenerative diseases and synaptic dysfunction. *Acta Naturae*, *8*(1), 58–73.

Pettegrew, J. W., Panchalingam, K., Hamilton, R. L., & Mcclure, R. J. (2001). Brain membrane phospholipid alterations in Alzheimer's disease. *Neurochemical Research*, *26*(7), 771–782. https://doi.org/10.1023/A:1011603916962.

Pfeiffer, K., Gohil, V., Stuart, R. A., Hunte, C., Brandt, U., Greenberg, M. L., et al. (2003). Cardiolipin stabilizes respiratory chain supercomplexes. *Journal of Biological Chemistry*, *278*(52), 52873–52880. https://doi.org/10.1074/jbc.M308366200.

Poirier, J., Baccichet, A., Dea, D., & Gauthier, S. (1993). Cholesterol synthesis and lipoprotein reuptake during synaptic remodelling in hippocampus in adult rats. *Neuroscience*, *55*(1), 81–90. https://doi.org/10.1016/0306-4522(93)90456-P.

Prasad, M., Kaur, J., Pawlak, K. J., Bose, M., Whittal, R. M., & Bose, H. S. (2015). Mitochondria-associated endoplasmic reticulum membrane (MAM) regulates steroidogenic activity via steroidogenic acute regulatory protein (StAR)-voltage-dependent anion channel 2 (VDAC2) interaction. *Journal of Biological Chemistry*, *290*(5), 2604–2616. https://doi.org/10.1074/jbc.M114.605808.

Puglielli, L., Ellis, B. C., Ingano, L. A. M., & Kovacs, D. M. (2004). Role of acyl-coenzyme A: Cholesterol acyltransferase activity in the processing of the amyloid precursor protein. *Journal of Molecular Neuroscience*, *24*(1), 093–096. https://doi.org/10.1385/jmn:24:1:093.

Puglielli, L., Konopka, G., Pack-Chung, E., Ingano, L. A. M. K., Berezovska, O., Hyman, B. T., et al. (2001). Acyl-coenzyme A: Cholesterol acyltransferase modulates the generation of the amyloid β-peptide. *Nature Cell Biology*, *3*(10), 905–912. https://doi.org/10.1038/ncb1001-905.

Puglielli, L., Tanzi, R. E., & Kovacs, D. M. (2003). Alzheimer's disease: The cholesterol connection. *Nature Neuroscience*, *6*(4), 345–351. https://doi.org/10.1038/nn0403-345.

Rea, T. D., Breitner, J. C., Psaty, B. M., Fitzpatrick, A. L., Lopez, O. L., Newman, A. B., et al. (2005). Statin use and the risk of incident dementia: The cardiovascular health study. *Archives of Neurology*, *62*(7), 1047–1051. https://doi.org/10.1001/archneur.62.7.1047.

Reddy, P. H. (2009). Amyloid beta, mitochondrial structural and functional dynamics in Alzheimer's disease. *Experimental Neurology*, *218*(2), 286–292. https://doi.org/10.1016/j.expneurol.2009.03.042.

Redjems-Bennani, N., Jeandel, C., Lefebvre, E., Blain, H., Vidailhet, M., & Gueant, J. L. (1998). Abnormal substrate levels that depend upon mitochondrial function in cerebrospinal fluid from Alzheimer patients. *Gerontology*, *44*(5), 300–304. https://doi.org/10.1159/000022031.

Reed, B., Villeneuve, S., Mack, W., DeCarli, C., Chui, H. C., & Jagust, W. (2014). Associations between serum cholesterol levels and cerebral amyloidosis. *JAMA Neurology*, *71*(2), 195–200. https://doi.org/10.1001/jamaneurol.2013.5390.

Refolo, L. M., Pappolla, M. A., Malester, B., LaFrancois, J., Bryant-Thomas, T., Wang, R., et al. (2000). Hypercholesterolemia accelerates the Alzheimer's amyloid pathology in a transgenic mouse model. *Neurobiology of Disease*, 7(4), 321–331. https://doi.org/10.1006/nbdi.2000.0304.

Reiserer, R. S., Harrison, F. E., Syverud, D. C., & McDonald, M. P. (2007). Impaired spatial learning in the APPSwe + PSEN1DeltaE9 bigenic mouse model of Alzheimer's disease. *Genes, Brain, and Behavior*, *6*(1), 54–65. https://doi.org/10.1111/j.1601-183X.2006.00221.x.

Riemer, J., & Kins, S. (2013). Axonal transport and mitochondrial dysfunction in alzheimer's disease. *Neurodegenerative Diseases*, *12*(3), 111–124. https://doi.org/10.1159/000342020.

Ringheim, G., & Szczepanik, A. (2006). Brain inflammation, cholesterol, and glutamate as interconnected participants in the pathology of Alzheimers disease. *Current Pharmaceutical Design*, *12*(6), 719–738. https://doi.org/10.2174/138161206775474215.

Rockenstein, E., Mante, M., Alford, M., Adame, A., Crews, L., Hashimoto, M., et al. (2005). High β-secretase activity elicits neurodegeneration in transgenic mice despite reductions in amyloid β levels. *Journal of Biological Chemistry*, *280*(38), 32957–32967. https://doi.org/10.1074/jbc.m507016200.

Rusiñol, A. E., Cui, Z., Chen, M. H., & Vance, J. E. (1994). A unique mitochondria-associated membrane fraction from rat liver has a high capacity for lipid synthesis and contains pre-Golgi secretory proteins including nascent lipoproteins. *Journal of Biological Chemistry*, *269*(44), 27494–27502.

Salminen, A., Kauppinen, A., Suuronen, T., Kaarniranta, K., & Ojala, J. (2009). ER stress in Alzheimer's disease: A novel neuronal trigger for inflammation and Alzheimer's pathology. *Journal of Neuroinflammation*, *6*, 1–13. https://doi.org/10.1186/1742-2094-6-41.

Sano, M., Bell, K. L., Galasko, D., Galvin, J. E., Thomas, R. G., Van Dyck, C. H., et al. (2011). A randomized, double-blind, placebo-controlled trial of simvastatin to treat Alzheimer disease. *Neurology*, 77(6), 556–563. https://doi.org/10.1212/WNL.0b013e318228bf11.

Santos, R. X., Correia, S. C., Wang, X., Perry, G., Smith, M. A., Moreira, P. I., et al. (2010). A synergistic dysfunction of mitochondrial fission/fusion dynamics and mitophagy in Alzheimer's disease. *Journal of Alzheimer's Disease*, *20*(Suppl. 2), S401–S412. https://doi.org/10.3233/JAD-2010-100666.

Saraiva, A. A., Borges, M. M., Madeira, M. D., Tavares, M. A., & Paula-Barbosa, M. M. (1985). Mitochondrial abnormalities in cortical dendrites from patients with Alzheimer's disease. *Journal of Submicroscopic Cytology*, *17*(3), 459–464.

Sarter, M., Hagan, J., & Dudchenko, P. (1992a). Behavioral screening for cognition enhancers: From indiscriminate to valid testing: Part I. *Psychopharmacology*, *107*, 144–159. Retrieved from https://link.springer.com/content/pdf/10.1007%2FBF02245132.pdf.

Sarter, M., Hagan, J., & Dudchenko, P. (1992b). Behavioral screening for cognition enhancers: From indiscriminate to valid testing: Part II. *Psychopharmacology*, *107*(4), 461–473. https://doi.org/10.1007/bf02245257.

Satoi, H., Tomimoto, H., Ohtani, R., Kitano, T., Kondo, T., Watanabe, M., et al. (2005). Astroglial expression of ceramide in Alzheimer's disease brains: A role during neuronal apoptosis. *Neuroscience*, *130*(3), 657–666. https://doi.org/10.1016/j.neuroscience.2004.08.056.

Saura, C. A., Choi, S.-Y., Beglopoulos, V., Malkani, S., Zhang, D., Shankaranarayana Rao, B. S., et al. (2004). Loss of presenilin function causes impairments of memory and synaptic plasticity followed by age-dependent neurodegeneration. *The Neuron*, *42*(1), 23–36. https://doi.org/10.1016/s0896-6273(04)00182-5.

Schlame, M., Rua, D., & Greenberg, M. L. (2000). The biosynthesis and functional role of cardiolipin. *Progress in Lipid Research*, *39*(3), 257–288. https://doi.org/10.1016/S0163-7827(00)00005-9.

Schlattner, U., Tokarska-Schlattner, M., Ramirez, S., Brückner, A., Kay, L., Polge, C., et al. (2009). Mitochondrial kinases and their molecular interaction with cardiolipin. *Biochimica et Biophysica Acta—Biomembranes*, *1788*(10), 2032–2047. https://doi.org/10.1016/j.bbamem.2009.04.018.

Schreiner, B., Hedskog, L., Wiehager, B., & Ankarcrona, M. (2015). Amyloid-β peptides are generated in mitochondria-associated endoplasmic reticulum membranes. *Journal of Alzheimer's Disease*, *43*(2), 369–374. https://doi.org/10.3233/JAD-132543.

Schumacher, M. M., Choi, J. Y., & Voelker, D. R. (2002). Phosphatidylserine transport to the mitochondria is regulated by ubiquitination. *Journal of Biological Chemistry*, *277*(52), 51033–51042. https://doi.org/10.1074/jbc.M205301200.

Scorrano, L., De Matteis, M. A., Emr, S., Giordano, F., Hajnóczky, G., Kornmann, B., et al. (2019). Coming together to define membrane contact sites. *Nature Communications*, *10*(1). https://doi.org/10.1038/s41467-019-09253-3.

Segatto, M., Di Giovanni, A., Marino, M., & Pallottini, V. (2013). Analysis of the protein network of cholesterol homeostasis in different brain regions: An age and sex dependent perspective. *Journal of Cellular Physiology*, *228*(7), 1561–1567. https://doi.org/10.1002/jcp.24315.

Seshadri, S., Fitzpatrick, A. L., Ikram, M. A., DeStefano, A. L., Gudnason, V., Boada, M., et al. (2010). Genome-wide analysis of genetic loci associated with Alzheimer disease. *JAMA: The Journal of the American Medical Association*, *303*(18), 1832–1840. https://doi.org/10.1001/jama.2010.574.

Sheng, Z.-H. (2017). The interplay of axonal energy homeostasis and mitochondrial trafficking and anchoring. *Trends in Cell Biology*, *27*(6), 403–416. https://doi.org/10.1016/j.physbeh.2017.03.040.

Shinohara, M., Sato, N., Shimamura, M., Kurinami, H., Hamasaki, T., Chatterjee, A., et al. (2014). Possible modification of Alzheimer's disease by statins in midlife: Interactions with genetic and non-genetic risk factors. *Frontiers in Aging Neuroscience*, 6(APR), 1–12. https://doi.org/10.3389/fnagi.2014.00071.

Shobab, L. A., Hsiung, G. R., & Feldman, H. H. (2005). *Cholesterol in Alzheimer's disease*. 4 (December).

Simmen, T., Aslan, J. E., Blagoveshchenskaya, A. D., Thomas, L., Wan, L., Xiang, Y., et al. (2005). PACS-2 controls endoplasmic reticulum-mitochondria communication and Bid-mediated apoptosis. *The EMBO Journal*, *24*(4), 717–729. https://doi.org/10.1038/sj.emboj.7600559.

Simons, M., Keller, P., De Strooper, B., Beyreuther, K., Dotti, C. G., & Simons, K. (1998). Cholesterol depletion inhibits the generation of β-amyloid in hippocampal neurons. *Proceedings of the National Academy of Sciences of the United States of America*, *95*(11), 6460–6464. https://doi.org/10.1073/pnas.95.11.6460.

Simons, K., & Vaz, W. L. C. (2004). Model systems, lipid rafts, and cell membranes. *Annual Review of Biophysics and Biomolecular Structure*, *33*(1), 269–295. https://doi.org/10.1146/annurev.biophys.32.110601.141803.

Sims, N. R., Finegan, J. M., & Blass, J. P. (1987). Altered metabolic properties of cultured skin fibroblasts in Alzheimer's disease. *Annals of Neurology*, *21*(5), 451–457. https://doi.org/10.1002/ana.410210507.

Söderberg, M., Edlund, C., Kristensson, K., & Dallner, G. (1991). Fatty acid composition of brain phospholipids in aging and in Alzheimer's disease. *Lipids*, *26*(6), 421–425. https://doi.org/10.1007/BF02536067.

Sorbi, S., Bird, E. D., & Blass, J. P. (1983). Decreased pyruvate dehydrogenase complex activity in Huntington and Alzheimer brain. *Annals of Neurology*, *13*(1), 72–78. https://doi.org/10.1002/ana.410130116.

Sparks, D. L., Scheff, S. W., Hunsaker, J. C., 3rd, Liu, H., Landers, T., & Gross, D. R. (1994). Induction of Alzheimer-like beta-amyloid immunoreactivity in the brains of rabbits with dietary cholesterol. *Experimental Neurology*, *126*(1), 88–94. https://doi.org/10.1006/exnr.1994.1044.

Spuch, C., Ortolano, S., & Navarro, C. (2012). New insights in the amyloid-beta interaction with mitochondria. *Journal of Aging Research*, *2012*, 9, Article ID 324968, https://doi.org/10.1155/2012/324968.

Stefani, M., & Liguri, G. (2009). Cholesterol in Alzheimers disease: Unresolved questions. *Current Alzheimer Research*, *6*(1), 15–29. https://doi.org/10.2174/156720509787313899.

Sterniczuk, R., Antle, M. C., Laferla, F. M., & Dyck, R. H. (2010). Characterization of the 3xTg-AD mouse model of Alzheimer's disease: Part 2. Behavioral and cognitive changes. *Brain Research*, *1348*, 149–155. https://doi.org/10.1016/j.brainres.2010.06.011.

Stokin, G. (2005). axonopathy and transport deficits in the pathogenesis of Alzheimer's disease. *Science*, *307*, 1282–1288. https://doi.org/10.1210/jcem-10-10-1361.

Stone, S. J., & Vance, J. E. (2000). Phosphatidylserine synthase-1 and -2 are localized to mitochondria-associated membranes. *Journal of Biological Chemistry*, *275*(44), 34534–34540. https://doi.org/10.1074/jbc.M002865200.

Su, B., Wang, X., Bonda, D., Perry, G., Smith, M., & Zhu, X. (2010). Abnormal mitochondrial dynamics-a novel therapeutic target for alzheimer's disease? *Molecular Neurobiology*, *41*(2–3), 87–96. https://doi.org/10.1007/s12035-009-8095-7.

Sumpter, P. Q., Mann, D. M., Davies, C. A., Yates, P. O., Snowden, J. S., & Neary, D. (1986). An ultrastructural analysis of the effects of accumulation of neurofibrillary tangle in pyramidal neurons of the cerebral cortex in Alzheimer's disease. *Neuropathology and Applied Neurobiology*, *12*(3), 305–319. https://doi.org/10.1111/j.1365-2990.1986.tb00142.x.

Sun, X., He, G., Qing, H., Zhou, W., Dobie, F., Cai, F., et al. (2006). Hypoxia facilitates Alzheimer's disease pathogenesis by up-regulating BACE1 gene expession. *Proceedings of the National Academy of Sciences of the United States of America*, *103*(49), 18727–18732. https://doi.org/10.1073/pnas.0606298103.

Svennerholm, L., Boström, K., Jungbjer, B., & Olsson, L. (1994). membrane lipids of adult human brain: Lipid composition of frontal and temporal lobe in subjects of age 20 to 100 years. *Journal of Neurochemistry*, *63*(5), 1802–1811. https://doi.org/10.1046/j.1471-4159.1994.63051802.x.

Swerdlow, R. H. (2011). Brain aging, Alzheimer's disease, and mitochondria. *Biochimica et Biophysica Acta—Biomembranes*, *1812*(12), 1630–1639. https://doi.org/10.1038/jid.2014.371.

Swerdlow, R. H., Burns, J. M., & Khan, S. M. (2014). The Alzheimer's disease mitochondrial cascade hypothesis: Progress and perspectives. *Biochimica et Biophysica Acta—Molecular Basis of Disease, 1842*(8), 1219–1231. https://doi.org/10.1016/j.bbadis.2013.09.010.

Swerdlow, R., Marcus, D. L., Landman, J., Kooby, D., Frey, W., & Freedman, M. L. (1994). Brain glucose metabolism in Alzheimer's disease. *American Journal of the Medical Sciences, 308*(3), 141–144. https://doi.org/10.1097/00000441-199409000-00003.

Szabadkai, G., Bianchi, K., Várnai, P., De Stefani, D., Wieckowski, M. R., Cavagna, D., et al. (2006). Chaperone-mediated coupling of endoplasmic reticulum and mitochondrial Ca2+ channels. *Journal of Cell Biology, 175*(6), 901–911. https://doi.org/10.1083/jcb.200608073.

Tamayev, R., & Dadamio, L. (2012). Inhibition of -secretase worsens memory deficits in a genetically congruous mouse model of Danish dementia. *Molecular Neurodegeneration*, 7(1), 1–7. https://doi.org/10.1186/1750-1326-7-19.

Tambini, M. D., Pera, M., Kanter, E., Yang, H., Guardia-Laguarta, C., Holtzman, D., et al. (2016). ApoE4 upregulates the activity of mitochondria-associated ER membranes. *The EMBO Reports, 17*(1), 27–36. https://doi.org/10.15252/embr.201540614.

Tully, A. M., Roche, H. M., Doyle, R., Fallon, C., Bruce, I., Lawlor, B., et al. (2003). Low serum cholesteryl ester-docosahexaenoic acid levels in Alzheimer's disease: A case–control study. *British Journal of Nutrition, 89*(4), 483–489. https://doi.org/10.1079/bjn2002804.

Umeda, T., Tomiyama, T., Kitajima, E., Idomoto, T., Nomura, S., Lambert, M. P., et al. (2012). Hypercholesterolemia accelerates intraneuronal accumulation of Aβ oligomers resulting in memory impairment in Alzheimer's disease model mice. *Life Sciences, 91*(23–24), 1169–1176. https://doi.org/10.1016/j.lfs.2011.12.022.

Unterberger, U., Höftberger, R., Gelpi, E., Flicker, H., Budka, H., & Voigtländer, T. (2006). Endoplasmic reticulum stress features are prominent in Alzheimer disease but not in prion diseases in vivo. *Journal of Neuropathology and Experimental Neurology, 65*(4), 348–357. https://doi.org/10.1097/01.jnen.0000218445.30535.6f.

Van Dam, D., Vloeberghs, E., Abramowski, D., Staufenbiel, M., & De Deyn, P. P. (2005). APP23 mice as a model of Alzheimer's disease: An example of a transgenic approach to modeling a CNS disorder. *CNS Spectrums, 10*(3), 207–222. https://doi.org/10.1017/S1092852900010051.

Van Der Kant, R., Langness, V. F., Herrera, C. M., Wagner, S. L., Bang, A. G., Goldstein, L. S. B., et al. (2019). Cholesterol metabolism is a druggable axis that independently regulates tau and amyloid-b in iPSC-derived Alzheimer's disease neurons. *Cell Stem Cell, 24*(3), 363–375.e9. https://doi.org/10.1016/j.stem.2018.12.013.

Van Meer, G., Voelker, D. R., & Feigenson, G. W. (2008). Membrane lipids: Where they are and how they behave. *Nature Reviews Molecular Cell Biology, 9*(2), 112–124. https://doi.org/10.1038/nrm2330.

van Vliet, A. R., Verfaillie, T., & Agostinis, P. (2014). New functions of mitochondria associated membranes in cellular signaling. *Biochimica et Biophysica Acta, Molecular Cell Research, 1843*(10), 2253–2262. https://doi.org/10.1016/j.bbamcr.2014.03.009.

Vance, J. E. (2003). Molecular and Cell Biology of Phosphatidylserine and Phosphatidylethanolamine Metabolism. *Progress in Nucleic Acid Research and Molecular Biology, 75*, 69–111. https://doi.org/10.1016/S0079-6603(03)75003-X.

Vance, J. E. (2014). MAM (mitochondria-associated membranes) in mammalian cells: Lipids and beyond. *Biochimica et Biophysica Acta, Molecular and Cell Biology of Lipids, 1841*, 595–609. https://doi.org/10.1016/j.bbalip.2013.11.014.

Vance, D. E., Walkey, C. J., & Cui, Z. (1997). Phosphatidylethanolamine N-methyltransferase from liver. *Biochimica et Biophysica Acta, Lipids and Lipid Metabolism, 1348*(1–2), 142–150. https://doi.org/10.1016/S0005-2760(97)00108-2.

Vetrivel, K. S., Cheng, H., Lin, W., Sakurai, T., Li, T., Nukina, N., et al. (2004). Association of γ-secretase with lipid rafts in post-golgi and endosome membranes. *Journal of Biological Chemistry, 279*(43), 44945–44954. https://doi.org/10.1074/jbc.M407986200.

Voelker, D. R. (2000). Interorganelle transport of aminoglycerophospholipids. *Biochimica et Biophysica Acta, 1486*(1), 97–107. https://doi.org/10.1016/s1388-1981(00)00051-2.

Voelker, D. R. (2005). Bridging gaps in phospholipid transport. *Trends in Biochemical Sciences, 30*(7), 396–404. https://doi.org/10.1016/j.tibs.2005.05.008.

Völgyi, K., Badics, K., Sialana, F. J., Gulyássy, P., Udvari, E. B., Kis, V., et al. (2018). Early presymptomatic changes in the proteome of mitochondria-associated membrane in the APP/PS1 mouse model of Alzheimer's disease. *Molecular Neurobiology, 55*(10), 7839–7857. https://doi.org/10.1007/s12035-018-0955-6.

Volgyi, K., Haden, K., Kis, V., Gulyassy, P., Badics, K., Gyorffy, B. A., et al. (2017). Mitochondrial proteome changes correlating with beta-amyloid accumulation. *Molecular Neurobiology, 54*(3), 2060–2078. https://doi.org/10.1007/s12035-015-9682-4.

Wang, J., Liu, X., Qiu, Y., Shi, Y., Cai, J., Wang, B., et al. (2018). Cell adhesion-mediated mitochondria transfer contributes to mesenchymal stem cell-induced chemoresistance on T cell acute lymphoblastic leukemia cells. *Journal of Hematology & Oncology, 11*(1), 1–13. https://doi.org/10.1186/s13045-018-0554-z.

Wang, X., Perry, G., Smith, M. A., & Zhu, X. (2010). Amyloid-β-derived diffusible ligands cause impaired axonal transport of mitochondria in neurons. *Neurodegenerative Diseases*, 7(1–3), 56–59. https://doi.org/10.1159/000283484.

Wang, X., Su, B., Fujioka, H., & Zhu, X. (2008a). Dynamin-like protein 1 reduction underlies mitochondrial morphology and distribution abnormalities in fibroblasts from sporadic Alzheimer's disease patients. *American Journal of Pathology, 173*(2), 470–482. https://doi.org/10.2353/ajpath.2008.071208.

Wang, X., Su, B., Siedlak, S. L., Moreira, P. I., Fujioka, H., Wang, Y., et al. (2008b). Amyloid-β overproduction causes abnormal mitochondrial dynamics via differential modulation of mitochondrial fission/fusion proteins. *Proceedings of the National Academy of Sciences of the United States of America, 105*(49), 19318–19323. https://doi.org/10.1073/pnas.0804871105.

Wang, X., Su, B., & Zheng, L. (2009a). The role of abnormal mitochondrial dynamics in the pathogenesis of Alzheimer's disease. *Journal of Neruchemistry, 109*(Suppl. 1), 153–159. https://doi.org/10.1111/j.1471-4159.2009.05867.x.The.

Wang, X., Su, B., Zheng, L., Perry, G., Smith, M. A., & Zhu, X. (2009b). The role of abnormal mitochondrial dynamics in the pathogenesis of Alzheimer's disease. *Journal of Neurochemistry, 109*(Suppl), 153–159. https://doi.org/10.1111/j.1471-4159.2009.05867.x.

Wang, X., Wang, W., Li, L., Perry, G., & Lee, H. gon, & Zhu, X. (2014). Oxidative stress and mitochondrial dysfunction in Alzheimer's disease. *Biochimica et Biophysica Acta—Molecular Basis of Disease, 1842*(8), 1240–1247. https://doi.org/10.1016/j.bbadis.2013.10.015.

Watanabe, S., Ilieva, H., Tamada, H., Nomura, H., Komine, O., Endo, F., et al. (2016). Mitochondria-associated membrane collapse is a common pathomechanism in SIGMAR 1 - and SOD 1 -linked ALS. *EMBO Molecular Medicine, 8*(12), 1421–1437. https://doi.org/10.15252/emmm.201606403.

Westerman, M. A., Cooper-Blacketer, D., Mariash, A., Kotilinek, L., Kawarabayashi, T., Younkin, L. H., et al. (2002). The relationship between Aβ and memory in the Tg2576 mouse model of Alzheimer's disease. *Journal of Neuroscience, 22*(5), 1858–1867. https://doi.org/10.1523/jneurosci.22-05-01858.2002.

Wilkins, H. M., & Swerdlow, R. H. (2017). Amyloid precursor protein processing and bioenergetics. *Brain Research Bulletin, 133*, 71–79. https://doi.org/10.1016/j.brainresbull.2016.08.009.

Wisniewski, H., Terry, R. D., & Hirano, A. (1970). Neurofibrillary pathology. *Journal of Neuropathology and Experimental Neurology*, *29*(2), 163–176.

Wolozin, B., Kellman, W., Ruosseau, P., Celesia, G. G., & Siegel, G. (2000). Decreased prevalence of Alzheimer disease associated with 3-hydroxy-3-methyglutaryl coenzyme A reductase inhibitors. *Archives of Neurology*, *57*(10), 1439–1443. https://doi.org/10.1001/archneur.57.10.1439.

Yang, L. B., Lindholm, K., Yan, R., Citron, M., Xia, W., Yang, X. L., et al. (2003). Elevated β-secretase expression and enzymatic activity detected in sporadic Alzheimer disease [1]. *Nature Medicine*, *9*, 3–4. https://doi.org/10.1038/nm0103-3.

Yao, J., Irwin, R. W., Zhao, L., Nilsen, J., Hamilton, R. T., & Brinton, R. D. (2009). Mitochondrial bioenergetic deficit precedes Alzheimer's pathology in female mouse model of Alzheimer's disease. *Proceedings of the National Academy of Sciences of the United States of America*, *106*(34), 14670–14675. https://doi.org/10.1073/pnas.0903563106. Epub 2009 Aug 10.

Young-Collier, K. J., McArdle, M., & Bennett, J. P. (2012). The dying of the light: Mitochondrial failure in Alzheimer's disease. *Journal of Alzheimer's Disease*, *28*(4), 771–781. https://doi.org/10.3233/JAD-2011-111487.

Yu, J., Novgorodov, S. A., Chudakova, D., Zhu, H., Bielawska, A., Bielawski, J., et al. (2007). JNK3 signaling pathway activates ceramide synthase leading to mitochondrial dysfunction. *Journal of Biological Chemistry*, *282*(35), 25940–25949. https://doi.org/10.1074/jbc.M701812200.

Zhang, X., & Le, W. (2010). Pathological role of hypoxia in Alzheimer's disease. *Experimental Neurology*, *223*(2), 299–303. https://doi.org/10.1016/j.expneurol.2009.07.033.

Zigdon, H., Kogot-Levin, A., Park, J. W., Goldschmidt, R., Kelly, S., Merrill, A. H., et al. (2013). Ablation of ceramide synthase 2 causes chronic oxidative stress due to disruption of the mitochondrial respiratory chain. *Journal of Biological Chemistry*, *288*(7), 4947–4956. https://doi.org/10.1074/jbc.M112.402719.

CHAPTER TEN

Experimental studies of mitochondrial and lysosomal function in *in vitro* and *in vivo* models relevant to Parkinson's disease genetic risk

Ria Thomas*, Penelope J. Hallett*, Ole Isacson*
Neuroregeneration Research Institute, Harvard Medical School/McLean Hospital, Belmont, MA, United States
*Corresponding authors: e-mail address: rthomas@mclean.harvard.edu; phallett@mclean.harvard.edu; isacson@mclean.harvard.edu

Contents

1. Introduction 280
2. Mitochondrial dysfunction in Parkinson's disease 281
   2.1 LRRK2-associated mitochondrial dysfunction in Parkinson's disease 282
3. Lysosomal dysfunction in aging and Parkinson's disease 291
   3.1 Effect of modulation of lysosomal glucocerebrosidase activity 293
4. Future considerations for using cell culture models for studying disease-relevant mechanisms *in vitro* 295
5. Conclusions and future directions 296
References 296

Abstract

Several studies have identified the involvement of mitochondrial and lysosomal dysfunction in Parkinson's disease (PD) pathology. In this review we discuss recent work that has identified deficits in mitophagy, mitochondrial network formation, increased sensitivity to mitochondrial stressors and alterations in proteins regulating mitochondrial fission and fusion associated with patient-derived fibroblasts harboring mutations in *LRRK2* gene and from sporadic PD patient cells. We further focus on alterations of lysosomal enzymes, in particular glucocerebrosidase activity, and resultant lipid dyshomeostasis in PD and aging, in human tissue and *in vivo* rodent models. Future studies aimed at understanding the convergence of mitochondrial and lysosomal pathways will be of essence for the identification of unique cellular defects in PD and for the development of new treatments.

International Review of Neurobiology, Volume 154
ISSN 0074-7742
https://doi.org/10.1016/bs.irn.2020.02.004

Abbreviations

ATP	adenosine triphosphate
ATP13A2	ATPase type 13A2
BafA1	bafilomycinA1
CBE	conduritol β-epoxide
DFP	defeiprone
GBA	glucosylceramidase beta
GCase	glucocerebrosidase
GFP	green fluorescent protein
GlcCer	glucosylceramide
GluSph	glucosylsphingosine
GSL	glycosphingolipid
HS	healthy subject
iPSC	induced pluripotent stem cells
LAMP1	lysosome-associated membrane glycoprotein1
LC3A and B	microtubule-associated proteins 1A/1B light chain 3A and B
LRRK2	leucine rich repeat kinase 2
mtDNA	mitochondrial DNA
PD	Parkinson's disease
PINK1	PTEN-induced putative kinase protein 1
RAB7	Ras-related protein 7
RAB10	Ras-related protein 10
RBD	REM sleep behavior disorder
RFP	red fluorescent protein
ROS	reactive oxygen species
SN	substantia nigra
SNc	substantia nigra pars compacta
Tom20	translocase of outer membrane 20 kDa subunit
WT	wild-type

1. Introduction

Parkinson's disease (PD) is a late onset neurodegenerative disease characterized by the focal degeneration of dopaminergic neurons in the substantia nigra pars compacta (SNc) resulting in deficits in the motor system and by the presence of intracellular Lewy bodies. It is the second most common neurodegenerative disorder after Alzheimer's disease, affecting about 2–3% of the population above the age of 65 (Poewe et al., 2017). Studies into gender specific differences identified that males exhibit a significantly higher incidence rate of the disease (around twofold) compared to females (Baldereschi et al., 2000; Hirsch et al., 2016). In addition to the motor symptoms, PD patients also exhibit a multitude of non-motor features, such as cognitive dysfunction, dementia, sleep disorders, constipation, fatigue, depression, hallucinations, sexual dysfunction, cardiovascular problems, hyposmia, etc. (Chaudhuri and Schapira, 2009; Hallett et al., 2012; Sánchez-Martínez et al., 2019; Schapira et al., 2017). With the presence of many non-motor symptoms that most often precede the cardinal motor features of the disease, PD is now rightfully considered a systemic disease and not just a motor disorder. Multiple pathways that underlie the molecular pathogenesis of the disease, including mitochondrial and lysosomal dysfunction, oxidative and endoplasmic reticulum stress, neuroinflammation, calcium and lipid homeostasis, vesicular transport, and α-synuclein proteostasis have been identified (Hallett et al., 2019; Poewe et al., 2017; Toulorge et al., 2016), and this review will primarily focus and expand on our work on mitochondrial and lysosomal defects in PD.

2. Mitochondrial dysfunction in Parkinson's disease

Mitochondria are dynamic intracellular organelles that primarily produce adenosine triphosphate (ATP), the energy currency of the cell. They also play an important role in calcium homeostasis, metabolism of lipids and amino acids, scavenging of free radicals, and apoptosis (Bose and Beal, 2016; Chen et al., 2019). Even though the human brain accounts for only 2% of the body weight, it consumes nearly 20% of the body's total energy production for its proper function (Rolfe and Brown, 1997; Yellen, 2018). This high energy demand also makes it especially vulnerable to any alterations in the mitochondrial energy production pathways. An uninterrupted cycle of mitochondrial turnover, which includes mitochondrial biogenesis and degradation of defective mitochondria through mitophagy, is essential for efficient functioning of the organelle. Besides this, mitochondrial dynamics that includes localization and transport of the organelle to energy demanding areas, network formation through fission and fusion, and presence of intact mitochondrial DNA (mtDNA) also play vital role in its proper function (Chen et al., 2019). All mutations and toxic insults that affect the aforementioned features of mitochondria can be detrimental, and several studies have identified defects in mitochondrial function in PD tissues and patient-derived cells. These include increased production of reactive oxygen species (ROS), respiratory chain deficiencies, depletion of ATP, increased mtDNA damage (Sanders et al., 2014), altered mitophagy and caspase release.

In postmortem studies, reduced activity and abundance of respiratory chain complex I and II subunits have been observed in the substantia nigra (SN) of PD brain tissue (Grünewald et al., 2016; Schapira et al., 1990). Studies have also identified more widespread decrease in respiratory complex I expression in putamen, hippocampus and pedunculopontine nucleus in PD patients (Flønes et al., 2018). The 16.6 kb double stranded mtDNA in humans comprises 37 genes, the majority of which encode for subunits of the respiratory chain machinery (Taanman, 1999). SN neurons from PD and aged brains exhibit large deletions (up to 50%) of mtDNA resulting in impaired cellular respiration and have been found to contribute to their selective neuronal loss (Bender et al., 2006; Dölle et al., 2016; Grünewald et al., 2016; Kraytsberg et al., 2006; Reeve et al., 2008). This mtDNA depletion could also be detected in peripheral blood cells and cerebrospinal fluid, and was suggested to be a potential biomarker for PD (Pyle et al., 2015, 2016). Impaired cellular respiration due to accumulation of somatic mtDNA deletions might, potentially, lead to reduced synthesis of ATP and excessive production of ROS. The resultant oxidative stress could result in oxidation of intracellular dopamine and cytotoxicity in the energy intense SNc dopaminergic neuron. With their naked DNA and limited antioxidant capacity, mitochondria themselves are particularly vulnerable to oxidative stress, amplifying the whole process to end up in a vicious cycle.

Several familial PD genes, either directly or indirectly, have been identified to regulate mitochondrial turnover and dynamics. These include PTEN-induced putative kinase protein 1 (*PINK1*), Parkin (*PRKN*), DJ-1 (*PARK7*), leucine rich repeat kinase 2 (*LRRK2*), ATPase type 13A2 (*ATP13A2*) and glucocerebrosidase (*GBA*).

2.1 LRRK2-associated mitochondrial dysfunction in Parkinson's disease

The LRRK2 protein is a large 286 kDa multidomain protein with four protein-protein interacting domains and two distinct enzymatic activity domains. The GTPase activity of the protein conferred by the Roc (Ras of complex proteins)-COR (C-terminal of Roc) domain helps in the hydrolysis of GTP to GDP, and the kinase activity conferred by the kinase domain is required for phosphorylation, leading to functional regulation of its multiple substrates (Berwick et al., 2019). Even though implicated in a multitude of pathways, including synaptic vesicle recycling (Inoshita et al., 2017; Piccoli et al., 2011), neurite outgrowth (Korecka et al., 2019a; Winner et al., 2011), neuroinflammation

(Kozina et al., 2018; Puccini et al., 2015), mitophagy (Hsieh et al., 2016; Korecka et al., 2019b), autophagy (Giaime et al., 2017; Ho et al., 2018), cytoskeletal regulation (Gandhi et al., 2008; Parisiadou et al., 2009), and translational control (Pellegrini et al., 2018), its exact physiological role is not completely understood. Mutations in *LRRK2* have been linked to autosomal dominant forms of PD and are the most known genetic cause of familial PD. While close to 100 non-sense and missense mutations have been identified in *LRRK2*, only a small number of them (R1441C/G/H, Y1699C, S1761R, I2012T, G2019S and I2020T), primarily located in the Roc-COR and kinase domains, have been identified to be pathogenic (Hernandez et al., 2016) and reported to increase the kinase activity of the protein. In this review we will focus on two studies from our group on mitochondrial dysfunction mediated by the *LRRK2* gene.

2.1.1 Mitochondrial phenotypes in fibroblasts derived from sporadic Parkinson's disease patients

Familial PD cases account for only about 10% of the overall incidence, and the remaining 90% constitute sporadic forms of the disease (Ascherio and Schwarzschild, 2016). About 19 PD causing genes, including 10 autosomal dominant and nine autosomal recessive ones have been identified so far (Deng et al., 2018). While the majority of the familial forms of the disease are monogenic, caused by mutations in a single gene, sporadic forms of the disease have unknown etiology and are likely to be the result of a combination of multiple factors, both genetic and environmental. The lack of a clear genetic basis in sporadic cases calls for the identification of alternate phenotypes that can be targeted for therapeutic intervention. With multiple etiologies causing sporadic PD, it is highly unlikely that a single drug is going to be successful in all these patients. Therefore, stratification of the sporadic patients into groups based on common cellular features would be important for the development of more specific drug treatments.

With these aims in mind, we used skin fibroblasts derived from sporadic PD patients, familial PD patients harboring mutations in the *LRRK2* (G2019S and R1441C) gene and age-matched healthy subject (HS) controls to study disease relevant mitochondrial phenotypes. Cellular phenotypes were studied at baseline and upon treatment with a mitochondrial stressor (valinomycin), and their potential rescue using a pharmacological LRRK2 kinase inhibitor (LRRK2-IN-1) was also tested (Smith et al., 2016).

Increased mitochondrial stress and reduced membrane potential have been observed in PD postmortem brain tissues (Grünewald et al., 2016; Schapira et al., 1990). To mimic this *in vitro* and to exacerbate the phenotypes, cells were treated with valinomycin, a K^+ ionophore that causes depolarization of mitochondria (Cooper et al., 2012; Fedorowicz et al., 2014; Rakovic et al., 2010). Sporadic fibroblasts treated with escalating doses of valinomycin from 10 to 200 μM exhibited varied levels of vulnerability as determined by lactate dehydrogenase cell toxicity assay. Based on the level of vulnerability, sporadic fibroblasts were grouped into two categories. The group of sporadic cells that showed heightened sensitivity to mid concentrations of valinomycin, similar to that observed in cells with LRRK2 G2019S mutation, were labeled as the "highly sensitive" group, and the sporadic cells that did not respond much to the stressor, similar to HS controls, were labeled "normally sensitive" group. All subsequent mitochondrial phenotype assays performed were aimed at finding fundamental differences between these two groups of sporadic cells.

LRRK2-IN-1 is a first generation kinase inhibitor that was shown to reduce LRRK2 kinase activity and lead to dephosphorylation of S910 and S935 (Deng et al., 2011). Treatment of the LRRK2 G2019S and highly sensitive sporadic group of cells with LRRK2-IN-1 reduced their increased vulnerability to valinomycin. This was accompanied by a reduction of LRRK2 phosphorylation at S935 in both these group of cells, indicative of a reduction in kinase activity, as expected. To assess the level of mitochondrial network connectivity and branching in these five groups of fibroblasts, form factor analysis was performed on translocase of outer membrane 20 kDa subunit (Tom20) stained fibroblasts (Fig. 1). At baseline, the highly sensitive group of fibroblasts exhibited a greater percentage of cells with higher form factor (>0.3), suggestive of collapsed, isolated mitochondria with less network (Fig. 1B). This could be explained, in part, by the higher protein expression of the mitochondrial fusion activating proteins Mfn1 and OPA1 and a reduced expression of the fission activating protein Dlp1 observed in these cells at baseline. Upon treatment with valinomycin, all PD groups of cells displayed increased percentage of cells with higher form factor (>0.3) suggestive of mitochondrial collapse and lack of connectivity (Fig. 1B), which was partially rescued with scores normalized to between 0.1 and 0.3 upon treatment with LRRK2-IN-1 (Fig. 1B). Analysis of mitochondrial membrane potential assessed using active mitochondrial stain MitoTracker Red FM did not show any genotype difference between the five groups of cells at baseline. This was reduced massively in a

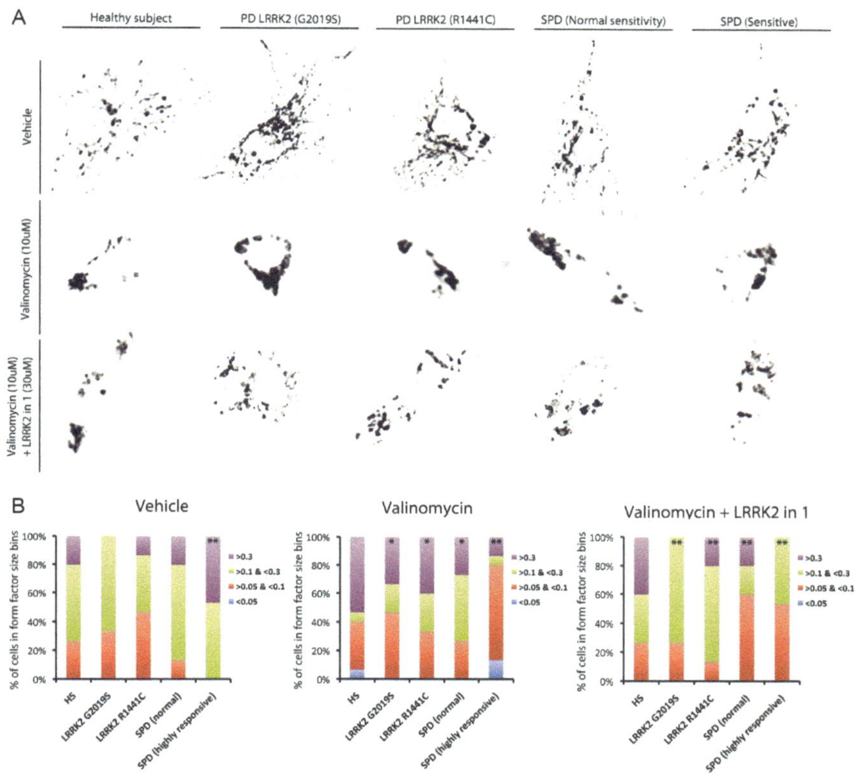

Fig. 1 Mitochondrial network connectivity and branching in PD patient-derived fibroblast cells (A) Representative images of mitochondrial connectivity in Tom20 stained fibroblasts derived from healthy subject controls, PD patients harboring LRRK2 G2019S and R1441C mutations and sporadic PD patients with normal and higher sensitivity to valinomycin stress. (B) Quantification of the images showed that, at baseline, sporadic PD (sPD) group with higher sensitivity had a significantly higher percentage of cells with higher form factor (>0.3). Upon treatment with valinomycin, all four PD groups of cells displayed increased percentage of cells with higher form factor (>0.3) compared to controls, suggestive of mitochondrial collapse and lack of connectivity. Treatment of cells with LRRK2-IN-1 increased the percentage of cells with lower form factor (between 0.1 and 0.3) indicative of a partial rescue and restoration of the mitochondrial network. Data are presented as mean ± SEM. Statistical analysis was performed by one-way ANOVA followed by Dunnett's *post hoc* analysis. $^{*}P<0.05$ and $^{**}P<0.01$. $n=4$–5 cell lines derived from different patients per group. *From Smith, G. A., Jansson, J., Rocha, E. M., Osborn, T., Hallett, P. J., & Isacson, O. (2016). Fibroblast biomarkers of sporadic Parkinson's disease and LRRK2 kinase inhibition.* Molecular Neurobiology, 53, *5161–5177. With permission.*

non-cell line specific manner upon treatment with valinomycin and was restored by LRRK2 kinase inhibition. Mitophagy was studied in these cells using co-localization of Tom20 with the lysosomal marker lysosome-associated membrane glycoprotein1 (LAMP1). Both *LRRK2* mutants (G2019S and R1441C) and the highly sensitive group of sporadic fibroblasts exhibited increased co-localization of Tom20 with LAMP1 compared to HS controls and the normal sensitive group of sporadic fibroblasts. This localization was completely lost upon valinomycin treatment in all cell lines suggestive of alterations in mitophagy, and was restored to normal basal levels with the application of LRRK2-IN-1. While the co-localization of Tom20 and LAMP1 indicates alterations in mitophagy in the *LRRK2* mutants and a subset of the sporadic cells, it is difficult to interpret the actual directionality of the pathway from this static assay.

One of the key findings from this work is that fibroblasts from sporadic PD patients can be classified into two statistically distinct groups based on the presence of shared mitochondrial phenotypes. While half of the sporadic PD lines did not exhibit any phenotype at baseline or upon stress with valinomycin, the other half responded to mitochondrial stress in a manner comparable to that of *LRRK2* mutant cells. This has relevance for potential stratification of patients for clinical trials as well as for pharmacodynamic testing of drugs. Interestingly, the sporadic PD fibroblast lines with heightened sensitivity to valinomycin exhibited several phenotypes that were similar to LRRK2 G2019S cells and most of them were rescued upon inhibition with LRRK2-IN-1, which suggests a role for increased LRRK2 kinase activity in the regulation of these mitochondrial phenotypes. In line with our data, a recent study identified that wild type (WT) LRRK2 in dopaminergic neurons of idiopathic PD SNc displays increased kinase activity in response to oxidative stress and mitochondrial impairment (Maio et al., 2018). It is plausible that the subset of highly sensitive sporadic PD cells that display mitochondrial phenotypes that can be ameliorated with LRRK2-IN-1 treatment might also have increased LRRK2 kinase activity.

2.1.2 Fibroblasts from Parkinson's disease patients harboring the LRRK2 G2019S mutation exhibit reduced mitophagy and maturation of autophagosomes

Work from our group and others have shown that LRRK2 protein regulates mitochondrial function (Cooper et al., 2012; Hsieh et al., 2016; Papkovskaia et al., 2012; Smith et al., 2016; Su and Qi, 2013). Using a static co-localization assay utilizing LAMP1 and Tom20, we have previously shown that fibroblasts from PD patients with the LRRK2 G2019S mutation exhibit aberrant co-localization of Tom20 and LAMP1, which was restored by pharmacological inhibition of LRRK2 kinase activity. Even though this indicates aberrant mitophagy in these cells, the dynamics of this process largely remain unknown. To investigate this better, we studied mitophagy in LRRK2 G2019S fibroblasts using a novel Rosella pH-sensitive, dual fluorescence biosensor probe (Korecka et al., 2019b). The bioprobe comprises mitochondrial target sequence of human ATP3 (mitochondrial membrane ATP synthase protein) coupled with a Rosella construct encoding a pH-sensitive, green fluorescent protein (GFP) variant, pHluorin fused with a pH-insensitive, red fluorescent protein (RFP) variant, DsRed (Rosado et al., 2008). Upon transduction into fibroblasts, the healthy mitochondria labeled by the bioprobe exhibit dual fluorescence. Upon induction of mitophagy, mitochondria associate with a phagophore vacuole before being enclosed into an autophagosome, which then matures and eventually fuses with the lysosome for complete degradation of the organelle. The pH of the various vacuoles enclosing mitochondria along this pathway decreases as it goes from the phagophores to the lysosomes, leading to quenching of the pH-sensitive GFP signal, thereby leaving the pH-insensitive RFP signal intact until complete degradation. From live cell imaging of the transduced cells, ratios of the green to red fluorescence can be calculated and serves as a measure of the rate of mitophagy. An increased green/red fluorescence ratio indicates a reduced rate of mitophagy, while a reduced ratio indicates increased mitophagy. To test the functionality of the probe, HEK293T cells were transfected with it and treated with compounds known to induce mitophagy, such as rotenone (a mitochondrial complex I inhibitor) and defeiprone (DFP), and with a compound that induces autophagy, rapamycin. As expected, treatment of cells with rotenone and DFP reduced the green/red fluorescence ratio when compared to vehicle-treated controls and treatment with rapamycin increased it, thus confirming its specificity and functionality.

Utilizing this optimized probe, we assessed rate of mitophagy in fibroblasts derived from PD patients carrying the LRRK2 G2019S mutation compared to age-matched, HS controls. Cells with the LRRK2 G2019S mutation exhibited a significantly increased green/red ratio compared to controls, indicative of a reduced rate of mitophagy (Fig. 2A). The LRRK2 G2019S mutation (in exon41) is located in the kinase domain and is known to increase the kinase activity of the protein (West et al., 2005). To test whether the kinase activity of the protein regulated

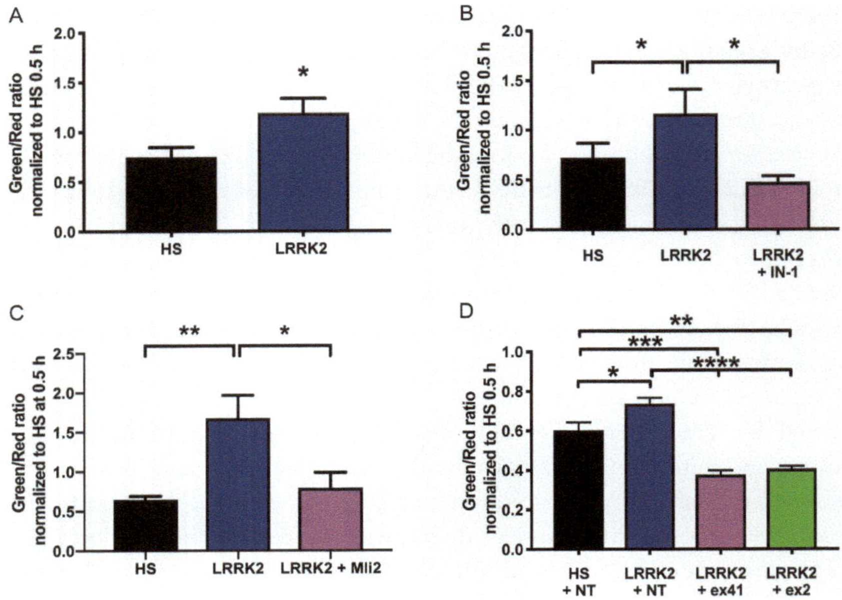

Fig. 2 Fibroblasts harboring the LRRK2 G2019S mutation exhibit reduced rate of mitophagy (A) PD patient-derived fibroblast cells harboring the LRRK2 G2019S mutation (LRRK2) exhibit increased green/red ratio indicative of reduced rate of mitophagy compared to age-matched, HS controls (unpaired *t*-test, $P=0.0201$, $n=11$ HS and $n=9$ LRRK2 cell lines derived from separate patients). This phenotype was rescued by the inhibition of LRRK2 kinase activity using the pharmacological inhibitors (B) LRRK2-IN-1 (one-way ANOVA $F(2,16)=4.136$, $P=0.0356$, $n=7$ HS and $n=6$ LRRK2), (C) Mli-2 (one-way ANOVA $F(2,28)=7.783$, $P=0.0021$, $n=11$ HS and $n=10$ LRRK2), and by genetic inhibition using (D) antisense oligonucleotides (ASO) targeting exon41 of the *LRRK2* gene (ex41) (one-way ANOVA $F(3,18)=27.22$, $P=0.0001$, $n=7$ HS and $n=5$ LRRK2). Non-target (NT) control ASO and ASO targeting exon2 of the *LRRK2* gene (ex2) were used as negative and positive controls respectively. All data represented as arithmetic mean $\pm$ SEM. $^{*}P<0.5$, $^{**}P<0.01$, $^{***}P<0.001$, $^{****}P<0.0001$. *From Korecka, J. A., Thomas, R., Christensen, D. P., Hinrich, A. J., Ferrari, E. J., Levy, S. A., et al. (2019). Mitochondrial clearance and maturation of autophagosomes are compromised in LRRK2 G2019S familial Parkinson's disease patient fibroblasts.* Human Molecular Genetics, 28, *3232–3243. With permission.*

mitophagy, the assay was repeated in the presence of LRRK2 kinase inhibitors. Two pharmacological inhibitors comprising the first generation LRRK2-IN-1, and the more recent, more specific inhibitor Mli-2 were used for the experiment. In addition to the chemical compounds, genetic inhibition of the LRRK2 kinase activity was performed using antisense oligonucleotides (ASOs) designed to target and induce skipping of exon41 of the *LRRK2* gene. Application of all the three methods of LRRK2 kinase inhibition significantly reduced the green/red ratio (Fig. 2B–D), thus leading to a functional rescue of mitophagy in these mutant cells. This suggests an important role for LRRK2 kinase activity in the regulation of intracellular mitophagic flux.

The process of mitochondrial degradation, including initiation of the phagophore, transport of the vesicles, maturation of autophagosomes, and degradation by the lysosomes is a complex pathway tightly regulated by several proteins. To assess the mechanism underlying the regulation of mitophagy by LRRK2 kinase, expression level of the key protein players of the pathway was analyzed between the mutant cells and HS controls at baseline and upon induction of mitochondrial stress with valinomycin. Mitophagy is a dynamic process with quick turnover of the protein and vesicles in the pathway. To investigate the expression level of its regulators, inhibition of the pathway at the degradative stage is important. This was achieved using the treatment of cells with bafilomycin A1 (bafA1), an inhibitor of vacuolar H^+ ATPase, which prevents the fusion of autophagosomes to the lysosomes. Among the various proteins analyzed, microtubule-associated proteins 1A/1B light chain 3A and B (LC3A II/I and B II/I) showed a significant reduction in *LRRK2* mutant cells compared to controls in bafA1-treated cells, and this effect was reversed upon treatment with the kinase inhibitor Mli-2. The LC3A and B proteins are present on the surface of autophagosomes. Upon maturation, LC3 I is lipidated and cleaved to form LC3 II, and ratio between the two forms have been used as an indicator of the content of mature autophagosomes within the cell. Decreased LC3A and B II/I ratio in *LRRK2* mutant cells suggests that these cells have a lower amount of mature autophagosomes compared to controls, and its reversal upon LRRK2 kinase inhibition suggests a functional involvement of LRRK2 in this process. No significant changes were observed in levels of proteins associated with other stages of the pathway suggesting that alterations in the rate of mitophagy in LRRK2 G2019S mutant cells is primarily associated with maturation of the autophagosomes and not the other stages of the pathway.

In line with our data, the LRRK2 G2019S mutation was reported to reduce autophagic flux and lead to accelerated dopaminergic neuron loss in *Caenorhabditis elegans* (Saha et al., 2015), and was associated with reduced mitophagy in human induced pluripotent stem cell (iPSC)-derived neurons and fibroblasts due to a failure to remove the transporter protein Miro from the mitochondria before its entry into the degradative pathway

(Hsieh et al., 2016). Reduced mitochondrial degradation can lead to accumulation of defective organelles within the cell. This, in turn, can increase oxidative stress which impedes mitochondrial respiration and metabolic output, and increase mtDNA mutations. All these factors can make the SNc dopaminergic neurons more vulnerable to genetic and environmental stress, eventually leading to their cell death.

2.1.3 Summary of LRRK2 associated mitochondrial dysfunction in Parkinson's disease

Key concepts from the above described studies (Korecka et al., 2019b; Smith et al., 2016) are discussed below:

(1) Patient-derived peripheral cells as an *in vitro* model system—for the identification of the causative pathways and development of successful therapeutic interventions, a good human *in vitro* model for PD is required. While patient specific iPSC-derived dopaminergic neurons are a great option for modeling the disease, this is a labor, skill and resource intensive technique, and not suitable for large scale therapeutic screens. Patient specific, non-neuronal, easy-to-obtain peripheral cells, such as blood cells and fibroblasts, have an advantage over iPSCs as equally good *in vitro* model systems and have been utilized by several groups to identify disease mechanisms (Auburger et al., 2012; Berenguer-Escuder et al., 2019; Juárez-Flores et al., 2018; Mutez et al., 2014). Even though PD is primarily a neuronal disorder, the mutations that cause the disease and the ensuing cellular phenotypes are present in all cells of the body. The SNc dopaminergic neurons tend to be more vulnerable to the effects of stressors (both genetic and environmental), thus leading to their degeneration in PD (Chung et al., 2005; Cooper et al., 2012; Surmeier et al., 2017).

(2) Presence of shared phenotypes between sporadic and genetic PD—devoid of genetic mutations, the cause of sporadic PD largely remains unknown. Identification of phenotypes shared between the genetic and sporadic cases can shed light into the key pathological mechanisms and convergent pathways that cause the disease. Presence of such phenotypes can be of use for the stratification of sporadic patients into subgroups for personalized therapies. Supporting this idea, a recent study identified that a subset of sporadic PD fibroblasts showed defect in removal of Miro from defective mitochondria which prevented it from entering the mitophagy pathway. Treatment of this subgroup with a specific molecule that aided in the reduction of Miro helped improve mitophagy in these cells (Hsieh et al., 2019).

(3) Utility of cellular biomarkers—monogenic forms of PD caused by genetic mutations can be screened using gene sequencing, which serves as a good genetic biomarker. However, a vast majority of PD is sporadic with no known etiology. Hence, there is a large need for identification of PD phenotypes, such as those associated with mitochondrial dysfunction as discussed here that can be utilized as cellular biomarkers for the disease. Our study shows that *LRRK2* mutant cells with increased LRRK2 kinase activity have a reduced rate of mitophagy (Korecka et al., 2019b). A recent study identified that idiopathic PD patients exhibit increased kinase activity despite the presence of a normal LRRK2 WT sequence (Maio et al., 2018).

3. Lysosomal dysfunction in aging and Parkinson's disease

Even though historically thought to be only involved in the breakdown of cytosolic macromolecules and defective cellular organelles, lysosomes have recently been identified to play an important role in nutrient sensing, storage of lipids, and calcium homeostasis (Audano et al., 2018). Many factors, including lysosomal pH, volume and activity of its constituent enzymes, govern lysosomal function. Its dysfunction in cells like neurons can have serious negative effects, as being post mitotic, these cells heavily depend on functional lysosomes for the clearance of defective organelles and autophagic cargoes.

Aging is one of the biggest risk factors for most of the neurodegenerative diseases, including PD (Collier et al., 2011; Hou et al., 2019). Interestingly, several pathological changes associated with genetic forms of PD resemble cellular changes occurring in normal aging. Mild to moderate loss of SN dopamine neurons and presence of Lewy bodies were observed in aged human brain (Buchman et al., 2012; Ma et al., 1999; Rudow et al., 2008). In addition to this, several other pathophysiological changes in PD, such as mitochondrial and lysosomal dysfunction, mtDNA lesions, protein aggregation and oxidative stress, are also observed in aging (Collier et al., 2011; Navarro and Boveris, 2010; Rodriguez et al., 2015). This commonality between the two situations led to the hypothesis that aging creates a "pre-PD" situation, which gets accentuated in the presence of external stressors to eventually lead to the disease (Collier et al., 2017).

Glucocerebrosidase (GCase) encoded by *GBA1* is a lysosomal hydrolase required for the metabolism of the glycolipids, glucosylceramide (GlcCer) and glucosylsphingosine (GluSph). Mutations causing haploinsufficiency of *GBA1* leading to loss-of-function of the enzyme activity is one of the most common genetic risk factors identified in PD (Clark et al., 2007; Sidransky et al., 2009) and homozygous mutations in this gene cause the lysosomal storage disorder, Gaucher's disease. To gain a better understanding of GCase activity levels in aging and sporadic PD, we utilized postmortem brain tissue from sporadic PD patients and compared them to age-matched, HS controls (Rocha et al., 2015a). We observed a reduction in GCase activity in several regions of the sporadic PD brain (SNc, putamen, hippocampus and cerebellum) and this was associated with a concomitant increase of GluSph level in hippocampus and SN. Interestingly, in addition to the changes in PD, a gradual, age-related decline in GCase activity was also observed in the SN and putamen of HS control samples indicating that the reduction of enzyme activity, and potentially lysosomal function, is a shared feature of both PD and aging. Supporting this data, reductions in GCase activity levels have also been described in blood, cerebrospinal fluid and blood-derived monocytes of PD patients with and without *GBA* mutations (Alcalay et al., 2015; Atashrazm et al., 2018; Parnetti et al., 2017). Further expanding on our study in human aging, we showed that glucocerebrosidase activities and several glycosphingolipids were altered upon aging in different strains of WT mice (Hallett et al., 2018). Compared to young animals (<3 months), aged whole mouse brain lysates (17–24 months) had increased levels of several glycosphingolipids, including glucosylceramide, lactosylceramide and the ganglioside GM1a, with a reduction in the level of GD1a, GD1b and GT1b gangliosides. Reduced activity of lysosomal GCase (*GBA1*), non-lysosomal GCase (*GBA2*) and β galactosidase activity, and increased neuraminidase activity was observed in aged mice compared to young animals. A subsequent study in two human sporadic PD cohorts replicated several of these findings (Huebecker et al., 2019). Substantia nigra lysates from sporadic PD patients showed reduction in the lysosomal and non-lysosomal GCase and β galactosidase activity levels compared to controls. In addition, reduced β hexosaminidase and α galactosidase enzyme activity levels were observed in aged (in their 80s) PD patients. These changes in enzyme activity levels were also accompanied by alterations in several glycosphingolipid (GSL) levels, including accumulation of GluCer and GluSph, and depletion of GM1a, GD1a, GD1b and GT1b. Additionally, the changes in GD1a, GD1b and GT1b levels were also reflected in ante mortem cerebrospinal fluid samples and reduction of GM1a and GD1a in serum samples from PD patients, suggesting their potential application as biomarkers for PD. An additional interesting finding was that the serum sample from a cohort of patients with REM sleep behavior disorder (RBD) who have a significantly high risk of developing PD (used as prodromal PD subjects here) also displayed several changes in GSL levels (reduction of lactosylceramide, Gb3, Gb4, GM3, GM2, GM1a and GD1a). While these changes in lipid levels in RBD patients might be characteristic of the disease itself, they might also play an important role in predisposing these patients to develop PD over time.

Glycosphingolipids are important components of cell membrane required for membrane organization and signal transduction. Membrane glycolipids are degraded in the lysosomes by catabolic enzymes to simpler compounds, which are then recycled back for synthesis of new sphingolipids (Grassi et al., 2019). Glucosylceramide is the final product of GSL catabolism and its increase in neuronopathic Gaucher's disease is accompanied by pathological changes and degeneration of neurons (Farfel-Becker et al., 2019; Nilsson and Svennerholm, 1982). Similarly, exogenous treatment of cholinergic neurons with GlcSph adversely affected neuronal outgrowth and led to a reduction of several lysosomal enzymes (Schueler et al., 2003). A general loss of lysosomal capacity in aging and sporadic PD brain due to reduced activity of the lysosomal enzymes can impair proper glycosphingolipid catabolism leading to adverse alterations in membrane composition and neuronal function.

3.1 Effect of modulation of lysosomal glucocerebrosidase activity

Studies from our group and others have identified a reduction of lysosomal GCase activity in aging and sporadic PD brain tissue and the ensuing lysosomal dysfunction and GSL dyshomeostasis could detrimentally affect neuronal function and increase its vulnerability to external stressors. One of the hallmarks of PD is the presence of intracellular proteinaceous Lewy bodies comprising undegraded α-synuclein. Recent ultrastructural analysis identified that in addition to α-synuclein, Lewy bodies are also made up of distorted organelles, membranes and lipids (Shahmoradian et al., 2019).

Accumulation of α-synuclein oligomers is considered pathogenic in PD and defects in the lysosomal degradation system could add to this. To test whether lysosomal dysfunction caused due to chronic inhibition of lysosomal GCase contribute to α-synuclein accumulation and aggregation, pharmacological inhibition of the enzyme activity was performed in a murine model. Inhibition of GBA was achieved by systemic administration of a selective and irreversible competitive inhibitor, CBE (conduritol β-epoxide) (Grabowski et al., 1986) for 28 consecutive days. As expected, CBE treatment reduced GCase activity and increased accumulation of GlcCer and GlcSph in the forebrain compared to vehicle injected controls. In addition to widespread neuroinflammation (cortical layer V, SN and striatum) indicated by the presence of Iba1-positive reactive microglia, inhibition of GCase activity also led to accumulation of proteinase-K resistant α-synuclein aggregates in the SN. Glucocerebrosidase inhibition also led to neuronal loss in motor cortex layers V and VI, and there was no obvious neuronal death elsewhere in the brain. Conversely, the overexpression of GBA1 in two PD rodent models (transgenic mice over-expressing WT α-synuclein, and a rat model over-expressing A53T α-synuclein in nigral dopaminergic neurons) was protective and prevented α-synuclein-induced, nigrostriatal dopamine degeneration (Rocha et al., 2015b).

Over-expression of human A53T α-synuclein in rat SN leads to widespread nigrostriatal degeneration preceded by changes in proteins related to synaptic and axonal transport and an increased neuroinflammatory response (Chung et al., 2009; Koprich et al., 2010). While the chronic GBA inhibition mouse model mentioned here recapitulated some of the pre-degenerative changes of α-synuclein over-expression model, including neuroinflammation, and alteration in axonal and synaptic transport proteins, nigrostriatal degeneration was absent in these animals. Strikingly, this GBA inhibition model caused a marked increase in high molecular weight α-synuclein species and aggregation of α-synuclein, reminiscent of pathological changes seen in PD. The CBE dosing paradigm used in this study leads to a 100% reduction of the enzyme activity, which mirrors a Gaucher's disease model more than sporadic PD (which is associated with only 30–50% reduction of GCase activity), and it is possible that nigrostriatal dopaminergic degeneration could occur later. Extending analysis to a later time point following GBA inhibition might be helpful to observe such changes. Supporting this hypothesis, studies using a subchronic CBE treatment paradigm showed reduced potassium chloride evoked dopamine release in the striatum indicative of GBA-induced regulation of dopaminergic transmission (Ginns et al., 2014), suggesting that lysosomal deficits caused by inhibition of GBA might, potentially, be able to recapitulate nigrostriatal pathway degeneration as that observed in PD. Targeting the lysosomal pathway using small molecules that reduce the accumulation of GSLs and chaperones to increase the folding, transport and activity of lysosomal enzymes are especially attractive, and are currently under development (Sardi et al., 2018).

4. Future considerations for using cell culture models for studying disease-relevant mechanisms *in vitro*

While a significant number of studies over the years have identified some of the cellular pathways and stressors that contribute to PD, much remains to be understood. For further studies into the etiology of the disease and for developing good therapeutic interventions, it is important to have access to good human specific cell systems. Cell culture is itself an artificial system and extreme care must be taken to keep all handling and culture conditions (media composition, incubation parameters, seeding density) as constant as possible. Fibroblasts upon repeated passaging *in vitro* can undergo senescence. Staining cells regularly to identify senescent cells and performing experiments with cells maintained within a specific range of early passages will be critical to control for this. Additionally, if left in culture for too long, these epithelial cells can also undergo contact inhibition and cell cycle arrest, the effect of which on the quality of cultures is not fully understood. Regular karyotyping to confirm that the cells are devoid of chromosomal aberrations and testing to ensure that they are not infected with mycoplasma are also critical for maintaining good quality of the cultures. If not controlled properly, cells could adapt and evolve to varying culture conditions, detrimentally affecting the phenotypes under study. Age-matched, HS cells are often used as controls for most of the studies. However, oftentimes they display a wide range of variability among themselves, which, in turn, can critically affect the result and directionality of the phenotypes under study. Increasing the overall number of HS cells used in an experiment or using well characterized isogenic controls (in the case of studies with genetically mutant cells) are important. While patient-derived, human cells constitute an invaluable tool in PD research, properly controlling for the experimental set up and validating the results in a secondary system (different type of cell system, human PD postmortem brain tissue or *in vivo* models) is critical to avoid artifactual data.

Even though patient-derived cells offer several advantages over *in vivo* models, such as easy maintenance, genetic manipulation and pharmacological intervention, there are several aspects that have to be considered carefully in terms of the experimental design and interpretation of results. While several types of patient-derived cells, such as iPSC-derived neuronal populations (dopaminergic neurons, cortical cells, etc.), peripheral cells such as blood cells and dermal fibroblasts are available to study disease pathology, no single *in vitro* cellular system that can recapitulate all features of the disease exists. Hence, choosing the right system depending on the specific aspect to study, such as neuronal cells to investigate synaptic and axonal transport, is important. While studying cell types such as neurons in isolation is interesting in terms of their response to a pharmacological treatment, their behavior might be different in the absence of other non-neuronal cells. Considering co-culture systems that recapitulate interactions between multiple cell types will be more relevant.

5. Conclusions and future directions

As outlined in this review, a significant number of studies, including ours, have identified the involvement of mitochondrial and lysosomal dysfunction in PD. While these organelle systems and biochemical pathways have usually been studied independently of each other, recent data point towards potential functional interplay between them. Physical interaction between the two organelles resulting in mitochondria-lysosome contact sites regulated by Ras-related protein 7 (RAB7) that aids in mitochondrial fission have been identified in mammalian cells (Wong et al., 2018). A more direct regulation of LRRK2 and GBA was identified in a recent study, which reported that GCase activity was negatively regulated by the LRRK2 kinase and this was mediated by one of its direct substrates, Ras-related protein 10 (Rab10) (Ysselstein et al., 2019). Thus, future studies aimed at understanding the interdependence of the two organelles, beyond just mitophagy, and how dysfunction of one might impact the functionality and dynamics of the other is of essence.

References

Alcalay, R. N., Levy, O. A., Waters, C. C., Fahn, S., Ford, B., Kuo, S.-H., et al. (2015). Glucocerebrosidase activity in Parkinson's disease with and without GBA mutations. *Brain: A Journal of Neurology*, *138*, 2648–2658.

Ascherio, A., & Schwarzschild, M. A. (2016). The epidemiology of Parkinson's disease: Risk factors and prevention. *The Lancet Neurology*, *15*, 1257–1272.

Atashrazm, F., Hammond, D., Perera, G., Dobson-Stone, C., Mueller, N., Pickford, R., et al. (2018). Reduced glucocerebrosidase activity in monocytes from patients with Parkinson's disease. *Scientific Reports*, *8*, 15446.
Auburger, G., Klinkenberg, M., Drost, J., Marcus, K., Morales-Gordo, B., Kunz, W. S., et al. (2012). Primary skin fibroblasts as a model of Parkinson's disease. *Molecular Neurobiology*, *46*, 20–27.
Audano, M., Schneider, A., & Mitro, N. (2018). Mitochondria, lysosomes, and dysfunction: Their meaning in neurodegeneration. *Journal of Neurochemistry*, *147*, 291–309.
Baldereschi, M., Di Carlo, A., Rocca, W. A., Vanni, P., Maggi, S., Perissinotto, E., et al. (2000). Parkinson's disease and parkinsonism in a longitudinal study: Two-fold higher incidence in men. ILSA Working Group. Italian Longitudinal Study on Aging. *Neurology*, *55*, 1358–1363.
Bender, A., Krishnan, K. J., Morris, C. M., Taylor, G. A., Reeve, A. K., Perry, R. H., et al. (2006). High levels of mitochondrial DNA deletions in substantia nigra neurons in aging and Parkinson disease. *Nature Genetics*, *38*, 515–517.
Berenguer-Escuder, C., Grossmann, D., Massart, F., Antony, P., Burbulla, L. F., Glaab, E., et al. (2019). Variants in Miro1 cause alterations of ER-mitochondria contact sites in fibroblasts from Parkinson's disease patients. *Journal of Clinical Medicine*, *8*, 2226.
Berwick, D. C., Heaton, G. R., Azeggagh, S., & Harvey, K. (2019). LRRK2 biology from structure to dysfunction: Research progresses, but the themes remain the same. *Molecular Neurodegeneration*, *14*, 49.
Bose, A., & Beal, M. F. (2016). Mitochondrial dysfunction in Parkinson's disease. *Journal of Neurochemistry*, *139*, 216–231.
Buchman, A. S., Shulman, J. M., Nag, S., Leurgans, S. E., Arnold, S. E., Morris, M. C., et al. (2012). Nigral pathology and parkinsonian signs in elders without Parkinson disease. *Annals of Neurology*, *71*, 258–266.
Chaudhuri, K. R., & Schapira, A. H. (2009). Non-motor symptoms of Parkinson's disease: Dopaminergic pathophysiology and treatment. *Lancet Neurology*, *8*, 464–474.
Chen, C., Turnbull, D. M., & Reeve, A. K. (2019). Mitochondrial dysfunction in Parkinson's disease—Cause or consequence? *Biology*, *8*, 38.
Chung, C. Y., Koprich, J. B., Siddiqi, H., & Isacson, O. (2009). Dynamic changes in presynaptic and axonal transport proteins combined with striatal neuroinflammation precede dopaminergic neuronal loss in a rat model of AAV α-synucleinopathy. *The Journal of Neuroscience*, *29*, 3365–3373.
Chung, C. Y., Seo, H., Sonntag, K. C., Brooks, A., Lin, L., & Isacson, O. (2005). Cell type-specific gene expression of midbrain dopaminergic neurons reveals molecules involved in their vulnerability and protection. *Human Molecular Genetics*, *14*, 1709–1725.
Clark, L. N., Ross, B. M., Wang, Y., Mejia-Santana, H., Harris, J., Louis, E. D., et al. (2007). Mutations in the glucocerebrosidase gene are associated with early-onset Parkinson disease. *Neurology*, *69*, 1270–1277.
Collier, T. J., Kanaan, N. M., & Kordower, J. H. (2011). Ageing as a primary risk factor for Parkinson's disease: Evidence from studies of non-human primates. *Nature Reviews Neuroscience*, *12*, 359–366.
Collier, T. J., Kanaan, N. M., & Kordower, J. H. (2017). Aging and Parkinson's disease: Different sides of the same coin? *Movement Disorders: Official Journal of the Movement Disorder Society*, *32*, 983–990.
Cooper, O., Seo, H., Andrabi, S., Guardia-Laguarta, C., Graziotto, J., Sundberg, M., et al. (2012). Pharmacological rescue of mitochondrial deficits in iPSC-derived neural cells from patients with familial Parkinson's disease. *Science Translational Medicine*, *4*, 141ra90.
Deng, X., Dzamko, N., Prescott, A., Davies, P., Liu, Q., Yang, Q., et al. (2011). Characterization of a selective inhibitor of the Parkinson's disease kinase LRRK2. *Nature Chemical Biology*, 7, 203–205.

Deng, H., Wang, P., & Jankovic, J. (2018). The genetics of Parkinson disease. *Ageing Research Reviews*, *42*, 72–85.

Dölle, C., Flønes, I., Nido, G. S., Miletic, H., Osuagwu, N., Kristoffersen, S., et al. (2016). Defective mitochondrial DNA homeostasis in the substantia nigra in Parkinson disease. *Nature Communications*, 7, 1–11.

Farfel-Becker, T., Roney, J. C., Cheng, X.-T., Li, S., Cuddy, S. R., & Sheng, Z.-H. (2019). Neuronal soma-derived degradative lysosomes are continuously delivered to distal axons to maintain local degradation capacity. *Cell Reports*, *28*, 51–64.e4.

Fedorowicz, M. A., de Vries-Schneider, R. L. A., Rüb, C., Becker, D., Huang, Y., Zhou, C., et al. (2014). Cytosolic cleaved PINK1 represses Parkin translocation to mitochondria and mitophagy. *EMBO Reports*, *15*, 86–93.

Flønes, I. H., Fernandez-Vizarra, E., Lykouri, M., Brakedal, B., Skeie, G. O., Miletic, H., et al. (2018). Neuronal complex I deficiency occurs throughout the Parkinson's disease brain, but is not associated with neurodegeneration or mitochondrial DNA damage. *Acta Neuropathologica*, *135*, 409–425.

Gandhi, P. N., Wang, X., Zhu, X., Chen, S. G., & Wilson-Delfosse, A. L. (2008). The Roc domain of leucine-rich repeat kinase 2 is sufficient for interaction with microtubules. *Journal of Neuroscience Research*, *86*, 1711–1720.

Giaime, E., Tong, Y., Wagner, L. K., Yuan, Y., Huang, G., & Shen, J. (2017). Age-dependent dopaminergic neurodegeneration and impairment of the autophagy-lysosomal pathway in LRRK-deficient mice. *Neuron*, *96*, 796–807.e6.

Ginns, E. I., Mak, S. K.-K., Ko, N., Karlgren, J., Akbarian, S., Chou, V. P., et al. (2014). Neuroinflammation and α-synuclein accumulation in response to glucocerebrosidase deficiency are accompanied by synaptic dysfunction. *Molecular Genetics and Metabolism*, *111*, 152–162.

Grabowski, G. A., Osiecki-Newman, K., Dinur, T., Fabbro, D., Legler, G., Gatt, S., et al. (1986). Human acid beta-glucosidase. Use of conduritol B epoxide derivatives to investigate the catalytically active normal and Gaucher disease enzymes. *The Journal of Biological Chemistry*, *261*, 8263–8269.

Grassi, S., Chiricozzi, E., Mauri, L., Sonnino, S., & Prinetti, A. (2019). Sphingolipids and neuronal degeneration in lysosomal storage disorders. *Journal of Neurochemistry*, *148*, 600–611.

Grünewald, A., Rygiel, K. A., Hepplewhite, P. D., Morris, C. M., Picard, M., & Turnbull, D. M. (2016). Mitochondrial DNA depletion in respiratory chain–deficient Parkinson disease neurons. *Annals of Neurology*, *79*, 366–378.

Hallett, P. J., Engelender, S., & Isacson, O. (2019). Lipid and immune abnormalities causing age-dependent neurodegeneration and Parkinson's disease. *Journal of Neuroinflammation*, *16*, 153.

Hallett, P. J., Huebecker, M., Brekk, O. R., Moloney, E. B., Rocha, E. M., Priestman, D. A., et al. (2018). Glycosphingolipid levels and glucocerebrosidase activity are altered in normal aging of the mouse brain. *Neurobiology of Aging*, *67*, 189–200.

Hallett, P. J., McLean, J. R., Kartunen, A., Langston, J. W., & Isacson, O. (2012). Alpha-synuclein overexpressing transgenic mice show internal organ pathology and autonomic deficits. *Neurobiology of Disease*, *47*, 258–267.

Hernandez, D. G., Reed, X., & Singleton, A. B. (2016). Genetics in Parkinson disease: Mendelian versus non-Mendelian inheritance. *Journal of Neurochemistry*, *139*, 59–74.

Hirsch, L., Jette, N., Frolkis, A., Steeves, T., & Pringsheim, T. (2016). The incidence of Parkinson's disease: A systematic review and meta-analysis. *Neuroepidemiology*, *46*, 292–300.

Ho, D. H., Kim, H., Nam, D., Sim, H., Kim, J., Kim, H. G., et al. (2018). LRRK2 impairs autophagy by mediating phosphorylation of leucyl-tRNA synthetase. *Cell Biochemistry and Function*, *36*, 431–442.

Hou, Y., Dan, X., Babbar, M., Wei, Y., Hasselbalch, S. G., Croteau, D. L., et al. (2019). Ageing as a risk factor for neurodegenerative disease. *Nature Reviews Neurology*, *15*, 565–581.

Hsieh, C.-H., Li, L., Vanhauwaert, R., Nguyen, K. T., Davis, M. D., Bu, G., et al. (2019). Miro1 marks Parkinson's disease subset and Miro1 reducer rescues neuron loss in Parkinson's models. *Cell Metabolism*, *30*, 1131–1140.e7.

Hsieh, C.-H., Shaltouki, A., Gonzalez, A. E., Bettencourt da Cruz, A., Burbulla, L. F., St Lawrence, E., et al. (2016). Functional impairment in Miro degradation and mitophagy is a shared feature in familial and sporadic Parkinson's disease. *Cell Stem Cell*, *19*, 709–724.

Huebecker, M., Moloney, E. B., van der Spoel, A. C., Priestman, D. A., Isacson, O., Hallett, P. J., et al. (2019). Reduced sphingolipid hydrolase activities, substrate accumulation and ganglioside decline in Parkinson's disease. *Molecular Neurodegeneration*, *14*, 40.

Inoshita, T., Arano, T., Hosaka, Y., Meng, H., Umezaki, Y., Kosugi, S., et al. (2017). Vps35 in cooperation with LRRK2 regulates synaptic vesicle endocytosis through the endosomal pathway in Drosophila. *Human Molecular Genetics*, *26*, 2933–2948.

Juárez-Flores, D. L., González-Casacuberta, I., Ezquerra, M., Bañó, M., Carmona-Pontaque, F., Catalán-García, M., et al. (2018). Exhaustion of mitochondrial and autophagic reserve may contribute to the development of LRRK2G2019S-Parkinson's disease. *Journal of Translational Medicine*, *16*, 160.

Koprich, J. B., Johnston, T. H., Reyes, M. G., Sun, X., & Brotchie, J. M. (2010). Expression of human A53T alpha-synuclein in the rat substantia nigra using a novel AAV1/2 vector produces a rapidly evolving pathology with protein aggregation, dystrophic neurite architecture and nigrostriatal degeneration with potential to model the pathology of Parkinson's disease. *Molecular Neurodegeneration*, *5*, 43.

Korecka, J. A., Talbot, S., Osborn, T. M., de Leeuw, S. M., Levy, S. A., Ferrari, E. J., et al. (2019a). Neurite collapse and altered ER Ca^{2+} control in human Parkinson disease patient iPSC-derived neurons with LRRK2 G2019S mutation. *Stem Cell Reports*, *12*, 29–41.

Korecka, J. A., Thomas, R., Christensen, D. P., Hinrich, A. J., Ferrari, E. J., Levy, S. A., et al. (2019b). Mitochondrial clearance and maturation of autophagosomes are compromised in LRRK2 G2019S familial Parkinson's disease patient fibroblasts. *Human Molecular Genetics*, *28*, 3232–3243.

Kozina, E., Sadasivan, S., Jiao, Y., Dou, Y., Ma, Z., Tan, H., et al. (2018). Mutant LRRK2 mediates peripheral and central immune responses leading to neurodegeneration in vivo. *Brain*, *141*, 1753–1769.

Kraytsberg, Y., Kudryavtseva, E., McKee, A. C., Geula, C., Kowall, N. W., & Khrapko, K. (2006). Mitochondrial DNA deletions are abundant and cause functional impairment in aged human substantia nigra neurons. *Nature Genetics*, *38*, 518–520.

Ma, S. Y., Röytt, M., Collan, Y., & Rinne, J. O. (1999). Unbiased morphometrical measurements show loss of pigmented nigral neurones with ageing. *Neuropathology and Applied Neurobiology*, *25*, 394–399.

Maio, R. D., Hoffman, E. K., Rocha, E. M., Keeney, M. T., Sanders, L. H., Miranda, B. R. D., et al. (2018). LRRK2 activation in idiopathic Parkinson's disease. *Science Translational Medicine*, *10*, eaar5429.

Mutez, E., Nkiliza, A., Belarbi, K., de Broucker, A., Vanbesien-Mailliot, C., Bleuse, S., et al. (2014). Involvement of the immune system, endocytosis and EIF2 signaling in both genetically determined and sporadic forms of Parkinson's disease. *Neurobiology of Disease*, *63*, 165–170.

Navarro, A., & Boveris, A. (2010). Brain mitochondrial dysfunction in aging, neurodegeneration, and Parkinson's disease. *Frontiers in Aging Neuroscience*, *2*, 34.

Nilsson, O., & Svennerholm, L. (1982). Accumulation of glucosylceramide and glucosylsphingosine (psychosine) in cerebrum and cerebellum in infantile and juvenile Gaucher disease. *Journal of Neurochemistry*, *39*, 709–718.

Papkovskaia, T. D., Chau, K.-Y., Inesta-Vaquera, F., Papkovsky, D. B., Healy, D. G., Nishio, K., et al. (2012). G2019S leucine-rich repeat kinase 2 causes uncoupling protein-mediated mitochondrial depolarization. *Human Molecular Genetics*, *21*, 4201–4213.

Parisiadou, L., Xie, C., Cho, H. J., Lin, X., Gu, X.-L., Long, C.-X., et al. (2009). Phosphorylation of ezrin/radixin/moesin proteins by LRRK2 promotes the rearrangement of actin cytoskeleton in neuronal morphogenesis. *The Journal of Neuroscience*, *29*, 13971–13980.

Parnetti, L., Paciotti, S., Eusebi, P., Dardis, A., Zampieri, S., Chiasserini, D., et al. (2017). Cerebrospinal fluid β-glucocerebrosidase activity is reduced in Parkinson's disease patients. *Movement Disorders: Official Journal of the Movement Disorder Society*, *32*, 1423–1431.

Pellegrini, L., Hauser, D. N., Li, Y., Mamais, A., Beilina, A., Kumaran, R., et al. (2018). Proteomic analysis reveals co-ordinated alterations in protein synthesis and degradation pathways in LRRK2 knockout mice. *Human Molecular Genetics*, *27*, 3257–3271.

Piccoli, G., Condliffe, S. B., Bauer, M., Giesert, F., Boldt, K., Astis, S. D., et al. (2011). LRRK2 controls synaptic vesicle storage and mobilization within the recycling Pool. *The Journal of Neuroscience*, *31*, 2225–2237.

Poewe, W., Seppi, K., Tanner, C. M., Halliday, G. M., Brundin, P., Volkmann, J., et al. (2017). Parkinson disease. *Nature Reviews Disease Primers*, *3*, 1–21.

Puccini, J. M., Marker, D. F., Fitzgerald, T., Barbieri, J., Kim, C. S., Miller-Rhodes, P., et al. (2015). Leucine-rich repeat kinase 2 modulates neuroinflammation and neurotoxicity in models of human immunodeficiency virus 1-associated neurocognitive disorders. *The Journal of Neuroscience*, *35*, 5271–5283.

Pyle, A., Anugrha, H., Kurzawa-Akanbi, M., Yarnall, A., Burn, D., & Hudson, G. (2016). Reduced mitochondrial DNA copy number is a biomarker of Parkinson's disease. *Neurobiology of Aging*, *38*, 216.e7–216.e10.

Pyle, A., Brennan, R., Kurzawa-Akanbi, M., Yarnall, A., Thouin, A., Mollenhauer, B., et al. (2015). Reduced cerebrospinal fluid mitochondrial DNA is a biomarker for early-stage Parkinson's disease. *Annals of Neurology*, *78*, 1000–1004.

Rakovic, A., Grünewald, A., Seibler, P., Ramirez, A., Kock, N., Orolicki, S., et al. (2010). Effect of endogenous mutant and wild-type PINK1 on Parkin in fibroblasts from Parkinson disease patients. *Human Molecular Genetics*, *19*, 3124–3137.

Reeve, A. K., Krishnan, K. J., Elson, J. L., Morris, C. M., Bender, A., Lightowlers, R. N., et al. (2008). Nature of mitochondrial DNA deletions in substantia nigra neurons. *American Journal of Human Genetics*, *82*, 228–235.

Rocha, E. M., Smith, G. A., Park, E., Cao, H., Brown, E., Hallett, P., et al. (2015a). Progressive decline of glucocerebrosidase in aging and Parkinson's disease. *Annals of Clinical Translational Neurology*, *2*, 433–438.

Rocha, E. M., Smith, G. A., Park, E., Cao, H., Brown, E., Hayes, M. A., et al. (2015b). Glucocerebrosidase gene therapy prevents α-synucleinopathy of midbrain dopamine neurons. *Neurobiology of Disease*, *82*, 495–503.

Rodriguez, M., Rodriguez-Sabate, C., Morales, I., Sanchez, A., & Sabate, M. (2015). Parkinson's disease as a result of aging. *Aging Cell*, *14*, 293–308.

Rolfe, D. F., & Brown, G. C. (1997). Cellular energy utilization and molecular origin of standard metabolic rate in mammals. *Physiological Reviews*, *77*, 731–758.

Rosado, C. J., Mijaljica, D., Hatzinisiriou, I., Prescott, M., & Devenish, R. J. (2008). Rosella: A fluorescent pH-biosensor for reporting vacuolar turnover of cytosol and organelles in yeast. *Autophagy*, *4*, 205–213.

Rudow, G., O'Brien, R., Savonenko, A. V., Resnick, S. M., Zonderman, A. B., Pletnikova, O., et al. (2008). Morphometry of the human substantia nigra in ageing and Parkinson's disease. *Acta Neuropathologica*, *115*, 461–470.

Saha, S., Ash, P. E. A., Gowda, V., Liu, L., Shirihai, O., & Wolozin, B. (2015). Mutations in LRRK2 potentiate age-related impairment of autophagic flux. *Molecular Neurodegeneration*, *10*, 26.

Sánchez-Martínez, C. M., Choreño-Parra, J. A., Placencia-Álvarez, N., Nuñez-Orozco, L., & Guadarrama-Ortiz, P. (2019). Frequency and dynamics of non-motor symptoms presentation in Hispanic patients with Parkinson disease. *Frontiers in Neurology*, *10*, 1197.

Sanders, L. H., Laganière, J., Cooper, O., Mak, S. K., Vu, B. J., Huang, Y. A., et al. (2014). LRRK2 mutations cause mitochondrial DNA damage in iPSC-derived neural cells from Parkinson's disease patients: Reversal by gene correction. *Neurobiology of Disease*, *62*, 381–386.

Sardi, S. P., Cedarbaum, J. M., & Brundin, P. (2018). Targeted therapies for Parkinson's disease: From genetics to the clinic. *Movement Disorders: Official Journal of the Movement Disorder Society*, *33*, 684–696.

Schapira, A. H. V., Chaudhuri, K. R., & Jenner, P. (2017). Non-motor features of Parkinson disease. *Nature Reviews Neuroscience*, *18*, 435–450.

Schapira, A. H. V., Cooper, J. M., Dexter, D., Clark, J. B., Jenner, P., & Marsden, C. D. (1990). Mitochondrial complex I deficiency in Parkinson's disease. *Journal of Neurochemistry*, *54*, 823–827.

Schueler, U. H., Kolter, T., Kaneski, C. R., Blusztajn, J. K., Herkenham, M., Sandhoff, K., et al. (2003). Toxicity of glucosylsphingosine (glucopsychosine) to cultured neuronal cells: A model system for assessing neuronal damage in Gaucher disease type 2 and 3. *Neurobiology of Disease*, *14*, 595–601.

Shahmoradian, S. H., Lewis, A. J., Genoud, C., Hench, J., Moors, T. E., Navarro, P. P., et al. (2019). Lewy pathology in Parkinson's disease consists of crowded organelles and lipid membranes. *Nature Neuroscience*, *22*, 1099–1109.

Sidransky, E., Nalls, M. A., Aasly, J. O., Aharon-Peretz, J., Annesi, G., Barbosa, E. R., et al. (2009). Multicenter analysis of glucocerebrosidase mutations in Parkinson's disease. *The New England Journal of Medicine*, *361*, 1651–1661.

Smith, G. A., Jansson, J., Rocha, E. M., Osborn, T., Hallett, P. J., & Isacson, O. (2016). Fibroblast biomarkers of sporadic Parkinson's disease and LRRK2 kinase inhibition. *Molecular Neurobiology*, *53*, 5161–5177.

Su, Y.-C., & Qi, X. (2013). Inhibition of excessive mitochondrial fission reduced aberrant autophagy and neuronal damage caused by LRRK2 G2019S mutation. *Human Molecular Genetics*, *22*, 4545–4561.

Surmeier, D. J., Obeso, J. A., & Halliday, G. M. (2017). Selective neuronal vulnerability in Parkinson disease. *Nature Reviews Neuroscience*, *18*, 101–113.

Taanman, J.-W. (1999). The mitochondrial genome: Structure, transcription, translation and replication. *Biochimica et Biophysica Acta*, *1410*, 103–123.

Toulorge, D., Schapira, A. H. V., & Hajj, R. (2016). Molecular changes in the postmortem parkinsonian brain. *Journal of Neurochemistry*, *139*, 27–58.

West, A. B., Moore, D. J., Biskup, S., Bugayenko, A., Smith, W. W., Ross, C. A., et al. (2005). Parkinson's disease-associated mutations in leucine-rich repeat kinase 2 augment kinase activity. *Proceedings of the National Academy of Sciences of the United States of America*, *102*, 16842–16847.

Winner, B., Melrose, H., Zhao, C., Hinkle, K., Yue, M., Kent, C., et al. (2011). Adult neurogenesis and neurite outgrowth are impaired in LRRK2 G2019S mice. *Neurobiology of Disease*, *41*, 706–716.

Wong, Y. C., Ysselstein, D., & Krainc, D. (2018). Mitochondria–lysosome contacts regulate mitochondrial fission via RAB7 GTP hydrolysis. *Nature*, *554*, 382–386.

Yellen, G. (2018). Fueling thought: Management of glycolysis and oxidative phosphorylation in neuronal metabolism. *The Journal of Cell Biology*, *217*, 2235–2246.

Ysselstein, D., Nguyen, M., Young, T. J., Severino, A., Schwake, M., Merchant, K., et al. (2019). LRRK2 kinase activity regulates lysosomal glucocerebrosidase in neurons derived from Parkinson's disease patients. *Nature Communications*, *10*, 1–9.

CHAPTER ELEVEN

Endosomal-lysosomal dysfunction in metabolic diseases and Alzheimer's disease

Michael F. Almeida[a], Ben A. Bahr[a], Stephen T. Kinsey[b,]*

[a]Biotechnology Research and Training Center, University of North Carolina—Pembroke, Pembroke, NC, United States
[b]Department of Biology and Marine Biology, University of North Carolina—Wilmington, Wilmington, NC, United States
*Corresponding author: e-mail address: kinseys@uncw.edu

Contents

1. Introduction 305
2. Metabolic energy balance 306
   2.1 Diabetes and obesity 306
   2.2 Exercise and exercise mimetics 308
3. Alzheimer's disease and related dementias 311
4. α-Synucleinopathies 315
5. Conclusion 318
References 318

Abstract

The endosomal-lysosomal pathways and related autophagic processes are responsible for proteostasis, involving complexes between lysosomes and autophagosomes. Lysosomes are a key component of homeostasis, involved in cell signaling, metabolism, and quality control, and they experience functional compromise in metabolic diseases, aging, and neurodegenerative diseases. Many genetic mutations and risk factor genes associated with proteinopathies, as well as with metabolic diseases like diabetes, negatively influence endocytic trafficking and autophagic clearance. In contrast, health-improving exercise induces autophagy-lysosomal degradation, perhaps promoting efficient digestion of injured organelles so that undamaged organelles ensure cellular healthiness. Reductions in lysosomal hydrolases are implicated in Alzheimer's, Parkinson's, and lysosomal storage diseases, as well as obesity-related pathology, and members of the cathepsin enzyme family are involved in clearing both Aβ42 and α-synuclein. Upregulation of cathepsin hydrolases improves synaptic and memory functions in models of dementia and in exercising humans, thus identifying lysosomal-related systems as vital for healthy cognitive aging.

International Review of Neurobiology, Volume 154
ISSN 0074-7742
https://doi.org/10.1016/bs.irn.2020.02.012

Abbreviations

ABCA1	ATP-binding cassette transporter subfamily A member 1
AD	Alzheimer's disease
AHN	adult hippocampal neurogenesis
AICAR	5-aminoimidazole-4-carboxamide ribonucleotide
ApoE4	apolipoprotein E4
ASAH1	*N*-acylsphingosine amidohydrolase 1
ATG7	autophagy related protein 7
ATP13A2	ATPase cation transporting 13A2
Aβ	amyloid β protein
BDNF	brain-derived neurotrophic factor
BECN1	Beclin 1
CMA	chaperone mediated autophagy
CREB	cAMP response element binding protein
CTSB	cathepsin B
CTSD	cathepsin D
FNDC5	fibronectin type III domain-containing protein 5
GALC	galactosylceramidase
GBA1	glucosylceramidase β1
GLA	galactosidase α
IRS1	insulin receptor substrate 1
LAMP1	lysosomal-associated membrane protein 1
LC3	microtubule-associated protein 1A/1B-light chain 3
LC3II	the lipid conjugated form of LC3
LRRK2	leucine-rich repeat kinase 2
MCI	mild cognitive impairment
mTORC1	mechanistic target of rapamycin complex 1
PD	Parkinson's disease
PI3K	phosphoinositide 3-kinase
PICALM	phosphatidylinositol-binding clathrin assembly protein
PINK1	phosphatase and tensin homolog (PTEN) induced putative kinase 1
PKA	protein kinase A
PRKN	Parkin RBR (RING between RING-RING) E3 ubiquitin protein ligase
SCARB2	scavenger receptor class B member 2
SIRT1	sirtuin 1
SMPD1	sphingomyelin phosphodiesterase 1
SQSTM1/p62	sequestosome 1/p62
T2DM	Type 2 diabetes mellitus
TFEB	transcription factor EB
TMEM175	lysosomal K^+ channel transmembrane protein 175
ULK1	UNC51-like autophagy activating kinase 1
VPS35	vacuolar protein sorting-associated protein 35 retromer complex component
β-GPA	β-guandinopropionic acid

1. Introduction

The endosomal-lysosomal pathways and related autophagic responses are critical for proteostatic regulation involving dynamic, membrane-bound organelles that function as the primary catabolic compartment of cells. The pathways maintain cell homeostasis through the control of vital processes (e.g., nutrient sensing, protein and other substrate trafficking, cellular migration), processes that experience compromise during metabolic diseases, aging, and neurodegenerative diseases (Condon and Sabatini, 2019; Wang et al., 2018; Xu and Ren, 2015). Table 1 provides a list of diseases that exhibit endosomal-lysosomal compromise and dysfunction. Lysosomal storage diseases (LSDs) are rare metabolic diseases that typically present at infancy and young ages by disrupting the function of one or several of the over 60 different hydrolytic enzymes that break down proteins, lipids, glycoconjugates, and nucleic acids (Parenti et al., 2015; Platt, 2018). The heterogeneous group of over 70 LSDs is characterized by lysosomal dysfunction and substrate accumulation, with progressive neurodegeneration common among LSDs, as well as frequent defects in other organ systems.

Table 1 Disease states that exhibit endosomal-lysosomal dysfunction.

• Lysosomal storage diseases
• Type 2 diabetes mellitus
• Obesity
• Aging
• Alzheimer's disease
• Parkinson's disease
• Huntington's disease
• Pick's disease
• Chronic traumatic encephalopathy
• Sporadic multiple system tauopathy with dementia
• Amyotrophic lateral sclerosis/frontotemporal dementia
• Frontotemporal dementia and parkinsonism linked to chromosome 17

More prevalent are age-related diseases of which numerous pathways are involved, regulating cell and system processes that influence aging and age-related risk factors for neurodegenerative diseases (see Carmona-Gutierrez et al., 2016; López-Otín et al., 2013; Mattson and Arumugam, 2018). Age-related neurodegenerative diseases, which increasingly are complicated by metabolic perturbations like obesity and diabetes, often involve intracellular or extracellular aggregate formations, due to dysfunction among the endo-lysosomal pathways and impaired clearance of biological polymers. The substrates are delivered to lysosomes through endocytic trafficking and pathways of autophagy, and the cross-talk between these trafficking routes provides efficient substrate clearance. Cross-talk between distinct protein clearance systems—the autophagy lysosomal and the proteasome pathways—also functions as a complimentary network to ensure cellular homeostasis, involving balanced regulation of the pathways and their compensatory responses to govern protein accumulation stress (see Farizatto et al., 2017; Pandey et al., 2007; Pohl and Dikic, 2019).

In this review, we will focus on endosomal-lysosomal dysfunction that plays a role in lifestyle-induced metabolic perturbations, as well as Alzheimer's disease (AD) and other protein accumulation pathologies.

2. Metabolic energy balance

In recent years, it has become clear that whole body metabolic state can profoundly influence the risk of acquiring and the progression of neurodegenerative diseases. Energy balance is of particular importance, where excessive caloric intake is a principle driver of obesity and diabetes, which can impede normal endosomal-lysosomal function in the brain. Enhancing energy expenditure via physical activity, however, has a restorative effect and offers insights into treatment options.

2.1 Diabetes and obesity

Insulin resistance, Type 2 diabetes mellitus (T2DM) and obesity lead to an increased risk of cognitive decline and development of AD (Jayaraman and Pike, 2014; Neth and Craft, 2017; Ryu et al., 2019; Thambisetty et al., 2013). In fact, the connection between T2DM and AD etiology overlaps to the extent that the term "type 3 diabetes" has been used to describe AD induced by insulin resistance and T2DM (Steen et al., 2005), although this nomenclature is not officially accepted by the World Health Organization or the American Diabetes Association. While there are conflicting studies, in general, obesity and T2DM tend to decrease autophagy, and a decrease in autophagic flux appears to facilitate the progression from obesity to T2DM (Levine and Kroemer, 2019). Mitochondrial dysfunction associated with reduced autophagy (including mitochondria-specific autophagy known as mitophagy), as well as mitochondrial fusion, fission, and biogenesis, appears to be a central driver of the responses to high circulating glucose and insulin resistance in the brain (Cardoso and Moreira, 2019).

Carvalho et al. (2015) found that the autophagy related 7 protein (ATG7) and the glycosylated form of the lysosomal-associated membrane protein 1 (LAMP1) were downregulated in brains of mouse models of both diet-induced T2DM and a transgenic model of AD. Furthermore, these disease models had compromised mitochondrial function and biogenesis, as well as synaptic damage. In non-obese diabetic mice, Santos et al. (2014) observed a decrease in autophagy in the brain, indicated by a reduction in the autophagy protein microtubule-associated protein 1A/1B-light chain 3 (LC3, particularly its lipid-modified form LC3II) and an increase in phosphorylation of the mechanistic target of rapamycin complex 1 (mTORC1). These findings are consistent with decreased autophagy, which Santos et al. (2014) proposed was part of an adaptive strategy to maintain mitochondrial quality and preserve cellular energetics during the early stages of T2DM. In contrast, shifting the balance toward net energy expenditure via calorie restriction and/or exercise can restore insulin sensitivity, increase brain-derived neurotrophic factor (BDNF) and cAMP response element binding protein (CREB), reduce mTORC1 activation, and thus restore autophagy and mitochondrial function (Mattson et al., 2018; also see Fernandez et al., 2018).

Tramutola et al. (2015) examined post-mortem human inferior parietal lobule in patients with pre-clinical AD, mild cognitive impairment (MCI), and late AD. There was an increase in the amyloid β protein (Aβ) in all three groups, which was associated with decreases in the autophagy-initiating proteins Beclin-1 and LC3, consistent with a reduction in autophagy. Furthermore, the PI3K/Akt/mTORC1 pathway, which inhibits autophagy, was hyper-activated and associated with increased insulin receptor substrate 1 (IRS1), an indicator of insulin resistance. Tramutola et al. (2015) proposed that even in pre-clinical AD the elevated Aβ levels induce insulin resistance that exacerbates disease progression. Consistent with these findings, Han et al. (2017) demonstrated that the neuroprotective agent, humanin, improved insulin sensitivity by inducing an upregulation of autophagy, which led to enhanced learning and memory in the APP/PS1 mouse model of AD.

Other studies have similarly found that treatment with insulin or insulin-sensitizing agents, such as metformin, leptin or glucagon-like peptide-1, restores autophagy in AD models and enhances cognition, and the PI3K/Akt/mTORC1 pathway appears to be a central mediator of these effects (de Mello et al., 2019; Gabbouj et al., 2019; Hölscher, 2019).

The central role of autophagy in both T2DM and AD was recently evaluated by Caberlotto et al. (2019). A transcriptomic analysis of post-mortem T2DM and AD human brains, followed by a network analysis revealed overlap of 158 genes among the two disease states. Within the network, overlap was a strong involvement of autophagy genes, and this was confirmed in AD mouse models, indicating the fundamental importance of autophagy dysregulation in both T2DM and AD (Caberlotto et al., 2019). While the causative mechanisms of insulin resistance in the brain are yet to be determined, the linkage to autophagy, cognitive impairment and AD has become well established (de Mello et al., 2019).

2.2 Exercise and exercise mimetics

Exercise has positive effects on synaptic plasticity, cognition and memory, and affords protection against neurodegeneration (Cotman et al., 2007; Mattson, 2012). In older adults, the level of physical activity and frailty influence learning and the progression of AD and other neuropathologies (Buchman et al., 2019; Wallace et al., 2019). Two recent clinical trials demonstrated that a healthy lifestyle, including physical activity, had positive outcomes for the progression of AD (Isaacson et al., 2019) and heritable frontotemporal degeneration (Casaletto et al., 2020). While it is well established that in skeletal muscle, exercise induces an increase in autophagy, and elevated autophagic flux and lysosomal activity appear to be a prerequisite for muscle protein and mitochondrial turnover associated with exercise training effects, the effects of exercise on brain autophagy and lysosome function are only recently coming into focus.

He et al. (2012) demonstrated that exercise induces autophagy in the brain, using wild-type and B-cell lymphoma 2 (BCL2) mutant mice that lacked the capacity for exercise- and starvation-inducible autophagy but not basal autophagy. In wild-type transgenic mice expressing green-fluorescent protein (GFP) conjugated to the autophagy marker LC3, fluorescence microscopy images revealed an increase in GFP-LC3 puncta following acute exercise in the cerebral cortex, which is consistent with elevated autophagic activity. In addition, sequestosome 1 (SQSTM1/p62) was decreased in cerebral cortex in response to acute exercise in wild-type mice, while BCL2 mutant mice lacked this response. SQSTM1/p62 is degraded during autophagy, so a decrease in protein levels is generally associated with elevated autophagic flux, while SQSTM1/p62 accumulation is indicative of autophagy impairment.

Zhao et al. (2018) found that exercise reduced Aβ plaque deposition in the APP/PS1 transgenic model of AD, which has reduced autophagy activity relative to wild-type mice as indicated by the combination of increases in LC3II and SQSTM1/p62, as well as an overabundance of the lysosomal protein LAMP1. While an increase in LC3II expression typically is thought to be associated with upregulation of autophagy, when it accumulates in concert with increases in SQSTM1/p62, it is indicative of a decrease in degradation of autophagosomal substrates (Yoshii and Mizushima, 2017). Furthermore, Zhao et al. (2018) argue that the elevated LAMP1 in APP/PS1 mice suggests an accumulation of lysosomes and reduced clearance of autophagic substrates and is consistent with prior findings in this AD model. In contrast, exercise in APP/PS1 mice led to a further increase in LC3II expression and decreases in both SQSTM1/p62 and LAMP1, which are results consistent with an increase in autophagic flux that likely explains, at least in part, the reduction in Aβ plaque accumulation.

The mechanisms by which exercise enhances synaptic plasticity and provides neuroprotective effects, and the linkage to autophagy remain to be fully elucidated. Recent work has identified the myokine, irisin, which is a cleavage product of fibronectin type III domain-containing protein 5 (FNDC5), as a key factor that is released from muscle during exercise and mediates some of the positive effects of exercise on brain function (Lourenco et al., 2019; Munoz et al., 2018).

Genetic and pharmacological reduction of FNDC5/irisin in mouse brain impairs synaptic plasticity and memory, while increasing FNDC5/irisin rescues these functions in wild-type and AD models (Lourenco et al., 2019). Swimming protected against Aβ oligomer-induced memory loss in an AD mouse model and increased FNDC5/irisin in the brain, whereas these beneficial effects were lost when irisin was inhibited. In addition, peripheral treatment with FNDC5/irisin had the same beneficial effect, suggesting potential therapeutic approaches to treating AD and related diseases. These authors further showed that irisin stimulated the cAMP-PKA-CREB signaling pathway, which is known to be involved with memory formation. In addition, CREB has been shown to induce autophagy by upregulating the autophagy genes *ATG7*, UNC51-like autophagy activating kinase 1 (*ULK1*) and transcription factor EB (*TFEB*) (Füllgrabe et al., 2016), providing a potential mechanistic link between exercise, FNDC5/irisin and autophagy.

FNDC5/irisin triggers the expression of BDNF, which is involved in neuronal growth, memory and learning (Wrann et al., 2013). Choi et al. (2018) evaluated the effects of genetic and pharmacological induction of adult hippocampal neurogenesis (AHN), elevations of BDNF, and exercise in a mouse AD model. Exercise induced AHN, as well as an increase in BDNF and FNDC5, all of which led to cognitive improvement. While artificial elevation of AHN or BDNF alone did not improve cognition, activating AHN and BDNF in concert led to cognitive improvements comparable to those seen following exercise. Moon et al. (2016) found that running exercise induces the release of secretory cathepsin B, a candidate exercise myokine, in mice and humans. In cathepsin B knockout mice the exercise-induced increases in BDNF, hippocampal neurogenesis, and spatial memory were blocked, suggesting that like FNDC5/irisin, cathepsin B is also an important mediator of the beneficial effects of exercise on brain health.

Neuroprotective effects of exercise can also be mediated through the AMP-activated protein kinase (AMPK)-sirtuin1 (SIRT1)-TFEB pathway, independently of mTORC1. Bayod et al. (2011) showed in rat brain that exercise reduced levels of phospho-tau, and this was associated with the activation of AMPK and upregulation of SIRT1. Kim and Leem (2016) found that stress-induced memory deficits could be offset by exercise, which led to an increase in BDNF in an AMPK-dependent manner. Huang et al. (2019) further demonstrated that exercise activated AMPK in brains of mice, and this was associated with phosphorylation of ULK1 and an increase in LC3II. The phosphorylated modification of AMPK, in turn, led to an increase in NAD^+ and activation of SIRT1, which increased TFEB in the brain and induced expression of lysosomal proteins.

While exercise has clear functional benefits in brain, some of which are mediated by autophagy enhancement, there is also evidence that exercise mimetic drugs can improve brain function, which may provide useful treatment options for patients who lack the capacity for physical exercise. The AMP analog 5-aminoimidazole-4-carboxamide ribonucleotide (AICAR) activates AMPK and leads to whole body training effects and improves endurance in mice. Kobilo et al. (2014) found that AICAR treatment led to improved spatial memory and motor function, and increased gene expression in brain similar to the patterns seen in muscle that lead to the endurance phenotype. The poor capacity for AICAR to cross the blood-brain barrier suggests that these effects are indirect, and skeletal muscle may be a key contributor to the effects seen in brain.

In a follow up study, Guerrieri and van Praag (2015) found that AICAR induced positive short-term effects on brain in an AMPK-dependent manner, but unlike in muscle, the AICAR treatment led to long-term cytotoxicity in the brain, suggesting that it is not a suitable substitute for exercise. Hill et al. (2016) showed that traumatic brain injury reduces AMPK activity, and pharmacological stimulation of AMPK with AICAR or metformin improved spatial memory. Likewise, the creatine analog and known AMPK activator, β-guanidinopropionic acid (β-GPA), led to an upregulation in brain mRNA for genes associated with mitochondrial biogenesis and oxidative stress, as well as behavioral changes (Gureev et al., 2018).

While these and other exercise mimetic drugs target AMPK, which can activate autophagy and lysosomal activity, the role of these treatments in autophagic processes and lysosomal activity is yet to be determined (Guerrieri et al., 2017).

3. Alzheimer's disease and related dementias

Alzheimer's disease is an irreversible and progressive diseases, noteworthy as possibly the biggest challenge in medicine today with millions of the aging population facing an impending epidemic if effective strategies to prevent or slow the disease are not discovered. Alzheimer's disease is one of many age-related protein accumulation diseases. Compromised clearance of protein aggregates is implicated in several proteinopathies, and lysosomal instability is a feature of brain aging that decreases clearing efficiency and thereby disrupts the balance between protein synthesis and protein degradation/recycling (Colacurcio and Nixon, 2016; Miquel et al., 2018; Peng et al., 2019). This altered balance increases the risk of AD and related dementias as abnormal accumulations of Aβ peptides and tau form long before signs of memory impairment (see Bateman et al., 2012).

Altered levels of lysosomes, autophagosomes, and autolysosomes occur in AD, and familial AD and polymorphisms linked to sporadic AD exhibit the indications of endo-lysosomal and autophagic disturbances (see Wang et al., 2018). The gradual, age-related lysosomal instability leads to gradual deposition of pathogenic protein species, slow disruption of microtubule-based transport processes, and progressive synaptic decline (see Bahr et al., 2012; Bendiske et al., 2002; Dourlen et al., 2019; Forner et al., 2017). This cascade of synaptopathology is notable as synapse loss stands out as a strong correlate of cognitive deterioration in both AD and aging.

Several risk factors for Alzheimer-type pathogenesis have also been linked to protein clearance machinery including autophagic-lysosomal functions. Two examples of AD-associated genes are involved in autophagic initiation and autophagosome formation, namely phosphatidylinositol-binding clathrin assembly protein (*PICALM*) and the Beclin-1 component of autophagy biogenesis (*BECN1*) (Bonam et al., 2019; Wang et al., 2018). Reduction in BECN1 in certain cells is linked to reduced trafficking of retromer, a complex involved in endo-lysosomal sorting which is impaired in familial AD (Caglayan et al., 2014). Apolipoprotein E4 (ApoE4) promotes the trapping of its lipidation controlling protein ATP-binding cassette transporter subfamily A member 1 (ABCA1) in late-endosomes and alters its membrane trafficking in astrocytes (Rawat et al., 2019). In addition to other avenues toward abnormal sorting, recycling, and trafficking that are linked to AD (see Wang et al., 2018), lysosomal organelles themselves exhibit impairment of their axonal transport, maturation, and hydrolase trafficking in AD mice (see Gowrishankar et al., 2015; Hwang et al., 2019), events that occur at early time points of the AD-type proteinopathy.

Lysosomes play a key role in governing AD-type protein accumulation events since increasing lysosomal biogenesis reduces amyloid deposits and neurofibrillary pathology in APP/PS1 and tauopathy models, respectively (Polito et al., 2014; Xiao et al., 2015). Altered tau secretion in tauopathies has been linked to impairment in the autophagy-lysosomal pathway, thereby enhancing the spreading of the protein (Chen et al., 2020). In addition, lysosomal cathepsins B and L, members of the papain-family of cysteine endopeptidases, have been found to be involved in the fragmentation and clearance of the Aβ peptide (Butler et al., 2011; Mueller-Steiner et al., 2006; Wang et al., 2012) and tau isoforms (Bendiske and Bahr, 2003; Wang et al., 2009). The autophagic-lysosomal and proteasomal systems for protein clearance display concerted regulation and cross-talk with regards to the governance of Aβ pathology (e.g., see Farizatto et al., 2017), perhaps comprising a complimentary network that maintains a balance between the two clearance pathways through compensatory responses. Evidence indicates that the lysosomal component of the compensatory action is compromised with age as shown by different studies (Gavilán et al., 2015; Mueller-Steiner et al., 2006; Torres et al., 2012), thus pointing to an unprotected protein clearance system as an aging risk factor for the development of proteinopathy.

Protective compensatory responses by the autophagic-lysosomal system are related, at least in part, to exercise-mediated induction of the autophagy-lysosomal degradation pathway as discussed above. Epidemiological studies reported that exercise reduces the risk of age-related cognitive decline, as well as the likelihood of developing AD (Hamer et al., 2018; Livingston et al., 2017). Eight of thirteen randomized controlled trials found that exercise improves cognitive function in people with AD (Du et al., 2018), and high levels of physical activity were associated with slower clinical decline in autosomal dominant frontotemporal dementia (Casaletto et al., 2020). In healthy humans and monkeys, regular treadmill exercise increased systemic levels of cathepsin B and improved learning and memory (Moon et al., 2016), thus further indicating the neuroprotective potential of the autophagy-lysosomal pathway for healthy brain aging (also see Bayod et al., 2011; Chen and Gan, 2019; He et al., 2012; Zhao et al., 2018). Endurance exercise was also found to be protective in a mouse model of MPTP-induced parkinsonism (Jang et al., 2018), similar to Moon et al. (2016) as being associated with increased autophagy and cathepsin proteins. Exercise also elicits new synapse formation/repair, which is clearly relevant since gradual synaptic decline is a significant correlate of cognitive deterioration (Chen et al., 2018; Spires-Jones and Hyman, 2014).

Broadening the importance of compensatory actions through the autophagy-lysosomal pathway, note the contrast between lysosomal dysfunction, including reduced cathepsin B and L activities recently linked to obesity-related pathology (Mizunoe et al., 2019) versus evident lysosomal enhancement by exercise-induced cathepsin B levels that correlate with improvements in the type of memory encoding lost in AD (see Moon et al., 2016). In addition to the hippocampal-dependent recall ability in humans, the treadmill training in the Moon et al. (2016) study also detected a positive correlation between aerobic fitness change and the change in cathepsin B plasma levels.

Cathepsin B has been shown to significantly lower levels of $A\beta_{42}$ amyloid peptide when positively modulated by genetic and pharmacological strategies in AD transgenic mouse models, leading to reductions in both neuronal and behavioral disease parameters (Butler et al., 2011; Hwang et al., 2019; Mueller-Steiner et al., 2006; Viswanathan et al., 2012; Wang et al., 2012; Yang et al., 2011). It is also of interest that, cilostazol, a positive modulator of the autophagy-lysosomal system, enhanced cathepsin B activity and reduced $A\beta_{42}$ accumulation via sirtuin 1 (SIRT1)-coupled AMPK signaling in neuroblastoma cells (Park et al., 2016)—note that SIRT1 is widely considered a crucial component for proteostasis. Cathepsin B indeed degrades

$A\beta_{42}$ through carboxy-terminal truncation, resulting in proteolytic detoxification of the peptide (Butler et al., 2011; Hwang et al., 2019; Mueller-Steiner et al., 2006; Wang et al., 2012). In addition, the hydrolase promotes the clearance of α-synuclein, including the A53T α-synuclein variant linked to familial Parkinson's disease (Hwang et al., 2019; McGlinchey and Lee, 2015). The McGlinchey–Lee study found that cathepsin B cleaves α-synuclein in such a way that circumvents pathogenic fibril formation.

The regulation of cathepsin B has been extensively studied and the enzyme shown to be modestly enhanced in response to different types of protein accumulation pathology (see Bonam et al., 2019; Farizatto et al., 2017) as compared to the level produced by direct pharmacological enhancement. Such findings suggest the involvement of endogenous compensatory signals that explain the slow nature of disease progression in many dementias. In addition to the synaptic and behavioral protection by cathepsin B upregulation in several transgenic AD models (see above), similar enhancement of the hydrolase improved memory performance and synaptic integrity in models of MCI (Hwang et al., 2019).

In striking contrast to the manner of protection the cysteine protease cathepsin B produces when enriched in lysosomes, inhibition of the non-lysosomal cysteine protease calpain has long been suggested as a neuroprotective avenue (see Hassen et al., 2017; Romine et al., 2017). Together, such protective avenues may act in a concerted manner since the calcium-activated calpain generates fragments of cytoskeletal components and membrane receptors during excitotoxic events, including the hyperexcitability and seizure episodes identified during early stages of AD and in a model of frontotemporal dementia (García-Cabrero et al., 2013; Vossel et al., 2013, 2016), thus exacerbating proteostatic stress on autophagic-lysosomal processes working to clear the accumulating breakdown products. It is also of interest here that neuronal excitability has been found to increase with age, and inhibition of excitatory neurotransmission increases longevity (Zullo et al., 2019).

Through the recent lines of research, a clearer understanding has evolved regarding the autophagy-lysosomal pathway and the role it plays in cellular quality control for healthy brain aging. Accordingly, regulators of protein clearance activity are being pursued to identify candidate therapeutic drugs, peptides, and natural products, following the proof-of-concept studies on both the positive modulation of cathepsins (Bendiske and Bahr, 2003; Butler et al., 2011; Mueller-Steiner et al., 2006), as well as the down-regulation of endogenous inhibitors of lysosomal cysteine proteases that leads to cathepsin-dependent reduction of proteinopathy (Wang et al., 2018; Yang et al., 2011). Exercise mimetics may be among the candidates as recent drug discovery efforts equaled the exercise-mediated benefits to neuronal networks and brain function, with exercise widely accepted as a health-improving lifestyle factor that can prevent a variety of diseases.

4. α-Synucleinopathies

The α-synuclein protein is the major component of proteinaceous inclusions found in neurodegenerative diseases called α-synucleinopathies, and it appears to contribute to the protein accumulation stress in 50% of AD patients. Parkinson's disease (PD) is the second most common, late-onset neurodegenerative disease, behind only AD. The accumulation of α-synuclein is the main pathological hallmark of PD and PD-related disorders. It is notable that α-synuclein is a regulator of the autophagic network (Winslow et al., 2010) in addition to being a substrate of the autophagy-lysosomal pathway (Hwang et al., 2019; McGlinchey and Lee, 2015).

Parkinson's disease-associated genes and evidence from model systems strongly implicate intracellular trafficking and protein clearance systems as playing a major role in α-synucleionopathy. Several PD-associated genes are related to endo-lysosomal/autophagy pathways, such as *ATP13A2*, *PINK1*, *PRKN*, *TMEM175*, *VPS35*, *LRRK2*, *GBA1*, *SMPD1*, *GALC*, *ASAH1*, *SCARB2*, *CTSD*, *CTSB*, and *GLA* (see Table 2), thus in support of dysfunction in these and related pathways as being a contributing or even primary cause of PD (Bellomo et al., 2019; Bento et al., 2016; Wallings et al., 2019; Wang et al., 2018). *GBA1* and *LRRK2* genes harbor some of the most common PD-associated mutations and they are linked to autophagic-lysosomal processing. Note that mutations in the glucocerebrosidase gene *GBA1* also cause Gaucher disease, an autosomal recessive lysosomal storage disease, and Mullin et al. (2020) recently reported that ambroxol, a drug known to increase glucocerebrosidase activity and reduce α-synuclein levels, can positively target the glucocerebrosidase pathway in PD patients with and without *GBA1* mutations.

Leucine-rich repeat kinase 2 (LRRK2) regulates autophagy, perhaps through vesicle sorting and/or lysosomal positioning. Using generated lines of *C. elegans* expressing LRRK2 and α-synuclein in dopaminergic neurons, PD-causing mutations in *LRRK2* were found to inhibit autophagy and accelerate age-related failure of autophagic function, while wild-type LRRK2 improved autophagy throughout the *C. elegans* lifespan

Table 2 Summary of genes with mutations associated with familial PD and linked to the autophagy-lysosomal pathway.

Symbol	Name
ATP13A2	ATPase cation transporting 13A2
PINK1	Phosphatase and tensin homolog (PTEN) induced kinase 1
PRKN	Parkin RBR E3 ubiquitin protein ligase
TMEM175	Transmembrane protein 175
VPS35	Vacuolar protein sorting-associated protein 35 retromer complex component
LRRK2	Leucine-rich repeat kinase 2
GBA1	Glucosylceramidase β1
SMPD1	Sphingomyelin phosphodiesterase 1
GALC	Galactosylceramidase
ASAH1	*N*-acylsphingosine amidohydrolase 1
SCARB2	Scavenger receptor class B member 2
CTSB	Cathepsin B
GLA	Galactosidase α

(Saha et al., 2015). The relationship between autophagic-lysosomal dysfunction and the pathogenesis of α-synuclein aggregation/propagation points to positive modulatory agents of the protein clearance pathway as a potential disease-modifying avenue.

Disturbances in the endo-lysosomal system play a prominent role in PD pathogenesis (see Wang et al., 2018), including the dysfunction of endo-lysosomal sorting and trafficking. Mazzulli et al. (2016) established that α-synuclein accumulation disrupts lysosomal hydrolase trafficking in human midbrain dopaminergic neurons, finding a significant reduction in cathepsin B activity in PD and Gaucher disease neurons. They also showed that overexpression of an endoplasmic reticulum-Golgi trafficking regulator improved hydrolase maturation and activity in patient neurons, and the lysosomal enhancement reduced pathological α-synuclein in neurons and improved their viability.

Note that a mutation in *VPS35*, a component of the large retromer complex for endo-lysosomal sorting and trafficking, causes late-onset PD (Zimprich et al., 2011) and thereby implicates retromer dysfunction in

PD, as well as AD (see Caglayan et al., 2014). A retromer-like heterotrimer called retriever was also identified and found to mediate recycling of α5β1 integrins and other signaling receptors from endosomes to the plasma membrane (McNally et al., 2017). Such transport machineries may be involved in compensatory responses important for neuronal homeostasis and synapse integrity, as a neurotoxic insult enhanced β1 integrin levels at synaptic contacts and promoted integrin-cofilin signaling (Farizatto et al., 2019).

The protein accumulation stress from α-synuclein inclusions and associated transport deficiency may explain the synaptic decline in transgenic mouse models of PD (Batey et al., 2005; Hwang et al., 2019). Enhancing the autophagy-lysosomal system with a small-molecule modulator (*Z*-phenylalanyl-alanyl-diazomethylketone), which appears to act as an inhibitory chaperone to promote transport to lysosomes and elution of active cathepsin B (Butler et al., 2011), was found to reduce α-synuclein levels in PD mice, similar to the decreased proteinopathy the positive modulator produced in AD mouse models (Butler et al., 2011; Hwang et al., 2019). Such enhancement led to the positive relationship between improved clearance of α-synuclein and the extent of synaptic marker recovery.

Related compensatory signaling was evident in response to protein accumulation stress and autophagic compromise in Huntington's disease mice, with activation of chaperone-mediated autophagy (CMA) resulting in constitutive upregulation (Koga et al., 2011). Interestingly, an autophagy-inducing peptide (Shoji-Kawata et al., 2013) and LC3 linker compounds that target mutant huntingtin protein for degradation using an autophagosome-tethering approach (Li et al., 2019) also reduced the polyglutamine proteinopathy related to Huntington's disease. Chaperone mediated autophagy works to control cellular energy through involvement in glucose metabolism and recycling amino acids from lysosomal degradation of substrate proteins. Tapping into CMA compensatory responses is key for therapeutic routes since dysfunctional CMA occurs with age, thus the development of CMA activators has been progressing. For example, in the lysosomal storage diseases cystinosis, which exhibits CMA defects and impaired trafficking of the CMA receptor LAMP2A, the developed small-molecule CMA activator QX77 (*N*-[4-(7-chloro-2H-1, 4-benzoxazin-3-yl)phenyl]acetamide) improved LAMP2A localization to lysosomes and increased cell survival (Zhang et al., 2017). As with lysosomal-targeted cathepsin modulation, CMA upregulation is a potential strategy to treat a variety of diseases exhibiting defective CMA.

5. Conclusion

Lysosomes are clearly a core component of proteostasis with multiple types of substrates of extracellular and intracellular origin being delivered for degradation through endocytic trafficking and autophagic events. Several related metabolic and genetic cascades influence aging and longevity, and the aging process and many disease states weaken the proteostasis network of mechanisms that works to maintain (i) the integrity of the proteome, (ii) complex glycoconjugates and glycans, and (iii) posttranslational modifications that influence functionality. Vital compensatory responses for the maintenance of cellular and subcellular homeostasis (to include pre- and postsynaptic moieties) involve lysosomal maturation, autophagic regulation, as well as cross-talk with the ubiquitin-proteasome system (Bahr et al., 2012; Bonam et al., 2019; Colacurcio and Nixon, 2016; Farizatto et al., 2017; Scrivo et al., 2018; Wang et al., 2012). These distinct restorative outlets are related due to shared components or due to having correlative activation events in response to proteinopathy. Together they represent an intimate circle of potential therapeutic targets for neurodegenerative diseases, with lysosomes at the hub of each avenue.

References

Bahr, B. A., Wisniewski, M. L., & Butler, D. (2012). Positive lysosomal modulation as a unique strategy to treat age-related protein accumulation diseases. *Rejuvenation Research*, *15*, 189–197.

Bateman, R. J., Xiong, C., Benzinger, T. L. S., Fagan, A. M., Goate, A., Fox, N. C., et al. (2012). Clinical and biomarker changes in dominantly inherited Alzheimer's disease. *The New England Journal of Medicine*, *367*, 795–804.

Batey, L., Chin, D. J., Karim, S., Karanian, D. A., Butler, D., & Bahr, B. A. (2005). Lysosomal modulation and autophagic induction: Compensatory processes in response to protein deposition pathology. In A. Fisher (Ed.), *New trends in Alzheimer and Parkinson related disorders, ADPD 2005* (pp. 55–58). Bologna, Italy: Medimond International Proceedings.

Bayod, S., del Valle, J., Canudas, A. M., Lalanza, J. F., Sanchez-Roige, S., Camins, A., et al. (2011). Long-term treadmill exercise induces neuroprotective molecular changes in rat brain. *Journal of Applied Physiology*, *111*, 1380–1390.

Bellomo, G., Paciotti, S., Gatticchi, L., & Parnetti, L. (2019). The vicious cycle between α-synuclein aggregation and autophagic-lysosomal dysfunction. *Movement Disorders*, *35*, 34–44.

Bendiske, J., & Bahr, B. A. (2003). Lysosomal activation is a compensatory response against protein accumulation and associated synaptopathogenesis—An approach for slowing Alzheimer disease? *Journal of Neuropathology and Experimental Neurology*, *62*, 451–463.

Bendiske, J., Caba, E., Brown, Q. B., & Bahr, B. A. (2002). Intracellular deposition, microtubule destabilization, and transport failure: An "early" pathogenic cascade leading to synaptic decline. *Journal of Neuropathology and Experimental Neurology*, *61*, 640–650.

Bento, C. F., Ashkenazi, A., Jimenez-Sanchez, M., & Rubinsztein, D. C. (2016). The Parkinson's disease-associated genes *ATP13A2* and *SYT11* regulate autophagy via a common pathway. *Nature Communications*, 7, 11803.

Bonam, S. R., Wang, F., & Muller, S. (2019). Lysosomes as a therapeutic target. *Nature Reviews Drug Discovery*, *18*, 923–948.

Buchman, A. S., Yu, L., Wilson, R. S., Lim, A., Dawe, R. J., Gaiteri, C., et al. (2019). Physical activity, common brain pathologies, and cognition in community-dwelling older adults. *Neurology*, *92*, 811–822.

Butler, D., Hwang, J., Estick, C., Nishiyama, A., Kumar, S. S., Baveghems, C., et al. (2011). Protective effects of positive lysosomal modulation in Alzheimer's disease transgenic mouse models. *PLoS One*, *6*, e20501.

Caberlotto, L., Nguyen, T. P., Lauria, M., Priami, C., Rimondini, R., Maioli, S., et al. (2019). Cross-disease analysis of Alzheimer's disease and type-2 diabetes highlights the role of autophagy in the pathophysiology of two highly comorbid diseases. *Scientific Reports*, *9*, 3965.

Caglayan, S., Takagi-Niidome, S., Liao, F., Carlo, A. S., Schmidt, V., Burgert, T., et al. (2014). Lysosomal sorting of amyloid-β by the SORLA receptor is impaired by a familial Alzheimer's disease mutation. *Science Translational Medicine*, *6*, 223ra20.

Cardoso, S., & Moreira, P. I. (2019). Diabesity and brain disturbances: A metabolic perspective. *Molecular Aspects of Medicine*, *66*, 71–79.

Carmona-Gutierrez, D., Hughes, A. L., Madeo, F., & Ruckenstuhl, C. (2016). The crucial impact of lysosomes in aging and longevity. *Ageing Research Reviews*, *32*, 2–12.

Carvalho, C., Santos, M. S., Oliveira, C. R., & Moreira, P. I. (2015). Alzheimer's disease and type 2 diabetes-related alterations in brain mitochondria, autophagy and synaptic markers. *Biochimica et Biophysica Acta*, *1852*, 1665–1675.

Casaletto, K. B., Staffaroni, A. M., Wolf, A., Appleby, B., Brushaber, D., Coppola, G., et al. (2020). Active lifestyles moderate clinical outcomes in autosomal dominant frontotemporal degeneration. *Alzheimer's & Dementia*, *16*, 91–105.

Chen, Z., Brodie, M. J., Liew, D., & Kwan, P. (2018). Treatment outcomes in patients with newly diagnosed epilepsy treated with established and new antiepileptic drugs: A 30-year longitudinal cohort study. *JAMA Neurology*, *75*, 279–286.

Chen, X., & Gan, L. (2019). An exercise-induced messenger boosts memory in Alzheimer's disease. *Nature Medicine*, *25*, 20–21.

Chen, X., Li, Y., Wang, C., Tang, Y., Mok, S.-A., Tsai, R. M., et al. (2020). Promoting tau secretion and propagation by hyperactive p300/CBP via autophagy-lysosomal pathway in tauopathy. *Molecular Neurodegeneration*, *15*, 2.

Choi, S. H., Bylykbashi, E., Chatila, Z. K., Lee, S. W., Pulli, B., Clemenson, G. D., et al. (2018). Combined adult neurogenesis and BDNF mimic exercise effects on cognition in an Alzheimer's mouse model. *Science*, *361*, eaan8821.

Colacurcio, D. J., & Nixon, R. A. (2016). Disorders of lysosomal acidification—The emerging role of v-ATPase in aging and neurodegenerative disease. *Ageing Research Reviews*, *32*, 75–88.

Condon, K. J., & Sabatini, D. M. (2019). Nutrient regulation of mTORC1 at a glance. *Journal of Cell Science*, *132*, jcs222570.

Cotman, C. W., Berchtold, N. C., & Christie, L.-A. (2007). Exercise builds brain health: Key roles of growth factor cascades and inflammation. *Trends in Neuroscience*, *30*, 464–472.

de Mello, N. P., Orellana, A. M., Mazucanti, C. H., de Morais Lima, G., Scavone, C., & Kawamoto, E. M. (2019). Insulin and autophagy in neurodegeneration. *Frontiers in Neuroscience*, *13*, 491.

Dourlen, P., Kilinc, D., Malmanche, N., Chapuis, J., & Lambert, J.-C. (2019). The new genetic landscape of Alzheimer's disease: From amyloid cascade to genetically driven synaptic failure hypothesis? *Acta Neuropathologica*, *138*, 221–236.

Du, Z., Li, Y., Li, J., Zhou, C., Li, F., & Yang, X. (2018). Physical activity can improve cognition in patients with Alzheimer's disease: A systematic review and meta-analysis of randomized controlled trials. *Clinical Interventions in Aging*, *13*, 1593–1603.

Farizatto, K. L. G., Almeida, M. F., Long, R. T., & Bahr, B. A. (2019). Early synaptic alterations and selective adhesion signaling in hippocampal dendritic zones following organophosphate exposure. *Scientific Reports*, *9*, 6532.

Farizatto, K. L. G., Ikonne, U. S., Almeida, M. F., Ferrari, M. F. R., & Bahr, B. A. (2017). $A\beta_{42}$-mediated proteasome inhibition and associated tau pathology in hippocampus are governed by a lysosomal response involving cathepsin B: Evidence for protective crosstalk between protein clearance pathways. *PLoS One*, *12*, e0182895.

Fernandez, A. M., Santi, A., & Torres Aleman, I. (2018). Insulin peptides as mediators of the impact of lifestyle in Alzheimer's disease. *Brain Plasticity*, *4*, 3–15.

Forner, S., Baglietto-Vargas, D., Martini, A. C., Trujillo-Estrada, L., & LaFerla, F. M. (2017). Synaptic impairment in Alzheimer's disease: A dysregulated symphony. *Trends in Neuroscience*, *40*, 347–357.

Füllgrabe, J., Ghislat, G., Cho, D.-H., & Rubinsztein, D. C. (2016). Transcriptional regulation of mammalian autophagy at a glance. *Journal of Cell Science*, *129*, 3059–3066.

Gabbouj, S., Ryhänen, S., Marttinen, M., Wittrahm, R., Takalo, M., Kemppainen, S., et al. (2019). Altered insulin signaling in Alzheimer's disease brain. Special emphasis on PI3K-Akt pathway. *Frontiers in Neuroscience*, *13*, 629.

García-Cabrero, A. M., Guerrero-López, R., Giráldez, B. G., Llorens-Martín, M., Avila, J., Serratosa, J. M., et al. (2013). Hyperexcitability and epileptic seizures in a model of frontotemporal dementia. *Neurobiology of Disease*, *58*, 200–208.

Gavilán, E., Pintado, C., Gavilan, M. P., Daza, P., Sánchez-Aguayo, I., Castaño, A., et al. (2015). Age-related dysfunctions of the autophagy lysosomal pathway in hippocampal pyramidal neurons under proteasome stress. *Neurobiology of Aging*, *36*, 1953–1963.

Gowrishankar, S., Yuan, P., Wu, Y., Schrag, M., Paradise, S., Grutzendler, J., et al. (2015). Massive accumulation of luminal protease-deficient axonal lysosomes at Alzheimer's disease amyloid plaques. *Proceedings of the National Academy of Sciences of the United States of America*, *112*, 3699–3708.

Guerrieri, D., Moon, H. Y., & van Praag, H. (2017). Exercise in a pill: The latest on exercise-mimetics. *Brain Plasticity*, *2*, 153–169.

Guerrieri, D., & van Praag, H. (2015). Exercise-mimetic AICAR transiently benefits brain function. *Oncotarget*, *6*, 18293–18313.

Gureev, A. P., Shaforostova, E. A., Starkov, A. A., & Popov, V. N. (2018). β-Guanidinopropionic acid stimulates brain mitochondria biogenesis and alters cognitive behavior in nondiseased mid-age mice. *Journal of Experimental Neuroscience*, *12*, 1179069518766524.

Hamer, M., Terrera, G. M., & Demakakos, P. (2018). Physical activity and trajectories in cognitive function: English longitudinal study of ageing. *Journal of Epidemiology and Community Health*, *72*, 477–483.

Han, K., Jia, N., Zhong, Y., & Shang, X. (2017). S14G-humanin alleviates insulin resistance and increases autophagy in neurons of APP/PS1 transgenic mouse. *Journal of Cellular Biochemistry*, *119*, 3111–3117.

Hassen, G. W., Kesner, L., Stracher, A., & Shulman, A. (2017). Calpain inhibition: A potential therapeutic target for neurodegenerative and neuromuscular disorders. In *Vol. 3. Frontiers in CNS drug discovery* (pp. 33–71). Bentham Science Publishers.

He, C., Sumpter, R., Jr., & Levine, B. (2012). Exercise induces autophagy in peripheral tissues and in the brain. *Autophagy*, *8*, 1548–1551.

Hill, J. L., Kobori, N., Zhao, J., Rozas, N. S., Hylin, M. J., Moore, A. N., et al. (2016). Traumatic brain injury decreases AMP-activated protein kinase activity and pharmacological enhancement of its activity improves cognitive outcome. *Journal of Neurochemistry*, *139*, 106–119.

Hölscher, C. (2019). Insulin signaling impairment in the brain as a risk factor in Alzheimer's disease. *Frontiers in Aging Neuroscience, 11*, 88.
Huang, J., Wang, X., Zhu, Y., Li, Z., Zhu, Y.-T., Wu, J.-C., et al. (2019). Exercise activates lysosomal function in the brain through AMPK-SIRT1-TFEB pathway. *CNS Neuroscience and Therapeutics, 25*, 796–807.
Hwang, J., Estick, C. M., Ikonne, U. S., Butler, D., Pait, M. C., Elliott, L. H., et al. (2019). The role of lysosomes in a broad disease-modifying approach evaluated across transgenic mouse models of Alzheimer's disease and Parkinson's disease and models of mild cognitive impairment. *International Journal of Molecular Sciences, 20*, 4432.
Isaacson, R. S., Hristov, H., Saif, N., Hackett, K., Hendrix, S., Melendez, J., et al. (2019). Individualized clinical management of patients at risk for Alzheimer's dementia. *Alzheimer's & Dementia, 15*, 1588–1602.
Jang, Y., Kwon, I., Song, W., Cosio-Lima, L. M., & Lee, Y. (2018). Endurance exercise mediates neuroprotection against MPTP-mediated Parkinson's disease via enhanced neurogenesis, antioxidant capacity, and autophagy. *Neuroscience, 379*, 292–301.
Jayaraman, A., & Pike, C. J. (2014). Alzheimer's disease and type 2 diabetes: Multiple mechanisms contribute to interactions. *Current Diabetes Reports, 14*, 476.
Kim, D. M., & Leem, Y. H. (2016). Chronic stress-induced memory deficits are reversed by regular exercise via AMPK-mediated BDNF induction. *Neuroscience, 324*, 271–285.
Kobilo, T., Guerrieri, D., Zhang, Y., Collica, S. C., Becker, K. G., & van Praag, H. (2014). AMPK agonist AICAR improves cognition and motor coordination in young and aged mice. *Learning and Memory, 21*, 119–126.
Koga, H., Martinez-Vicente, M., Arias, E., Kaushik, S., Sulzer, D., & Cuervo, A. M. (2011). Constitutive upregulation of chaperone-mediated autophagy in Huntington's disease. *The Journal of Neuroscience, 31*, 18492–18505.
Levine, B., & Kroemer, G. (2019). Biological functions of autophagy genes: A disease perspective. *Cell, 176*, 11–42.
Li, Z., Wang, C., Wang, Z., Zhu, C., Li, J., Sha, T., et al. (2019). Allele-selective lowering of mutant HTT protein by HTT-LC3 linker compounds. *Nature, 575*, 203–209.
Livingston, G., Sommerlad, A., Orgeta, V., Costafreda, S. G., Huntley, J., Ames, D., et al. (2017). Dementia prevention, intervention, and care. *The Lancet, 390*, 2673–2734.
López-Otín, C., Blasco, M. A., Partridge, L., Serrano, M., & Kroemer, G. (2013). The hallmarks of aging. *Cell, 153*, 1194–1217.
Lourenco, M. V., Frozza, R. L., de Freitas, G. B., Zhang, H., Kincheski, G. C., Ribeiro, F. C., et al. (2019). Exercise-linked FNDC5/irisin rescues synaptic plasticity and memory defects in Alzheimer's models. *Nature Medicine, 25*, 165–175.
Mattson, M. P. (2012). Energy intake and exercise as determinants of brain health and vulnerability to injury and disease. *Cell Metabolism, 16*, 706–722.
Mattson, M. P., & Arumugam, T. V. (2018). Hallmarks of brain aging: Adaptive and pathological modification by metabolic states. *Cell Metabolism, 27*, 1176–1199.
Mattson, M. P., Moehl, K., Ghena, N., Schmaedick, M., & Cheng, A. (2018). Intermittent metabolic switching, neuroplasticity and brain health. *Nature Reviews Neuroscience, 19*, 63–80.
Mazzulli, J. R., Zunke, F., Isacson, O., Studer, L., & Krainc, D. (2016). α-Synuclein-induced lysosomal dysfunction occurs through disruptions in protein trafficking in human midbrain synucleinopathy models. *Proceedings of the National Academy of Sciences of the United States of America, 113*, 1931–1936.
McGlinchey, R. P., & Lee, J. C. (2015). Cysteine cathepsins are essential in lysosomal degradation of α-synuclein. *Proceedings of the National Academy of Sciences of the United States of America, 11*, 9322–9327.
McNally, K. E., Faulkner, R., Steinberg, F., Gallon, M., Ghai, R., Pim, D., et al. (2017). Retriever is a multiprotein complex for retromer-independent endosomal cargo recycling. *Nature Cell Biology, 19*, 1214–1225.

Miquel, S., Champ, C., Day, J., Aarts, E., Bahr, B. A., Bakker, M., et al. (2018). Poor cognitive ageing: Vulnerabilities, mechanisms and the impact of nutritional interventions. *Ageing Research Reviews*, *42*, 40–55.

Mizunoe, Y., Kobayashi, M., Tagawa, R., Nakagawa, Y., Shimano, H., & Higami, Y. (2019). Association between lysosomal dysfunction and obesity-related pathology: A key knowledge to prevent metabolic syndrome. *International Journal of Molecular Sciences*, *20*, 3688.

Moon, H. Y., Becke, A., Berron, D., Becker, B., Sah, N., Benoni, G., et al. (2016). Running-induced systemic cathepsin B secretion is associated with memory function. *Cell Metabolism*, *24*, 332–340.

Mueller-Steiner, S., Zhou, Y., Arai, H., Roberson, E. D., Sun, B., Chen, J., et al. (2006). Antiamyloidogenic and neuroprotective functions of cathepsin B: Implications for Alzheimer's disease. *Neuron*, *51*, 703–714.

Mullin, S., Smith, L., Lee, K., D'Souza, G., Woodgate, P., Elflein, J., et al. (2020). Ambroxol for the treatment of patients with Parkinson disease with and without glucocerebrosidase gene mutations: A nonrandomized, noncontrolled trial. *JAMA Neurology*, *77*, e194611. https://doi.org/10.1001/jamaneurol.2019.4611.

Munoz, I. Y. M., Camarillo Romero, E. D. S., & Garduno Garcia, J. J. (2018). Irisin a novel metabolic biomarker: Present knowledge and future directions. *International Journal of Endocrinology*, Article ID 7816806. https://doi.org/10.1155/2018/7816806.

Neth, B. J., & Craft, S. (2017). Insulin resistance and Alzheimer's disease: Bioenergetic linkages. *Frontier in Aging Neuroscience*, *9*, 345.

Pandey, U. B., Nie, Z., Batlevi, Y., McCray, B. A., Ritson, G. P., Nedelsky, N. B., et al. (2007). HDAC6 rescues neurodegeneration and provides an essential link between autophagy and the UPS. *Nature*, *447*, 859–863.

Parenti, G., Andria, G., & Ballabio, A. (2015). Lysosomal storage diseases: From pathophysiology to therapy. *Annual Review of Medicine*, *66*, 471–486.

Park, S. Y., Lee, H. R., Lee, W. S., Shin, H. K., Kim, H. Y., Hong, K. W., et al. (2016). Cilostazol modulates autophagic degradation of β-amyloid peptide via SIRT1-coupled LKB1/AMPKα signaling in neuronal cells. *PLoS One*, *11*, e0160620.

Peng, W., Minakaki, G., Nguyen, M., & Krainc, D. (2019). Preserving lysosomal function in the aging brain: Insights from neurodegeneration. *Neurotherapeutics*, *16*, 611–634.

Platt, F. M. (2018). Emptying the stores: Lysosomal diseases and therapeutic strategies. *Nature Reviews Drug Discovery*, *17*, 133–150.

Pohl, C., & Dikic, I. (2019). Cellular quality control by the ubiquitin-proteasome system and autophagy. *Science*, *366*, 818–822.

Polito, V. A., Li, H., Martini-Stoica, H., Wang, B., Yang, L., Xu, Y., et al. (2014). Selective clearance of aberrant tau proteins and rescue of neurotoxicity by transcription factor EB. *EMBO Molecular Medicine*, *6*, 1142–1160.

Rawat, V., Wang, S., Sima, J., Bar, R., Liraz, O., Gundimeda, U., et al. (2019). ApoE4 alters ABCA1 membrane trafficking in astrocytes. *The Journal of Neuroscience*, *39*, 9611–9622.

Romine, H., Rentschler, K. M., Smith, K., Edwards, A., Colvin, C., Farizatto, K. L. G., et al. (2017). Potential Alzheimer's disease therapeutics among weak cysteine protease inhibitors exhibit mechanistic differences regarding extent of cathepsin B up-regulation and ability to block calpain. *European Scientific Journal*, *13*, 38–59.

Ryu, J. C., Zimmer, E. R., Rosa-Neto, P., & Yoon, S. O. (2019). Consequences of metabolic disruption in Alzheimer's disease pathology. *Neurotherapeutics*, *16*, 600–610.

Saha, S., Ash, P. E. A., Gowda, V., Liu, L., Shirihai, O., & Wolozin, B. (2015). Mutations in LRRK2 potentiate age-related impairment of autophagic flux. *Molecular Neurodegeneration*, *10*, 26.

Santos, R. X., Correia, S. C., Cardoso, S., Carvalho, C., Seiça, R., Santos, M. S., et al. (2014). Mitochondrial quality control systems sustain brain mitochondrial bioenergetics in early stages of type 2 diabetes. *Molecular and Cellular Biochemistry*, *394*, 13–22.

Scrivo, A., Bourdenx, M., Pampliega, O., & Cuervo, A. M. (2018). Selective autophagy as a potential therapeutic target for neurodegenerative disorders. *The Lancet Neurology*, *17*, 802–815.

Shoji-Kawata, S., Sumpter, R., Jr., Leveno, M., Campbell, G. R., Zou, Z., Kinch, L., et al. (2013). Identification of a candidate therapeutic autophagy-inducing peptide. *Nature*, *494*, 201–206.

Spires-Jones, T. L., & Hyman, B. T. (2014). The intersection of amyloid-β and tau at synapses in Alzheimer's disease. *Neuron*, *82*, 756–771.

Steen, E., Terry, B. M., Rivera, E. J., Cannon, J. L., Neely, T. R., Tavares, R., et al. (2005). Impaired insulin and insulin-like growth factor expression and signaling mechanisms in Alzheimer's disease—Is this type 3 diabetes? *Journal of Alzheimer's Disease*, 7, 63–80.

Thambisetty, M., Jeffrey Metter, E., Yang, A., Dolan, H., Marano, C., Zonderman, A. B., et al. (2013). Glucose intolerance, insulin resistance, and pathological features of Alzheimer disease in the Baltimore longitudinal study of aging. *JAMA Neurology*, *70*, 1167–1172.

Torres, M., Jimenez, S., Sanchez-Varo, R., Navarro, V., Trujillo-Estrada, L., Sanchez-Mejias, E., et al. (2012). Defective lysosomal proteolysis and axonal transport are early pathogenic events that worsen with age leading to increased APP metabolism and synaptic Aβ in transgenic APP/PS1 hippocampus. *Molecular Neurodegeneration*, 7, 59.

Tramutola, A., Triplett, J. C., Di Domenico, F., Niedowicz, D. M., Murphy, M. P., Coccia, R., et al. (2015). Alteration of mTOR signaling occurs early in the progression of Alzheimer disease (AD): Analysis of brain from subjects with pre-clinical AD, amnestic mild cognitive impairment and late-stage AD. *Journal of Neurochemistry*, *133*, 739–749.

Viswanathan, K., Hoover, D. J., Hwang, J., Wisniewski, M. L., Ikonne, U. S., Bahr, B. A., et al. (2012). Nonpeptidic lysosomal modulators derived from Z-Phe-Ala-diazomethylketone for treating protein accumulation diseases. *ACS Medicinal Chemistry Letters*, *3*, 920–924.

Vossel, K. A., Beagle, A. J., Rabinovici, G. D., Shu, H., Lee, S. E., Naasan, G., et al. (2013). Seizures and epileptiform activity in the early stages of Alzheimer disease. *JAMA Neurology*, *70*, 1158–1166.

Vossel, K. A., Ranasinghe, K. G., Beagle, A. J., Mizuiri, D., Honma, S. M., Dowling, A. F., et al. (2016). Incidence and impact of subclinical epileptiform activity in Alzheimer's disease. *Annals of Neurology*, *80*, 858–870.

Wallace, L. M. K., Theou, O., Godin, J., Andrew, M. K., Bennett, D. A., & Rockwood, K. (2019). Investigation of frailty as a moderator of the relationship between neuropathology and dementia in Alzheimer's disease: A cross-sectional analysis of data from the rush memory and aging project. *Lancet Neurology*, *18*, 177–184.

Wallings, R. L., Humble, S. W., Ward, M. E., & Wade-Martins, R. (2019). Lysosomal dysfunction at the centre of Parkinson's disease and frontotemporal dementia/amyotrophic lateral sclerosis. *Trends in Neuroscience*, *42*, 899–912.

Wang, Y., Martinez-Vicente, M., Krüger, U., Kaushik, S., Wong, E., Mandelkow, E.-M., et al. (2009). Tau fragmentation, aggregation and clearance: The dual role of lysosomal processing. *Human Molecular Genetics*, *18*, 4153–4170.

Wang, C., Sun, B., Zhou, Y., Grubb, A., & Gan, L. (2012). Cathepsin B degrades amyloid-β in mice expressing wild-type human amyloid precursor protein. *The Journal of Biological Chemistry*, *287*, 39834–39841.

Wang, C., Telpoukhovskaia, M. A., Bahr, B. A., Chen, X., & Gan, L. (2018). Endo-lysosomal dysfunction: A converging mechanism in neurodegenerative diseases. *Current Opinion in Neurobiology*, *48*, 52–58.

Winslow, A. R., Chen, C.-W., Corrochano, S., Acevedo-Arozena, A., Gordon, D. E., Peden, A. A., et al. (2010). α-Synuclein impairs macroautophagy: Implications for Parkinson's disease. *The Journal of Cell Biology*, *190*, 1023–1037.

Wrann, C. D., White, J. P., Salogiannnis, J., Laznik-Bogoslavski, D., Wu, J., Ma, D., et al. (2013). Exercise induces hippocampal BDNF through a PGC-1α/FNDC5 pathway. *Cell Metabolism*, *18*, 649–659.

Xiao, Q., Yan, P., Ma, X., Liu, H., Perez, R., Zhu, A., et al. (2015). Neuronal-targeted TFEB accelerates lysosomal degradation of APP, reducing Aβ generation and amyloid plaque pathogenesis. *The Journal of Neuroscience*, *35*, 12137–12151.

Xu, H., & Ren, D. (2015). Lysosomal physiology. *Annual Review of Physiology*, *77*, 57–80.

Yang, D.-S., Stavrides, P., Mohan, P. S., Kaushik, S., Kumar, A., Ohno, M., et al. (2011). Reversal of autophagy dysfunction in the TgCRND8 mouse model of Alzheimer's disease ameliorates amyloid pathologies and memory deficits. *Brain*, *134*, 258–277.

Yoshii, S. R., & Mizushima, N. (2017). Monitoring and measuring autophagy. *International Journal of Molecular Sciences*, *18*, 1865.

Zhang, J., Johnson, J. L., He, J., Napolitano, G., Ramadass, M., Rocca, C., et al. (2017). Cystinosin, the small GTPase Rab11, and the Rab7 effector RILP regulate intracellular trafficking of the chaperone-mediated autophagy receptor LAMP2A. *Journal of Biological Chemistry*, *292*, 10328–10346.

Zhao, N., Zhang, X., Song, C., Yang, Y., He, B., & Xu, B. (2018). The effects of treadmill exercise on autophagy in hippocampus of APP/PS1 transgenic mice. *Neuroreport*, *29*, 819–825.

Zimprich, A., Benet-Pagès, A., Struhal, W., Graf, E., Eck, S. H., Offman, M. N., et al. (2011). A mutation in VPS35, encoding a subunit of the retromer complex, causes late-onset Parkinson disease. *The American Journal of Human Genetics*, *89*, 168–175.

Zullo, J. M., Drake, D., Aron, L., O'Hern, P., Dhamne, S. C., Davidsohn, N., et al. (2019). Regulation of lifespan by neural excitation and REST. *Nature*, *574*, 359–364.

CHAPTER TWELVE

Neuroinflammation, microglial activation, and glucose metabolism in neurodegenerative diseases

Paul Edison[a,b,*]

[a]Department of Brain Sciences, Faculty of Medicine, Imperial College London, London, United Kingdom
[b]Cardiff University, Cardiff, United Kingdom
[*]Corresponding author: e-mail address: paul.edison@imperial.ac.uk

Contents

1. Introduction 327
2. Microglia 327
3. Type 2 diabetes mellitus 329
4. Obesity 331
5. Hypercholesterolemia 334
6. Role of systemic inflammation in central inflammation 335
7. Blood-brain barrier and systemic inflammation 337
8. Neuroinflammation and its major players 339
9. Conclusion 340

Acknowledgment and conflicts 340
References 341

Abstract

Alzheimer's disease is characterized by aggregated amyloid beta plaques and neurofibrillary tangles. Apart from the plaques and tangles, microglial activation plays a significant role in neurodegeneration and neuronal function. This review discusses the way in which microglial activation influences neurodegeneration and how systemic inflammation, type 2 diabetes mellitus, obesity and hypercholesterolemia influence neuroinflammation. Also reviewed is how systemic inflammation influences microglial activation along with the relationship between microglial activation and glucose metabolism.

International Review of Neurobiology, Volume 154
ISSN 0074-7742
https://doi.org/10.1016/bs.irn.2020.03.017

Abbreviations

Aβ	amyloid-β
AD	Alzheimer's disease
ADNI	Alzheimer's Disease Neuroimaging Initiative
AGEs	advanced glycation end-products
APP	amyloid precursor protein
BACE1	beta-site amyloid precursor protein cleaving enzyme 1
BBB	Blood-brain barrier
BMI	body mass index
CCL2	chemokine CC motif ligand 2
CDK5	cyclin-dependent kinase 5
CHI3L1/YKL-40	chitinase-3-like-1
COX	cyclo-oxygenases
CRP	C-reactive protein
CSF	cerebral spinal fluid
GMCSF	granulocyte-macrophage colony-stimulating factor
GSK-3	glycogen synthase kinase-3
GSK-3β	glycogen synthase kinase 3 beta
HDL	high-density lipoproteins
HFD	high-fat diet
IDE	insulin degrading enzyme
IGF-1	insulin-like growth factor 1
iNOS	inducible nitric oxide synthase
IFNγ	interferon gamma
IgG-Fc	immunoglobulin G Fc
IL-3	interleukin-3
IL-4	interleukin 4
IL-6	interleukin-6
IL-10	interleukin-10
IL-12	interleukin-12
IL-13	interleukin-13
IL-15	interleukin-15
IL-1β	interleukin-1 beta
IR	insulin receptors
JNK	c-Jun NH2-terminal kinase
LDL	low-density lipoproteins
LPS	lipopolysaccharide
MAPK	mitogen-activated protein kinase
MCI	mild cognitive impairment
MCP-1	monocyte chemoattractant protein-1
MHC	major histocompatibility complex
MIF	macrophage migration inhibitory factor
NFTs	neurofibrillary tangles
NF-κB	nuclear factor kappa-light-chain-enhancer of activated B cells
NMDA	*N*-methyl-D-aspartate
NO	nitric oxide
PAI-1	plasminogen activator inhibitor-1

PGD2	prostaglandin D2
PGE2	prostaglandin E2
PHF	paired helical filaments
PKB/AKT	protein kinase B
PKC	protein kinase C
PI3K	phosphatidylinositide 3-kinases
RAGE	receptor for advanced glycation end products
RBP4	retinol binding protein 4
ROS	reactive oxygen species
SOCS3	suppressor of cytokine signaling 3
TLR	toll-like receptor
TNF-α	tumor necrosis factor alpha
TNFR1	tumor necrosis factor receptor 1
T2DM	type 2 diabetes mellitus

1. Introduction

Alzheimer's disease (AD) is characterized by extracellular amyloid-β (Aβ) plaques formed from aggregated amyloid β peptide and intraneuronal deposits of neurofibrillary tangles (NFTs), which are paired helical filaments (PHF) constituted by hyperphosphorylated tau proteins. Abnormal tau deposition leads to destabilization of microtubules, membrane degeneration and neuronal damage. The NFTs are found in hippocampal formations, the parahippocampal gyrus, and temporal association cortex. The density of NFTs is highest in the entorhinal cortex, the CA1 and subiculum of hippocampus, and amygdala. Compared to the entorhinal cortex, the density of NFTs is lower in superior temporal sulcus or inferior temporal gyrus, with much lower density in the primary cortices. The extracellular deposits are constituted by Aβ that aggregates and forms amyloid plaques. The distribution of amyloid plaques predominantly starts in the basal forebrain and spreads to the rest of the cortex (Arnold et al., 1991).

2. Microglia

Microglia are resident macrophages in the central nervous system (CNS) and play a vital role in brain homeostasis. In addition to amyloid plaques and NFTs, microglial activation plays a significant role in neurodegenerative diseases. In response to a toxic insult to the brain, microglia are activated in an attempt to protect the brain from toxic damage (Calsolaro and Edison, 2016). The initial response of the microglia is thought to be protective. As the toxic insult continues, however, unabated microglial activation occurs, which manifests as a detrimental phenotype, which releases pro-inflammatory cytokines. This excessive release of pro-inflammatory cytokines can subsequently cause neuronal damage (Lyman et al., 2014).

Microglia have the capacity to phagocytose products and can act as antigen-presenting cells to also release toxic factors (Morales et al., 2014). In the resting stage, the microglia scan the brain milieu, and are usually in the ramified stage. These microglia can scan regions close to their ramified process without interfering with neurons and neuronal activities, a process that occurs every few hours (Stence et al., 2001). When activated, microglia can pass through different morphological phases, from an extremely ramified morphology and changing their shape to achieve an activated amoeboid and mobile morphology, the goal being to engulf the toxic material (Hanisch and Kettenmann, 2007). It is also possible that these microglia can remain in the activated phase for a long time; if microglia persist in their activated state, this may cause significant neuronal damage (Liu and Hong, 2003). Microglial cells can have both a pro-inflammatory and an anti-inflammatory phenotype, previously referred to as M1 and M2 (Calsolaro and Edison, 2016). The pro-inflammatory phenotype response is stimulated by lipopolysaccharide (LPS) along with interferon gamma (IFNγ) and produces a massive inflammatory response associated with the release of various cytokines, including interleukin-1 beta (IL-β), interleukin-12 (IL-12), tumor necrosis factor alpha (TNF-α). The other phenotype is M2, where three different subtypes have been described previously. M2a usually responds to IL-4 and IL-13, with an anti-inflammatory profile, M2b is stimulated by immune complex formation and toll-like receptor (TLR) or IL-1β activation, M2c represents the deactivated macrophages and contributes to the suppression of pro-inflammatory cytokines (Cunningham, 2013; Lyman et al., 2014). Several pro-inflammatory cytokines and anti-inflammatory cytokines have also been identified. Pro-inflammatory cytokines are a group of proteins, which include IL-1, TNF-α, IL-6, IL-8 and IFNγ, and are important for promoting inflammatory responses. Their action, aimed to be protective, usually regulates inflammation and damages tissues. Conversely, anti-inflammatory cytokines, including IL-1 receptor antagonist, IL-4, IL-6, IL-10, IL-11, and IL-13, are regulatory molecules that control the pro-inflammatory response. The third microglial phenotype represents deactivated macrophages and contributes to suppression of pro-inflammatory cytokines (Cunningham, 2013; Lyman et al., 2014).

Different factors inhibit microglial activation, including neurotrophic factors that inhibit the expression of major histocompatibility complex (MHC) class II molecules on microglial cells (Perry and Teeling, 2013). It has also been suggested that the switch from the anti-inflammatory to pro-inflammatory phenotype is a dynamic process, which makes targeting specific phenotypes in disease modification extremely difficult, and the switching of the phenotype can be influenced by peripheral inflammation (Lyman et al., 2014). Additionally, microglia can switch from an anti-inflammatory to pro-inflammatory phenotype during the course of the disease, especially in the case of neurodegenerative diseases. *In vitro* stimulation of microglial cells using LPS has shown that the anti-inflammatory phenotypes can be switched to the pro-inflammatory aggressive phenotype, which leads to increased production of IL-1β, TNF-α and interleukin-6 (IL-6), and subsequent neuronal damage and neurodegeneration (Cunningham et al., 2005).

Endothelial cells, pericytes and vascular smooth muscle maintain the intact function of the blood-brain barrier (BBB) and neurovascular unit. It is possible that microglial activation and neuroinflammation could be triggered by microvascular dysfunction of the blood-brain barrier leading to increased permeability and leakage, which in turn leads to increased neural inflammation. Interestingly, it has been shown that the microvessels from the brains of individuals with AD secrete higher amounts of pro-inflammatory cytokines, which leads to neuronal damage (Zlokovic, 2011). It has been shown that the neuroinflammatory marker YKL-40 (chitinase-3-like-1 [CHI3L1], human cartilage glycoprotein-39) is elevated in the patients (Alcolea et al., 2015).

Systemic inflammation plays an important factor in many diseases, such as type 2 diabetes mellitus (T2DM), atherosclerosis, rheumatoid arthritis, and several other systemic disorders.

3. Type 2 diabetes mellitus

Emerging evidence suggests a close link between T2DM and AD. It has been shown that there is greater prevalence of T2DM in the AD population compared to healthy controls (34.6% vs. 18.1%; $P < 0.05$). Additionally, there is evidence of a higher incidence of AD among T2DM patients (Holscher, 2011), as revealed by longitudinal population-based studies.

Insulin could prevent Aβ accumulation and plaque formation by promoting brain Aβ clearance. The insulin receptors (IR) are abundant in the brain and particularly present in both astrocytes and neurons in the hippocampus, hypothalamus and cerebral cortex (Abbott et al., 1999; Kandimalla et al., 2017). The insulin receptor signaling pathways are involved in synaptic plasticity, neuroprotection, and energy metabolism (Chiu et al., 2008; van der Heide et al., 2006). It has been suggested that neurotoxicity in AD is mediated by oligomeric Aβ, and insulin degrading enzyme (IDE) is one of the regulators of Aβ levels in neuronal cells. Insulin degrading enzyme is responsible for degradation of insulin, Aβ and amylin, and a reduction of IDE could lead to Aβ deposition.

Insulin resistance, defined as the inability of a known quantity of exogenous or endogenous insulin to increase glucose uptake, induces Aβ oligomerization and activates the downstream c-Jun NH2-terminal kinase (JNK) pathway. It has also been shown that insulin resistance is associated with decreased IDE levels and increased amyloidosis (Ho et al., 2005). Additionally, Aβ oligomers could act on hypothalamus and trigger peripheral insulin resistance via nuclear factor kappa-light-chain-enhancer of activated B cells (NF-κB) signaling. Insulin resistance can also promote neurodegeneration and facilitate the onset of AD (Kandimalla et al., 2017).

Studies have suggested a role for insulin resistance in tau phosphorylation and aggregation. This is mediated by the influence of various kinases and phosphatases, including JNK, phosphatidylinositide 3-kinases (PI3K), protein kinase B (PKB/AKT) and glycogen synthase kinase-3 (GSK-3). Chronic hyperglycaemia and insulin resistance are thought to induce JNK activity in diabetic patients and cause oxidative stress and pancreatic beta cell apoptosis. Induction of the JNK and PI3K/PKB/GSK-3 pathways is crucial in AD development because these promotes abnormal tau hyperphosphorylation. Glycogen synthase kinase 3 beta (GSK-3β) is widely considered to be the primary kinase responsible for the phosphorylation of tau, and is modulated by insulin via the PKB/AKT pathway. It has been suggested that GSK-3β contributes to AD-type neurodegeneration (Petit-Paitel, 2010).

Different studies have shown an increased prevalence of dementia in individuals with T2DM. In the Rotterdam study, patients had a doubled risk of developing AD if they had T2DM (Ott et al., 1999), and in the Framingham study, a relationship between T2DM and dementia was found to be more evident in individuals older than 75 years (Akomolafe et al., 2006).

Apart from soluble Aβ lowering the response of insulin receptors to insulin, advanced glycation end-products (AGEs) also play a pivotal role in the relationship between T2DM and AD-type dementia (Zhao et al., 2008). The AGEs are proteins or lipids that become glycated after exposure to sugars, and they are also present in different neurodegenerative diseases.

Excessively high levels of AGEs occur in the circulation and promote oxidative stress and inflammation by binding with cell surface receptors or cross-linking with body proteins, altering their structure and function. It has also been shown that AGEs occur in the cytosol of neurons, especially in the hippocampus and parahippocampal gyrus. Levels of AGEs and the receptor for advanced glycation end products (RAGE) are both significantly higher in AD patients and are associated with amyloid plaques and NFTs.

4. Obesity

Epidemiological studies have shown that the probability of developing AD and other forms of dementia is associated with obesity and metabolic disorders (Nguyen et al., 2014). A study in Sweden reported an increased risk of AD of 36% associated with every unit increase in body mass index (BMI) in female, but not male, subjects. One study, including both male and female subjects, showed a greater risk in subjects having a midlife BMI $>30\,kg/m^2$. A longitudinal population-based study, in which subjects were followed up after 27 years, demonstrated that overweight subjects ($BMI = 25–29.9\,kg/m^2$) aged 40–45 years had a 35% increased risk of developing dementia, whereas obese individuals (BMI $\geq 30\,kg/m^2$) had a 74% increased risk of dementia (Whitmer et al., 2005).

Obesity leads to an aberrant inflammatory response, and it has been suggested that subclinical inflammation of adipose tissue may interact with the cerebral inflammatory response causing neurodegeneration. Adipocytes produce pro-inflammatory adipokines, cytokines and chemokines including leptin, CRP, MCP-1, MIF, PAI-1, RBP4 (Monteiro and Azevedo, 2010). Interestingly, amyloid precursor protein is upregulated in adipocytes in obesity. In obese patients, adipocytes are infiltrated by immune cells, including macrophages, monocytes, natural killer cells and lymphocytes. In obesity, there is increased release of pro-inflammatory cytokines, such as IL-6 and TNF-α, which produce a large number of reactive oxygen species (ROS). These molecules could cross the BBB and cause central neuroinflammation (Cai, 2013) and resultant activation of microglial cells. This could be mediated by microglial activation relating to pro-inflammatory cytokines, complement proteins, and other inflammatory mediators in the brain, leading to tau phosphorylation and neurodegenerative changes (Eikelenboom et al., 2002).

Metabolic syndrome is characterized by a high fasting glucose level, regarded to be an important precursor of T2DM. Furthermore, decreased insulin-stimulated glucose transport exhibits as insulin resistance, and is commonly observed in obese individuals. It has been suggested that, in a state of peripheral insulin resistance in obesity, insulin may cross the BBB and stimulate overproduction of IL-1β, IL-6 and TNF-α. These inflammatory cytokines induce trafficking of amyloid precursor protein (APP) and lead to increased deposition of Aβ. Moreover, Aβ itself contributes to the increased production of pro-inflammatory cytokines by means of a vicious cycle.

As indicated above, evolving evidence suggests that changes in body weight are associated with an increased risk of AD. Thus, factors that regulate body weight are expected to influence the development and progression of AD. The adipocyte-derived hormone, leptin, is a foremost regulator of body weight, chiefly by activating hypothalamic neural circuits. As well as being expressed in the hypothalamus, leptin receptors are broadly expressed throughout the brain, including the hippocampus and neocortex. It has been suggested that hyperleptinemia may be associated with cognitive deficits. In obese subjects, hyperleptinemia may result from different changes, for example the down-regulation of leptin receptors, an increase in circulating C-reactive protein (CRP), oxidative stress and overexpression of suppressor of cytokine signaling 3 (SOCS3) (Beltowski et al., 2002). C-reactive protein reduces the physiological roles of leptin and contributes to leptin resistance (Singh et al., 2007). Furthermore, leptin may itself stimulate the production of CRP, alluding to the existence of feedback between leptinemia and the inflammatory response. *In vivo*, it has been shown that neuronal MyD88-dependent signaling functions as a key regulator of diet-induced leptin and insulin resistance (Konner and Bruning, 2012).

Inconsistencies between age, obesity and the probability of developing AD have been reported in previous studies. These varied results could possibly be explained by the length of the studies. Certainly, there is much evidence to suggest that obesity or a high-fat diet (HFD) has a negative effect on cognitive performance. Conversely, one study demonstrated that an inverse association between BMI and dementia risk over time. A retrospective cohort study also showed that, over 20 years, being underweight in middle age and old age conveys an increased risk of dementia (Qizilbash et al., 2015). The association between BMI and dementia has been postulated to be attributable to two different processes: a harmful effect of higher BMI, which is apparent in long follow-up, and a reverse-causation effect that makes a higher BMI appear to be protective when the follow-up is short. Evidently, further investigation is required to fully comprehend exactly how obesity relates to AD pathology, but it can be established that there is a relationship between obesity and cognitive performance.

While there is substantial evidence linking metabolic syndrome (visceral obesity, dyslipidemia, hyperglycemia, and hypertension) with increased probability of developing AD, the exact mechanisms by which this, and in particular obesity, affect AD remain unclear. It has been suggested that visceral obesity and insulin resistance may arise from low grade inflammation of the adipose tissue. There is a general agreement that the aberrant inflammatory response underlying metabolic syndrome may be a result of a dysregulation of the endocrine homeostasis of adipose tissue. It has been suggested that subclinical inflammation of adipose tissue may interact with the impaired central inflammatory response, resulting in neurodegeneration. It has also been proposed that subjects with metabolic syndrome are likely to develop cognitive impairment within 5 years, and that high levels of inflammation, as demonstrated by high levels of IL-6 and CRP, contribute to the cognitive deterioration. With the appearance of abdominal obesity, there is a switch in the populations of immune cells, resulting in increased predominance of M1 macrophages that release pro-inflammatory cytokines, such as IL-6 or TNF-α, and create a large number of ROS. These could cross the BBB and cause central neuroinflammation (Cai, 2013).

In AD brains, pro-inflammatory cytokines and other inflammatory mediators co-localize with Aβ in senile plaques. The modified amyloid cascade/neuroinflammation hypothesis suggests that Aβ formation activates microglial cells which, in turn, release potentially neurotoxic substances, such as nitric oxide, pro-inflammatory cytokines, complement proteins, and other inflammatory mediators, resulting in tau phosphorylation and neurodegenerative changes (Akiyama et al., 2000; Eikelenboom et al., 2002; McGeer and McGeer, 2001).

Other studies have shown an association between metabolic alterations, such obesity, insulin resistance and T2DM and the release of inflammatory mediators, resulting in a chronic inflammatory status (Ho et al., 2005; Lee et al., 2008; Misiak et al., 2012). It is acknowledged that obesity can promote neurodegeneration, inducing both pro-inflammatory cytokine release and insulin/insulin-like growth factor 1 (IGF-1) resistance (Spielman et al., 2014). Adipocytes produce pro-inflammatory cytokines that can penetrate the BBB by a range of mechanisms and cause neuronal damage. Leptin, an adipose-derived hormone, is able to bind to receptors expressed in the hippocampus and has demonstrated some protective effects on the CNS. However, in conditions of hyperleptinemia and leptin resistance, this hormone can act as a neuroinflammatory signal (Misiak et al., 2012). Lee et al. (2008) demonstrated that APP is up-regulated in the adipocytes of obese subjects, with levels linked to insulin resistance and to the pro-inflammatory phenotype of adipocytes. Amyloid precursor protein appears to be processed in Aβ40 and Aβ42 fragments, as in the brain, and is it possible that circulating Aβ42 could impair CNS clearance of amyloid fragments. Aβ40 and Aβ42 may also contribute to hyperinsulinemia and/or insulin resistance, indirectly increasing the risk of dementia. A number of prospective studies have also demonstrated that obesity in middle age as a risk factor for AD (Lee et al., 2008). Insulin resistance, together with visceral obesity, has a key function in the pathogenesis of metabolic syndrome, and can also play a role in neurodegeneration. In conditions of peripheral insulin resistance, insulin can cross the BBB and result in the production of various cytokines. Previous studies have demonstrated that, in AD patients, the brain can become resistant to the activity of insulin (Misiak et al., 2012). Talbot et al. (2012) demonstrated brain insulin resistance in hippocampal formation and cerebellar cortex subjects with AD but not diabetes. The response to insulin and to IGF-1 in the signaling pathway appeared to be impaired in those subjects, suggesting that insulin resistance is a common feature of AD (Talbot et al., 2012).

Rodent models of obesity induced by the consumption of a HFD are characterized by inflammation, in both peripheral tissues and hypothalamic areas critical for energy homeostasis (Magnuson et al., 2020; McLean et al., 2019). It has been shown that, in inflammation in peripheral tissues that develops as a consequence of obesity, hypothalamic inflammatory signaling was evident in both rats and mice within 1–3 days of HFD onset prior to substantial weight gain. Furthermore, both reactive gliosis and markers suggestive of neuronal injury were evident in the hypothalamic arcuate nucleus of rats and mice within the first week of HFD feeding. Reactive gliosis in human hypothalamus was highly correlated with BMI. These data suggest that obesity is associated with central inflammation. However, cortical inflammation in humans has not been evaluated to date.

5. Hypercholesterolemia

Clinical studies indicate that middle-aged individuals with increased cholesterol have a greater risk of developing AD. This is supported by studies indicating that diet-induced hypercholesterolemia increases Aβ accumulation in white rabbits (Sparks et al., 2000) and in transgenic mouse models of AD (Levin-Allerhand et al., 2002). Numerous studies suggest that cholesterol can influence the development of AD. One Japanese study tested cholesterol levels of nearly 2600 people between the ages of 40 and 79 who had no signs of AD. They found formation of senile plaques in 86% of subjects with high cholesterol, whereas plaques were present in only 62% of people with low cholesterol. A more recent study demonstrated that cholesterol could act as a catalyst and accelerate the aggregation of Aβ. It was estimated that the presence of cholesterol caused Aβ plaques to develop 20-fold more rapidly. Therefore, the use of cholesterol-lowering medication, such as statins, has given rise to optimism for treating or preventing AD. It should be noted that statins do not simply inhibit the synthesis of cholesterol, but also inhibit effects on inflammation (Wolozin et al., 2007). However, study results were inconsistent: some studies found beneficial effects, while others did not.

Cholesterol is mainly carried by two different types of lipoproteins: low-density lipoproteins (LDL) and high-density lipoproteins (HDL). It is important to note that the effect of LDL cholesterol and HDL cholesterol on the likelihood of developing neurodegenerative diseases, including AD, is not yet well-established. Elevated LDL concentration in mid-life was shown to increase AD risk in later life. Increased oxidative modification and nitration of LDL were observed in AD-type dementia and hypercholesterolemia in comparison with control subjects. On the other hand, higher levels of HDL were associated with a decreased risk of both probable and possible AD. However, the connection between high LDL and low HDL to AD risk is not definite and requires further investigation.

Changes in cholesterol levels have also been found to cause alterations in cell membrane properties. Lipid rafts are cholesterol-rich and sphingomyelin-rich membrane micro-domains in which beta-site amyloid precursor protein cleaving enzyme 1 (BACE1) and γ-secretases are located. This increase in cholesterol stimulates the binding of APP to the lipid rafts, leading to the production of Aβ following cleavage by BACE1 and γ-secretase (Beel et al., 2010). If the level of intracellular cholesterol decreases, this may result in a decrease of catalytic activity of BACE1 and γ-secretase (Grimm et al., 2008).

6. Role of systemic inflammation in central inflammation

There is further evidence that systemic inflammation can lead on to central inflammation resulting in neurodegenerative processes. In AD and other neurodegenerative diseases, as well as the presence of microglia in the brain, increased levels of pro-inflammatory cytokines occur in the cerebrospinal fluid (CSF), and limited studies have also demonstrated that there are increased inflammatory markers in the periphery (Morales et al., 2014). Apart from a direct neurotoxic effect, activated microglia and astrocytes can promote Aβ deposition: studies in mouse models have shown that amyloid deposition is increased under inflammatory conditions (Guo et al., 2002). Furthermore, cytokines are able to up-regulate β-secretase mRNA protein and enzymatic activity (BACE1) (Sastre et al., 2003), a key enzyme in neuronal Aβ formation (Sastre et al., 2006) and whose transcription also appears to be increased by a TNF-α-activated NF-κB signal, leading to increased Aβ production (Chen et al., 2012).

A study conducted in three independent cohorts of subjects (University of Pennsylvania, Philadelphia; Washington University, St. Louis; Alzheimer's Disease Neuroimaging Initiative) found a panel of 17 plasma analytes that were associated with AD, mild cognitive impairment (MCI) and other dementias. Of the cytokines, IL-10 and IL-15 had a positive association in Penn and WU data, while interleukin-3 (IL-3) showed the opposite correlation. Alzheimer's Disease Neuroimaging Initiative (ADNI) data have shown that, of the analytes evaluated, IL-3 was strongly associated with AD, MCI and other forms of dementia. The association between these plasma analytes and CSF biomarkers warrants further study (Hu et al., 2012).

Caspases are proteases known to regulate cellular apoptosis. Some have a known role in inflammation; for example, caspase 1 is acknowledged to be essential for maturation of pro-inflammatory cytokines. It has been shown that caspases 3, 7 and 8 can regulate microglial activation through a protein kinase C (PKC)-δ-dependent signaling (Burguillos et al., 2011). Chemokines and complement factors are also important mediators in neuroinflammatory processes. Chemokines have been shown to be upregulated in the CNS during neuroinflammation, stimulating astrocytes and microglial chemotaxis. It is known that Aβ activates complement in the alternative pathway and that complement components C3a, C3b, C5a play a role in the chemotaxis and phagocytic functions of microglial cells (Lyman et al., 2014).

Prostaglandins are additional players in the modulation of neuroinflammation. These 20-carbon saturated fatty acids are produced by the cyclo-oxygenase pathway in response to a stimulus. It has been shown that the cyclo-oxygenases COX-1 and COX-2 normally are expressed in the brain, and their role in neurodegeneration has been recognized. In particular, expression of COX2 is induced in both neurons and glial cells in neurodegenerative disorders. When inflammation develops, high levels of prostaglandins such as PGE2, PGD2 and J2 prostaglandins (the most abundant prostaglandins in the brain, derived from PGD2) are observed. In particular, PGD2 levels were found to be increased in frontal cortex of AD patients compared with healthy control subjects (de Leeuw et al., 2017). In both AD patients and in mouse models, the levels of PGD2 synthase and the PGD2 receptor led to activated microglia and astrocytes in senile plaques (Figueiredo-Pereira et al., 2014).

The CSF has been studied extensively in AD and the levels of Aβ and tau are routinely used as an indicator of AD pathology. Several studies have been conducted evaluating the levels of cytokines in the CNS in AD and healthy control subjects. The results have been inconclusive, with some studies showing comparable levels of certain cytokines (IL-10, IL-12, IL-6, IL-8, IL-1β), while others found increased concentration of TNF-α, IL-6, IL-8 and granulocyte-macrophage colony-stimulating factor, (GM-CSF) in AD subjects compared to controls (Llano et al., 2012). Rosen et al. (2014) recently evaluated the association between different inflammatory markers in the CNS and the development of AD. The first of these is chitotriosidase, an enzyme secreted by activated macrophages and known to be elevated in AD subjects. The second is the glycoprotein chitinase-3-like protein 1 (YKL-40), expressed in astrocytes and found to be increased in AD patients in some of the studies reported. The final one was the monocyte chemo-attractant protein-1 [MCP-1 or chemokine CC motif ligand 2 (CCL2)], previously studied with controversial results. In that study, comparing 25 AD and 25 controls, it was found that chitotriosidase and YKL-40 were increased in the CSF of AD subjects, while MCP-1 was comparable between AD and controls (Rosen et al., 2014). In conclusion, peripheral inflammation plays a significant role in neuroinflammation and neurodegeneration.

7. Blood-brain barrier and systemic inflammation

The "immune privilege" status of the CNS has been recognized since 1960 by Sir Peter Medawar and Sir Frank Macfarlane Burnet. It was previously thought that the BBB completely disconnects the CNS from the peripheral immune system and that inflammatory cytokines cannot cross the BBB. However, several previous studies have described evidence of the influence of cytokines on the integrity of the BBB, which is able to release and transmit these mediators and allow leucocyte migration into the brain. Several paths allow communication between the systemic circulation and the CNS (Lyman et al., 2014). One of these routes is around circumventricular organs, where the BBB is not intact and permits cytokine transport. Moreover, cytokines, in the presence of intact BBB, can stimulate the endothelium, which signals to perivascular macrophages, which in turn are able to activate microglia. The third route involves the direct influence of the vagal stimulus by peripheral inflammation (Goehler et al., 1999; Lyman et al., 2014).

It is known that intraperitoneal administration of IL-1β induces brain-mediated sickness symptoms that can be blocked by subdiaphragmatic vagotomy, and that perivagal immune cells are induced to express IL-1β immunoreactivity following intraperitoneal administration of LPS (Goehler et al., 1999). Furthermore, sensory fibers of the vagus nerve express receptors that can be stimulated by IL-1 (Kurosawa et al., 1998). A further route is an active transport system by which cytokines cross the BBB that can allow their migration into the brain (Lyman et al., 2014). In addition, TNF-α, IL-6 and IL-1β have the capacity to damage the integrity of the BBB, modifying the resistance of the tight junctions of brain vessels (Lyman et al., 2014). Chemokines, such as CCL10 and CCL21, permit the movement of leucocytes across the BBB, enabling the adhesion of T-cells to the BBB, and are involved in T-lymphocyte migration into the immune privileged CNS during immunosurveillance and chronic inflammation (Lyman et al., 2014).

Various clinical studies carried out in the last decades showed a possible role of systemic inflammation in AD pathogenesis. A study of twins in 1999 showed that a patient with a history of serious systemic inflammation developed AD with earlier onset than their twin sibling (Eikelenboom et al., 2012). A large case-control study in 2005 showed an association between dementia and infectious disease in older patients (Dunn et al., 2005). Finally, a study recruiting 300 subjects with mild to severe dementia between November 2003 and May 2006 showed that acute systemic inflammation with increased levels of pro-inflammatory cytokines, such as TNF-α), was associated with an increased rate of cognitive decline over a 6-month period. Indeed, high baseline levels of TNF-α were associated with a fourfold increase in the rate of cognitive deterioration, which was not seen in patients with lower initial levels of TNF-α (Holmes et al., 2009).

Several factors could explain the high susceptibility of the brain to systemic inflammation: in AD patients, pro-inflammatory changes are already apparent in the early stages of the disease when microglial cells are activated and produce pro-inflammatory mediators. Additionally, acetylcholine, which is known to have a role in controlling glial activation and cytokine production, is reduced in AD (Eikelenboom et al., 2012). Common cardiovascular risk factors, such as hypertension, hypercholesterolemia, and diabetes, could disrupt BBB function and result in endothelial cell activation leading to microvessel leakage of high levels of neurotoxic and inflammatory factors (Eikelenboom et al., 2012). There is evidence that the presence of systemic inflammation has an impact on CNS. Studies in animal models of sepsis showed CNS inflammation, microglial inducible nitric oxide synthase (iNOS) expression and neuronal death with cognitive decline after challenge with LPS or other stimuli that mimic systemic viral infection and, moreover, via immunoglobulin G Fc (IgG-Fc) receptors or activation of complement (Cunningham, 2013). Apart from acute inflammation and the relationship with cognitive impairment, several studies have shown an association between various types of chronic inflammation (e.g., chronic periodontitis) and the risk of AD. It has been demonstrated that increased plasma antibodies against periodontal bacteria, and TNF-α are independently associated with AD (Kamer et al., 2009).

Evidence in the last 10 years showed a connection, especially in late life dementia, between vascular disease and dementia. Small vessel disease, caused by high blood pressure, T2DM, obesity, and an unhealthy lifestyle, can lead to neuronal loss in a more complex network of pathological events, as Fotuhi et al. (2009) hypothesized in a "dynamic polygon hypothesis." According to this hypothesis, late onset dementia results from an interaction of several cardiovascular risk factors along with different other pathologies like amyloid and tau. In particular, amyloid plaques and NFTs are part of the pathological processes that impair cortex and hippocampal volume, such as inflammation, synucleinopathy, strokes and other environmental factors, including education, exercise, leisure activities, or obesity (Fotuhi et al., 2009). Tau hyperphosphorylation also appears to be facilitated by neuroinflammation: IL-6 via a cyclin-dependent kinase 5 (CDK5) pathway, IL-1 via mitogen-activated protein kinase (MAPK) pathway and nitric oxide (NO) can deal with this process (Lyman et al., 2014). Moreover, dead neurons release tau aggregates, which can activate microglial cells (Morales et al., 2014).

8. Neuroinflammation and its major players

Neuroinflammation commonly refers to an inflammatory response within the CNS, which is caused by several pathological insults, including infection, trauma, ischemia and toxins. The process is marked by pro-inflammatory cytokines (including IL-1β, IL-6, IL-18 and TNF-α), chemokines (CCL1, CCL5, CXCL1), small molecule messengers (prostaglandins, NO) and ROS, which are produced by innate immune cells in the CNS (DiSabato et al., 2016). The innate immune cells involved in the process are primarily microglia and astrocytes, but capillary endothelial cells and infiltrating blood cells also contribute to neuroinflammation, particularly when the BBB suffers biochemical or mechanic damage (DiSabato et al., 2016; Heneka et al., 2014).

The release of pro-inflammatory molecules could result in further synaptic dysfunction, neuronal death and inhibition of neurogenesis (Lyman et al., 2014). It has been shown that IL-1β induces synaptic loss by increasing PGE2 production, which further leads to presynaptic glutamate release and postsynaptic *N*-methyl-D-aspartate (NMDA) receptor activation (Mishra et al., 2012), while TNF-α could cause neuronal death by activating TNF receptor 1 (TNFR1) and recruiting caspase 8 when the NF-κB pathway was inhibited (Micheau and Tschopp, 2003). In addition, the complement system is activated in the process and promotes a phagocytic role of microglia, which could lead to inappropriate pruning of synapses (Hong et al., 2016). Conversely, anti-inflammatory cytokines, including IL-1 receptor antagonist, IL-4, IL-10 and IL-11, are also produced in neuroinflammatory processes, which could be part of the sophisticated mechanism to prevent excessive neuroinflammation (Calsolaro and Edison, 2016). However, in the context of neurodegenerative disease, neuroinflammation tends to be a chronic process that does not resolve spontaneously, and is considered to be a vital driver of neurodegenerative diseases (Heneka et al., 2015).

9. Conclusion

Neuroinflammation plays a significant role in neurodegeneration. This could be the result of different pathological substrates associated with neurodegenerative diseases along with systemic inflammation and factors influencing insulin resistance, such as T2DM, obesity and hypercholesterolemia. Treatment of these conditions could lead to improved neuronal function and prevent neurodegeneration.

Acknowledgment and conflicts

P.E. was funded by the Medical Research Council and now by Higher Education Funding Council for England (HEFCE). He has also received grants from Alzheimer's Research, UK, Alzheimer's Drug Discovery Foundation, Alzheimer's Society, UK, Medical Research Council. P.E. is a consultant to Pfizer and Novo Nordisk. He has received speaker fees from Novo Nordisk, Pfizer, Nordea, Piramal Life Sciences. He has received educational and research grants from GE Healthcare, Novo Nordisk, Piramal Life Sciences/Life Molecular Imaging, Avid Radiopharmaceuticals and Eli Lilly. He is an external consultant to Novo Nordisk and a member of their Scientific Advisory Board.

References

Abbott, M. A., Wells, D. G., & Fallon, J. R. (1999). The insulin receptor tyrosine kinase substrate p58/53 and the insulin receptor are components of CNS synapses. *The Journal of Neuroscience*, *19*(17), 7300–7308.

Akiyama, H., Barger, S., Barnum, S., Bradt, B., Bauer, J., Cole, G. M., et al. (2000). Inflammation and Alzheimer's disease. *Neurobiology of Aging*, *21*(3), 383–421.

Akomolafe, A., Beiser, A., Meigs, J. B., Au, R., Green, R. C., Farrer, L. A., et al. (2006). Diabetes mellitus and risk of developing Alzheimer disease: Results from the Framingham study. *Archives of Neurology*, *63*(11), 1551–1555.

Alcolea, D., Martinez-Lage, P., Sanchez-Juan, P., Olazaran, J., Antunez, C., Izagirre, A., et al. (2015). Amyloid precursor protein metabolism and inflammation markers in preclinical Alzheimer disease. *Neurology*, *85*(7), 626–633.

Arnold, S. E., Hyman, B. T., Flory, J., Damasio, A. R., & Van Hoesen, G. W. (1991). The topographical and neuroanatomical distribution of neurofibrillary tangles and neuritic plaques in the cerebral cortex of patients with Alzheimer's disease. *Cerebral Cortex*, *1*(1), 103–116.

Beel, A. J., Sakakura, M., Barrett, P. J., & Sanders, C. R. (2010). Direct binding of cholesterol to the amyloid precursor protein: An important interaction in lipid-Alzheimer's disease relationships? *Biochimica et Biophysica Acta*, *1801*(8), 975–982.

Beltowski, J., Wojcicka, G., & Borkowska, E. (2002). Human leptin stimulates systemic nitric oxide production in the rat. *Obesity Research*, *10*(9), 939–946.

Burguillos, M. A., Deierborg, T., Kavanagh, E., Persson, A., Hajji, N., Garcia-Quintanilla, A., et al. (2011). Caspase signalling controls microglia activation and neurotoxicity. *Nature*, *472*(7343), 319–324.

Cai, D. (2013). Neuroinflammation and neurodegeneration in overnutrition-induced diseases. *Trends in Endocrinology and Metabolism*, *24*(1), 40–47.

Calsolaro, V., & Edison, P. (2016). Neuroinflammation in Alzheimer's disease: Current evidence and future directions. *Alzheimers Dement*, *12*(6), 719–732.

Chen, C. H., Zhou, W., Liu, S., Deng, Y., Cai, F., Tone, M., et al. (2012). Increased NF-kappaB signalling up-regulates BACE1 expression and its therapeutic potential in Alzheimer's disease. *The International Journal of Neuropsychopharmacology*, *15*(1), 77–90.

Chiu, C. C., Su, K. P., Cheng, T. C., Liu, H. C., Chang, C. J., Dewey, M. E., et al. (2008). The effects of omega-3 fatty acids monotherapy in Alzheimer's disease and mild cognitive impairment: A preliminary randomized double-blind placebo-controlled study. *Progress in Neuro-Psychopharmacology & Biological Psychiatry*, *32*(6), 1538–1544.

Cunningham, C. (2013). Microglia and neurodegeneration: The role of systemic inflammation. *Glia*, *61*(1), 71–90.

Cunningham, C., Wilcockson, D. C., Campion, S., Lunnon, K., & Perry, V. H. (2005). Central and systemic endotoxin challenges exacerbate the local inflammatory response and increase neuronal death during chronic neurodegeneration. *The Journal of Neuroscience*, *25*(40), 9275–9284.

de Leeuw, F. A., Peeters, C. F. W., Kester, M. I., Harms, A. C., Struys, E. A., Hankemeier, T., et al. (2017). Blood-based metabolic signatures in Alzheimer's disease. *Alzheimer's & dementia: diagnosis, assessment & disease monitoring*, *8*, 196–207.

DiSabato, D. J., Quan, N., & Godbout, J. P. (2016). Neuroinflammation: The devil is in the details. *Journal of Neurochemistry*, *139*(Suppl. 2), 136–153.

Dunn, N., Mullee, M., Perry, V. H., & Holmes, C. (2005). Association between dementia and infectious disease: Evidence from a case-control study. *Alzheimer Disease and Associated Disorders*, *19*(2), 91–94.

Eikelenboom, P., Bate, C., Van Gool, W. A., Hoozemans, J. J., Rozemuller, J. M., Veerhuis, R., et al. (2002). Neuroinflammation in Alzheimer's disease and prion disease. *Glia*, *40*(2), 232–239.

Eikelenboom, P., Hoozemans, J. J., Veerhuis, R., van Exel, E., Rozemuller, A. J., & van Gool, W. A. (2012). Whether, when and how chronic inflammation increases the risk of developing late-onset Alzheimer's disease. *Alzheimer's Research & Therapy*, *4*(3), 15.

Ek, M., Kurosawa, M., Lundeberg, T., & Ericsson, A. (1998). Activation of vagal afferents after intravenous injection of interleukin-1beta: Role of endogenous prostaglandins. *The Journal of Neuroscience*, *18*(22), 9471–9479.

Figueiredo-Pereira, M. E., Rockwell, P., Schmidt-Glenewinkel, T., & Serrano, P. (2014). Neuroinflammation and J2 prostaglandins: Linking impairment of the ubiquitin-proteasome pathway and mitochondria to neurodegeneration. *Frontiers in Molecular Neuroscience*, 7, 104.

Fotuhi, M., Hachinski, V., & Whitehouse, P. J. (2009). Changing perspectives regarding late-life dementia. *Nature Reviews Neurology*, *5*(12), 649–658.

Goehler, L. E., Gaykema, R. P., Nguyen, K. T., Lee, J. E., Tilders, F. J., Maier, S. F., et al. (1999). Interleukin-1beta in immune cells of the abdominal vagus nerve: A link between the immune and nervous systems? *The Journal of Neuroscience*, *19*(7), 2799–2806.

Grimm, M. O., Grimm, H. S., Tomic, I., Beyreuther, K., Hartmann, T., & Bergmann, C. (2008). Independent inhibition of Alzheimer disease beta- and gamma-secretase cleavage by lowered cholesterol levels. *The Journal of Biological Chemistry*, *283*(17), 11302–11311.

Guo, J. T., Yu, J., Grass, D., de Beer, F. C., & Kindy, M. S. (2002). Inflammation-dependent cerebral deposition of serum amyloid a protein in a mouse model of amyloidosis. *The Journal of Neuroscience*, *22*(14), 5900–5909.

Hanisch, U. K., & Kettenmann, H. (2007). Microglia: Active sensor and versatile effector cells in the normal and pathologic brain. *Nature Neuroscience*, *10*(11), 1387–1394.

Heneka, M. T., Carson, M. J., El Khoury, J., Landreth, G. E., Brosseron, F., Feinstein, D. L., et al. (2015). Neuroinflammation in Alzheimer's disease. *Lancet Neurology*, *14*(4), 388–405.

Heneka, M. T., Kummer, M. P., & Latz, E. (2014). Innate immune activation in neurodegenerative disease. *Nature Reviews Immunology*, *14*(7), 463–477.

Ho, G. J., Drego, R., Hakimian, E., & Masliah, E. (2005). Mechanisms of cell signaling and inflammation in Alzheimer's disease. *Current Drug Targets. Inflammation and Allergy*, *4*(2), 247–256.

Holmes, C., Cunningham, C., Zotova, E., Woolford, J., Dean, C., Kerr, S., et al. (2009). Systemic inflammation and disease progression in Alzheimer disease. *Neurology*, *73*(10), 768–774.

Holscher, C. (2011). Diabetes as a risk factor for Alzheimer's disease: Insulin signalling impairment in the brain as an alternative model of Alzheimer's disease. *Biochemical Society Transactions*, *39*(4), 891–897.

Hong, S., Beja-Glasser, V. F., Nfonoyim, B. M., Frouin, A., Li, S. M., Ramakrishnan, S., et al. (2016). Complement and microglia mediate early synapse loss in Alzheimer mouse models. *Science*, *352*(6286), 712–716.

Hu, W. T., Holtzman, D. M., Fagan, A. M., Shaw, L. M., Perrin, R., Arnold, S. E., et al. (2012). Plasma multianalyte profiling in mild cognitive impairment and Alzheimer disease. *Neurology*, *79*(9), 897–905.

Kamer, A. R., Craig, R. G., Pirraglia, E., Dasanayake, A. P., Norman, R. G., Boylan, R. J., et al. (2009). TNF-alpha and antibodies to periodontal bacteria discriminate between Alzheimer's disease patients and normal subjects. *Journal of Neuroimmunology*, *216*(1–2), 92–97.

Kandimalla, R., Thirumala, V., & Reddy, P. H. (2017). Is Alzheimer's disease a type 3 diabetes? A critical appraisal. *Biochimica et Biophysica Acta, Molecular Basis of Disease*, *1863*(5), 1078–1089. https://doi.org/10.1016/j.bbadis.2016.08.018.

Konner, A. C., & Bruning, J. C. (2012). Selective insulin and leptin resistance in metabolic disorders. *Cell Metabolism*, *16*(2), 144–152.

Lee, Y. H., Tharp, W. G., Maple, R. L., Nair, S., Permana, P. A., & Pratley, R. E. (2008). Amyloid precursor protein expression is upregulated in adipocytes in obesity. *Obesity (Silver Spring)*, *16*(7), 1493–1500.

Levin-Allerhand, J. A., Lominska, C. E., & Smith, J. D. (2002). Increased amyloid-levels in APPSWE transgenic mice treated chronically with a physiological high-fat high-cholesterol diet. *The Journal of Nutrition, Health & Aging*, *6*(5), 315–319.

Liu, B., & Hong, J. S. (2003). Role of microglia in inflammation-mediated neurodegenerative diseases: Mechanisms and strategies for therapeutic intervention. *The Journal of Pharmacology and Experimental Therapeutics*, *304*(1), 1–7.

Llano, D. A., Li, J. H., Waring, J. F., Ellis, T., Devanarayan, V., Witte, D. G., et al. (2012). Cerebrospinal fluid cytokine dynamics differ between Alzheimer disease patients and elderly controls. *Alzheimer Disease & Associated Disorders*, *26*(4), 322–328.

Lyman, M., Lloyd, D. G., Ji, X., Vizcaychipi, M. P., & Ma, D. (2014). Neuroinflammation: The role and consequences. *Neuroscience Research*, *79*, 1–12.

Magnuson, A. M., Regan, D. P., Booth, A. D., et al. (2020). High-fat diet induced central adiposity (visceral fat) is associated with increased fibrosis and decreased immune cellularity of the mesenteric lymph node in mice. *European Journal of Nutrition*, *59*, 1641–1654.

McGeer, P. L., & McGeer, E. G. (2001). Inflammation, autotoxicity and Alzheimer disease. *Neurobiology of Aging*, *22*(6), 799–809.

McLean, F. H., Campbell, F. M., Langston, R. F., Sergi, D., Resch, C., Grant, C., et al. (2019). A high-fat diet induces rapid changes in the mouse hypothalamic proteome. *Nutrition & Metabolism (London)*, *16*, 26.

Micheau, O., & Tschopp, J. (2003). Induction of TNF receptor I-mediated apoptosis via two sequential signaling complexes. *Cell*, *114*(2), 181–190.

Mishra, A., Kim, H. J., Shin, A. H., & Thayer, S. A. (2012). Synapse loss induced by interleukin-1beta requires pre- and post-synaptic mechanisms. *Journal of Neuroimmune Pharmacology*, 7(3), 571–578.

Misiak, B., Leszek, J., & Kiejna, A. (2012). Metabolic syndrome, mild cognitive impairment and Alzheimer's disease—The emerging role of systemic low-grade inflammation and adiposity. *Brain Research Bulletin*, *89*(3–4), 144–149.

Monteiro, R., & Azevedo, I. (2010). Chronic inflammation in obesity and the metabolic syndrome. *Mediators of Inflammation*, *2010*289645.

Morales, I., Guzman-Martinez, L., Cerda-Troncoso, C., Farias, G. A., & Maccioni, R. B. (2014). Neuroinflammation in the pathogenesis of Alzheimer's disease. A rational framework for the search of novel therapeutic approaches. *Frontiers in Cellular Neuroscience*, *8*, 112.

Nguyen, J. C., Killcross, A. S., & Jenkins, T. A. (2014). Obesity and cognitive decline: Role of inflammation and vascular changes. *Frontiers in Neuroscience*, *8*, 375.

Ott, A., Stolk, R. P., van Harskamp, F., Pols, H. A., Hofman, A., & Breteler, M. M. (1999). Diabetes mellitus and the risk of dementia: The Rotterdam study. *Neurology*, *53*(9), 1937–1942.

Perry, V. H., & Teeling, J. (2013). Microglia and macrophages of the central nervous system: The contribution of microglia priming and systemic inflammation to chronic neurodegeneration. *Seminars in Immunopathology*, *35*(5), 601–612.

Petit-Paitel, A. (2010). GSK-3beta: A central kinase for neurodegenerative diseases? *Medical Science (Paris)*, *26*(5), 516–521.

Qizilbash, N., Gregson, J., Johnson, M. E., Pearce, N., Douglas, I., Wing, K., et al. (2015). BMI and risk of dementia in two million people over two decades: A retrospective cohort study. *The Lancet Diabetes and Endocrinology*, *3*(6), 431–436.

Rosen, C., Andersson, C. H., Andreasson, U., Molinuevo, J. L., Bjerke, M., Rami, L., et al. (2014). Increased levels of chitotriosidase and YKL-40 in cerebrospinal fluid from patients with Alzheimer's disease. *Dementia and geriatric cognitive disorders extra*, *4*(2), 297–304.

Sastre, M., Dewachter, I., Landreth, G. E., Willson, T. M., Klockgether, T., van Leuven, F., et al. (2003). Nonsteroidal anti-inflammatory drugs and peroxisome proliferator-activated receptor-gamma agonists modulate immunostimulated processing of amyloid precursor protein through regulation of beta-secretase. *The Journal of Neuroscience, 23*(30), 9796–9804.

Sastre, M., Klockgether, T., & Heneka, M. T. (2006). Contribution of inflammatory processes to Alzheimer's disease: Molecular mechanisms. *International Journal of Developmental Neuroscience, 24*(2–3), 167–176.

Singh, P., Hoffmann, M., Wolk, R., Shamsuzzaman, A. S., & Somers, V. K. (2007). Leptin induces C-reactive protein expression in vascular endothelial cells. *Arteriosclerosis, Thrombosis, and Vascular Biology, 27*(9), e302–e307.

Sparks, D. L., Kuo, Y. M., Roher, A., Martin, T., & Lukas, R. J. (2000). Alterations of Alzheimer's disease in the cholesterol-fed rabbit, including vascular inflammation. Preliminary observations. *Annals of the New York Academy of Sciences, 903*, 335–344.

Spielman, L. J., Little, J. P., & Klegeris, A. (2014). Inflammation and insulin/IGF-1 resistance as the possible link between obesity and neurodegeneration. *Journal of Neuroimmunology, 273*(1–2), 8–21.

Stence, N., Waite, M., & Dailey, M. E. (2001). Dynamics of microglial activation: A confocal time-lapse analysis in hippocampal slices. *Glia, 33*(3), 256–266.

Talbot, K., Wang, H. Y., Kazi, H., Han, L. Y., Bakshi, K. P., Stucky, A., et al. (2012). Demonstrated brain insulin resistance in Alzheimer's disease patients is associated with IGF-1 resistance, IRS-1 dysregulation, and cognitive decline. *The Journal of Clinical Investigation, 122*(4), 1316–1338.

van der Heide, L. P., Ramakers, G. M., & Smidt, M. P. (2006). Insulin signaling in the central nervous system: Learning to survive. *Progress in Neurobiology, 79*(4), 205–221.

Whitmer, R. A., Gunderson, E. P., Barrett-Connor, E., Quesenberry, C. P., Jr., & Yaffe, K. (2005). Obesity in middle age and future risk of dementia: A 27 year longitudinal population based study. *BMJ, 330*(7504), 1360.

Wolozin, B., Wang, S. W., Li, N. C., Lee, A., Lee, T. A., & Kazis, L. E. (2007). Simvastatin is associated with a reduced incidence of dementia and Parkinson's disease. *BMC Medicine, 5*, 20.

Zhao, W. Q., De Felice, F. G., Fernandez, S., Chen, H., Lambert, M. P., Quon, M. J., et al. (2008). Amyloid beta oligomers induce impairment of neuronal insulin receptors. *The FASEB Journal, 22*(1), 246–260.

Zlokovic, B. V. (2011). Neurovascular pathways to neurodegeneration in Alzheimer's disease and other disorders. *Nature Reviews. Neuroscience, 12*(12), 723–738.

CHAPTER THIRTEEN

The NLRP3 inflammasome as a bridge between neuro-inflammation in metabolic and neurodegenerative diseases

Grażyna Söderbom[a,*], Bai-Yun Zeng[b]

[a]Klipspringer AB, Sävedalen, Sweden

[b]Institute of Pharmaceutical Science, Faculty of Life Sciences & Medicine, King's College London, London, United Kingdom

[*]Corresponding author: e-mail address: grazyna@klipspringer.se

Contents

1. Introduction 348
2. The NLRP3 inflammasome and peripheral metabolic disease 348
   2.1 NLRP3 inflammasome formation requires priming and activation 348
   2.2 What triggers the NLRP3 inflammasome? 350
3. The NLRP3 inflammasome, chronic inflammation and endothelial dysfunction 354
4. Neuro-inflammation in Alzheimer's disease 355
   4.1 Pathology of Alzheimer's disease 355
   4.2 Chronic inflammation and endothelial dysfunction contribute to blood-brain barrier dysfunction in Alzheimer's disease 356
   4.3 Metabolic changes in Alzheimer's disease lead to aberrant protein accumulation 358
   4.4 Brain insulin resistance in Alzheimer's disease leads to aberrant protein accumulation 359
   4.5 Neuroinflammation and Alzheimer's disease: The NLRP3 inflammasome 360
   4.6 Chronic inflammation and neuroinflammation in Alzheimer's disease: The NLRP3 inflammasome as a major contributor 362
5. Neuro-inflammation in Parkinson's disease 363
   5.1 Pathology of Parkinson's disease 363
   5.2 Chronic inflammation is common to metabolic disorders and Parkinson's disease 364
   5.3 Neuroinflammation, Parkinson's disease, and the NLRP3 inflammasome 366
6. The NLRP3 inflammasome: A target for novel therapeutic approaches in Alzheimer's and Parkinson's diseases? 369
   6.1 Targeting mitochondrial dysfunction and ATP as DAMPs 369
   6.2 Targeting metabolic drivers of the NLRP3 inflammasome 369
   6.3 Targeting pro-inflammatory cytokines 371
   6.4 Targeting DAMPs like amyloid-β, tau and α-synuclein 372

International Review of Neurobiology, Volume 154
ISSN 0074-7742
https://doi.org/10.1016/bs.irn.2020.03.023

7. Conclusion 372
7.1 Type 2 diabetes is a risk factor for neurodegenerative diseases 372
7.2 Metabolic and neurodegenerative diseases share common risk factors 373
7.3 Metabolic and neurodegenerative diseases share common pathology 373
7.4 What does this mean for Alzheimer's and Parkinson's diseases? 374
References 375

Abstract

Evidence increasingly suggests that type 2 diabetes mellitus (T2DM) is a risk factor for neurodegenerative diseases (NDDs), such as Alzheimer's disease (AD) and Parkinson's disease (PD). These diseases share many pathological processes, including oxidative stress, local inflammation/neuroinflammation and chronic, low-grade (systemic) inflammation, which are exacerbated by aging, a common risk factor for T2DM and NDDs. Here, we focus on the link between chronic inflammation driven by peripheral metabolic disease and how this may impact neurodegeneration in AD and PD. We review the relationship between these common pathological processes in AD and PD from the perspective of the "pro-inflammatory" signaling of the nucleotide-binding oligomerization domain (NOD)-, leucine-rich repeat- (LRR)-, and pyrin domain-containing protein 3 (NLRP3) inflammasome complex. Since the need for effective disease-modifying therapies in T2DM, AD and PD is significant, the relationship between these diseases is important as a positive clinical impact on one may benefit the others. We briefly consider how novel strategies may target neuro-inflammation and provide potential therapies for AD and PD.

List of Abbreviations

AD Alzheimer's disease
ADS antioxidant defense system
AGE advanced glycation end product
AMPK AMP-activated protein kinase
APOE gene encoding apolipoprotein E
APOE-ε4 apolipoprotein E-ε4 gene variant
ASC Apoptosis-associated Speck-like protein containing a Caspase activating and recruitment domain
ATP adenosine triphosphate
BBB blood-brain barrier
BCSFB blood-cerebrospinal fluid barrier
CARD caspase activation and recruitment domain
CSF cerebrospinal fluid
DAMPs danger-associated molecular patterns
DPP-4 dipeptidyl peptidase-4
GLP-1 glucagon-like peptide-1
GLUT1 glucose transporter 1
IAPP pro-islet amyloid polypeptide
IBD inflammatory bowel disease
IL-18 interleukin-18
IL1β gene encoding IL-1β

IL-1β	interleukin-1β
IL-6	interleukin-6
IL-1α	interleukin-1α
IMM	inner mitochondrial membrane
iNOS	inducible nitric oxide synthase
ISF	interstitial fluid
LPS	lipopolysaccharide
***LRRK2* (*PARK8*)**	gene encoding leucine-rich repeat kinase 2
MAMPs	microbial-associated molecular patterns
MAPK	mitogen-activated protein kinase
MCI	mild cognitive impairment
MetS	metabolic syndrome
MHC-II	major histocompatibility complex class II
MPP^+	1-methyl-4-phenylpyridinium
MPTP	1-methyl-4-phenyl-1,2,3,6-tetrahydropyridine
mtDNA	mitochondrial DNA
MyD88	myeloid differentiation primary response 88 protein
NDDs	neurodegenerative diseases
NF-κB	nuclear factor kappa-light-chain-enhancer of activated B cells
NLRP1, 3	genes encoding NLRP1 and NLRP3, respectively
NLRP3 inflammasome	nucleotide-binding oligomerization domain (NOD)-, leucine-rich repeat- (LRR)-, and pyrin domain-containing protein 3 (NLRP3) inflammasome
NLRs	nucleotide-binding oligomerization domain (NOD)-like receptors
NOX	NADPH oxidase
$O_2^{\bullet-}$	superoxide radical anion
OMM	outer mitochondrial membrane
$ONOO^-$	peroxynitrite anion
OXPHOS	oxidative phosphorylation
OxS	oxidative stress
P2X7R	purinergic P2X7 receptor
PAMPs	pathogen-associated molecular patterns
PD	Parkinson's disease
PI3K-AKT	phosphatidylinositol-3-kinase/protein kinase B
PI3K/AKT/mTOR	phosphatidylinositol-3-kinase/protein kinase B/mechanistic target of rapamycin
PPRs	pattern recognition receptors
RAGE	receptor for AGE products
RNS	reactive nitrogen species
ROS	reactive oxygen species
RSPO3	R-spondin 3
SGLT-2	sodium/glucose cotransporter 2
SIRT2	sirtuin-2
SIRT3	sirtuin-3
SMOCs	supramolecular organizing centers
T2DM	type 2 diabetes mellitus
TLRs	toll-like receptors
TNF1/TNF2	tumor necrosis factor-α receptors
TNF-α	tumor necrosis factor-α

ULK1	unc-51-like kinase-1
α-Syn	α-synuclein
βHB	β-hydroxybutyrate

1. Introduction

A strong link exists between type 2 diabetes mellitus (T2DM) and neurodegenerative diseases (NDDs), such as Alzheimer's disease (AD) and Parkinson's disease (PD). These diseases share many pathological processes, including oxidative stress (OxS), local inflammation/neuroinflammation and chronic, low-grade (systemic) inflammation. Metabolic disorders cause OxS, which triggers inflammatory responses via activation of redox sensitive transcription factors and stress signaling pathways that feedback positively to exacerbate OxS, inflammation and peripheral tissue damage. Consequent chronic inflammation damages the vascular endothelium and disrupts the blood-brain barrier (BBB). This increases the brain's vulnerability to systemic oxidative, inflammatory and other, potentially toxic, mediators. These processes are exacerbated by aging, a common risk factor for T2DM, AD and PD.

Here, we review the relationship between OxS and chronic inflammation and neuroinflammation in T2DM, AD and PD from the perspective of innate immune signaling, nucleotide-binding oligomerization domain (NOD)-, leucine-rich repeat- (LRR)-, and pyrin domain-containing protein 3 (NLRP3) inflammasome. We will review how peripheral metabolic diseases and chronic inflammation can impact the brain, the role of the NLRP3 inflammasome in AD and PD, and how a therapeutic impact on any of these diseases might offer mutual benefits based on their mechanistically intertwined pathological processes.

2. The NLRP3 inflammasome and peripheral metabolic disease

2.1 NLRP3 inflammasome formation requires priming and activation

The inflammasomes are important components of the innate immune system. First described almost two decades ago by Martinon, Burns, and

Tschopp (2002), the large, multi-protein inflammasome complexes are pattern recognition receptors (PPRs) and innate immune cell sensors for Pathogen-Associated Molecular Patterns (PAMPs), Microbial-Associated Molecular Patterns (MAMPs) and Danger-Associated Molecular Patterns (DAMPs) associated with cellular dysfunction, and trigger cytokine gene transcription. The PPRs include membrane-bound toll-like receptors (TLRs) that induce pro-inflammatory cytokine secretion by triggering stress signaling pathways, such as those involving nuclear factor kappa-light-chain-enhancer of activated B cells (NF-κB) and mitogen-activated protein kinase (MAPK) via the myeloid differentiation primary response 88 protein (MyD88; a signaling adaptor protein that mediates TLR signaling) (Brown et al., 2011). They also include the nucleotide-binding oligomerization domain (NOD)-like receptors (NLRs), including those of the NLRP family, such as NLRP3 expressed by macrophages, glial cells (primarily microglia), neurons and endothelial cells (Gustin et al., 2015; Heid et al., 2013; von Herrmann et al., 2018; Xia et al., 2014). Upon activation, the NLRP3 protein becomes part of a complex structure called the NLRP3 inflammasome that drives interleukin-1β (IL-1β) and interleukin-18 (IL-18) pro-inflammatory responses.

Formation of an active, cytosolic NLRP3 inflammasome complex depends upon a coordinated, two-step process of priming and activation (Groslambert and Py, 2018; Jo et al., 2016; Song et al., 2017). The priming step involves transcriptional activation of the auto-inhibited NLRP3 protein mediated by activation of cell surface receptors, such as TLRs by PAMPs and/or DAMPs [e.g., TLR4 activation by lipopolysaccharide (LPS) expressed by gram-negative bacteria], or cytokine receptors by pro-inflammatory cytokines [e.g., tumor necrosis factor-α receptors (TNF1/TNF2) by tumor necrosis factor-α (TNF-α)] (Groslambert and Py, 2018). Activation of TLRs causes NF-κB disinhibition and translocation into the nucleus via the MyD88 signaling pathway, and subsequent transcription of cytokines and chemokines, as well as expression of NLRP3, pro-IL-1β and pro-IL-18 (Bauernfeind et al., 2009; Franchi et al., 2009).

The NLRP3 sensor is triggered by DAMPs indicative of metabolic or cellular stress [e.g., mitochondrial or lysosomal damage induced by amyloid-beta (Aβ)] (Abais et al., 2015; Abderrazak et al., 2015). Expressed NLRP3 protein associates with the adaptor protein, Apoptosis-associated Speck-like protein containing a Caspase activating and recruitment domain (ASC), which mediates the recruitment of pro-caspase-1 via its Caspase Activation and Recruitment Domain (CARD) to the complex (Srinivasula et al., 2002). Pro-caspase-1 undergoes auto-proteolytic conversion to its active form, caspase-1, a protease that subsequently converts inactive pro-IL-1β and pro-IL-18 into pro-inflammatory IL-1β and IL-18, respectively, as well as the pro-form of gasdermin D into its active form. Gasdermin D is a cytosolic protein associated with pyroptosis, a highly inflammatory form of programmed cell death that protects against intracellular bacterial infection. Active gasdermin D causes pyroptotic lysis of the cell plasma membrane to allow canonical secretion of active IL-1β and IL-18, as well as non-lytic, non-canonical IL-1β release (He et al., 2015; Heilig et al., 2018; Russo et al., 2016; Schneider et al., 2017). Plasma membrane rupture also allows the release of the other pro-inflammatory cytokines, such as interleukin-6 (IL-6) and TNF-α produced in response to TLR-mediated activation of NF-κB. This complex, interlinked cascade of pro-inflammatory events and resultant mediators subsequently drive and feedback to amplify inflammation in both the periphery and brain.

2.2 What triggers the NLRP3 inflammasome?

Many factors can trigger the formation and activation of the NLRP3 inflammasome. However, both the "priming" step mediated by TLRs and/or cytokine signaling and the "formation and activation" step triggered by DAMPs are prerequisite for an NLRP3 inflammasome-mediated inflammatory response (Abais et al., 2015; Groslambert and Py, 2018).

2.2.1 Danger-associated molecular patterns linked to mitochondrial dysfunction

Mitochondrial dysfunction is an early pathological feature common to T2DM, AD and PD. Oxidative phosphorylation (OXPHOS) in mitochondria generates endogenous reactive oxygen species (ROS; e.g., $O_2^{\bullet -}$) and reactive nitrogen species (RNS; e.g., $ONOO^-$), particularly at complexes I and III of the electron transport chain (Halliwell and Gutteridge, 2015). Pancreatic mitochondrial complex I hyperactivity associated with OxS and pancreatic β-cell failure has been observed in T2DM mice (Wu et al., 2017), while early dysfunctions in complexes IV and I are observed in human AD and PD, respectively (Holper et al., 2019; Mutisya et al., 1994; Schapira et al., 1990).

Although mitochondria deploy an antioxidant defense system (ADS) to inactivate these highly reactive ROS/RNS, inefficient "deactivation" overwhelms ADS capacity and/or evades a sub-optimal (deficient or defective) ADS and tips the $NADH/NAD^+$ redox balance into a pathological state of

OxS. Oxidative stress occurs in T2DM, AD and PD, causing oxidative/peroxidative and nitrosative stress and damage to cellular macromolecules, nucleic acids and mitochondrial DNA (mtDNA). Pancreatic β-cells are vulnerable and subject to OxS, which is exacerbated by hyperglycemia in T2DM by increasing the mitochondrial glucose load and ROS production (Miki et al., 2018). The brain also shows clear evidence of oxidative damage in AD and PD (Cenini et al., 2019; Cioffi et al., 2019; Toulorge et al., 2016; Zabel et al., 2018).

Mitochondria-generated ROS are DAMPs. They trigger NLRP3 inflammasome formation and activation (He et al., 2016) to cause NLRP3 dependent-lysosomal membrane permeabilization and subsequent mitochondrial breakdown (Heid et al., 2013). Mitochondrial DNA released into the cytosol, particularly oxidized mtDNA released during apoptosis, is another DAMP (Shimada et al., 2012). Oxidative stress and mitochondrial dysfunction can impair apoptosis and autophagy/mitophagy. Mitochondrial fission and fusion proteins regulate the assembly of respiratory complexes, such that any disruption in the dynamics of mitochondria may contribute to their dysfunction in NDDs (Cenini et al., 2019). Consequent accumulation of damaged and dysfunctional, ROS-generating mitochondria activate the NLRP3 inflammasome, indicating its ability to sense cellular/neuronal damage and resulting in inflammation (Park et al., 2015; Zhou et al., 2011). However, NF-κB can delay accumulation of the autophagy receptor p62/SQSTM1 associated with mitophagic clearance of damaged mitochondria to inhibit NLRP3-inflammasome activation and limit the inflammatory response (Zhong et al., 2016). These studies demonstrate a complex interaction between lysosomes, mitochondrial dysfunction and ROS/OxS in the induction of NLRP3 inflammasome-mediated inflammatory responses.

Interestingly, Elliott et al. (2018) have proposed that mitochondria are supramolecular organizing centers (SMOCs) for the NLRP3 inflammasome. Using mouse-derived cells, they showed that inflammasome priming requires NLRP3 and caspase-1 migration to the mitochondria and their association with the outer mitochondrial membrane (OMM). This is crucial for subsequent association of ASC with mitochondria, which depends on elevated cytosolic calcium levels and activates the NLRP3 inflammasome (Elliott et al., 2018). Cardiolipin, a phospholipid made and located in the inner mitochondrial membrane (IMM), is involved in mitochondrial bioenergetic and apoptotic processes, and in mitochondrial membrane stability and dynamics (Paradies et al., 2014). Elliott et al. (2018) also showed that cardiolipin moves from the IMM to the OMM following inflammasome priming; this, as with the movement of NLRP3 and caspase 1 to the OMM, is ROS dependent. At the OMM, cardiolipin directly associates with NLRP3 and caspase-1, at which point caspase-1 undergoes autocatalysis and the NLRP3 inflammasome is active.

Mitochondria are, therefore, vital for the formation and activation of the NLRP3 inflammasome, serving as SMOCs for the inflammasome components. Inflammasome activation is ROS dependent and mitochondria are the primary generators of ROS, not only during normal OXPHOS, but in a state of OxS associated with metabolic and neurodegenerative diseases. Mitochondrial dysfunction occurs upstream of OxS and neuroinflammation, suggesting a central role of mitochondria in a complex, pathological, OxS-inflammatory amplification loop associated with the NLRP3 inflammasome.

2.2.2 Danger-associated molecular patterns linked to metabolic dysfunction

Metabolic syndrome (MetS) is defined as a cluster of three or more cardiometabolic pathologies, including obesity, insulin resistance, hyperglycemia, dyslipidemia and hypertension; each component *per se* or as a MetS cluster are risk factors for T2DM. Type 2 diabetes mellitus is a risk factor for AD and PD, and the NLRP3 inflammasome may link these diseases since it can detect metabolic dysregulation and stress (Zhou et al., 2011), which are common to all three diseases.

Inflammation can cause insulin resistance and T2DM development, and the NLRP3 inflammasome has been implicated in T2DM pathogenesis (Abderrazak et al., 2015). Individuals with a combined increase in IL-6 and IL-β1 levels have an increased risk of developing T2DM (Spranger et al., 2003), and levels of NLRP3 protein and mRNA in peripheral blood mononuclear cells and IL-1β in plasma are increased in T2DM (Wan et al., 2019). The NLRP3 inflammasome contributes to T2DM inflammation as a consequence of hypoxia-induced ROS and OxS (Sokolova et al., 2018). Moreover, LPS can exacerbate injury caused by high glucose and hypoxia/re-oxygenation in H9C2 cardiomyocytes *in vitro*; this promotes ROS production that induces NLRP3 inflammasome-mediated pyroptosis (Qiu et al., 2019).

The NLRP3 inflammasome also contributes to inflammation associated with diet-induced obesity and progression to T2DM (Youm et al., 2011). Saturated fatty acids are metabolic triggers that prime the NLRP3 inflammasome via TLR4 activation causing adipocyte inflammation and insulin resistance (Reynolds et al., 2012). Palmitate appears to induce NLRP3 inflammasome-mediated pro-inflammatory responses involving IL-1β and IL-18 associated with mitochondrial ROS, AMP-activated protein kinase (AMPK) and the unc-51-like kinase-1 (ULK1) autophagy signaling cascade (Wen et al., 2011; Xing et al., 2019). In contrast, mice fed a high-fat diet in which saturated fatty acids were replaced with monounsaturated fatty acids showed improved insulin sensitivity with a concurrent reduction in pro-IL-1β priming, reduced IL-1β secretion from adipose tissue and sustained adipose AMPK activation (Finucane et al., 2015). Moreover, obesity-induced lipids (e.g., ceramide) can trigger the NLRP3 sensor to promote IL-1β-induced inflammation and reduced plasma glucose and insulin sensitivity in obese individuals with T2DM (Vandanmagsar et al., 2011).

Type 2 diabetes mellitus *per se* can trigger NLRP3-inflammasome-mediated inflammation. In T2DM, β-cells produce high levels of pro-islet amyloid polypeptide (IAPP) (Montane et al., 2012). This accumulates with other toxic, pro-apoptotic oligomers in amyloid plaques in the pancreatic islets (Jurgens et al., 2011; Wang et al., 2001) leading to OxS and consequent β-cell apoptosis (Zraika et al., 2009). Together with these pro-apoptotic oligomers, IAPP can trigger NLRP3 inflammasome activation and IL-1β production, and subsequent macrophage recruitment exacerbates the local inflammation-OxS-β-cell dysfunction loop (Masters et al., 2010; Morikawa et al., 2018; Vandanmagsar et al., 2011; Youm et al., 2011).

One consequence of T2DM is diabetic nephropathy. This is associated with chronic inflammation also mediated by the NLRP3 inflammasome, which is triggered by mitochondrial ROS in glucose- or advanced glycation end product (AGE)-stressed podocytes (Shahzad et al., 2015). Thus, the NLRP3 inflammasome is implicated in chronic inflammatory responses and peripheral damage associated with T2DM, and may also negatively impact other organs, such as the brain. Indeed, such T2DM-associated DAMPs, including ROS/RNS, ceramides, AGEs, and cytokines associated with chronic inflammation, can access and negatively impact the brain (Lee et al., 2018; Maciejczyk et al., 2019). Ceramide, for example, positively modulates NLRP3 inflammasome assembly and IL-1β secretion, as demonstrated in microglia primed *in vitro* with sodium palmitate to induce *de novo* ceramide synthesis (Scheiblich et al., 2017).

Importantly, mitochondrial dysfunction associated with T2DM and NDDs may be associated with a sustained anabolic state (Cheng and Almeida, 2014; Esterline et al., 2018; Yan et al., 2013). Chronic hyperglycemia and/or hyperlipidemia induce metabolic overload/mitochondrial stress in pancreatic β-cell function (Fu et al., 2017) and a metabolic switch from OXPHOS to β-oxidation of fuel substrates and the accumulation of incompletely β-oxidized metabolites and mitochondrial failure (Befroy et al., 2007; Koves et al., 2008).

2.2.3 Pathogen- and microbial-associated molecular patterns and the NLRP3 inflammasome

Gut dysbiosis associated with MetS/T2DM can contribute to the chronic inflammation seen in T2DM, AD and PD. Microbiome-generated metabolic products like short-chain fatty acids (e.g., propionic acid), and bacterial components (e.g., LPS, lipoteichoic acids, and peptidoglycans) are all PAMPs/DAMPs. By triggering PPRs, they increase cellular secretion of pro-inflammatory cytokines, including TNF-α, IL-6 and IL-1β, which contribute to a pathological, chronic inflammatory loop and may negatively impact the brain (Hirschberg et al., 2019; Saad et al., 2016).

3. The NLRP3 inflammasome, chronic inflammation and endothelial dysfunction

Pro-inflammatory cytokines are associated with the development of T2DM and consequent chronic inflammation that contributes to endothelial dysfunction. This may involve the NLRP3 inflammasome. The NLRP3, ASC and caspase-1 proteins are expressed in cultured mouse microvascular endothelial cells (Xia et al., 2014). C-reactive protein, another T2DM-associated cytokine, has been shown to upregulate NF-κB activity, induce expression of pro-IL-1β and NLRP3, and activate the NLRP3 inflammasome in endothelial cells (Bian et al., 2019). Activation of the NLRP3 inflammasome by visfatin (an obesity-associated adipokine) or palmitate leads to endothelial inflammation and dysfunction in mice (Xia et al., 2014; Xing et al., 2019). Endothelial dysfunction contributes to BBB dysfunction associated with AD and PD, and so links chronic inflammation associated with peripheral metabolic disorders to NDDs.

Interleukin-1β can increase vascular permeability by inducing the expression of pro-permeability factor R-spondin 3 (RSPO3) in vascular endothelial cells (Skaria et al., 2017). This is important because RSPO3 enhances the permeability of human primary endothelial cells and synergizes with IL-1β to exacerbate IL-1β-induced permeability of the vascular endothelium (Skaria et al., 2018). Exposure of human brain microvessel endothelial cell cultures to cytokines (e.g., IL-1β) or LPS is associated with increased permeability of the tight junctions between the cells (Wong et al., 2004). Interestingly, the sodium/glucose cotransporter 2 (SGLT-2) inhibitor empagliflozin reduces vascular dysfunction and cognitive impairment in T2DM mice by attenuating OxS and inflammation (Lin et al., 2014) and can protect the neurovascular unit from abnormal remodeling in the brain (Hayden et al., 2019). Clearly, there is a link between metabolic dysfunction and the brain partly mediated by inflammation with the NLRP3 inflammasome as a key contributor.

In summary, MetS/T2DM is associated with mitochondrial dysfunction and consequent OxS and local and chronic inflammation. Oxidative stress and neuro-inflammation are both cause and consequence of one another; they mutually exacerbate mitochondrial dysfunction and positively drive the OxS-inflammation loop associated with cell death and neurodegeneration. The NLRP3 inflammasome contributes to pro-inflammatory pathology via IL-1β and IL-18 in response to various DAMPs/PAMPs associated with MetS/T2DM, which can amplify the inflammation. The pathological OxS-inflammation loop involving the NLRP3 inflammasome in T2DM is clearly demonstrated in vascular endothelium, causing endothelial dysfunction. This links chronic inflammation to neuroinflammation and neurodegeneration as endothelial dysfunction contributes to BBB dysfunction.

4. Neuro-inflammation in Alzheimer's disease

4.1 Pathology of Alzheimer's disease

Alzheimer's disease starts years before clinical symptoms of progressive cognitive impairment and episodic memory loss appear (Lane et al., 2018). It progresses through three stages: in preclinical AD, only biomarkers are evident (Sperling et al., 2011); in mild cognitive impairment (MCI) cognitive deficits, but no functional impairment are apparent (Albert et al., 2011); and, in AD dementia a decline in two or more cognitive domains has reached a stage where functioning at work or daily activities are impaired (McKhann et al., 2011).

Pathological changes in the brain defining AD are abundant; extracellular Aβ plaques and intracellular neurofibrillary tangles consisting of abnormally hyperphosphorylated tau protein are accompanied by synaptic and neuronal loss in the cerebral cortex and specific subcortical regions (Alonson et al., 2001; Goedert et al., 1991; Selkoe, 1986). These pathological changes cause gross atrophy of affected regions, including the temporal and parietal lobes, and parts of the frontal cortex and cingulate gyrus (Li et al., 2016; Wenk, 2003).

The cause and progression of AD are still unclear, but may involve neuro-inflammation. Inflammation occurs early at the MCI stage of AD (King et al., 2018), indicated by increased peripheral levels of pro-inflammatory cytokines (e.g., IL-1β, IL-18, IL-6, and TNF-α) that are also associated with AD (Brosseron et al., 2014; Italiani et al., 2018; Lai et al., 2017). Evidence suggests that T2DM is a risk factor for AD (Hooshmand et al., 2019; Hughes et al., 2017; Ott et al., 1999; Priyadarshini et al., 2012; Wang et al., 2017) and chronic inflammation is common to both diseases. Serum levels of pro-inflammatory cytokines, including IL-1β, are significantly higher in individuals with AD or T2DM compared to age-matched controls (Bozluocay et al., 2016). Midlife onset of T2DM has been associated with brain atrophy and causally linked to a negative impact on late-life cognition in elderly individuals without dementia (Roberts et al., 2014), while cortical atrophy associated with T2DM resembles patterns seen in preclinical AD (Moran et al., 2013). This suggests that cognitive deterioration may be accelerated in cognitively normal, but genetically and/or metabolically at risk individuals by a negative impact on brain metabolism and volume, partially involving chronic inflammation.

Aging is the major risk factor for AD. Increased systemic levels of pro-inflammatory cytokines are associated with cognitive impairment in older adults (Mooijaart et al., 2013; Schram et al., 2007; Simpson et al., 2013; Warren et al., 2018), brain atrophy in middle-aged individuals (Marsland et al., 2015) and the elderly (Gu et al., 2017; Satzibal et al., 2012), and a higher risk of AD (Brosseron et al., 2014). Normal aging is associated with BBB dysfunction in the hippocampus (Montagne et al., 2015). Pro-inflammatory drivers and chronic and central inflammation disrupt the BBB, which increases brain exposure to the systemic environment in the elderly and in AD (Blennow et al., 1990). Individuals with MCI have more pronounced BBB dysfunction than controls, which may contribute to early cognitive dysfunction irrespective of aberrant protein changes (Montagne et al., 2015). Thus, chronic inflammation may be associated with brain atrophy during normal aging with negative consequences for the BBB and cognitive function.

4.2 Chronic inflammation and endothelial dysfunction contribute to blood-brain barrier dysfunction in Alzheimer's disease

Metabolic syndrome and T2DM are vascular risk factors for cerebral small vessel disease, BBB leakage, and AD (Kivipelto et al., 2018; Nelson et al., 2016;

Skoog and Gustafson, 2006). The BBB consists of the neurovascular unit, which includes neurons, glial cells and pericytes that allows bi-directional movement of nutrients (e.g., glucose), oxygen, insulin, immune system components and other molecules and ions into and out of the brain, while preventing influx of systemic DAMPs/PAMPs into the brain (Obermeier et al., 2013). Cerebral endothelial cells are vital in controlling molecular ion transport across the BBB and regulating cerebral blood flow. Thus, endothelial dysfunction associated with MetS/T2DM causes hypoperfusion, reduced nutrient transport into the brain and a negative impact on normal BBB function (Hawkins et al., 2007; Hayden, 2019; Zhao et al., 2015).

The brain is no longer considered to be immuno-privileged. Interleukin-6 can cross the BBB (Banks et al., 1994) and both IL-6 and TNF-α have been shown to increase BBB permeability linked to NADPH oxidase (NOX)-dependent ROS generation in microvascular endothelial cells from human brain (Rochfort et al., 2014). In AD, pro-inflammatory mediators are associated with endothelial dysfunction. Inducible nitric oxide synthase (iNOS) can be generated by glia-associated inflammation, contributing to OxS, endothelial dysfunction and causing neurodegeneration in conditions of hypoxia and active NOX (Brown and Neher, 2010; Nagareddy et al., 2005). Chronic inflammation and iNOS induction in rats can lead to impaired short- and long-term memory (Eckel et al., 2011). Intracerebral administration of LPS in a mouse model of prion disease (ME7-infected C57BL/6 mice) increases the expression of iNOS and IL-1β levels, as well as neutrophil infiltration into brain parenchyma that can also occur in response to chronic inflammation evoked by LPS associated with increased neuronal apoptosis (Cunningham et al., 2005).

Blood-brain barrier dysfunction may increase the mutual exchange of pro-inflammatory drivers between brain and periphery, and the vulnerability of the brain to systemic influence increases when the body is in a diseased state, e.g., T2DM (Nation et al., 2019; Starr et al., 2003). Blood-brain barrier dysfunction correlates partly with peripheral endothelial dysfunction and is associated with AD, dementia with Lewy bodies or PD dementia, and T2DM (Janelidze et al., 2017). Blood-cerebrospinal fluid barrier (BCSFB) permeability also occurs, independently of, and possibly before both Aβ and tau pathologies, in specific AD populations (Chalbot et al., 2011). In individuals with MCI, decreased BCSFB permeability and transport capacity has been associated with expression of a specific pro-inflammatory protein signature in serum and CSF (Ott et al., 2018).

This supports a negative role of chronic inflammation and neuroinflammation on BCSFB integrity early in AD development. It also suggests that inflammation increases brain vulnerability to both systemic neurotoxins and metabolic dysfunction due to reduced active transport of nutrients from the systemic circulation (Ott et al., 2018).

An important consequence of chronic inflammation is the movement of Aβ across the BBB. Systemic IL-6 has been associated with increased influx and decreased efflux of Aβ across the BBB and increased neuronal synthesis of Aβ in the brain in a murine model of AD (Jaeger et al., 2009). This may contribute to Aβ-associated pathology in AD. The *APOE-ε4* genotype is a risk factor for late-onset AD and vascular pathology (i.e., BBB dysfunction and inflammation). Blood-brain barrier dysfunction and increased IL-6 levels are associated with greater AD pathology in *APOE-ε4* carriers, which codes for the apolipoprotein E cholesterol carrier (Riphagen et al., 2020). Importantly, *APOE-ε4* correlates with amyloid deposition in AD cerebral vasculature (Nelson et al., 2013) and is linked to increased vascular stiffness and higher vascular Aβ deposition in elderly individuals without dementia, which may decrease cerebrospinal fluid (CSF) flow and reduce Aβ clearance from the brain (Hughes et al., 2014). The *APOE-ε4* gene is associated with disruption of the tight junctions of the BBB via an inflammatory effect (Bell et al., 2012; Nishitsuji et al., 2011). Thus, chronic inflammation associated with T2DM and mediated by pro-inflammatory cytokines may exacerbate genetic risk factors that predispose some individuals to AD and other pathological triggers.

4.3 Metabolic changes in Alzheimer's disease lead to aberrant protein accumulation

Cerebral hypoperfusion restricts oxygen and nutrient delivery to the brain and can impact the bioenergetic capacity of the brain (de la Monte and Tong, 2014). Hypoperfusion is also associated with accelerated cognitive decline in the general population (Wolters et al., 2017) and lower memory function in cognitively healthy, late middle-aged adults with MetS (Birdsill et al., 2013). Abnormal hemodynamics and a "leaky" BBB in affected cerebral vasculature will reduce Aβ clearance from the brain and instead promote its accumulation.

Another early step associated with BBB dysfunction as a consequence of peripheral metabolic disease is cerebral glucose hypometabolism. This occurs before AD symptoms are apparent and may be related to the glucose transporter 1 (GLUT1) that, in addition to transporting glucose into the brain, is required for maintenance of brain capillary networks, blood flow and BBB integrity, and neuronal function and structure (Winkler et al., 2015). Cerebrovascular pathology and BBB dysfunction have been linked to a GLUT1 deficiency in the endothelium. At 2 weeks of age, GLUT1 deficiency in AD mice (Slc2a1 +/−APPSw/0) show cerebral microvascular degeneration and BBB dysfunction associated with reduced brain perfusion and glucose uptake into the brain. By 6 months of age, these mice exhibit neuronal dysfunction and Aβ accumulation in the brain (Winkler et al., 2015).

Severe glucose hypometabolism occurs in the cerebral cortex in AD (Friedland et al., 1985; Koss et al., 1985), correlates with symptom severity, and precedes disease onset in at-risk individuals, including those with MCI or carrying mutations predisposing them to AD (Mosconi et al., 2008, 2014). In MCI, a lower metabolic rate of glucose utilization, particularly in the parietal cortex, posterior cingulate and thalamus has been associated with an energy substrate switch from glucose to ketones [e.g., acetoacetate and β-hydroxybutyrate (βHB)] in the brain (Castellano et al., 2015). Thus, a combination of abnormal cerebral hemodynamics, a dysfunctional BBB and glucose hypometabolism can all drive aberrant Aβ accumulation in the brain. Chronic inflammation may be an upstream driver of these metabolic changes.

4.4 Brain insulin resistance in Alzheimer's disease leads to aberrant protein accumulation

Hyperglycemia and hyperinsulinemia are a consequence of peripheral insulin resistance, and both are associated with AD (Macauley et al., 2015; Stanley et al., 2016a; Willette et al., 2015b). Most brain insulin appears to be made peripherally (Banks et al., 2012) and crosses into the brain via the cerebral microvasculature/BBB at the endothelium mediated by an insulin-receptor-specific, vesicle-mediated transport process in brain endothelial cells at physiological concentrations of insulin (Gray et al., 2017; Meijer et al., 2016). Insulin may also be produced locally in the brain (Molnár et al., 2014), although this has been challenged (Arnold et al., 2018).

Peripheral hyperinsulinemia increases plasma and interstitial fluid (ISF) levels of Aβ in APP/PS1 AD mice with no corresponding increase in brain, ISF or CSF insulin or brain insulin signaling (Stanley et al., 2016b). Similarly, peripheral hyperglycemia increases ISF Aβ levels in young and old APP/PS1 AD mice (Macauley et al., 2015). Insulin-degrading enzyme is responsible for insulin degradation, as well as other targets like Aβ-peptide, and can regulate proteasomal degradation (Farris et al., 2003). Its ability to degrade Aβ, however, is impaired by hyperinsulinemia (Selkoe, 2002) or OxS-induced damage (de Dios et al., 2019). Thus, any peripheral metabolic disease, such as T2DM that is associated with hyperinsulinemia and hyperglycemia or BBB dysfunction, may increase Aβ influx into the brain and impair its breakdown and/or elimination from the brain.

Brain insulin resistance occurs in AD (de la Monte and Tong, 2014; Steen et al., 2005) and is associated with significant regional reductions in cerebral glucose metabolism (Willette et al., 2015a). Together with increased glycogen synthase kinase-3β activity and gene expression of amyloid precursor protein in AD brain (Steen et al., 2005), subsequent aberrant accumulation of Aβ in the brain may trigger or exacerbate AD pathology.

4.5 Neuroinflammation and Alzheimer's disease: The NLRP3 inflammasome

4.5.1 *Aberrant Aβ accumulation and neuroinflammation in Alzheimer's disease*

Neuroinflammation is primarily driven by accumulation of damaged neurons, neurofibrillary tangles, and insoluble Aβ deposits in hippocampal neurons from AD onset to its terminal stage (Iadecola, 2004) and post-mortem studies show that excess Aβ expression in the plaques of AD brain is a main pathological feature of AD (Reale et al., 2012).

Chronic inflammation induces IR and T2DM in the periphery, and the link between AD, T2DM and neuroinflammation has been demonstrated in APP $\pm$ ob/ob mice (transgenic APP23 AD mice crossed with diabetic ob/ob mice). Cognitive impairment in these mice was akin to that of AD with evidence of cerebrovascular inflammation and cerebral amyloid angiopathy. Increased levels of the receptor for AGE products (RAGE) and IL-6 in the cerebral vasculature preceded the appearance of cerebral amyloid angiopathy, which is associated with small vessel disease and contributes to vascular dysfunction in AD (Takeda et al., 2011). Thus, chronic inflammation appears to precede Aβ deposition in cerebral vasculature, which together with changes in brain insulin signaling may be critical upstream processes linked to cognitive impairment.

Importantly, Tejera et al. (2019) showed that in a transgenic murine AD model (APP/PS1), chronic inflammation induced by LPS exacerbated Aβ deposition in the brain by reducing microglial uptake and clearance of Aβ. Knockout of the NLRP3 inflammasome blocked this effect, suggesting that the inflammasome mediates Aβ deposition in AD brain. Similarly, the

NLRP3 inflammasome inhibitor JC-124 inhibited the increased formation and activation of the NLRP3 inflammasome compared to non-transgenic controls, and decreased brain Aβ deposition accompanied by reduced activation of microglia in CRND8 APP transgenic mice (TgCRND8T) (Yin et al., 2018).

4.5.2 The NLRP3 inflammasome is triggered by Aβ and contributes to neuroinflammation

The NLRP3 inflammasome is closely associated with AD, as are genetic alterations in the *NLRP3* gene. Han Chinese individuals with a genetic variation in *NLRP3*, particularly the 5′-flanking rs2027432 polymorphism are at increased risk of late-onset AD (Tan et al., 2013) and single nucleotide variations in the *NLRP1* gene in AD brain have been associated with AD risk (Pontillo et al., 2012). In human AD, intraneuronal inflammasome activation appears to be linked to caspase 1-generated neuroinflammation mediated by IL-1β and caspase 6-mediated axonal degeneration (Kaushal et al., 2015). Saresella et al. (2016) were the first to show to that the NLRP3 inflammasome is activated in Aβ-stimulated peripheral immune cells of individuals with AD. Up-regulated mRNA expression of inflammasome components (NLRP1, NLRP3, PYCARD, caspase 1, 5 and 8), as well as IL-1β and IL-18 were found in AD monocytes, suggesting that at least two different inflammasome complexes are activated in AD-associated inflammation.

Cytokine release and neurotoxic events occurring downstream of Aβ accumulation are likely to be closely associated in AD since Aβ-induced astrocyte activation can increase release of inflammatory mediators (e.g., IL-1β, IL-6, TNF-α) from astrocytes causing neurotoxicity (Forloni et al., 1997; Garwood et al., 2011). Microglia, however, are the primary glial cells containing NLRP3 receptors in the brain and release IL-1β in response to NLRP3 inflammasome formation and activation by DAMPs, such as Aβ peptides and adenosine triphosphate (ATP) (Gustin et al., 2015; Wang et al., 2015). Amyloid-β monomers, oligomers and fibrils can all interact with NLRP3, but only oligomerized Aβ directly interacts with NLRP3 to stimulate inflammasome activation and IL-1β release (Halle et al., 2008; Nakanishi et al., 2018). The Aβ oligomers cause ROS-mediated lysosomal damage to induce the NLRP3 inflammasome in microglia in AD (Parajuli et al., 2013). Moreover, as Aβ deposits accumulate in AD brain, they progressively impair the ability of microglial cells to phagocytose Aβ, which exacerbates the neuro-inflammatory response and leads to neurodegeneration (Sarlus and Heneka, 2017).

Ising et al. (2019) have shown that tau can also activate the NLRP3 inflammasome, while intracerebral injection of fibrillar Aβ-containing brain homogenates induced tau pathology in an NLRP3-dependent manner, suggesting that neurofibrillary tangles develop downstream of Aβ-induced microglial activation. Importantly, the NLRP3 inflammasome stimulates an inflammatory response to mitochondrial stress, as seen in Aβ APP/PS1 mice exposed to paraquat, where its activation by mitochondrial ROS is associated with inflammation and exacerbated cognition impairment (Chen et al., 2015).

4.6 Chronic inflammation and neuroinflammation in Alzheimer's disease: The NLRP3 inflammasome as a major contributor

Increasing evidence suggests that neuro-inflammation may mediate the risk relationship between T2DM and AD with the NLRP3 inflammasome as a key player. The oxidative and inflammatory consequences of T2DM on systemic vasculature are such that BBB integrity is disrupted and T2DM-associated DAMPs and/or PAMPs can more easily access the brain and trigger microglia to release pro-inflammatory molecules that drive and/or contribute to neuroinflammation and neurodegenerative pathology. This is at least partly associated with the NLRP3 inflammasome. A dysfunctional BBB would also enable components of both the innate and adaptive immune system to enter the brain and trigger microglia, and also potentially contribute to mitochondrial dysfunction (Parodi-Rullán et al. 2019). Moreover, Saresella et al. (2016) have suggested that peripheral monocytes migrate across the BBB and contribute to AD-associated neuroinflammation.

Endothelial dysfunction is caused by chronic inflammation associated with T2DM, which may drive cerebral hypoperfusion, reduce brain oxygenation and contribute to BBB dysfunction. A dysfunctional BBB may, therefore, also disrupt cerebral metabolic function by impairing glucose uptake and insulin transport into the brain. Such cerebral hypometabolism and brain insulin resistance might occur decades before visible cognitive impairment.

Chronic inflammation occurs in T2DM long before symptoms appear and is also associated with AD. It is interesting to speculate that peripheral metabolic dysfunction may be an early trigger of BBB dysfunction in AD. It might, therefore, contribute to early pathological processes like brain metabolic dysfunction, Aβ and tau accumulation, and neuroinflammation mediated by the NLRP3 inflammasome. Hokama et al. (2014), however, have suggested that AD pathology *per se* impairs brain insulin signaling to increase the brain's vulnerability to metabolic or inflammatory insults that are exacerbated by peripheral insulin resistance or T2DM. Since the NLRP3 inflammasome causes insulin resistance and T2DM development in the periphery, it might also do so in the brain, contributing to AD pathology.

Chronic, low-grade, peripheral inflammation and the local OxS-inflammatory loop mediated by microglia is fuelled by a wide range of DAMPs that are both cause and consequence of pathology associated with MetS/T2DM. Many of these DAMPs are downstream of mitochondrial dysfunction, but they can also feedback to trigger or exacerbate mitochondrial dysfunction, such as Aβ, tau and α-synuclein (α-Syn) (Béraud et al., 2011; Halle et al., 2008; Martinon et al., 2002; Stancu et al., 2019). Aberrant accumulation of these proteins is associated with AD and PD. All of these factors either alone or in combination, could trigger NLRP3 inflammasome priming, formation and activation in NDDs. Moreover, the potential negative impact of the gut microbiome on BBB integrity requires further study, particularly as chronic inflammation and the microbiome are both associated with AD (Sochocka et al., 2019).

5. Neuro-inflammation in Parkinson's disease

5.1 Pathology of Parkinson's disease

Parkinson's disease is complex, progressive and influenced by genetic and environmental factors as yet of unknown etiology (Schapira and Jenner, 2011). Idiopathic PD is considered a movement disorder, characterized by bradykinesia, resting tremor, and muscles and joint rigidity. The severe loss of dopaminergic neurons in the substantia nigra and subsequent depletion of dopamine in the striatum are often associated with development of motor symptoms (Hornykiewicz, 2001; Obeso et al., 2008). These tend to be preceded and/or accompanied by non-motor symptoms, including autonomic dysfunction, gastrointestinal symptoms, sleep disorders, neuropsychiatric disturbances, and cognitive dysfunction and dementia, in the majority of individuals with PD (Aarsland et al., 2017; Chaudhuri et al., 2006).

The pathogenesis of PD is associated with OxS, impaired mitochondrial function, inflammation, apoptosis, loss of neurotrophic factors, and α-Syn toxicity (Hirsch et al., 2013; Toulorge et al., 2016). Such pathology is similar to that associated with aging, the major risk factor for PD (Jenner et al., 2013). α-Synuclein is found in neurons and glia of the substantia nigra, thalamus, neocortex and hippocampus (Braak and Del Tredici, 2009; Taguchi et al., 2016) and damages mitochondria, for example, by disrupting mitophagy and mitochondrial fusion (Mullin and Schapira, 2013). Parkinson's disease may even be a prion-like synucleinopathy associated with the propagation of misfolded α-Syn (Visanji et al., 2013), the pathology of which may originate in the enteric nervous system or brainstem and entorhinal cortex and ascend retrogradely into the brain (Kim et al., 2019).

5.2 Chronic inflammation is common to metabolic disorders and Parkinson's disease

Type 2 diabetes mellitus is a risk factor for PD (Cereda et al., 2011; De Pablo-Fernandez et al., 2018; Horvath and Wittung-Stafshede, 2016; Hu et al., 2007; Pagano et al., 2018) and metabolic dysfunction has been demonstrated in PD. Individuals with PD show a significant loss of insulin receptor mRNA levels and immunoreactivity in the substantia nigra pars compacta (Moroo et al., 1994; Takahashi et al., 1996). Wang et al. (2014) have demonstrated impaired insulin signaling not only in the pancreas and livers of preclinical T2DM models, but also in their midbrain. Expressions of monomeric and oligomeric α-Syn, as well as endoplasmic reticulum stress markers (CHOP and GRP78) were significantly elevated in both pancreas and midbrain of T2DM mice and accompanied by increased NLRP3 activation. Dopaminergic neurons in these mice were also more vulnerable to MPTP (1-methyl-4-phenyl-1,2,3,6-tetrahydropyridine, pro-drug to neurotoxic 1-methyl-4-phenylpyridinium, MPP^+), suggesting that metabolic inflammation exacerbates dopaminergic neurodegeneration in PD progression (Wang et al., 2014). The potential neuroprotective effects of glucagon-like peptide-1 (GLP-1) analogues in PD that may have anti-inflammatory properties (Athauda and Foltynie, 2016), also support an inflammatory link between T2DM and PD.

Chronic inflammation associated with T2DM may contribute to the overall chronic inflammatory burden in PD. As discussed for AD, T2DM-associated chronic inflammation and consequent BBB dysfunction may expose the PD brain not only to DAMP bi-products of T2DM, but systemic cytokines and PAMPs that trigger microglia to induce neuroinflammation and subsequent neurodegeneration. Indeed, chronic inflammation induced by systemic TNF-α and LPS administration can activate brain microglia to generate chronically elevated mRNA and protein levels of pro-inflammatory factors (e.g., TNF-α and IL-1β) associated with progressive dopaminergic neuron loss in the substantia nigra of adult, wild-type mice, but not in TNF-α receptor (TNFR1/R2) knockout mice. Thus, TNF-α associated with chronic inflammation is transferred to the brain via TNF-α receptors to induce self-driving neuroinflammation and neurodegeneration (Qin et al., 2007). As levels of IL-1β were also increased in the brain in this study, this would indicate that the neuroinflammation induced by TNF-α in the brain was mediated by the NLRP3 inflammasome.

The peripheral immune system may be associated with PD pathology and some model studies have linked this to α-Syn pathology. Chronic gut inflammation accelerates brain neuropathology and motor dysfunction in α-Syn mutant mice (Kishimoto et al., 2019). Intraperitoneal LPS administration prior to systemic administration of pathogenic strains of α-Syn increases levels of microglia in and leukocyte recruitment to the brain and spinal cord in wild-type mice (Peralta Ramos et al., 2019). Interestingly, radiolabelled α-Syn can move bi-directionally across the BBB (Sui et al., 2014) such that the BBB may be an important access point of peripheral α-Syn to the brain.

Evidence from human studies also links PD with chronic inflammation. α-Synuclein monomers increase IL-1β and IL-18 production in cultured blood mononuclear cells from individuals with PD, suggesting that α-Syn may be involved in NLRP3-inflammasome-related, peripheral cytokine production in PD (White et al., 2018). A recent cohort study has shown a significant association between inflammatory bowel disease (IBD) and later occurrence of PD with *LRRK2* as a common genetic link (Hui et al., 2018; Villumsen et al., 2019). An *NLRP3* polymorphism has been associated with IBD (Opipari and Franchi, 2015) and retrospective analyses indicate that anti-TNF-α therapy used to treat IBD also reduces the incidence of PD (Peter et al., 2018). Preclinical studies have shown that gut microbiota may regulate BBB integrity (Braniste et al., 2014), and a dysfunctional microbiota-gut-brain axis associated with a gut-derived, pro-inflammatory milieu contributes to PD-like pathology and neuroinflammation (Dodiya et al., 2020). Moreover, pre-existing neuroinflammation caused by a bacterial endotoxin and mediated by IL-1β can increase dopaminergic neuron susceptibility to degeneration in the 6-OHDA rat model of PD (Koprich et al., 2008).

Interestingly, a toxic protease (gingipain R1, RgpA) associated with the bacterium *Porphyromonas gingivalis* that causes chronic periodontitis has been detected systemically in a PD population (Adams et al., 2019). Both *Porphyromonas gingivalis* and gingipain antigens have also been found in AD brains at post mortem (Dominy et al., 2019), suggesting that systemic bacterial inflammagens may be associated with both NDDs by way of being PAMP/MAMPs.

Systemic, CSF and brain levels of IL-β1 and IL-6 are elevated in PD (Blum-Degen et al., 1995; Brodacki et al., 2008; Karpenko et al., 2018; Lindqvist et al., 2012; Mogi et al., 1996; Pereira et al., 2016; Veselý et al., 2018) linking the NLRP3 inflammasome to PD pathogenesis. A meta-analysis has also shown that peripheral concentrations of IL-1β and IL-6 are increased in PD (Qin et al., 2016). Thus, peripheral metabolic and inflammatory factors may impact PD, particularly since IL-6 levels are significantly increased in obesity-associated T2DM (Goyal et al., 2012). These pro-inflammatory factors can trigger the NLRP3 inflammasome, neuroinflammation and dopaminergic neurodegeneration or increase the vulnerability of dopaminergic neurons to degeneration. Interestingly, the peripherally administered metabolic toxin, rotenone, is also linked to NLRP3-independent neuroinflammation (Martinez et al., 2017). Since environmental toxins with peripheral access to the body have been implicated in PD, this suggests that systemic factors may lead to neuro-inflammatory damage in the brain mediated by the NLRP3 inflammasome.

5.3 Neuroinflammation, Parkinson's disease, and the NLRP3 inflammasome

5.3.1 α-Synuclein provokes microglial inflammatory responses in the brain

Peripheral α-Syn, which can move bi-directionally across the BBB, may influence neuroinflammation and PD. Grassi et al. (2018) have identified a non-fibrillar, phosphorylated α-Syn species (termed pα-Syn*) in human PD brain that induces mitochondrial toxicity and fission, energetic stress, and mitophagy. Notably, these processes would lead to the formation of DAMPs. In PD, erythrocytes contain levels of α-Syn about 1000-fold higher than the CSF and generate α-Syn-rich extracellular vesicles that cross the BBB, especially in an inflammatory milieu induced by peripheral LPS administration. The vesicles co-localize with microglia in the cerebral nuclei, cortex, midbrain and substantia nigra, and elicit microglial inflammatory responses (Matsumoto et al., 2017). α-Synuclein can also be internalized by peripheral, LPS-primed monocytes and subsequently taken up into the brain and spinal cord, suggesting that chronic inflammation-induced recruitment of peripheral leucocytes into the CNS may contribute to PD onset and/or progression (Peralta Ramos et al., 2019).

In the brain, α-Syn can activate microglia. Neuronal α-Syn can enter glial cells and induce secretion of pro-inflammatory cytokines [e.g., interleukin-1α (IL-α), IL-1β, and IL-6] and chemokines (e.g., platelet-derived chemokine CXCL) in direct proportion to its intracellular levels (Lee et al., 2010). Toll-like receptors, specifically TLR1/2 can be activated by misfolded α-Syn, which induce pro-inflammatory cytokine secretion by triggering NF-κB and MAPK pathways via MyD88 (Daniele et al., 2015). Extracellular α-Syn can trigger glial cell-induced neuroinflammation leading to neurodegeneration (Si et al., 2017) and the increased expression of striatal IL-1α, TNF-α and interferon-γ after bilateral infusion of human recombinant α-Syn monomers into mouse striatum indicates that α-Syn has induced microglial activation that contributes to dopaminergic neurotoxicity (Sznejder-Pachołek et al., 2017).

Intrastriatal injection of pre-formed α-Syn fibrils has been shown to cause widespread accumulation of phosphorylated α-Syn inclusions in brain areas that innervate the striatum and subsequent bilateral loss of dopaminergic neurons of the substantia nigra pars compacta in male Fischer rats within 5–6 months. Peak accumulation of α-Syn inclusions was closely associated with major histocompatibility complex class II (MHC-II)-immunoreactive microglia expression in the substantia nigra 3–4 months before neurodegeneration occurred, suggesting that reactive microglia may contribute to the vulnerability of nigral neurons (Duffy et al., 2018). Since microglial MHC-II expression is associated with co-expression of pro-inflammatory genes, including *IL1β*, neuroinflammation mediated by the NLRP3 inflammasome may have been involved, but this was not directly evaluated in this study.

5.3.2 The NLRP3 inflammasome is associated with Parkinson's disease

Elevated NLRP3 immunoreactivity has been demonstrated in mesencephalic neurons in a post-mortem PD study. Multiple *NLRP3* single-nucleotide polymorphisms were identified, including rs7525979 that was associated with a significantly reduced risk of developing PD since it caused the accumulation of an ubiquitinated, insoluble, and thus inactive NLRP3 form (von Herrmann et al., 2018). Importantly, this study also suggested that dopaminergic neurons are a cell-of-origin for NLRP3 inflammasome activity in PD implicating it in disease progression.

5.3.3 The NLRP3 inflammasome is triggered by α-Syn and contributes to neuroinflammation in Parkinson's disease

The activation of microglia by α-Syn is mediated by the NLRP3 inflammasome leading to IL-1β production, neuroinflammation and degeneration of dopaminergic neurons (Codolo et al., 2013; Zhou et al., 2016). *In vitro*, α-Syn can activate human umbilical cord blood-derived cultured mast cells to release cytokines IL-1β and TNF-α, as well as the chemokine interleukin-8 (Kempuraj et al., 2015), indicating that α-Syn in the brain could activate mast cells in inflammatory conditions *in vivo*. Activation of microglia by intranigral LPS administration causes α-Syn aggregation and loss of dopaminergic neurons in an interleukin-1-dependent manner with consequent, PD-like behavioral impairment in mice (Tanaka et al., 2013). Moreover, microglial NLRP3 inflammasome activation in the substantia nigra is important to dopaminergic neuronal loss and subsequent motor deficits in an MPTP mouse model of PD (Lee et al., 2019). *In vitro* studies have shown that MPTP/ATP treatment promotes NLRP3 inflammasome formation and activation mediated by mitochondrial ROS in primary microglia and mixed glial cell cultures. Moreover, NLRP3 inflammasome-active, microglia-induced neuronal death occurs in a microglia-neuron co-culture model (Lee et al., 2019).

Importantly, chemical inhibition of the NLRP3 inflammasome enzyme caspase-1 by a potent caspase-1 inhibitor VX765 or by silencing the caspase-1 gene with short hairpin RNA that decreased wild-type α-Syn truncation into its highly aggregation-prone form, improves the survival of neurons in a neuronal cell model of PD (Wang et al., 2016). Thus, activation of the NLRP3 inflammasome with known stimuli (e.g., OxS and LPS) induces α-Syn truncation to generate toxic species of α-Syn, which can exacerbate the neuro-inflammatory loop. The activation of TLR-4 has been shown to induce neuroinflammation associated with dopaminergic neurodegeneration mediated via OxS, activation of glia, NF-κB and the NLRP3 inflammasome pathways in both a murine MPTP model of PD (Campolo et al., 2019) and the chronic MPTP/probenecid murine model of PD (Shao et al., 2019). Since TLR4 recognizes LPS (and cytokines), this links PAMPs to neuro-inflammatory damage mediated by the NLRP3 inflammasome in PD.

6. The NLRP3 inflammasome: A target for novel therapeutic approaches in Alzheimer's and Parkinson's diseases?

Evidence clearly demonstrates pathological commonalities and metabolic disturbances in T2DM and NDDs that may be mediated by OxS and neuro-inflammation. This opens the field to novel strategies to treat NDDs.

6.1 Targeting mitochondrial dysfunction and ATP as DAMPs

The targeting of mitochondrial dysfunction, redox imbalance and OxS in PD has been reviewed elsewhere (Söderbom, 2020). Targeting mitochondrial dysfunction and mitophagy may also be a valid therapeutic approach to reduce the DAMP triggers for NLRP3-inflammasome-mediated neuro-inflammation in AD (Aman et al., 2020). As a DAMP, ATP can trigger the NLRP3 inflammasome. The effects of extracellular ATP are mediated by purinergic P2 plasma receptors, and the purinergic P2X7 receptor (P2X7R) is an ATP-gated ion channel that mediates NLRP3 inflammasome activation in response to LPS and ATP and subsequent cytokine and chemokine release (Bauernfeind et al., 2009; Di Virgilio et al., 2017; Franchi et al., 2007). The P2X7 receptor is also necessary for caspase-1 activation triggered by ATP in TNF-α-stimulated macrophages (Franchi et al., 2009). Thus, the role of the P2X7 receptor in neuroinflammation and pathophysiology of AD and PD is being evaluated in an observational study with individuals with newly-diagnosed, as yet untreated PD or AD (NCT03918616) and may provide a novel target to modulate NLRP3-mediated neuroinflammation in NDDs.

6.2 Targeting metabolic drivers of the NLRP3 inflammasome

Antidiabetic therapies, such as the GLP-1 analogues, demonstrated as beneficial in early-phase studies may impact OxS and neuroinflammation in AD and PD (Athauda and Foltynie, 2016; Green et al., 2019; Hölscher, 2014). In AD, intranasal insulin administration has been evaluated in individuals with MCI and early AD and found to improve memory and cognitive performance (Claxton et al., 2015; Craft et al., 2012). This may be mediated by the phosphatidylinositol-3-kinase/protein kinase B/mechanistic target of rapamycin (PI3K/AKT/mTOR) signaling pathway through which insulin increases neuronal growth and synapse formation (Gabbouj et al., 2019; Manning and Toker, 2017). Peroxisome proliferator-activated receptor-delta/gamma agonists are also being evaluated, such as T3D-959, which is being assessed for anti-inflammatory effects in preclinical AD (de la Monte et al., 2017; Tong et al., 2016). As NDDs have complex etiologies involving aberrant proteins as both cause and consequence of OxS and neuroinflammation and involving NLRP3 inflammasome activation, an approach combining immunotherapies with insulin-sensitizing therapies may be valuable (Takamatsu et al., 2017).

Preclinical studies with other antidiabetic agents indicated for T2DM have shown anti-inflammatory effects mediated by the NLRP3 inflammasome. In high-fat diet-fed rats which develop obesity-driven insulin resistance and cognitive decline, the SGLT-2 inhibitor dapagliflozin and the dipeptidyl peptidase-4 (DPP-4) inhibitor vildagliptin equally improved mitochondrial function, insulin signaling, apoptosis and prevented cognitive decline. Only dapagliflozin also improved hippocampal synaptic plasticity. In combination, the efficacy of these drugs in improving both insulin sensitivity and reducing OxS in the brain was increased (Sa-Nguanmoo et al., 2017). Based on the inter-related pathological pathways associated with mitochondrial dysfunction, OxS and inflammation, this effect may involve the inflammasome since dapagliflozin reduces NLRP3 inflammasome activation associated with diabetic cardiomyopathy (Ye et al., 2017) and nephropathy (Birnbaum et al., 2018) in experimental models. Saxagliptin, another DPP-4 inhibitor also inhibits NLRP3 inflammasome activation, diabetic nephropathy progression, and reduces inflammatory cytokine levels (e.g., IL-6 and IL-1β) (Birnbaum et al., 2016) in experimental models. As mentioned, empagliflozin also positively impacts vascular dysfunction and cognitive impairment in T2DM mice associated with ameliorating OxS and inflammation (Lin et al., 2014). As with the GLP1 analogues, and assuming they can cross the BBB, these agents may also translate to NDDs based on the common metabolic and inflammatory pathologies shared by T2DM and NDDs.

Interestingly, prolonged fasting in mice reduces NLRP3 inflammasome activation by NAD-dependent deacetylase and metabolic sensor sirtuin-3 (SIRT3)-mediated activation of mitochondrial, manganese superoxide dismutase (Traba et al., 2017). This would support a lifestyle approach to mitigating inflammation in T2DM and NDDs. The NLRP3 inflammasome is involved in the pathogenesis of ischemic stroke. Mice fed a ketogenic diet prior to middle cerebral artery occlusion/reperfusion-injury showed increased tolerance to injury, potentially associated with its suppression of endoplasmic reticular stress and protecting the mitochondrial integrity. Both of these factors inhibited NLRP3 inflammasome activation (Guo et al., 2018). A ketogenic diet, as well as starvation, caloric restriction and high-intensity exercise, elevates levels of βHB. Youm et al. (2015) have shown that βHB suppresses NLRP3 inflammasome activation and associated IL-1β and IL-18 production in response to stimuli, including lipotoxic fatty acids and ATP. β-Hydroxybutyrate also reduces ROS generation to improve mitochondrial respiration and inhibits NFκB activation and the NLRP3 inflammasome, all of which may be valuable in AD (Pinto et al., 2018) and PD (Norwitz et al., 2019).

6.3 Targeting pro-inflammatory cytokines

Cross-talk between cytokines occurs in the complex inflammatory cascades in T2DM, AD and PD. These diseases are also associated with increased systemic levels of TNF-α and IL-6 and as these cytokines can prime the formation and activation of the NLRP3 inflammasome via NF-κB (Groslambert and Py, 2018), targeting them may indirectly inhibit formation and activation of the NLRP3 inflammasome in NDDs.

Interleukin-6 employs two signaling pathways; IL-6 trans-signaling is pro-inflammatory, while the classic pathway is associated with protective anti-inflammatory responses. Trans-signaling appears to be a dominant pathway associated with IL-6 pathology in the brain and Rose-John (2017) has led the development of sgp130Fc, a protein that selectively inhibits IL-6 trans-signaling and may represent a novel therapeutic strategy for NDDs associated with neuroinflammation (Campbell et al., 2014).

Interleukin-6 release can be induced by soluble TNF-α, which has also been associated with AD. Inhibiting soluble TNF signaling in models of AD prevents AD-like pathology induced by chronic inflammation (McAlpine et al., 2009). Tumor necrosis factor-α is also a priming trigger for the NLRP3 inflammasome, formation and activation of which is subsequently induced by LPS and ATP (Franchi et al., 2009). Thus, inhibition of TNF-α production or signaling may prevent its downstream pro-inflammatory effects on glial cells and NLRP3 inflammasome activation. A novel, second-generation inhibitor of TNF is XPro1595 that selectively neutralizes soluble TNF and inhibits TNF1 signaling, reduces microglial activation in aged rats (Sama et al., 2012); prevents synaptic deficits in AD mice (Cavanagh et al., 2016); and reduces age-dependent increases in brain immune cell infiltration, neuroinflammation, and prevents/delays neuronal dysfunction in AD mice (MacPherson et al., 2017). This inhibitor is now in Phase 1b clinical trials in AD investigating biomarkers of inflammation (NCT03943264). Targeting pro-inflammatory cytokines associated with BBB dysfunction and neuroinflammation may, therefore, have beneficial effects in NDDs affected by NLRP3 inflammasome-mediated neuroinflammation.

As regards IL-1β, Kitazawa et al. (2011) were the first to show that inhibiting IL-1β signaling with an anti-IL-1 receptor blocking antibody in an AD mouse model reduced AD pathology. Thus, antibodies targeting IL-1β or IL-18, the pro-inflammatory effectors of the NLRP3 inflammasome, may provide novel disease modifying opportunities in AD and PD. The ability of MCC950, a small molecule inhibitor of NLRP3 inflammasome activation, to ameliorate fibrillar α-Syn-mediated inflammasome activation and extracellular ASC adaptor protein release is under investigation in preclinical PD studies (Gordon et al., 2018). Other pharmacological NLRP3 inflammasome inhibitors are reviewed by Zahid et al. (2019).

Other agents that can positively impact neuroinflammation and so offer a novel disease-modifying benefit in AD and PD include minocycline and dihydromyricetin. Minocycline is a tetracycline antibiotic that reduces LPS-induced cognitive impairments in mice, potentially via its ability to suppress neuroinflammation associated with glia activation and normalize brain-derived neurotrophic factor expression (Hou et al., 2016). Dihydromyricetin is a plant flavonoid compound. Its intraperitoneal injection into APP/PS1 transgenic mice is associated with reduced memory and cognition deficits, as well as the number of activated microglia in the hippocampus and cortex. Notably, dihydromyricetin also reduces NLRP3 inflammasome activation and expression of its components, and enhances microglial clearance of Aβ and its aggregates. Thus, the ability of dihydromyricetin to prevent AD pathology by inhibiting NLRP3 inflammasome-based, microglia-mediated neuroinflammation warrants further investigation (Feng et al., 2018).

6.4 Targeting DAMPs like amyloid-β, tau and α-synuclein

Many novel immunotherapies and other approaches designed to reduce aberrant accumulation or increase clearance of α-Syn are being evaluated in PD (Brundin et al., 2017; Mullin et al., 2020). This would reduce the neuroinflammation triggers in the brain that are linked to the NLRP3 inflammasome and consequent neurodegeneration. Notably, similar approaches based on the amyloid hypothesis in AD have not achieved successful clinical outcomes (Yiannopoulou et al., 2019).

7. Conclusion

7.1 Type 2 diabetes is a risk factor for neurodegenerative diseases

Evidence suggests that T2DM is a risk factor associated with AD and PD and this is supported by the beneficial impact of antidiabetic therapies that are under evaluation in AD and PD.

7.2 Metabolic and neurodegenerative diseases share common risk factors

Type 2 diabetes mellitus has a complex and multifactorial etiology, as do AD and PD. Aging is the major risk factor for AD (Xia et al., 2018) and PD (Jenner et al., 2013), and a major risk factor for T2DM (Palmer et al., 2019). The classical hallmarks of aging (e.g., mitochondrial dysfunction and redox imbalance, increased OxS and damage, impaired lysosome and proteasome function, systemic metabolic dysfunction, and chronic inflammation and neuroinflammation) (Mattson and Arumugam, 2018) reflect T2DM-, AD- and PD-associated pathology. Cellular senescence associated with ROS and pro-inflammatory cytokines, such as IL-1β, also contributes to age-associated inflammation (Palmer et al., 2019). These factors may exacerbate chronic inflammation associated with these diseases. In macrophages, NLRP3 can be acetylated to enable NLRP3 inflammasome activation and de-acetylated by sirtuin-2 (SIRT2) to reduce aging-associated inflammation and insulin resistance. Thus, dysregulation of NLRP3 inflammasome acetylation may trigger age-associated chronic inflammation and insulin resistance (He et al., 2020). This would position the NLRP3 inflammasome at the center of T2DM, AD, and PD pathogenesis.

7.3 Metabolic and neurodegenerative diseases share common pathology

Oxidative stress and neuro-inflammation are both cause and consequence of one another, and mutually exacerbate disease pathology progression. Complex interactions between factors, such as ROS, RNS, AGE products, lipid-induced OxS and chronic inflammation and local/neuroinflammation occur in T2DM (Hanley et al., 2004; Miyazaki et al., 2007), AD (Holmes et al., 2009; Rojas-Gutierrez et al., 2017) and PD (Chiang et al., 2017; de Farias et al., 2017). In T2DM, insulin resistance tends to precede OxS and inflammation, while OxS and neuroinflammation often precede and contribute to cell/neuronal death in AD and PD.

An upstream commonality is mitochondrial dysfunction. This can occur early in T2DM (Fex et al., 2018), AD (de la Monte and Wands, 2006; Martín-Maestro et al., 2017; Weidling and Swerdlow, 2019) and PD (Toulorge et al., 2016) and progresses with disease severity. The NLRP3 inflammasome may be an important common link in these pathogenic processes that can be exacerbated by cellular senescence and age-associated inflammation, particularly as neuro-inflammation occurs early in disease. Moreover, the Aβ peptide and IAPP deposits found in post-mortem AD brain and pancreatic islets cells in T2DM, respectively, indicate that amyloid formation is a pathological marker common to AD and T2DM (Dobson, 2001; Rasool et al., 2014), which is important as they are DAMPs for the NLRP3 inflammasome.

Insulin resistance is also common to these diseases and is important as it may lead to neuro-inflammation via OxS. Chronic inflammation can also lead to insulin resistance and T2DM so neuro-inflammation may contribute to brain insulin resistance. Alzheimer's disease is associated with brain insulin deficiency and insulin resistance, so-called "type-3 diabetes" (de la Monte and Tong, 2014), and there is also evidence of insulin resistance in PD that may contribute to neuroinflammation (Athauda and Foltynie, 2016). Ongoing investigations on the repurposing of antidiabetic therapies for AD and PD support this link to insulin resistance (Athauda et al., 2017; Claxton et al., 2015; Hölscher, 2014).

Overall, this highlights the link between T2DM, AD and PD with neuro-inflammation contributing to the complex, mechanistically intertwined, pathological processes underlying these diseases.

7.4 What does this mean for Alzheimer's and Parkinson's diseases?

Current clinical therapies for AD and PD are limited to symptomatic strategies. Neither AD nor PD can be cured, reversed, or halted with current therapies. Targeting hyperglycemia has been relatively successful in T2DM and repurposing hyperglycemic drugs, such as the GLP-1 receptor analogues in AD and PD, may provide disease-modifying, therapeutic breakthroughs crucial to these NDDs. While their neuroprotective/disease-modifying actions have not yet been conclusively demonstrated in the clinic, they may be partly related to anti-inflammatory mechanisms. This is important as neuro-inflammation is an early step in T2DM, AD and PD pathology, and T2DM is a risk factor for AD and PD.

The suggestion that NDDs may also be affected by chronic inflammatory conditions associated with metabolic dysfunction and/or pathogens requires further investigation. This is important because chronic inflammation contributes to BBB dysfunction, which allows systemic cytokines and DAMPs/PAMPS to enter the brain and trigger/exacerbate neurodegeneration. Genetic drivers predispose some individuals to central mitochondrial/metabolic stress and NDDs. Aging and age-related inflammation may exacerbate NDD pathology, while peripheral and central metabolic dysfunction may trigger these vulnerabilities. Cumulative damage over a lifetime might culminate in neurodegeneration as AD or PD. Metabolic dysfunction may, however, be more important than age as a pathological trigger since young individuals with T2DM who were neither demented nor aged had similar brain structural changes as seen in AD (Arnold et al., 2018).

More than one causative factor is likely to be involved in these complex, multifactorial diseases. Targeting neuroinflammation may contribute to ameliorating neurodegeneration in AD and PD. Specific targeting of pathological, pro-inflammatory signals is crucial, however, since some cytokines (e.g., IL-18 and IL-6) can induce both pro-inflammatory and anti-inflammatory responses. Neuroinflammation in AD and PD also involves glia, particularly the microglia, which express NLRP3 and release IL1-β and IL-18 in response to formation and activation of the NLRP3 inflammasome. The complex role of glia in neuroinflammation and neurodegeneration should be investigated further.

While chronic inflammation and/or neuroinflammation may bridge T2DM and AD/PD pathology, it remains to be seen if targeting these pathogenic processes and the NLRP3 inflammasome will provide the disease-modifying breakthroughs needed in AD and PD. It is increasingly clear, however, that it may be important to view AD and PD as systemic diseases, which may provide the necessary therapeutic breakthrough.

References

Aarsland, D., Creese, B., Politis, M., Chaudhuri, K. R., Ffytche, D. H., Weintraub, D., et al. (2017). Cognitive decline in Parkinson disease. *Nature Reviews. Neurology, 13*(4), 217–231.

Abais, J. M., Xia, M., Zhang, Y., Boini, K. M., & Li, P. L. (2015). Redox regulation of NLRP3 inflammasomes: ROS as trigger or effector? *Antioxidants & Redox Signaling, 22*(13), 1111–1129.

Abderrazak, A., Syrovets, T., Couchie, D., El Hadri, K., Friguet, B., Simmet, T., et al. (2015). NLRP3 inflammasome: From a danger signal sensor to a regulatory node of oxidative stress and inflammatory diseases. *Redox Biology, 4*, 296–307.

Adams, B., Nunes, J. M., Page, M. J., Roberts, T., Carr, J., Nell, T. A., et al. (2019). Parkinson's disease: A systemic inflammatory disease accompanied by bacterial inflammagens. *Frontiers in Aging Neuroscience, 11*, 210.

Albert, M. S., DeKosky, S. T., Dickson, D., Dubois, B., Feldman, H. H., Fox, N. C., et al. (2011). The diagnosis of mild cognitive impairment due to Alzheimer's disease: Recommendations from the National Institute on Aging-Alzheimer's Association workgroups on diagnostic guidelines for Alzheimer's disease. *Alzheimer's & Dementia, 7*(3), 270–279.

Alonso, A., Zaidi, T., Novak, M., Grundke-Iqbal, I., & Iqbal, K. (2001). Hyperphosphorylation induces self-assembly of tau into tangles of paired helical filaments/straight filaments. *Proceedings of the National Academy of Sciences of the United States of America, 98*(12), 6923–6928.

Aman, Y., Ryan, B., Torsetnes, D., Knapskog, A. B., Watne, L. O., McEwan, W. A., et al. (2020). Enhancing mitophagy as a therapeutic approach for neurodegenerative disorders. *International Review of Neurobiology, 155*.

Arnold, S. E., Arvanitakis, Z., Macauley-Rambach, S. L., Koenig, A. M., Wang, H. Y., et al. (2018). Brain insulin resistance in type 2 diabetes and Alzheimer disease: Concepts and conundrums. *Nature Reviews. Neurology*, *14*(3), 168–181.

Athauda, D., & Foltynie, T. (2016). The glucagon-like peptide 1 (GLP) receptor as a therapeutic target in Parkinson's disease: Mechanisms of action. *Drug Discovery Today*, *21*(5), 802–818.

Athauda, D., Maclagan, K., Skene, S. S., Bajwa-Joseph, M., Letchford, D., Chowdhury, K., et al. (2017). Exenatide once weekly versus placebo in Parkinson's disease: A randomised, double-blind, placebo-controlled trial. *Lancet*, *390*(10103), 1664–1675.

Banks, W. A., Kastin, A. J., & Gutierrez, E. G. (1994). Penetration of interleukin-6 across the murine blood-brain barrier. *Neuroscience Letters*, *179*(1–2), 53–56.

Banks, W. A., Owen, J. B., & Erickson, M. A. (2012). Insulin in the brain: There and back again. *Pharmacology & Therapeutics*, *136*(1), 82–93.

Bauernfeind, F. G., Horvath, G., Stutz, A., Alnemri, E. S., MacDonald, K., Speert, D., et al. (2009). Cutting edge: NF-kappaB activating pattern recognition and cytokine receptors license NLRP3 inflammasome activation by regulating NLRP3 expression. *Journal of Immunology*, *183*(2), 787–791.

Befroy, D. E., Petersen, K. F., Dufour, S., Mason, G. F., de Graaf, R. A., Rothman, D. L., et al. (2007). Impaired mitochondrial substrate oxidation in muscle of insulin-resistant offspring of type 2 diabetic patients. *Diabetes*, *56*(5), 1376–1381.

Bell, R. D., Winkler, E. A., Singh, I., Sagare, A. P., Deane, R., Wu, Z., et al. (2012). Apolipoprotein E controls cerebrovascular integrity via cyclophilin A. *Nature*, *485*(7399), 512–516.

Béraud, D., Twomey, M., Bloom, B., Mittereder, A., Ton, V., Neitzke, K., et al. (2011). α-Synuclein alters toll-like receptor expression. *Frontiers in Neuroscience*, *5*, 80.

Bian, F., Yang, X. Y., Xu, G., Zheng, T., & Jin, S. (2019). CRP-induced NLRP3 inflammasome activation increases LDL transcytosis across endothelial cells. *Frontiers in Pharmacology*, *10*, 40.

Birdsill, A. C., Carlsson, C. M., Willette, A. A., Okonkwo, O. C., Johnson, S. C., Xu, G., et al. (2013). Low cerebral blood flow is associated with lower memory function in metabolic syndrome. *Obesity (Silver Spring)*, *21*(7), 1313–1320.

Birnbaum, Y., Bajaj, M., Qian, J., & Ye, Y. (2016). Dipeptidyl peptidase-4 inhibition by Saxagliptin prevents inflammation and renal injury by targeting the Nlrp3/ASC inflammasome. *BMJ Open Diabetes Research & Care*, *4*(1), e000227.

Birnbaum, Y., Bajaj, M., Yang, H. C., & Ye, Y. (2018). Combined SGLT2 and DPP4 inhibition reduces the activation of the Nlrp3/ASC inflammasome and attenuates the development of diabetic nephropathy in mice with type 2 diabetes. *Cardiovascular Drugs and Therapy*, *32*(2), 135–145.

Blennow, K., Wallin, A., Fredman, P., Karlsson, I., Gottfries, C. G., & Svennerholm, L. (1990). Blood-brain barrier disturbance in patients with Alzheimer's disease is related to vascular factors. *Acta Neurologica Scandinavica*, *81*(4), 323–326.

Blum-Degen, D., Müller, T., Kuhn, W., Gerlach, M., Przuntek, H., & Riederer, P. (1995). Interleukin-1 beta and interleukin-6 are elevated in the cerebrospinal fluid of Alzheimer's and de novo Parkinson's disease patients. *Neuroscience Letters*, *202*(1–2), 17–20.

Bozluolcay, M., Andican, G., Fırtına, S., Erkol, G., & Konukoglu, D. (2016). Inflammatory hypothesis as a link between Alzheimer's disease and diabetes mellitus. *Geriatrics & Gerontology International*, *16*(10), 1161–1166.

Braak, H., & Del Tredici, K. (2009). Neuroanatomy and pathology of sporadic Parkinson's disease. *Advances in Anatomy, Embryology, and Cell Biology*, *201*, 1–119.

Braniste, V., Al-Asmakh, M., Kowal, C., Anuar, F., Abbaspour, A., Tóth, M., et al. (2014). The gut microbiota influences blood-brain barrier permeability in mice. *Science Translational Medicine*, *6*(263), 263ra158.

Brodacki, B., Staszewski, J., Toczyłowska, B., Kozłowska, E., Drela, N., Chalimoniuk, M., et al. (2008). Serum interleukin (IL-2, IL-10, IL-6, IL-4), TNFalpha, and INFgamma concentrations are elevated in patients with atypical and idiopathic parkinsonism. *Neuroscience Letters*, *441*(2), 158–162.

Brosseron, F., Krauthausen, M., Kummer, M., & Heneka, M. T. (2014). Body fluid cytokine levels in mild cognitive impairment and Alzheimer's disease: A comparative overview. *Molecular Neurobiology*, *50*(2), 534–544.

Brown, G. C., & Neher, J. J. (2010). Inflammatory neurodegeneration and mechanisms of microglial killing of neurons. *Molecular Neurobiology*, *41*(2–3), 242–247.

Brown, J., Wang, H., Hajishengallis, G. N., & Martin, M. (2011). TLR-signaling networks: An integration of adaptor molecules, kinases, and cross-talk. *Journal of Dental Research*, *90*(4), 417–427.

Brundin, P., Dave, K. D., & Kordower, J. H. (2017). Therapeutic approaches to target alpha-synuclein pathology. *Experimental Neurology*, *298*(Pt. B), 225–235.

Campbell, I. L., Erta, M., Lim, S. L., Frausto, R., May, U., Rose-John, S., et al. (2014). Trans-signaling is a dominant mechanism for the pathogenic actions of interleukin-6 in the brain. *The Journal of Neuroscience*, *34*(7), 2503–2513.

Campolo, M., Paterniti, I., Siracusa, R., Filippone, A., Esposito, E., & Cuzzocrea, S. (2019). TLR4 absence reduces neuroinflammation and inflammasome activation in Parkinson's diseases in vivo model. *Brain, Behavior, and Immunity*, *76*, 236–247.

Castellano, C. A., Nugent, S., Paquet, N., Tremblay, S., Bocti, C., Lacombe, G., et al. (2015). Lower brain 18F-fluorodeoxyglucose uptake but normal ^{11}C-acetoacetate metabolism in mild Alzheimer's disease dementia. *Journal of Alzheimer's Disease*, *43*(4), 1343–1353.

Cavanagh, C., Tse, Y. C., Nguyen, H. B., Krantic, S., Breitner, J. C., Quirion, R., et al. (2016). Inhibiting tumor necrosis factor-α before amyloidosis prevents synaptic deficits in an Alzheimer's disease model. *Neurobiology of Aging*, *47*, 41–49.

Cenini, G., Lloret, A., & Cascella, R. (2019). Oxidative stress in neurodegenerative diseases: From a mitochondrial point of view. *Oxidative Medicine and Cellular Longevity*, *2019*, 2105607.

Cereda, E., Barichella, M., Pedrolli, C., Klersy, C., Cassani, E., Caccialanza, R., et al. (2011). Diabetes and risk of Parkinson's disease: A systematic review and meta-analysis. *Diabetes Care*, *34*(12), 2614–2623.

Chalbot, S., Zetterberg, H., Blennow, K., Fladby, T., Andreasen, N., Grundke-Iqbal, I., et al. (2011). Blood-cerebrospinal fluid barrier permeability in Alzheimer's disease. *Journal of Alzheimer's Disease*, *25*(3), 505–515.

Chaudhuri, K. R., Healy, D. G., & Schapira, A. H. (2006). National Institute for Clinical Excellence. Non-motor symptoms of Parkinson's disease: Diagnosis and management. *Lancet Neurology*, *5*(3), 235–245.

Chen, L., Na, R., Boldt, E., & Ran, Q. (2015). NLRP3 inflammasome activation by mitochondrial reactive oxygen species plays a key role in long-term cognitive impairment induced by paraquat exposure. *Neurobiology of Aging*, *36*(9), 2533–2543.

Cheng, Z., & Almeida, F. A. (2014). Mitochondrial alteration in type 2 diabetes and obesity: An epigenetic link. *Cell Cycle*, *13*(6), 890–897.

Chiang, P. L., Chen, H. L., Lu, C. H., Chen, P. C., Chen, M. H., Yang, I. H., et al. (2017). White matter damage and systemic inflammation in Parkinson's disease. *BMC Neuroscience*, *18*(1), 48.

Cioffi, F., Adam, R. H. I., & Broersen, K. (2019). Molecular mechanisms and genetics of oxidative stress in Alzheimer's disease. *Journal of Alzheimer's Disease*, *72*(4), 981–1017.

Claxton, A., Baker, L. D., Hanson, A., Trittschuh, E. H., Cholerton, B., Morgan, A., et al. (2015). Long-acting intranasal insulin detemir improves cognition for adults with mild cognitive impairment or early-stage Alzheimer's disease dementia. *Journal of Alzheimer's Disease*, *44*(3), 897–906.

Codolo, G., Plotegher, N., Pozzobon, T., Brucale, M., Tessari, I., Bubacco, L., et al. (2013). Triggering of inflammasome by aggregated α-synuclein, an inflammatory response in synucleinopathies. *PLoS One*, *8*(1)e55375.

Craft, S., Baker, L. D., Montine, T. J., Minoshima, S., Watson, G. S., Claxton, A., et al. (2012). Intranasal insulin therapy for Alzheimer disease and amnestic mild cognitive impairment: A pilot clinical trial. *Archives of Neurology*, *69*(1), 29–38.

Cunningham, C., Wilcockson, D. C., Campion, S., Lunnon, K., & Perry, V. H. (2005). Central and systemic endotoxin challenges exacerbate the local inflammatory response and increase neuronal death during chronic neurodegeneration. *The Journal of Neuroscience*, *25*(40), 9275–9284.

Daniele, S. G., Béraud, D., Davenport, C., Cheng, K., Yin, H., & Maguire-Zeiss, K. A. (2015). Activation of MyD88-dependent TLR1/2 signaling by misfolded α-synuclein, a protein linked to neurodegenerative disorders. *Science Signaling*, *8*(376), ra45.

de Dios, C., Bartolessis, I., Roca-Agujetas, V., Barbero-Camps, E., Mari, M., Morales, A., et al. (2019). Oxidative inactivation of amyloid beta-degrading proteases by cholesterol-enhanced mitochondrial stress. *Redox Biology*, *26*, 101283.

de Farias, C. C., Maes, M., Bonifacio, K. L., Matsumoto, A. K., Bortolasci, C. C., Nogueira, A. S., et al. (2017). Parkinson's disease is accompanied by intertwined alterations in iron metabolism and activated immune-inflammatory and oxidative stress pathways. *CNS & Neurological Disorders Drug Targets*, *16*(4), 484–491.

de la Monte, S. M., & Tong, M. (2014). Brain metabolic dysfunction at the core of Alzheimer's disease. *Biochemical Pharmacology*, *88*(4), 548–559.

de la Monte, S. M., Tong, M., Schiano, I., & Didsbury, J. (2017). Improved brain insulin/IGF signaling and reduced neuroinflammation with T3D-959 in an experimental model of sporadic Alzheimer's disease. *Journal of Alzheimer's Disease*, *55*(2), 849–864.

de la Monte, S. M., & Wands, J. R. (2006). Molecular indices of oxidative stress and mitochondrial dysfunction occur early and often progress with severity of Alzheimer's disease. *Journal of Alzheimer's Disease*, *9*(2), 167–181.

De Pablo-Fernandez, E., Goldacre, R., Pakpoor, J., Noyce, A. J., & Warner, T. T. (2018). Association between diabetes and subsequent Parkinson disease: A record-linkage cohort study. *Neurology*, *91*(2), e139–e142.

Di Virgilio, F., Dal Ben, D., Sarti, A. C., Giuliani, A. L., & Falzoni, S. (2017). The P2X7 receptor in infection and inflammation. *Immunity*, *47*(1), 15–31.

Dobson, C. M. (2001). The structural basis of protein folding and its links with human disease. *Philosophical Transactions of the Royal Society of London. Series B, Biological Sciences*, *356*(1406), 133–145.

Dodiya, H. B., Forsyth, C. B., Voigt, R. M., Engen, P. A., Patel, J., Shaikh, M., et al. (2020). Chronic stress-induced gut dysfunction exacerbates Parkinson's disease phenotype and pathology in a rotenone-induced mouse model of Parkinson's disease. *Neurobiology of Disease*, *135*, 104352.

Dominy, S. S., Lynch, C., Ermini, F., Benedyk, M., Marczyk, A., Konradi, A., et al. (2019). Porphyromonas gingivalis in Alzheimer's disease brains: Evidence for disease causation and treatment with small-molecule inhibitors. *Science Advances*, *5*(1), eaau3333.

Duffy, M. F., Collier, T. J., Patterson, J. R., Kemp, C. J., Luk, K. C., Tansey, M. G., et al. (2018). Lewy body-like alpha-synuclein inclusions trigger reactive microgliosis prior to nigral degeneration. *Journal of Neuroinflammation*, *15*(1), 129.

Eckel, B., Ohl, F., Bogdanski, R., Kochs, E. F., & Blobner, M. (2011). Cognitive deficits after systemic induction of inducible nitric oxide synthase: A randomised trial in rats. *European Journal of Anaesthesiology*, *28*(9), 655–663.

Elliott, E. I., Miller, A. N., Banoth, B., Iyer, S. S., Stotland, A., Weiss, J. P., et al. (2018). Cutting edge: Mitochondrial assembly of the NLRP3 inflammasome complex is initiated at priming. *Journal of Immunology*, *200*(9), 3047–3052.

Esterline, R. L., Vaag, A., Oscarsson, J., & Vora, J. (2018). Mechanisms in endocrinology: SGLT2 inhibitors: Clinical benefits by restoration of normal diurnal metabolism? *European Journal of Endocrinology, 178*(4), R113–R125.

Farris, W., Mansourian, S., Chang, Y., Lindsley, L., Eckman, E. A., Frosch, M. P., et al. (2003). Insulin-degrading enzyme regulates the levels of insulin, amyloid beta-protein, and the beta-amyloid precursor protein intracellular domain in vivo. *Proceedings of the National Academy of Sciences of the United States of America, 100*(7), 4162–4167.

Feng, J., Wang, J. X., Du, Y. H., Liu, Y., Zhang, W., Chen, J. F., et al. (2018). Dihydromyricetin inhibits microglial activation and neuroinflammation by suppressing NLRP3 inflammasome activation in APP/PS1 transgenic mice. *CNS Neuroscience & Therapeutics, 24*(12), 1207–1218.

Fex, M., Nicholas, L. M., Vishnu, N., Medina, A., Sharoyko, V. V., Nicholls, D. G., et al. (2018). The pathogenetic role of β-cell mitochondria in type 2 diabetes. *The Journal of Endocrinology, 236*(3), R145–R159.

Finucane, O. M., Lyons, C. L., Murphy, A. M., Reynolds, C. M., Klinger, R., Healy, N. P., et al. (2015). Monounsaturated fatty acid-enriched high-fat diets impede adipose NLRP3 inflammasome-mediated IL-1β secretion and insulin resistance despite obesity. *Diabetes, 64*(6), 2116–2128.

Forloni, G., Mangiarotti, F., Angeretti, N., Lucca, E., & De Simoni, M. G. (1997). Beta-amyloid fragment potentiates IL-6 and TNF-alpha secretion by LPS in astrocytes but not in microglia. *Cytokine, 9*, 759–762.

Franchi, L., Eigenbrod, T., & Núñez, G. (2009). Cutting edge: TNF-alpha mediates sensitization to ATP and silica via the NLRP3 inflammasome in the absence of microbial stimulation. *Journal of Immunology, 183*(2), 792–796.

Franchi, L., Kanneganti, T. D., Dubyak, G. R., & Núñez, G. (2007). Differential requirement of P2X7 receptor and intracellular K+ for caspase-1 activation induced by intracellular and extracellular bacteria. *The Journal of Biological Chemistry, 282*(26), 18810–18818.

Friedland, R. P., Budinger, T. F., Koss, E., & Ober, B. A. (1985). Alzheimer's disease: Anteriorposterior and lateral hemispheric alterations in cortical glucose utilization. *Neuroscience Letters, 53*, 235.

Fu, J., Cui, Q., Yang, B., Hou, Y., Wang, H., Xu, Y., et al. (2017). The impairment of glucose-stimulated insulin secretion in pancreatic β-cells caused by prolonged glucotoxicity and lipotoxicity is associated with elevated adaptive antioxidant response. *Food and Chemical Toxicology, 100*, 161–167.

Gabbouj, S., Ryhänen, S., Marttinen, M., Wittrahm, R., Takalo, M., Kemppainen, S., et al. (2019). Altered insulin signaling in Alzheimer's disease brain—Special emphasis on PI3K-Akt pathway. *Frontiers in Neuroscience, 13*, 629.

Garwood, C. J., Pooler, A. M., Atherton, J., Hanger, D. P., & Noble, W. (2011). Astrocytes are important mediators of aβ-induced neurotoxicity and tau phosphorylation in primary culture. *Cell Death & Disease, 2*e167.

Goedert, M., Spillantini, M. G., & Crowther, R. A. (1991). Tau proteins and neurofibrillary degeneration. *Brain Pathology, 1*(4), 279–286.

Gordon, R., Albornoz, E. A., Christie, D. C., Langley, M. R., Kumar, V., Mantovani, S., et al. (2018). Inflammasome inhibition prevents α-synuclein pathology and dopaminergic neurodegeneration in mice. *Science Translational Medicine, 10*(465), pii: eaah4066.

Goyal, R., Faizy, A. F., Siddiqui, S. S., & Singhai, M. (2012). Evaluation of TNF-α and IL-6 levels in obese and non-obese diabetics: Pre- and postinsulin effects. *North American Journal of Medical Sciences, 4*(4), 180–184.

Grassi, D., Howard, S., Zhou, M., Diaz-Perez, N., Urban, N. T., Guerrero-Given, D., et al. (2018). Identification of a highly neurotoxic α-synuclein species inducing mitochondrial damage and mitophagy in Parkinson's disease. *Proceedings of the National Academy of Sciences of the United States of America, 115*(11), E2634–E2643.

Gray, S. M., Aylor, K. W., & Barrett, E. J. (2017). Unravelling the regulation of insulin transport across the brain endothelial cell. *Diabetologia, 60*(8), 1512–1521.

Green, H., Tsitsi, P., Markaki, I., Aarsland, D., & Svenningsson, P. (2019). Novel treatment opportunities against cognitive impairment in Parkinson's disease with an emphasis on diabetes-related pathways. *CNS Drugs, 33*(2), 143–160.

Groslambert, M., & Py, B. F. (2018). Spotlight on the NLRP3 inflammasome pathway. *Journal of Inflammation Research, 11*, 359–374.

Gu, Y., Vorburger, R., Scarmeas, N., Luchsinger, J. A., Manly, J. J., Schupf, N., et al. (2017). Circulating inflammatory biomarkers in relation to brain structural measurements in a non-demented elderly population. *Brain, Behavior, and Immunity, 65*, 150–160.

Guo, M., Wang, X., Zhao, Y., Yang, Q., Ding, H., Dong, Q., et al. (2018). Ketogenic diet improves brain ischemic tolerance and inhibits NLRP3 inflammasome activation by preventing Drp1-mediated mitochondrial fission and endoplasmic reticulum stress. *Frontiers in Molecular Neuroscience, 11*, 86.

Gustin, A., Kirchmeyer, M., Koncina, E., Felten, P., Losciuto, S., Heurtaux, T., et al. (2015). NLRP3 inflammasome is expressed and functional in mouse brain microglia but not in astrocytes. *PLoS One, 10*(6), e0130624.

Halle, A., Hornung, V., Petzold, G. C., Stewart, C. R., Monks, B. G., Reinheckel, T., et al. (2008). The NALP3 inflammasome is involved in the innate immune response to amyloid-beta. *Nature Immunology, 9*(8), 857–865.

Halliwell, B., & Gutteridge, J. M. C. (2015). Redox chemistry: The essentials. Chapter 2, pages 30–76 In *Free radicals in biology and medicine*. (5th ed.). Oxford University Press. ISBN: 978-0-19-871748-5.

Hanley, A. J., Festa, A., D'Agostino, R. B., Jr., Wagenknecht, L. E., Savage, P. J., Tracy, R. P., et al. (2004). Metabolic and inflammation variable clusters and prediction of type 2 diabetes: Factor analysis using directly measured insulin sensitivity. *Diabetes, 53*(7), 1773–1781.

Hawkins, B. T., Lundeen, T. F., Norwood, K. M., Brooks, H. L., & Egleton, R. D. (2007). Increased blood-brain barrier permeability and altered tight junctions in experimental diabetes in the rat: Contribution of hyperglycaemia and matrix metalloproteinases. *Diabetologia, 50*, 202–211.

Hayden, M. R. (2019). Type 2 diabetes mellitus increases the risk of late-onset Alzheimer's disease: Ultrastructural remodeling of the neurovascular unit and diabetic gliopathy. *Brain Sciences, 9*(10) pii: E262.

Hayden, M. R., Grant, D. G., Aroor, A. R., & DeMarco, V. G. (2019). Empagliflozin ameliorates type 2 diabetes-induced ultrastructural remodeling of the neurovascular unit and neuroglia in the female db/db mouse. *Brain Sciences, 9*(3), pii: E57.

He, M., Chiang, H. H., Luo, H., Zheng, Z., Qiao, Q., Wang, L., et al. (2020). An acetylation switch of the NLRP3 inflammasome regulates aging-associated chronic inflammation and insulin resistance. *Cell Metabolism, 31*(3), 580–591.e5.

He, Y., Hara, H., & Núñez, G. (2016). Mechanism and regulation of NLRP3 inflammasome activation. *Trends in Biochemical Sciences, 41*(12), 1012–1021.

He, W. T., Wan, H., Hu, L., Chen, P., Wang, X., Huang, Z., et al. (2015). Gasdermin D is an executor of pyroptosis and required for interleukin-1β secretion. *Cell Research, 25*(12), 1285–1298.

Heid, M. E., Keyel, P. A., Kamga, C., Shiva, S., Watkins, S. C., & Salter, R. D. (2013). Mitochondrial reactive oxygen species induces NLRP3-dependent lysosomal damage and inflammasome activation. *Journal of Immunology, 191*(10), 5230–5238.

Heilig, R., Dick, M. S., Sborgi, L., Meunier, E., Hiller, S., & Broz, P. (2018). The Gasdermin-D pore acts as a conduit for IL-1β secretion in mice. *European Journal of Immunology, 48*(4), 584–592.

Hirsch, E. C., Jenner, P., & Przedborski, S. (2013). Pathogenesis of Parkinson's disease. *Movement Disorders*, *28*(1), 24–30.

Hirschberg, S., Gisevius, B., Duscha, A., & Haghikia, A. (2019). Implications of diet and the Gut microbiome in neuroinflammatory and neurodegenerative diseases. *International Journal of Molecular Sciences*, *20*(12), pii: E3109.

Hokama, M., Oka, S., Leon, J., Ninomiya, T., Honda, H., Sasaki, K., et al. (2014). Altered expression of diabetes-related genes in Alzheimer's disease brains: The Hisayama study. *Cerebral Cortex*, *24*(9), 2476–2488.

Holmes, C., Cunningham, C., Zotova, E., Woolford, J., Dean, C., Kerr, S., et al. (2009). Systemic inflammation and disease progression in Alzheimer disease. *Neurology*, *73*(10), 768–774.

Holper, L., Ben-Shachar, D., & Mann, J. J. (2019). Multivariate meta-analyses of mitochondrial complex I and IV in major depressive disorder, bipolar disorder, schizophrenia, Alzheimer disease, and Parkinson disease. *Neuropsychopharmacology*, *44*(5), 837–849.

Hölscher, C. (2014). Drugs developed for treatment of diabetes show protective effects in Alzheimer's and Parkinson's diseases. *Sheng Li Xue Bao*, *66*(5), 497–510.

Hooshmand, B., Rusanen, M., Ngandu, T., Leiviskä, J., Sindi, S., von Arnim, C. A. F., et al. (2019). Serum insulin and cognitive performance in older adults: A longitudinal study. *The American Journal of Medicine*, *132*(3), 367–373.

Hornykiewicz, O. (2001). Chemical neuroanatomy of the basal ganglia—Normal and in Parkinson's disease. *Journal of Chemical Neuroanatomy*, *22*(1–2), 3–12.

Horvath, I., & Wittung-Stafshede, P. (2016). Cross-talk between amyloidogenic proteins in type-2 diabetes and Parkinson's disease. *Proceedings of the National Academy of Sciences of the United States of America*, *113*(44), 12473–12477.

Hou, Y., Xie, G., Liu, X., Li, G., Jia, C., Xu, J., et al. (2016). Minocycline protects against lipopolysaccharide-induced cognitive impairment in mice. *Psychopharmacology*, *233*(5), 905–916.

Hu, G., Jousilahti, P., Bidel, S., Antikainen, R., & Tuomilehto, J. (2007). Type 2 diabetes and the risk of Parkinson's disease. *Diabetes Care*, *30*(4), 842–847.

Hughes, T. M., Craft, S., Baker, L. D., Espeland, M. A., Rapp, S. R., Sink, K. M., et al. (2017). Changes in metabolic risk factors over 10 years and their associations with late-life cognitive performance: The Multi-Ethnic Study of Atherosclerosis. *Alzheimers Dement (Amst)*, *8*, 18–25.

Hughes, T. M., Kuller, L. H., Barinas-Mitchell, E. J., McDade, E. M., Klunk, W. E., Cohen, A. D., et al. (2014). Arterial stiffness and β-amyloid progression in nondemented elderly adults. *JAMA Neurology*, *71*(5), 562–568.

Hui, K. Y., Fernandez-Hernandez, H., Hu, J., Schaffner, A., Pankratz, N., Hsu, N. Y., et al. (2018). Functional variants in the LRRK2 gene confer shared effects on risk for Crohn's disease and Parkinson's disease. *Science Translational Medicine*, *10*(423), pii: eaai7795.

Iadecola, C. (2004). Neurovascular regulation in the normal brain and in Alzheimer's disease. *Nature Reviews. Neuroscience*, *5*(5), 347–360.

Ising, C., Venegas, C., Zhang, S., Scheiblich, H., Schmidt, S. V., Vieira-Saecker, A., et al. (2019). NLRP3 inflammasome activation drives tau pathology. *Nature*, *575*(7784), 669–673.

Italiani, P., Puxeddu, I., Napoletano, S., Scala, E., Melillo, D., Manocchio, S., et al. (2018). Circulating levels of IL-1 family cytokines and receptors in Alzheimer's disease: New markers of disease progression? *Journal of Neuroinflammation*, *15*(1), 342.

Jaeger, L. B., Dohgu, S., Sultana, R., Lynch, J. L., Owen, J. B., Erickson, M. A., et al. (2009). Lipopolysaccharide alters the blood-brain barrier transport of amyloid beta protein: A mechanism for inflammation in the progression of Alzheimer's disease. *Brain, Behavior, and Immunity*, *23*(4), 507–517.

Janelidze, S., Hertze, J., Nägga, K., Nilsson, K., Nilsson, C., Swedish BioFINDER Study Group, et al. (2017). Increased blood-brain barrier permeability is associated with dementia and diabetes but not amyloid pathology or APOE genotype. *Neurobiology of Aging*, *51*, 104–112.

Jenner, P., Morris, H. R., Robbins, T. W., Goedert, M., Hardy, J., Ben-Shlomo, Y., et al. (2013). Parkinson's disease—The debate on the clinical phenomenology, aetiology, pathology and pathogenesis. *Journal of Parkinson's Disease*, *3*(1), 1–11.

Jo, E. K., Kim, J. K., Shin, D. M., & Sasakawa, C. (2016). Molecular mechanisms regulating NLRP3 inflammasome activation. *Cellular & Molecular Immunology*, *13*(2), 148–159.

Jurgens, C. A., Toukatly, M. N., Fligner, C. L., Udayasankar, J., Subramanian, S. L., Zraika, S., et al. (2011). β-cell loss and β-cell apoptosis in human type 2 diabetes are related to islet amyloid deposition. *The American Journal of Pathology*, *178*(6), 2632–2640.

Karpenko, M. N., Vasilishina, A. A., Gromova, E. A., Muruzheva, Z. M., Miliukhina, I. V., & Bernadotte, A. (2018). Interleukin-1β, interleukin-1 receptor antagonist, interleukin-6, interleukin-10, and tumor necrosis factor-α levels in CSF and serum in relation to the clinical diversity of Parkinson's disease. *Cellular Immunology*, *327*, 77–82.

Kaushal, V., Dye, R., Pakavathkumar, P., Foveau, B., Flores, J., Hyman, B., et al. (2015). Neuronal NLRP1 inflammasome activation of Caspase-1 coordinately regulates inflammatory interleukin-1-beta production and axonal degeneration-associated Caspase-6 activation. *Cell Death and Differentiation*, *22*, 1676–1686.

Kempuraj, D., Thangavel, R., Yang, E., Pattani, S., Zaheer, S., Santillan, D. A., et al. (2015). Dopaminergic toxin 1-methyl-4-phenylpyridinium, proteins α-synuclein and glia maturation factor activate mast cells and release inflammatory mediators. *PLoS One*, *10*(8), e0135776.

Kim, S., Kwon, S. H., Kam, T. I., Panicker, N., Karuppagounder, S. S., Lee, S., et al. (2019). Transneuronal propagation of pathologic α-synuclein from the Gut to the brain models Parkinson's disease. *Neuron*, *103*(4), 627–641.e7.

King, E., O'Brien, J. T., Donaghy, P., Morris, C., Barnett, N., Olsen, K., et al. (2018). Peripheral inflammation in prodromal Alzheimer's and Lewy body dementias. *Journal of Neurology, Neurosurgery, and Psychiatry*, *89*(4), 339–345.

Kishimoto, Y., Zhu, W., Hosoda, W., Sen, J. M., & Mattson, M. P. (2019). Chronic mild gut inflammation accelerates brain neuropathology and motor dysfunction in α-synuclein mutant mice. *Neuromolecular Medicine*, *21*(3), 239–249.

Kitazawa, M., Cheng, D., Tsukamoto, M. R., Koike, M. A., Wes, P. D., Vasilevko, V., et al. (2011). Blocking IL-1 signaling rescues cognition, attenuates tau pathology, and restores neuronal β-catenin pathway function in an Alzheimer's disease model. *Journal of Immunology*, *187*(12), 6539–6549.

Kivipelto, M., Mangialasche, F., & Ngandu, T. (2018). Lifestyle interventions to prevent cognitive impairment, dementia and Alzheimer disease. *Nature Reviews. Neurology*, *14*(11), 653–666.

Koprich, J. B., Reske-Nielsen, C., Mithal, P., & Isacson, O. (2008). Neuroinflammation mediated by IL-1beta increases susceptibility of dopamine neurons to degeneration in an animal model of Parkinson's disease. *Journal of Neuroinflammation*, *5*, 8.

Koss, E., Friedland, R. P., Ober, B. A., & Jagust, W. (1985). Differences in lateral hemispheric asymmetries of glucose utilization between early- and late-onset Alzheimer type dementia. *The American Journal of Psychiatry*, *142*, 638.

Koves, T. R., Ussher, J. R., Noland, R. C., Slentz, D., Mosedale, M., Ilkayeva, O., et al. (2008). Mitochondrial overload and incomplete fatty acid oxidation contribute to skeletal muscle insulin resistance. *Cell Metabolism*, 7(1), 45–56.

Lai, K. S. P., Liu, C. S., Rau, A., Lanctôt, K. L., Köhler, C. A., Pakosh, M., et al. (2017). Peripheral inflammatory markers in Alzheimer's disease: A systematic review and meta-analysis of 175 studies. *Journal of Neurology, Neurosurgery, and Psychiatry*, *88*(10), 876–882.

Lane, C. A., Hardy, J., & Schott, J. M. (2018). Alzheimer's disease. *European Journal of Neurology*, *25*(1), 59–70.
Lee, E., Hwang, I., Park, S., Hong, S., Hwang, B., Cho, Y., et al. (2019). MPTP-driven NLRP3 inflammasome activation in microglia plays a central role in dopaminergic neurodegeneration. *Cell Death and Differentiation*, *26*(2), 213–228.
Lee, H. J., Suk, J. E., Patrick, C., Bae, E. J., Cho, J. H., Rho, S., et al. (2010). Direct transfer of alpha-synuclein from neuron to astroglia causes inflammatory responses in synucleinopathies. *The Journal of Biological Chemistry*, *285*(12), 9262–9272.
Lee, Y. S., Wollam, J., & Olefsky, J. M. (2018). An integrated view of immunometabolism. *Cell*, *172*(1–2), 22–40.
Li, W., Risacher, S. L., Huang, E., & Saykin, A. J. (2016). Alzheimer's disease neuroimaging initiative. Type 2 diabetes mellitus is associated with brain atrophy and hypometabolism in the ADNI cohort. *Neurology*, *87*(6), 595–600.
Lin, B., Koibuchi, N., Hasegawa, Y., Sueta, D., Toyama, K., Uekawa, K., et al. (2014). Glycemic control with empagliflozin, a novel selective SGLT2 inhibitor, ameliorates cardiovascular injury and cognitive dysfunction in obese and type 2 diabetic mice. *Cardiovascular Diabetology*, *13*, 148.
Lindqvist, D., Kaufman, E., Brundin, L., Hall, S., Surova, Y., & Hansson, O. (2012). Non-motor symptoms in patients with Parkinson's disease—Correlations with inflammatory cytokines in serum. *PLoS One*, 7(10)e47387.
Macauley, S. L., Stanley, M., Caesar, E. E., Yamada, S. A., Raichle, M. E., Perez, R., et al. (2015). Hyperglycemia modulates extracellular amyloid-β concentrations and neuronal activity in vivo. *The Journal of Clinical Investigation*, *125*(6), 2463–2467.
Maciejczyk, M., Żebrowska, E., & Chabowski, A. (2019). Insulin resistance and oxidative stress in the brain: What's new? *International Journal of Molecular Sciences*, *20*(4), pii: E874.
MacPherson, K. P., Sompol, P., Kannarkat, G. T., Chang, J., Sniffen, L., Wildner, M. E., et al. (2017). Peripheral administration of the soluble TNF inhibitor XPro1595 modifies brain immune cell profiles, decreases beta-amyloid plaque load, and rescues impaired long-term potentiation in 5xFAD mice. *Neurobiology of Disease*, *102*, 81–95.
Manning, B. D., & Toker, A. (2017). AKT/PKB signaling: Navigating the network. *Cell*, *169*(3), 381–405.
Marsland, A. L., Gianaros, P. J., Kuan, D. C., Sheu, L. K., Krajina, K., & Manuck, S. B. (2015). Brain morphology links systemic inflammation to cognitive function in midlife adults. *Brain, Behavior, and Immunity*, *48*, 195–204.
Martinez, E. M., Young, A. L., Patankar, Y. R., Berwin, B. L., Wang, L., von Herrmann, K. M., et al. (2017). Nlrp3 Is required for inflammatory changes and nigral cell loss resulting from chronic intragastric rotenone exposure in mice. *Toxicological Sciences*, *159*(1), 64–75.
Martín-Maestro, P., Gargini, R., García, E., Perry, G., Avila, J., & García-Escudero, V. (2017). Slower dynamics and aged mitochondria in sporadic Alzheimer's disease. *Oxidative Medicine and Cellular Longevity*, *2017*, 9302761.
Martinon, F., Burns, K., & Tschopp, J. (2002). The inflammasome: A molecular platform triggering activation of inflammatory caspases and processing of proIL-beta. *Molecular Cell*, *10*(2), 417–426.
Masters, S. L., Dunne, A., Subramanian, S. L., Hull, R. L., Tannahill, G. M., Sharp, F. A., et al. (2010). Activation of the NLRP3 inflammasome by islet amyloid polypeptide provides a mechanism for enhanced IL-1β in type 2 diabetes. *Nature Immunology*, *11*(10), 897–904.
Matsumoto, J., Stewart, T., Sheng, L., Li, N., Bullock, K., Song, N., et al. (2017). Transmission of α-synuclein-containing erythrocyte-derived extracellular vesicles across the blood-brain barrier via adsorptive mediated transcytosis: Another mechanism for initiation and progression of Parkinson's disease? *Acta Neuropathologica Communications*, *5*(1), 71.

Mattson, M. P., & Arumugam, T. V. (2018). Hallmarks of brain aging: Adaptive and pathological modification by metabolic states. *Cell Metabolism*, *27*(6), 1176–1199.

McAlpine, F. E., Lee, J. K., Harms, A. S., Ruhn, K. A., Blurton-Jones, M., Hong, J., et al. (2009). Inhibition of soluble TNF signaling in a mouse model of Alzheimer's disease prevents pre-plaque amyloid-associated neuropathology. *Neurobiology of Disease*, *34*(1), 163–177.

McKhann, G. M., Knopman, D. S., Chertkow, H., Hyman, B. T., Jack, C. R., Jr., Kawas, C. H., et al. (2011). The diagnosis of dementia due to Alzheimer's disease: Recommendations from the National Institute on Aging-Alzheimer's Association workgroups on diagnostic guidelines for Alzheimer's disease. *Alzheimer's & Dementia*, 7(3), 263–269.

Meijer, R. I., Gray, S. M., Aylor, K. W., & Barrett, E. J. (2016). Pathways for insulin access to the brain: The role of the microvascular endothelial cell. *American Journal of Physiology. Heart and Circulatory Physiology*, *311*(5), H1132–H1138.

Miki, A., Ricordi, C., Sakuma, Y., Yamamoto, T., Misawa, R., Mita, A., et al. (2018). Divergent antioxidant capacity of human islet cell subsets: A potential cause of beta-cell vulnerability in diabetes and islet transplantation. *PLoS One*, *13*(5)e0196570.

Miyazaki, Y., Kawano, H., Yoshida, T., Miyamoto, S., Hokamaki, J., Nagayoshi, Y., et al. (2007). Pancreatic B-cell function is altered by oxidative stress induced by acute hyperglycaemia. *Diabetic Medicine*, *24*(2), 154–160.

Mogi, M., Harada, M., Narabayashi, H., Inagaki, H., Minami, M., & Nagatsu, T. (1996). Interleukin (IL)-1 beta, IL-2, IL-4, IL-6 and transforming growth factor-alpha levels are elevated in ventricular cerebrospinal fluid in juvenile Parkinsonism and Parkinson's disease. *Neuroscience Letters*, *211*(1), 13–16.

Molnár, G., Faragó, N., Kocsis, Á. K., Rózsa, M., Lovas, S., Boldog, E., et al. (2014). GABAergic neurogliaform cells represent local sources of insulin in the cerebral cortex. *The Journal of Neuroscience*, *34*(4), 1133–1137.

Montagne, A., Barnes, S. R., Sweeney, M. D., Halliday, M. R., Sagare, A. P., Zhao, Z., et al. (2015). Blood-brain barrier breakdown in the aging human hippocampus. *Neuron*, *85*(2), 296–302.

Montane, J., Klimek-Abercrombie, A., Potter, K. J., Westwell-Roper, C., & Bruce Verchere, C. (2012). Metabolic stress, IAPP and islet amyloid. *Diabetes, Obesity & Metabolism*, *14*(Suppl. 3), 68–77.

Mooijaart, S. P., Sattar, N., Trompet, S., Lucke, J., Stott, D. J., Ford, I., et al. (2013). Circulating interleukin-6 concentration and cognitive decline in old age: The PROSPER study. *Journal of Internal Medicine*, *274*(1), 77–85.

Moran, C., Phan, T. G., Chen, J., Blizzard, L., Beare, R., Venn, A., et al. (2013). Brain atrophy in type 2 diabetes: Regional distribution and influence on cognition. *Diabetes Care*, *36*(12), 4036–4042.

Morikawa, S., Kaneko, N., Okumura, C., Taguchi, H., Kurata, M., Yamamoto, T., et al. (2018). IAPP/amylin deposition, which is correlated with expressions of ASC and IL-1β in β-cells of Langerhans' islets, directly initiates NLRP3 inflammasome activation. *International Journal of Immunopathology and Pharmacology*, *32* 2058738418788749.

Moroo, I., Yamada, T., Makino, H., Tooyama, I., McGeer, P. L., McGeer, E. G., et al. (1994). Loss of insulin receptor immunoreactivity from the substantia nigra pars compacta neurons in Parkinson's disease. *Acta Neuropathologica*, *87*, 343–348.

Mosconi, L., Murray, J., Tsui, W. H., Li, Y., Spector, N., Goldowsky, A., et al. (2014). Brain imaging of cognitively normal individuals with 2 parents affected by late-onset AD. *Neurology*, *82*(9), 752–760.

Mosconi, L., Pupi, A., & De Leon, M. J. (2008). Brain glucose hypometabolism and oxidative stress in preclinical Alzheimer's disease. *Annals of the New York Academy of Sciences*, *1147*, 180–195.

Mullin, S., & Schapira, A. (2013). α-Synuclein and mitochondrial dysfunction in Parkinson's disease. *Molecular Neurobiology*, *47*(2), 587–597.
Mullin, S., Smith, L., Lee, K., D'Souza, G., Woodgate, P., Elflein, J., et al. (2020). Ambroxol for the treatment of patients with Parkinson disease with and without glucocerebrosidase gene mutations: A nonrandomized, noncontrolled trial. *JAMA Neurology*. https://doi.org/10.1001/jamaneurol.2019.4611.
Mutisya, E. M., Bowling, A. C., & Beal, M. F. (1994). Cortical cytochrome oxidase activity is reduced in Alzheimer's disease. *Journal of Neurochemistry*, *63*(6), 2179–2184.
Nagareddy, P. R., Xia, Z., McNeill, J. H., & MacLeod, K. M. (2005). Increased expression of iNOS is associated with endothelial dysfunction and impaired pressor responsiveness in streptozotocin-induced diabetes. *American Journal of Physiology. Heart and Circulatory Physiology*, *289*(5), H2144–H2152.
Nakanishi, A., Kaneko, N., Takeda, H., Sawasaki, T., Morikawa, S., Zhou, W., et al. (2018). Amyloid β directly interacts with NLRP3 to initiate inflammasome activation: Identification of an intrinsic NLRP3 ligand in a cell-free system. *Inflammation and Regeneration*, *38*, 27.
Nation, D. A., Sweeney, M. D., Montagne, A., Sagare, A. P., D'Orazio, L. M., Pachicano, M., et al. (2019). Blood-brain barrier breakdown is an early biomarker of human cognitive dysfunction. *Nature Medicine*, *25*(2), 270–276.
Nelson, P. T., Pious, N. M., Jicha, G. A., Wilcock, D. M., Fardo, D. W., Estus, S., et al. (2013). APOE-ε2 and APOE-ε4 correlate with increased amyloid accumulation in cerebral vasculature. *Journal of Neuropathology and Experimental Neurology*, *72*(7), 708–715.
Nelson, A. R., Sweeney, M. D., Sagare, A. P., & Zlokovic, B. V. (2016). Neurovascular dysfunction and neurodegeneration in dementia and Alzheimer's disease. *Biochimica et Biophysica Acta*, *1862*(5), 887–900.
Nishitsuji, K., Hosono, T., Nakamura, T., Bu, G., & Michikawa, M. (2011). Apolipoprotein E regulates the integrity of tight junctions in an isoform-dependent manner in an in vitro blood-brain barrier model. *The Journal of Biological Chemistry*, *286*(20), 17536–17542.
Norwitz, N. G., Hu, M. T., & Clarke, K. (2019). The mechanisms by which the ketone body D-β-hydroxybutyrate may improve the multiple cellular pathologies of Parkinson's disease. *Frontiers in Nutrition*, *6*, 63.
Obermeier, B., Daneman, R., & Ransohoff, R. M. (2013). Development, maintenance and disruption of the blood-brain barrier. *Nature Medicine*, *19*(12), 1584–1596.
Obeso, J. A., Marin, C., Rodriguez-Oroz, C., Blesa, J., Benitez-Temiño, B., et al. (2008). The basal ganglia in Parkinson's disease: Current concepts and unexplained observations. *Annals of Neurology*, *64*(Suppl. 2), S30–S36.
Opipari, A., & Franchi, L. (2015). Role of inflammasomes in intestinal inflammation and Crohn's disease. *Inflammatory Bowel Diseases*, *21*, 173–181.
Ott, B. R., Jones, R. N., Daiello, L. A., de la Monte, S. M., Stopa, E. G., Johanson, C. E., et al. (2018). Blood-cerebrospinal fluid barrier gradients in mild cognitive impairment and Alzheimer's disease: Relationship to inflammatory cytokines and chemokines. *Frontiers in Aging Neuroscience*, *10*, 245.
Ott, A., Stolk, R. P., van Harskamp, F., Pols, H. A., Hofman, A., & Breteler, M. M. (1999). Diabetes mellitus and the risk of dementia: The Rotterdam Study. *Neurology*, *53*(9), 1937–1942.
Pagano, G., Polychronis, S., Wilson, H., Giordano, B., Ferrara, N., Niccolini, F., et al. (2018). Diabetes mellitus and Parkinson disease. *Neurology*, *90*(19), e1654–e1662.
Palmer, A. K., Gustafson, B., Kirkland, J. L., & Smith, U. (2019). Cellular senescence: At the nexus between ageing and diabetes. *Diabetologia*, *62*(10), 1835–1841.
Paradies, G., Paradies, V., De Benedictis, V., Ruggiero, F. M., & Petrosillo, G. (2014). Functional role of cardiolipin in mitochondrial bioenergetics. *Biochimica et Biophysica Acta*, *1837*(4), 408–417.

Parajuli, B., Sonobe, Y., Horiuchi, H., Takeuchi, H., Mizuno, T., & Suzumura, A. (2013). Oligomeric amyloid β induces IL-1β processing via production of ROS: Implication in Alzheimer's disease. *Cell Death & Disease, 4,* e975.

Park, S., Won, J. H., Hwang, I., Hong, S., Lee, H. K., & Yu, J. W. (2015). Defective mitochondrial fission augments NLRP3 inflammasome activation. *Scientific Reports, 5,* 15489.

Parodi-Rullán, R., Sone, J. Y., & Fossati, S. (2019). Endothelial mitochondrial dysfunction in cerebral amyloid angiopathy and Alzheimer's disease. *Journal of Alzheimer's Disease, 72*(4), 1019–1039.

Peralta Ramos, J. M., Iribarren, P., Bousset, L., Melki, R., Baekelandt, V., & Van der Perren, A. (2019). Peripheral inflammation regulates CNS immune surveillance through the recruitment of inflammatory monocytes upon systemic α-synuclein administration. *Frontiers in Immunology, 10,* 80.

Pereira, J. R., Santos, L. V. D., Santos, R. M. S., Campos, A. L. F., Pimenta, A. L., de Oliveira, M. S., et al. (2016). IL-6 serum levels are elevated in Parkinson's disease patients with fatigue compared to patients without fatigue. *Journal of the Neurological Sciences, 370,* 153–156.

Peter, I., Dubinsky, M., Bressman, S., Park, A., Lu, C., Chen, N., et al. (2018). Anti-tumor necrosis factor therapy and incidence of Parkinson disease among patients with inflammatory bowel disease. *JAMA Neurology, 75*(8), 939–946.

Pinto, A., Bonucci, A., Maggi, E., Corsi, M., & Businaro, R. (2018). Anti-oxidant and anti-inflammatory activity of ketogenic diet: New perspectives for neuroprotection in Alzheimer's disease. *Antioxidants (Basel),* 7(5) pii: E63.

Pontillo, A., Catamo, E., Arosio, B., Mari, D., & Crovella, S. (2012). NALP1/NLRP1 genetic variants are associated with Alzheimer disease. *Alzheimer Disease and Associated Disorders, 26*(3), 277–281.

Priyadarshini, M., Kamal, M. A., Greig, N. H., Reale, M., Abuzenadah, A. M., Chaudhary, A. G., et al. (2012). Alzheimer's disease and type 2 diabetes: Exploring the association to obesity and tyrosine hydroxylase. *CNS & Neurological Disorders Drug Targets, 11*(4), 482–489.

Qin, L., Wu, X., Block, M. L., Liu, Y., Breese, G. R., Hong, J. S., et al. (2007). Systemic LPS causes chronic neuroinflammation and progressive neurodegeneration. *Glia, 55*(5), 453–462.

Qin, X. Y., Zhang, S. P., Cao, C., Loh, Y. P., & Cheng, Y. (2016). Aberrations in peripheral inflammatory cytokine levels in Parkinson disease: A systematic review and meta-analysis. *JAMA Neurology, 73*(11), 1316–1324.

Qiu, Z., He, Y., Ming, H., Lei, S., Leng, Y., & Xia, Z. Y. (2019). Lipopolysaccharide (LPS) aggravates high glucose- and hypoxia/reoxygenation-induced injury through activating ROS-dependent NLRP3 inflammasome-mediated pyroptosis in H9C2 cardiomyocytes. *Journal Diabetes Research, 2019,* 8151836.

Rasool, M., Malik, A., Qazi, A. M., Sheikh, I. A., Manan, A., Shaheen, S., et al. (2014). Current view from Alzheimer disease to type 2 diabetes mellitus. *CNS & Neurological Disorders Drug Targets, 13*(3), 533–542.

Reale, M., Kamal, M. A., Velluto, L., Gambi, D., Di Nicola, M., & Greig, N. H. (2012). Relationship between inflammatory mediators, Aβ levels and ApoE genotype in Alzheimer disease. *Current Alzheimer Research, 9*(4), 447–457.

Reynolds, C. M., McGillicuddy, F. C., Harford, K. A., Finucane, O. M., Mills, K. H., & Roche, H. M. (2012). Dietary saturated fatty acids prime the NLRP3 inflammasome via TLR4 in dendritic cells-implications for diet-induced insulin resistance. *Molecular Nutrition & Food Research, 56*(8), 1212–1222.

Riphagen, J. M., Ramakers, I. H. G. M., Freeze, W. M., Pagen, L. H. G., Hanseeuw, B. J., Verbeek, M. M., et al. (2020). Linking APOE-ε4, blood-brain barrier dysfunction, and inflammation to Alzheimer's pathology. *Neurobiology of Aging, 85,* 96–103.

Roberts, R. O., Knopman, D. S., Przybelski, S. A., Mielke, M. M., Kantarci, K., Preboske, G. M., et al. (2014). Association of type 2 diabetes with brain atrophy and cognitive impairment. *Neurology*, *82*(13), 1132–1141.
Rochfort, K. D., Collins, L. E., Murphy, R. P., & Cummins, P. M. (2014). Downregulation of blood-brain barrier phenotype by proinflammatory cytokines involves NADPH oxidase-dependent ROS generation: Consequences for interendothelial adherens and tight junctions. *PLoS One*, *9*(7), e101815.
Rojas-Gutierrez, E., Muñoz-Arenas, G., Treviño, S., Espinosa, B., Chavez, R., Rojas, K., et al. (2017). Alzheimer's disease and metabolic syndrome: A link from oxidative stress and inflammation to neurodegeneration. *Synapse*, *71*(10), e21990.
Rose-John, S. (2017). The soluble interleukin 6 receptor: Advanced therapeutic options in inflammation. *Clinical Pharmacology and Therapeutics*, *102*(4), 591–598.
Russo, H. M., Rathkey, J., Boyd-Tressler, A., Katsnelson, M. A., Abbott, D. W., & Dubyak, G. R. (2016). Active caspase-1 induces plasma membrane pores that precede pyroptotic lysis and are blocked by lanthanides. *Journal of Immunology*, *197*(4), 1353–1367.
Saad, M. J., Santos, A., & Prada, P. O. (2016). Linking gut microbiota and inflammation to obesity and insulin resistance. *Physiology (Bethesda, Md)*, *31*(4), 283–293.
Sama, D. M., Mohmmad Abdul, H., Furman, J. L., Artiushin, I. A., Szymkowski, D. E., et al. (2012). Inhibition of soluble tumor necrosis factor ameliorates synaptic alterations and Ca2+ dysregulation in aged rats. *PLoS One*, 7(5), e38170.
Sa-Nguanmoo, P., Tanajak, P., Kerdphoo, S., Jaiwongkam, T., Pratchayasakul, W., Chattipakorn, N., et al. (2017). SGLT2-inhibitor and DPP-4 inhibitor improve brain function via attenuating mitochondrial dysfunction, insulin resistance, inflammation, and apoptosis in HFD-induced obese rats. *Toxicology and Applied Pharmacology*, *333*, 43–50.
Saresella, M., La Rosa, F., Piancone, F., Zoppis, M., Marventano, I., Calabrese, E., et al. (2016). The NLRP3 and NLRP1 inflammasomes are activated in Alzheimer's disease. *Molecular Neurodegeneration*, *11*, 23.
Sarlus, H., & Heneka, M. T. (2017). Microglia in Alzheimer's disease. *The Journal of Clinical Investigation*, *127*, 3240–3249.
Satizabal, C. L., Zhu, Y. C., Mazoyer, B., Dufouil, C., & Tzourio, C. (2012). Circulating IL-6 and CRP are associated with MRI findings in the elderly: The 3C-Dijon Study. *Neurology*, *78*(10), 720–727.
Schapira, A. H., Cooper, J. M., Dexter, D., Clark, J. B., Jenner, P., & Marsden, C. D. (1990). Mitochondrial complex I deficiency in Parkinson's disease. *Journal of Neurochemistry*, *54*(3), 823–827.
Schapira, A. H., & Jenner, P. (2011). Etiology and pathogenesis of Parkinson's disease. *Movement Disorders*, *26*(6), 1049–1055.
Scheiblich, H., Schlütter, A., Golenbock, D. T., Latz, E., Martinez-Martinez, P., & Heneka, M. T. (2017). Activation of the NLRP3 inflammasome in microglia: The role of ceramide. *Journal of Neurochemistry*, *143*(5), 534–550.
Schneider, K. S., Groß, C. J., Dreier, R. F., Saller, B. S., Mishra, R., Gorka, O., et al. (2017). The inflammasome drives GSDMD-independent secondary pyroptosis and IL-1 release in the absence of caspase-1 protease activity. *Cell Reports*, *21*(13), 3846–3859.
Schram, M. T., Euser, S. M., de Craen, A. J., Witteman, J. C., Frölich, M., Hofman, A., et al. (2007). Systemic markers of inflammation and cognitive decline in old age. *Journal of the American Geriatrics Society*, *55*(5), 708–716.
Selkoe, D. J. (1986). Altered structural proteins in plaques and tangles: What do they tell us about the biology of Alzheimer's disease? *Neurobiology of Aging*, 7(6), 425–432.
Selkoe, D. J. (2002). Alzheimer's disease is a synaptic failure. *Science*, *298*(5594), 789–791.
Shahzad, K., Bock, F., Dong, W., Wang, H., Kopf, S., Kohli, S., et al. (2015). Nlrp3-inflammasome activation in non-myeloid-derived cells aggravates diabetic nephropathy. *Kidney International*, *87*(1), 74–84.

Shao, Q. H., Chen, Y., Li, F. F., Wang, S., Zhang, X. L., Yuan, Y. H., et al. (2019). TLR4 deficiency has a protective effect in the MPTP/probenecid mouse model of Parkinson's disease. *Acta Pharmacologica Sinica*, *40*(12), 1503–1512.

Shimada, K., Crother, T. R., Karlin, J., Dagvadorj, J., Chiba, N., Chen, S., et al. (2012). Oxidized mitochondrial DNA activates the NLRP3 inflammasome during apoptosis. *Immunity*, *36*(3), 401–414.

Si, X., Pu, J., & Zhang, B. (2017). Structure, distribution, and genetic profile of α-synuclein and their potential clinical application in Parkinson's disease. *Journal of Movement Disorders*, *10*(2), 69–79.

Simpson, E. E., Hodkinson, C. F., Maylor, E. A., McCormack, J. M., Rae, G., Strain, S., et al. (2013). Intracellular cytokine production and cognition in healthy older adults. *Psychoneuroendocrinology*, *38*(10), 2196–2208.

Skaria, T., Bachli, E., & Schoedon, G. (2017). Wnt5A/Ryk signaling critically affects barrier function in human vascular endothelial cells. *Cell Adhesion & Migration*, *11*(1), 24–38.

Skaria, T., Bachli, E., & Schoedon, G. (2018). RSPO3 impairs barrier function of human vascular endothelial monolayers and synergizes with pro-inflammatory IL-1. *Molecular Medicine*, *24*(1), 45.

Skoog, I., & Gustafson, D. (2006). Update on hypertension and Alzheimer's disease. Hypertension may also lead to dysfunction in the blood-brain barrier, an event associated with the genesis of AD by previously discussed mechanisms. *Neurological Research*, *28*(6), 605–611.

Sochocka, M., Donskow-Łysoniewska, K., Diniz, B. S., Kurpas, D., Brzozowska, E., & Leszek, J. (2019). The gut microbiome alterations and inflammation-driven pathogenesis of Alzheimer's disease—A critical review. *Molecular Neurobiology*, *56*(3), 1841–1851.

Söderbom, G. (2020). Status and future directions of clinical trials in Parkinson's disease. *International Review of Neurobiology*, *154*, 153–188.

Sokolova, M., Sahraoui, A., Høyem, M., Øgaard, J., Lien, E., Aukrust, P., et al. (2018). NLRP3 inflammasome mediates oxidative stress-induced pancreatic islet dysfunction. *American Journal of Physiology. Endocrinology and Metabolism*, *315*(5), E912–E923.

Song, L., Pei, L., Yao, S., Wu, Y., & Shang, Y. (2017). NLRP3 Inflammasome in neurological diseases, from functions to therapies. *Frontiers in Cellular Neuroscience*, *11*, 63.

Sperling, R. A., Aisen, P. S., Beckett, L. A., Bennett, D. A., Craft, S., Fagan, A. M., et al. (2011). Toward defining the preclinical stages of Alzheimer's disease: Recommendations from the National Institute on Aging-Alzheimer's Association workgroups on diagnostic guidelines for Alzheimer's disease. *Alzheimer's & Dementia*, 7(3), 280–292.

Spranger, J., Kroke, A., Möhlig, M., Hoffmann, K., Bergmann, M. M., Ristow, M., et al. (2003). Inflammatory cytokines and the risk to develop type 2 diabetes: Results of the prospective population-based European Prospective Investigation into Cancer and Nutrition (EPIC)-Potsdam Study. *Diabetes*, *52*(3), 812–817.

Srinivasula, S. M., Poyet, J. L., Razmara, M., Datta, P., Zhang, Z., & Alnemri, E. S. (2002). The PYRIN-CARD protein ASC is an activating adaptor for caspase-1. *The Journal of Biological Chemistry*, *277*(24), 21119–21122.

Stancu, I. C., Cremers, N., Vanrusselt, H., Couturier, J., Vanoosthuyse, A., Kessels, S., et al. (2019). Aggregated Tau activates NLRP3-ASC inflammasome exacerbating exogenously seeded and non-exogenously seeded Tau pathology in vivo. *Acta Neuropathologica*, *137*(4), 599–617.

Stanley, M., Macauley, S. L., Caesar, E. E., Koscal, L. J., Moritz, W., Robinson, G. O., et al. (2016a). The effects of peripheral and central high insulin on brain insulin signaling and amyloid-β in young and Old APP/PS1 mice. *The Journal of Neuroscience*, *36*(46), 11704–11715.

Stanley, M., Macauley, S. L., & Holtzman, D. M. (2016b). Changes in insulin and insulin signaling in Alzheimer's disease: Cause or consequence? *The Journal of Experimental Medicine*, *213*(8), 1375–1385.

Starr, J. M., Wardlaw, J., Ferguson, K., MacLullich, A., Deary, I. J., & Marshall, I. (2003). Increased blood-brain barrier permeability in type II diabetes demonstrated by gadolinium magnetic resonance imaging. *Journal of Neurology, Neurosurgery, and Psychiatry*, *74*(1), 70–76.

Steen, E., Terry, B. M., Rivera, E. J., Cannon, J. L., Neely, T. R., Tavares, R., et al. (2005). Impaired insulin and insulin-like growth factor expression and signaling mechanisms in Alzheimer's disease—Is this type 3 diabetes? *Journal of Alzheimer's Disease*, 7, 63.

Sui, Y. T., Bullock, K. M., Erickson, M. A., Zhang, J., & Banks, W. A. (2014). Alpha synuclein is transported into and out of the brain by the blood-brain barrier. *Peptides*, *62*, 197–202.

Sznejder-Pachołek, A., Joniec-Maciejak, I., Wawer, A., Ciesielska, A., & Mirowska-Guzel, D. (2017). The effect of α-synuclein on gliosis and IL-1α, TNFα, IFNγ, TGFβ expression in murine brain. *Pharmacological Reports*, *69*(2), 242–251.

Taguchi, K., Watanabe, Y., Tsujimura, A., & Tanaka, M. (2016). Brain region-dependent differential expression of alpha-synuclein. *The Journal of Comparative Neurology*, *524*, 1236–1258.

Takahashi, M., Yamada, T., Tooyama, I., Moroo, I., Kimura, H., Yamamoto, T., et al. (1996). Insulin receptor mRNA in the substantia nigra in Parkinson's disease. *Neuroscience Letters*, *204*, 201–204.

Takamatsu, Y., Ho, G., Koike, W., Sugama, S., Takenouchi, T., Waragai, M., et al. (2017). Combined immunotherapy with "anti-insulin resistance" therapy as a novel therapeutic strategy against neurodegenerative diseases. *NPJ Parkinsons Disease*, *3*, 4.

Takeda, S., Sato, N., Rakug, H., & Morishita, R. (2011). Molecular mechanisms linking diabetes mellitus and Alzheimer disease: Beta-amyloid peptide, insulin signaling, and neuronal function. *Molecular BioSystems*, 7(6), 1822–1827.

Tan, M. S., Yu, J. T., Jiang, T., Zhu, X. C., Wang, H. F., Zhang, W., et al. (2013). NLRP3 polymorphisms are associated with late-onset Alzheimer's disease in Han Chinese. *Journal of Neuroimmunology*, *265*(1–2), 91–95.

Tanaka, S., Ishii, A., Ohtaki, H., Shioda, S., Yoshida, T., & Numazawa, S. (2013). Activation of microglia induces symptoms of Parkinson's disease in wild-type, but not in IL-1 knockout mice. *Journal of Neuroinflammation*, *10*, 143.

Tejera, D., Mercan, D., Sanchez-Caro, J. M., Hanan, M., Greenberg, D., Soreq, H., et al. (2019). Systemic inflammation impairs microglial Aβ clearance through NLRP3 inflammasome. *The EMBO Journal*, *38*(17), e101064.

Tong, M., Deochand, C., Didsbury, J., & de la Monte, S. M. (2016). T3D-959: A multi-faceted disease remedial drug candidate for the treatment of Alzheimer's disease. *Journal of Alzheimer's Disease*, *51*(1), 123–138.

Toulorge, D., Schapira, A. H., & Hajj, R. (2016). Molecular changes in the postmortem parkinsonian brain. *Journal of Neurochemistry*, *139*(Suppl. 1), 27–58.

Traba, J., Geiger, S. S., Kwarteng-Siaw, M., Han, K., Ra, O. H., Siegel, R. M., et al. (2017). Prolonged fasting suppresses mitochondrial NLRP3 inflammasome assembly and activation via SIRT3-mediated activation of superoxide dismutase 2. *The Journal of Biological Chemistry*, *292*(29), 12153–12164.

Vandanmagsar, B., Youm, Y. H., Ravussin, A., Galgani, J. E., Stadler, K., Mynatt, R. L., et al. (2011). The NLRP3 inflammasome instigates obesity-induced inflammation and insulin resistance. *Nature Medicine*, *17*(2), 179–188.

Veselý, B., Dufek, M., Thon, V., Brozman, M., Királová, S., Halászová, T., et al. (2018). Interleukin 6 and complement serum level study in Parkinson's disease. *Journal of Neural Transmission (Vienna)*, *125*(5), 875–881.

Villumsen, M., Aznar, S., Pakkenberg, B., Jess, T., & Brudek, T. (2019). Inflammatory bowel disease increases the risk of Parkinson's disease: A Danish nationwide cohort study 1977–2014. *Gut*, *68*(1), 18–24.

Visanji, N. P., Brooks, P. L., Hazrati, L. N., & Lang, A. E. (2013). The prion hypothesis in Parkinson's disease: Braak to the future. *Acta Neuropathologica Communications*, *1*, 2.

von Herrmann, K. M., Salas, L. A., Martinez, E. M., Young, A. L., Howard, J. M., Feldman, M. S., et al. (2018). NLRP3 expression in mesencephalic neurons and characterization of a rare NLRP3 polymorphism associated with decreased risk of Parkinson's disease. *NPJ Parkinsons Disease*, *4*, 24.

Wan, Z., Fan, Y., Liu, X., Xue, J., Han, Z., Zhu, C., et al. (2019). NLRP3 inflammasome promotes diabetes-induced endothelial inflammation and atherosclerosis. *Diabetes, Metabolic Syndrome and Obesity*, *12*, 1931–1942.

Wang, F., Hull, R. L., Vidal, J., Cnop, M., & Kahn, S. E. (2001). Islet amyloid develops diffusely throughout the pancreas before becoming severe and replacing endocrine cells. *Diabetes*, *50*(11), 2514–2520.

Wang, L., Zhai, Y. Q., Xu, L. L., Qiao, C., Sun, X. L., Ding, J. H., et al. (2014). Metabolic inflammation exacerbates dopaminergic neuronal degeneration in response to acute MPTP challenge in type 2 diabetes mice. *Experimental Neurology*, *251*, 22–29.

Wang, W. Y., Tan, M. S., Yu, J. T., & Tan, L. (2015). Role of pro-inflammatory cytokines released from microglia in Alzheimer's disease. *Annals of Translational Medicine*, *3*(10), 136.

Wang, W., Nguyen, L. T., Burlak, C., Chegini, F., Guo, F., Chataway, T., et al. (2016). Caspase-1 causes truncation and aggregation of the Parkinson's disease-associated protein α-synuclein. *Proceedings of the National Academy of Sciences of the United States of America*, *113*(34), 9587–9592.

Wang, X. F., Lin, X., Li, D. Y., Zhou, R., Greenbaum, J., Chen, Y. C., et al. (2017). Linking Alzheimer's disease and type 2 diabetes: Novel shared susceptibility genes detected by cFDR approach. *Journal of the Neurological Sciences*, *380*, 262–272.

Warren, K. N., Beason-Held, L. L., Carlson, O., Egan, J. M., An, Y., Doshi, J., et al. (2018). Elevated markers of inflammation are associated with longitudinal changes in brain function in older adults. *The Journals of Gerontology Series A, Biological Sciences and Medical Sciences*, *73*(6), 770–778.

Weidling, I., & Swerdlow, R. H. (2019). Mitochondrial dysfunction and stress responses in Alzheimer's disease. *Biology (Basel)*, *8*(2), pii: E39.

Wen, H., Gris, D., Lei, Y., Jha, S., Zhang, L., Huang, M. T., et al. (2011). Fatty acid-induced NLRP3-ASC inflammasome activation interferes with insulin signaling. *Nature Immunology*, *12*(5), 408–415.

Wenk, G. L. (2003). Neuropathologic changes in Alzheimer's disease. *The Journal of Clinical Psychiatry*, *64*(Suppl. 9), 7–10.

White, A. J., Wijeyekoon, R. S., Scott, K. M., Gunawardana, N. P., Hayat, S., Solim, I. H., et al. (2018). The peripheral inflammatory response to alpha-synuclein and endotoxin in Parkinson's disease. *Frontiers in Neurology*, *9*, 946.

Willette, A. A., Bendlin, B. B., Starks, E. J., Birdsill, A. C., Johnson, S. C., Christian, B. T., et al. (2015a). Association of insulin resistance with cerebral glucose uptake in late middle-aged adults at risk for Alzheimer disease. *JAMA Neurology*, *72*(9), 1013–1020.

Willette, A. A., Johnson, S. C., Birdsill, A. C., Sager, M. A., Christian, B., Baker, L. D., et al. (2015b). Insulin resistance predicts brain amyloid deposition in late middle-aged adults. *Alzheimer's & Dementia*, *11*(5), 504–510.e1.

Winkler, E. A., Nishida, Y., Sagare, A. P., Rege, S. V., Bell, R. D., Perlmutter, D., et al. (2015). GLUT1 reductions exacerbate Alzheimer's disease vasculo-neuronal dysfunction and degeneration. *Nature Neuroscience*, *18*(4), 521–530.

Wolters, F. J., Zonneveld, H. I., Hofman, A., van der Lugt, A., Koudstaal, P. J., Vernooij, M. W., et al. (2017). Cerebral perfusion and the risk of dementia: A population-based study. *Circulation*, *136*(8), 719–728.

Wong, D., Dorovini-Zis, K., & Vincent, S. R. (2004). Cytokines, nitric oxide, and cGMP modulate the permeability of an in vitro model of the human blood-brain barrier. *Experimental Neurology*, *190*(2), 446–455.

Wu, J., Luo, X., Thangthaeng, N., Sumien, N., Chen, Z., Rutledge, M. A., et al. (2017). Pancreatic mitochondrial complex I exhibits aberrant hyperactivity in diabetes. *Biochemistry and Biophysics Reports*, *11*, 119–129.

Xia, M., Boini, K. M., Abais, J. M., Xu, M., Zhang, Y., & Li, P. L. (2014). Endothelial NLRP3 inflammasome activation and enhanced neointima formation in mice by adipokine visfatin. *The American Journal of Pathology*, *184*(5), 1617–1628.

Xia, X., Jiang, Q., McDermott, J., & Han, J. J. (2018). Aging and Alzheimer's disease: Comparison and associations from molecular to system level. *Aging Cell*, *17*(5), e12802.

Xing, J. H., Li, R., Gao, Y. Q., Wang, M. Y., Liu, Y. Z., Hong, J., et al. (2019). NLRP3 inflammasome mediate palmitate-induced endothelial dysfunction. *Life Sciences*, *239*, 116882.

Yan, M. H., Wang, X., & Zhu, X. (2013). Mitochondrial defects and oxidative stress in Alzheimer disease and Parkinson disease. *Free Radical Biology & Medicine*, *62*, 90–101.

Ye, Y., Bajaj, M., Yang, H. C., Perez-Polo, J. R., & Birnbaum, Y. (2017). SGLT-2 inhibition with dapagliflozin reduces the activation of the Nlrp3/ASC inflammasome and attenuates the development of diabetic cardiomyopathy in mice with type 2 diabetes. Further augmentation of the effects with saxagliptin, a DPP4 inhibitor. *Cardiovascular Drugs and Therapy*, *31*(2), 119–132.

Yiannopoulou, K. G., Anastasiou, A. I., Zachariou, V., & Pelidou, S. H. (2019). Reasons for failed trials of disease-modifying treatments for Alzheimer disease and their contribution in recent research. *Biomedicines*, 7(4), pii: E97.

Yin, J., Zhao, F., Chojnacki, J. E., Fulp, J., Klein, W. L., Zhang, S., et al. (2018). NLRP3 inflammasome inhibitor ameliorates amyloid pathology in a mouse model of Alzheimer's disease. *Molecular Neurobiology*, *55*(3), 1977–1987.

Youm, Y. H., Adijiang, A., Vandanmagsar, B., Burk, D., Ravussin, A., & Dixit, V. D. (2011). Elimination of the NLRP3-ASC inflammasome protects against chronic obesity-induced pancreatic damage. *Endocrinology*, *152*(11), 4039–4045.

Youm, Y. H., Nguyen, K. Y., Grant, R. W., Goldberg, E. L., Bodogai, M., Kim, D., et al. (2015). The ketone metabolite β-hydroxybutyrate blocks NLRP3 inflammasome-mediated inflammatory disease. *Nature Medicine*, *21*(3), 263–269.

Zabel, M., Nackenoff, A., Kirsch, W. M., Harrison, F. E., Perry, G., & Schrag, M. (2018). Markers of oxidative damage to lipids, nucleic acids and proteins and antioxidant enzymes activities in Alzheimer's disease brain: A meta-analysis in human pathological specimens. *Free Radical Biology & Medicine*, *115*, 351–360.

Zahid, A., Li, B., Kombe, A. J. K., Jin, T., & Tao, J. (2019). Pharmacological Inhibitors of the NLRP3 Inflammasome. *Frontiers in Immunology*, *10*, 2538.

Zhao, Z., Nelson, A. R., Betsholtz, C., & Zlokovic, B. V. (2015). Establishment and dysfunction of the blood-brain barrier. *Cell*, *163*, 1064–1078.

Zhong, Z., Umemura, A., Sanchez-Lopez, E., Liang, S., Shalapour, S., Wong, J., et al. (2016). NF-κB restricts inflammasome activation via elimination of damaged mitochondria. *Cell*, *164*(5), 896–910.

Zhou, Y., Lu, M., Du, R. H., Qiao, C., Jiang, C. Y., Zhang, K. Z., et al. (2016). MicroRNA-7 targets Nod-like receptor protein 3 inflammasome to modulate neuroinflammation in the pathogenesis of Parkinson's disease. *Molecular Neurodegeneration*, *11*, 28.

Zhou, R., Yazdi, A. S., Menu, P., & Tschopp, J. (2011). A role for mitochondria in NLRP3 inflammasome activation. *Nature*, *469*(7329), 221–225.

Zraika, S., Hull, R. L., Udayasankar, J., Aston-Mourney, K., Subramanian, S. L., Kisilevsky, R., et al. (2009). Oxidative stress is induced by islet amyloid formation and time-dependently mediates amyloid-induced beta cell apoptosis. *Diabetologia*, *52*(4), 626–635.

CHAPTER FOURTEEN

Impact of circadian and diurnal rhythms on cellular metabolic function and neurodegenerative diseases

S. Kendall Smith[a], Erik S. Musiek[b,*]
[a]Department of Anesthesiology, Washington University School of Medicine, St. Louis, MO, United States
[b]Department of Neurology, Washington University School of Medicine, St. Louis, MO, United States
*Corresponding author: e-mail address: musieke@wustl.edu

Contents

1. Introduction to the circadian clock 394
2. Evidence of sleep and circadian disruption in neurodegenerative diseases 396
3. Does circadian disruption influence the pathogenesis of neurodegenerative diseases? 398
4. Circadian rhythms orchestrate cellular metabolism and bioenergetics in the periphery 400
5. Circadian rhythms orchestrate cellular metabolism and bioenergetics in the CNS 402
6. Conclusion and future directions 403
References 404

Abstract

The 24-h rotational period of the earth has driven evolution of biological systems that serve to synchronize organismal physiology and behavior to this predictable environmental event. In mammals, the circadian (*circa*, "about" and *dia*, "a day") clock keeps 24-h time at the organismal and cellular level, optimizing biological function for a given time of day. The most obvious circadian output is the sleep-wake cycle, though countless bodily functions, ranging from hormone levels to cognitive function, are influenced by the circadian clock. Here we discuss the regulation of metabolic pathways by the circadian clock, discuss the evidence implicating circadian and sleep disruption in neurodegenerative diseases, and suggest some possible connections between the clock, metabolism, and neurodegenerative disease.

International Review of Neurobiology, Volume 154
ISSN 0074-7742
https://doi.org/10.1016/bs.irn.2020.02.005

Abbreviations

AD	Alzheimer's disease
ALS	amyotrophic lateral sclerosis
AMPK	adenosine monophosphate kinase
ARNTL	aryl hydrocarbon receptor nuclear translocator-like protein 1
Aβ	amyloid beta
BMAL1	brain and muscle ARNT-like 1
CLOCK	circadian locomotor output cycles kaput
CSF	cerebrospinal fluid
DLB	dementia with Lewy bodies
HD	Huntington's disease
mTOR	mammalian target of rapamycin
NAD^+	nicotinamide adenine dinucleotide
NAMPT	nicotinamide phosphoribosyltransferase
NPAS2	neuronal PAS domain protein 2
PD	Parkinson's disease
PGC1-α	peroxisome proliferator-activated receptor gamma coactivator 1-alpha
RBD	REM behavior disorder
REM	rapid eye movement
SCN	suprachiasmatic nucleus
SIRT	sirtuin

1. Introduction to the circadian clock

The mammalian circadian system is hierarchical, with the suprachiasmatic nucleus (SCN) in the hypothalamus serving as the master clock of the body. The SCN is a complex nucleus that receives direct input from the retina, and where light exposure is transduced into patterns of clock gene expression and ultimately neuronal firing (Colwell, 2011; Herzog et al., 2017). Light-sensitive retinal ganglion cells, including unique non-image forming melanopsin-expressing cells (Gooley et al., 2001; Hattar et al., 2002), synapse on SCN neurons and synchronize oscillation of a molecular circadian clock which drives organized firing of the SCN with a 24-h rhythm that is entrained to the external light-dark cycle (Mohawk et al., 2012). The molecular clock that maintains cellular circadian activity is present not only in SCN neurons, but in most cells of the body, and regulates the expression of 10–50% of all transcripts, depending on cell type (Mohawk et al., 2012; Zhang et al., 2014). Rhythmic SCN activity serves to synchronize the molecular clocks of cells throughout the body, which then regulate cell- and tissue-specific functions in circadian synchrony.

The molecular circadian clock is transcriptional-translational feedback loop consisting of conserved clock proteins. The basic helix-loop helix PER-ARNT-SIM (bHLH-PAS) transcription factor brain and muscle ARNT-like 1 (BMAL1) is the primary positive transcriptional mediator of the clock, though it must heterodimerize with circadian locomotor output cycles kaput (CLOCK) or neuronal PAS domain protein 2 (NPAS2) in order to bind to E-box motifs throughout the genome, driving transcription. BMAL1 drives transcription of its own several negative feedback regulators, including PERIOD 1–3 (*Per1*, *2*, and *3*), CRYPTOCHROME 1 and 2 (*Cry1* and *Cry2*), and REV-ERBα and β (*Nr1d1* and *Nr1d2*) proteins, which inhibit BMAL1-mediated transcription via multiple mechanisms (Mohawk et al., 2012). This feedback loop is tuned to a 24-h period by a complex set of posttranslational regulatory events (Herzog et al., 2017). Once synchronized, cells can sustain near 24-h rhythms in clock gene expression and transcription with no external stimuli. Clock proteins themselves have functions outside of their immediate function in maintaining rhythms. For example, BMAL1 can directly regulate protein translation (Lipton et al., 2015) and mediates developmental effects (Yang et al., 2016), while PERIOD and CRYPTOCHROME genes can regulate carcinogenesis (Fu et al., 2002; Huber et al., 2016). Thus, disrupting clock genes is associated with a variety of striking phenotypes in rodents that are highly dependent on the particular clock gene manipulated (Yu and Weaver, 2011).

Notably, the timing of sleep is strongly influenced by circadian clock genes, as mutations in clock genes in humans often manifest as familial sleep syndromes (Hsu et al., 2015). In mice, deletion of *Bmal1* leads to arrhythmic, fragmented sleep, though the total amount and relative abundance of sleep stages are relatively normal (Laposky et al., 2005). Conversely, sleep deprivation can decrease DNA binding of BMAL1 and CLOCK (Mongrain et al., 2011), and can modestly blunt circadian oscillation in some clock genes, though rhythms in melatonin generally persist (Ackermann et al., 2013; Cedernaes et al., 2015; Lech et al., 2016; Möller-Levet et al., 2013). In the brain, sleep deprivation can cause long-lasting alteration in circadian transcription and blunting of clock gene expression (Hor et al., 2019), and profoundly reduces rhythmic protein expression and phosphorylation at synapses (Brüning et al., 2019; Noya et al., 2019). Thus, the circadian clock appears to function upstream of sleep, though sleep may exert critical mild feedback on the clock function, particularly in the brain outside the SCN.

2. Evidence of sleep and circadian disruption in neurodegenerative diseases

Changes in the sleep-wake pattern and in circadian function seem to be common, conserved symptoms of many neurodegenerative diseases. This issue has been reviewed in detail previously (Leng et al., 2019; Videnovic et al., 2014a), but we will present an overview of the evidence for sleep and circadian disruptions in several neurodegenerative conditions below.

Alzheimer's disease (AD) is the most common neurodegenerative disease in people older than 65, though early-onset and familial forms can strike much earlier. It is characterized by gradual, progressive hippocampal and cortical neurodegeneration, leading to dementia. Pathologically, the aggregation of amyloid beta (Aβ) peptide into extracellular amyloid plaques is an early event, followed years later by intraneuronal aggregation of hyperphosphorylated tau protein, yielding neurofibrillary tangles.

Several studies have shown fragmentation of sleep-wake behavior and circadian activity rhythms in AD dementia patients, who exhibit increased wakefulness and activity at night and increased napping during the day (Harper et al., 2008; Hatfield et al., 2004; Ju et al., 2013; Satlin et al., 1995; Vitiello and Prinz, 1989). Alzheimer's disease patients also have suppressed rhythmic release of melatonin, a circadian hormone critical for sleep timing (Mishima et al., 1999; Skene and Swaab, 2003; Uchida et al., 1996). In addition to this decrease in the amplitude of the overall circadian rhythm, a shift in the time of peak activity toward later hours (phase delay) has been described (Ancoli-Israel et al., 1997; Gehrman et al., 2005), and is hypothesized to potential explain the phenomenon of "sundowning," when dementia patients become more confused and agitated in the evening hours (Volicer et al., 2001).

Several studies show that AD patients have decreased amounts of both deep, restorative non-REM, as well as dream-related REM sleep (Martin et al., 1986; Peter-Derex et al., 2014). People with presymptomatic AD pathology report poor sleep (Benedict et al., 2015; Hahn et al., 2014; Spira et al., 2013; Sprecher et al., 2017) and exhibit decreased sleep efficiency and increased napping (Ju et al., 2013). Mild fragmentation of the circadian activity rhythm can also be detected in presymptomatic AD patients, suggesting that sleep and circadian changes begin very early in the disease process (Musiek et al., 2018). Moreover, poor sleep efficiency (Diem et al., 2016), blunting of circadian rhythms (Tranah et al., 2011) and sleep fragmentation (Lim et al., 2013) are associated with increased risk of incident dementia, suggesting that sleep-wake fragmentation may be an indication of the impending transition from preclinical to symptomatic AD.

On a pathological level, post-mortem studies of AD reveal significant loss of critical neurons in the SCN (Stopa et al., 1999; Swaab et al., 1985; Wang et al., 2015; Zhou et al., 1995). Degeneration of other sleep-related nuclei, including the ventrolateral preoptic area/intermediate nucleus (Lim et al., 2014) and the locus coeruleus (Oh et al., 2019) also occurs, suggesting that AD impacts both the circadian and sleep circuitry. Altered methylation of the *Bmal1* promoter may also play a role in AD-related circadian deficits (Cronin et al., 2017). Finally, AD patients have a loss of melanopsin-containing retinal ganglion cells (La Morgia et al., 2016). These cells relay circadian light information from the retina to the SCN, suggesting that there may be a loss of proper synchronizing input to the SCN in AD.

Parkinson's disease (PD) is the most common neurodegenerative movement disorder. Parkinson's disease is characterized by extrapyramidal motor symptoms, including bradykinesia, resting tremor, shuffling gait, and postural instability. The presence of Lewy bodies, pathological aggregations of α-synuclein protein, as well as degeneration of dopaminergic neurons in the substantia nigra, are hallmarks of disease (Obeso et al., 2017).

However, PD may begin with degeneration of brainstem nuclei. The most striking sleep symptom of PD, and its fellow synucleinopathy, dementia with Lewy bodies (DLB), is REM behavior disorder (RBD). This is characterized by the absence of normal muscle atonia during REM sleep, which leads to patients acting out their dreams, often violently (St Louis and Boeve, 2017). REM behavior disorder has been strongly linked to PD/DLB, with over 80% of RBD patients eventually developing PD or DLB on one cohort (Schenck et al., 2013). However, RBD can precede PD/DLB by decades in some cases, suggesting that RBD is a very early marker of brainstem neurodegeneration in these diseases (Borghammer and Van Den Berge, 2019).

Aside from RBD, PD/DLB patients often experience changes in sleep quality and excessive daytime sleepiness (Chahine et al., 2017; Videnovic and Golombek, 2013). Parkinson's disease-related changes in circadian rhythms are also prominent, including dampened amplitude of activity and melatonin secretion (Breen et al., 2014; Videnovic and Golombek, 2013; Videnovic et al., 2014b; Whitehead et al., 2008). Rhythmic expression of circadian clock genes in blood has also been described in PD patients, with dampened amplitude of several transcripts and decreased *Bmal1* expression (Breen et al., 2014; Cai et al., 2010). Pathologically, Lewy bodies have been reported in the SCN of PD patients, suggesting possible SCN degeneration as a mechanism of circadian dysfunction (De Pablo-Fernandez et al., 2018).

Huntington's disease (HD) is an autosomal-dominant neurodegenerative disorder caused by trinucleotide repeat expansion in the *huntingtin* gene. Huntington's disease is characterized by progressive degeneration of the striatum in early middle age, causing choreiform movements, neuropsychiatric symptoms, and dementia (Roos, 2010).

Several studies have shown sleep and circadian disturbances in patients with HD. Decreased sleep quality (as assessed by sleep questionnaire) and sleep-wake fragmentation (as assessed by actigraphy) occur very early in the disease (Bellosta Diago et al., 2017; Lazar et al., 2015) and worsen with disease progression (Morton et al., 2005). Circadian melatonin rhythms are also blunted in HD (Kalliolia et al., 2014).

While human studies are limited, mouse models of HD, which express CAG-expanded human *huntingtin*, recapitulate the circadian dysfunction. These mice experience nearly total arrhythmicity as they age, with disrupted SCN function (Kudo et al., 2011; Morton et al., 2005). Importantly, re-establishing a day-night rhythm in HD mice by inducing sleep with the benzodiazepine alprazolam each morning was sufficient to improve cognitive function (Pallier et al., 2007), while daily blue light therapy to stimulate the SCN mitigated pathological transcriptional changes (Wang et al., 2017). Time restricted feeding in a mouse HD model during the active phase only, which helps align circadian and metabolic function, also improved rhythms in behavior and clock gene expression, as well as motor function (Whittaker et al., 2018).

3. Does circadian disruption influence the pathogenesis of neurodegenerative diseases?

While sleep and circadian disruption is a well-established symptom of neurodegenerative disease, the evidence that sleep and or circadian disruption causes or accelerates the neurodegenerative process is less firm. The strongest evidence comes from the AD field. In mice, levels of Aβ peptide in the interstitial fluid of the hippocampus exhibit clear diurnal rhythms which correspond closely with the sleep-wake cycle, as Aβ levels increase during wake and decrease during sleep (Kang et al., 2009).

This is consistent with data showing that Aβ is released from neurons during synaptic activity (Cirrito et al., 2005). Sleep deprivation in mice keeps Aβ levels elevated, while disruption of the circadian clock abrogates these rhythms in both sleep-wake cycle and Aβ (Kang et al., 2009; Kress et al., 2018). Sleep deprivation in mouse models of AD which express human APP and PS1 transgenes, and thus develop amyloid plaque pathology, greatly accelerates the accumulation of plaques. Conversely, sleep enhancement using the orexin receptor antagonist drug almorexant, or by genetic deletion of the sleep-modulating neuropeptide orexin, decreases plaque accumulation (Kang et al., 2009; Roh et al., 2014).

In humans, diurnal rhythms in cerebrospinal fluid (CSF) Aβ can be observed, and acute sleep deprivation causes increases in CSF Aβ over the course of the night (Huang et al., 2012; Lucey et al., 2017; Ooms et al., 2014). Positron emission tomography imaging of amyloid plaques in humans suggests that a single night of sleep deprivation may slightly increase amyloid deposition (Shokri-Kojori et al., 2018). Sleep in mice is also associated with striking increases in glia-mediated fluid flow through the brain, termed "glymphatic" flow, which appears to mediate increased bulk clearance of Aβ from the brain (Xie et al., 2013). In humans, non-REM sleep is associated with waves of CSF movement which are coupled to vascular oscillations and may serve to remove metabolites and other waste products from the brain (Fultz et al., 2019).

Like Aβ, tau protein levels in the interstitial fluid also show diurnal oscillations and are increased by sleep deprivation in mice (Holth et al., 2019). Tau is also released from neurons during synaptic activity (Yamada et al., 2014), and sleep deprivation exacerbates tau pathology in human tau transgenic mice (Di Meco et al., 2014; Holth et al., 2019; Rothman et al., 2013; Zhu et al., 2018). Moreover, acute sleep deprivation in humans causes increases in CSF tau levels, while chronic fragmented sleep is also associated with higher CSF tau (Holth et al., 2019; Ju et al., 2017).

Thus, sleep and the circadian system have striking effects on both Aβ and tau levels in humans and mice, and on plaque and tangle pathology in AD mouse models, strongly suggesting that decreased or fragmented sleep may influence AD pathogenesis.

Evidence of a role for the circadian system in disease pathogenesis in other neurodegenerative conditions is less clear, and generally restricted to mouse experiments. Circadian disruption by exposure to altered light: dark cycles (20 h:4 h light:dark) increases α-synuclein pathology in a mouse PD model (Lauretti et al., 2017). Similarly, exposure of mice expressing human mutant SOD1, which causes familial amyotrophic lateral sclerosis (ALS), to 20 h:4 h light:dark cycle accelerated motor deficits, gliosis, and death (Huang et al., 2018). The ALS-associated gene fused in sarcoma (FUS), which encodes an RNA binding protein, shows strong circadian oscillation in the brain, is regulated by the clock protein REV-ERBα, and can inhibit PERIOD 2 expression, thereby altering core clock function (Jiang et al., 2018). Rats expressing mutant FUS also show marked circadian and sleep deficits which occur very early in the disease course (Zhang et al., 2018), though a causative role for circadian dysfunction has not been proven.

Epidemiological data is sparse, with one study suggesting that shift work (which causes circadian disruption and sleep abnormalities, among other things) during middle age may increase risk of late-life dementia (Bokenberger et al., 2018), but not PD (Chen et al., 2006; Schernhammer et al., 2015). A few small studies have suggested associations between polymorphisms in the circadian gene CLOCK (Chen et al., 2013a, 2013b; Yang et al., 2013) or BMAL1 (Chen et al., 2015) and risk of AD. Polymorphisms in BMAL1 or PER1 have also been associated with PD risk (Gu et al., 2015).

Thus, while current data is suggestive, particularly for AD, definitive evidence that sleep/circadian disruption is a causative agent in neurodegeneration is still lacking. Furthermore, specific cellular mechanisms linking the clock to neurodegeneration are still poorly understood. Circadian regulation of metabolism may represent such a link.

4. Circadian rhythms orchestrate cellular metabolism and bioenergetics in the periphery

It is well known that the circadian clock orchestrates cellular metabolism in peripheral cells thereby allowing oxidative and reductive cycles to synchronize with the sleep-wake cycle (Bass, 2016). The bidirectional relationship between the clock and metabolism has been reviewed extensively elsewhere (Bass and Takahashi, 2010; Eckel-Mahan and Sassone-Corsi, 2013; Panda, 2016). One major mechanism is through control of nicotinamide adenine dinucleotide (NAD^+), which is canonically known as a coenzyme that mediates many cellular redox reactions. For example, cellular NAD^+ levels in mouse hepatocytes oscillate as a result of clock-mediated transcriptional control of nicotinamide phosphoribosyltransferase (NAMPT), the rate-limiting enzyme in NAD^+ biosynthesis (Nakahata et al., 2009;

Ramsey et al., 2009). NAD^+ deficiency is implicated in the aging process, and increasing NAD^+ levels remains a major therapeutic anti-aging strategy (Imai and Guarente, 2014; Imai and Johnson, 2018).

Sirtuins, which are highly conserved NAD^+-dependent deacetylases, have also been implicated in the regulation of many metabolic pathways influencing aging and neurodegeneration, and have strong interaction with the circadian clock (Houtkooper et al., 2012; Imai and Guarente, 2014). Sirtuin 1 (SIRT1), the most studied sirtuin protein, regulates transcription of multiple core clock genes (*Bmal1*, Rorγ, *Per2*, *Cry1*) in mouse hepatocytes or fibroblasts (Asher et al., 2008; Nakahata et al., 2008). SIRT1 positively regulates circadian clock function via deacetylation of the clock protein PERIOD 2, as well as peroxisome proliferator-activated receptor gamma coactivator 1-alpha (PGC-1α), a master regulator of mitochondrial biogenesis (Foteinou et al., 2018), and may influence chromatic remodeling (Nakahata et al., 2008).

Mitochondrial function exhibits circadian regulation at multiple levels, some of which are NAD^+ dependent. For example, alterations in SIRT3 activity in mouse liver cells, regulated by NAD^+ levels, results in an oscillatory pattern of mitochondrial oxidative function at the appropriate times in the solar cycle (Peek et al., 2013). Additionally, mitochondrial fission and mitophagy have been shown to be regulated by the clock protein BMAL1 in mouse hepatocytes (Jacobi et al., 2015) and by the clock protein and nuclear receptor REV-ERBα in mouse muscle (Woldt et al., 2013). While we know that autophagy and mitophagy are critical in neurodegeneration (Menzies et al., 2015), direct links to the circadian clock remain less well defined. Mitochondrial enzymes responsible for carbohydrate and lipid metabolism appear to accumulate in a diurnal manner dependent on clock proteins PER1/2 (Neufeld-Cohen et al., 2016). It is also noteworthy that the circadian clock machinery is also implicated in mouse mitochondrial regulation via reversible acetylation, reinforcing the notion that circadian machinery also governs through post-translational mechanisms (Masri et al., 2013).

The nutrient sensor mammalian target of rapamycin (mTOR) also interacts with the circadian clock. Insulin signaling and other nutrient-related pathways regulate mTOR activation, which is turn controls numerous cellular metabolic processes, including autophagy, oxidative stress responses, and protein synthesis (Liu and Sabatini, 2020). In the SCN, mTOR regulates entrainment of neuronal firing to light and helps to maintain synchrony of circadian output, as mTOR-deficient mice have a longer circadian period when kept in constant darkness (Cao et al., 2010; Liu et al., 2018). At a cellular level, mTOR regulates transcription of multiple clock genes, thereby directly influencing core clock function (Ramanathan et al., 2018). Insulin itself can alter core clock function by inducing PERIOD proteins via an mTOR-dependent mechanism (Crosby et al., 2019). Additionally, adenosine monophosphate kinase (AMPK), another nutrient-sensing kinase which functions upstream of mTOR, regulates the circadian clock via phosphorylation of CRYPTOCHROME (Lamia et al., 2009). Thus, nutrient sensing and cellular metabolism are intertwined with circadian clock function.

5. Circadian rhythms orchestrate cellular metabolism and bioenergetics in the CNS

While many studies demonstrate that the circadian clock regulates many aspects of metabolism in peripheral tissues, a smaller but growing body of literature suggests that the interplay between the circadian clock machinery and metabolism translates to the CNS. In the mouse cortex and hippocampus, deletion of BMAL1 results in age-dependent astrogliosis, synaptic degeneration, neuronal oxidative damage and decreased expression of several redox defense genes (Lananna et al., 2018; Musiek et al., 2013). Other lines of evidence support the concept that synaptic gene expression and protein phosphorylation are strongly influenced by the clock and sleep. In mice, circadian gene expression patterns are present in neuronal cell bodies, while striking rhythms in protein expression and phosphorylation are observed only in synapses (Brüning et al., 2019; Noya et al., 2019). Rhythmic proteins expressed in synapses during the mouse active phase (dark) were generally involved in neuronal metabolism, and these rhythms were abrogated by sleep deprivation (Noya et al., 2019), suggesting that sleep and the circadian system interact to support synaptic metabolic function.

Circadian regulation of NAD^+, sirtuins, mTOR, and PGC-1α function may also play a significant role within the CNS, although this must still be inferred. Deletion of PGC-1α promotes neurodegeneration, while overexpression is protective (St-Pierre et al., 2006). Genetic or pharmacological activation of SIRT1 is neuroprotective in mouse AD and HD models, while SIRT1 deletion exacerbates degeneration (Jiang et al., 2011; Kim et al., 2007). Interaction between the clock and SIRT1 has also been demonstrated in the mouse SCN, where SIRT1 directly regulates central circadian control by activating BMAL1 and CLOCK via deacetylation of PGC-1α (Bonda et al., 2011; Chang and Guarente, 2013). Furthermore, aging was associated with decreased SIRT1 levels and decreased activation of BMAL1 and PER2, resulting in weakened circadian control (Chang and Guarente, 2013). Moreover, inhibition of mTOR in the brain is linked to autophagy and degradation of pathogenic proteins (Caccamo et al., 2013; Ravikumar et al., 2004; Tsvetkov et al., 2010), and mTOR inhibition has been proposed as a novel therapeutic strategy for neurodegenerative diseases (Bové et al., 2011). Inhibition of mTOR has also been shown to increase lifespan in mice (Harrison et al., 2009). Taken together, these studies suggest metabolic regulation in the brain by the circadian clock could influence brain aging and neurodegeneration.

The circadian clock has been shown to interact with a number of other metabolic pathways, which are critical for brain health and neurodegeneration. The clock controls redox homeostasis through regulation of the Nrf2 pathway (Pekovic-Vaughan et al., 2014), glutathione homeostasis (Beaver et al., 2012), and interaction with the pentose phosphate pathway (Rey et al., 2016). Autophagy is also influenced by clock proteins, including REV-ERBα (Huang et al., 2016; Woldt et al., 2013). Circadian regulation of glial function could have profound effects on brain metabolism and neurodegeneration (Lananna et al., 2018; Marpegan et al., 2011). Thus, metabolic regulation by the circadian clock could influence neurodegeneration via many mechanisms, although at this point few have been thoroughly explored in the brain.

6. Conclusion and future directions

A deep understanding of how sleep and circadian clocks influence the panoply of brain metabolism and neurodegeneration could illuminate new therapeutic strategies. However, much is still unknown. The exact molecular pathways controlled by the circadian systems in different cell types of the brain, as well as the effects of clock perturbation in disease models are not well understood. Optimizing circadian function with clock-targeted drugs could, in theory, synergize multiple processes, including brain metabolism, inflammation, redox defenses, proteostasis, and synaptic function, to fight neurodegeneration on multiple fronts. Only with continued discovery can we potentially reach this goal and advance from initial efforts such as morning light exposure and melatonin supplementation to more nuanced strategies, harnessing specific clock-controlled pathways to protect the brain.

References

Ackermann, K., et al. (2013). Effect of sleep deprivation on rhythms of clock gene expression and melatonin in humans. *Chronobiology International, 30*, 901–909. https://doi.org/10.3109/07420528.2013.784773.

Ancoli-Israel, S., et al. (1997). Variations in circadian rhythms of activity, sleep, and light exposure related to dementia in nursing-home patients. *Sleep, 20*(1), 18–23.

Asher, G., et al. (2008). SIRT1 regulates circadian clock gene expression through PER2 deacetylation. *Cell, 134*(2), 317–328. https://doi.org/10.1016/j.cell.2008.06.050.

Bass, J. (2016). Circadian mechanisms in bioenergetics and cell metabolism. In P. Sassone-Corsi & Y. Christen (Eds.), *A time for metabolism and hormones* (pp. 25–32). Springer. https://doi.org/10.1007/978-3-319-27069-2.

Bass, J., & Takahashi, J. S. (2010). Circadian integration of metabolism and energetics. *Science, 330*(6009), 1349–1354. Available at: http://www.ncbi.nlm.nih.gov/entrez/query.fcgi?cmd=Retrieve&db=PubMed&dopt=Citation&list_uids=21127246.

Beaver, L. M., et al. (2012). Circadian regulation of glutathione levels and biosynthesis in *Drosophila melanogaster. PLoS One*, 7(11), e50454. https://doi.org/10.1371/journal.pone.0050454 PONE-D-12-24156 [pii].

Bellosta Diago, E., et al. (2017). Circadian rhythm and autonomic dysfunction in presymptomatic and early Huntington's disease. *Parkinsonism & Related Disorders, 44*, 95–100. https://doi.org/10.1016/j.parkreldis.2017.09.013. Elsevier Ltd.

Benedict, C., et al. (2015). Self-reported sleep disturbance is associated with Alzheimer's disease risk in men. *Alzheimers Dement, 11*(9), 1090–1097. https://doi.org/10.1016/j.jalz.2014.08.104.

Bokenberger, K., et al. (2018). Shift work and risk of incident dementia: A study of two population-based cohorts. *European Journal of Epidemiology, 33*, 977–987. https://doi.org/10.1007/s10654-018-0430-8.

Bonda, D. J., et al. (2011). The critical role of the sirtuin pathway in aging and Alzheimer disease: Mechanistic and therapeutic considerations. *Lancet Neurology, 10*(3), 275–279. https://doi.org/10.1016/S1474-4422(11)70013-8.

Borghammer, P., & Van Den Berge, N. (2019). Brain-first versus gut-first Parkinson's disease: A hypothesis. *Journal of Parkinson's Disease, 9*, S281–S295. https://doi.org/10.3233/JPD-191721.

Bové, J., Martínez-Vicente, M., & Vila, M. (2011). Fighting neurodegeneration with rapamycin: Mechanistic insights. *Nature Reviews Neuroscience, 12*, 437–452. https://doi.org/10.1038/nrn3068.

Breen, D. P., et al. (2014). Sleep and circadian rhythm regulation in early Parkinson disease. *JAMA Neurology, 71*(5), 589–595. https://doi.org/10.1001/jamaneurol.2014.65.

Brüning, F., et al. (2019). Sleep-wake cycles drive daily dynamics of synaptic phosphorylation. *Science, 366*(6462), eaav3617, American Association for the Advancement of Science. https://doi.org/10.1126/science.aav3617.

Caccamo, A., et al. (2013). mTOR regulates tau phosphorylation and degradation: Implications for Alzheimer's disease and other tauopathies. *Aging Cell, 12*, 370–380. https://doi.org/10.1111/acel.12057.

Cai, Y., et al. (2010). Expression of clock genes Per1 and Bmal1 in total leukocytes in health and Parkinson's disease. *European Journal of Neurology, 17*(4), 550–554. https://doi.org/10.1111/j.1468-1331.2009.02848.x.

Cao, R., et al. (2010). Mammalian target of rapamycin signaling modulates photic entrainment of the suprachiasmatic circadian clock. *Journal of Neuroscience, 30*, 6302–6314. https://doi.org/10.1523/JNEUROSCI.5482-09.2010.

Cedernaes, J., et al. (2015). Acute sleep loss induces tissue-specific epigenetic and transcriptional alterations to circadian clock genes in men. *The Journal of Clinical Endocrinology and Metabolism, 100*(9), E1255–E1261. https://doi.org/10.1210/jc.2015-2284.

Chahine, L. M., Amara, A. W., & Videnovic, A. (2017). A systematic review of the literature on disorders of sleep and wakefulness in Parkinson's disease from 2005 to 2015. *Sleep Medicine Reviews*, *35*, 33–50. https://doi.org/10.1016/j.smrv.2016.08.001.
Chang, H. C., & Guarente, L. (2013). XSIRT1 mediates central circadian control in the SCN by a mechanism that decays with aging. *Cell*, *153*(7), 1448. https://doi.org/10.1016/j.cell.2013.05.027. Elsevier Inc.
Chen, H., et al. (2006). A prospective study of night shift work, sleep duration, and risk of Parkinson's disease. *American Journal of Epidemiology*, *163*, 726–730. https://doi.org/10.1093/aje/kwj096.
Chen, H. F., et al. (2013a). Polymorphism of CLOCK gene rs 4580704 C > G is associated with susceptibility of Alzheimer's disease in a Chinese population. *Archives of Medical Research*, *44*(3), 203–207, S0188-4409(13)00036-2 [pii]. https://doi.org/10.1016/j.arcmed.2013.01.002.
Chen, Q., et al. (2013b). Functional CLOCK gene rs1554483 G/C polymorphism is associated with susceptibility to Alzheimer's disease in the Chinese population. *The Journal of International Medical Research*, *41*(2), 340–346, 41/2/340 [pii]. https://doi.org/10.1177/0300060513476430.
Chen, Q., et al. (2015). Association between ARNTL (BMAL1) rs2278749 polymorphism T >C and susceptibility to Alzheimer disease in a Chinese population. *Genetics and Molecular Research*, *14*(4), 18515–18522. https://doi.org/10.4238/2015.December.23.39.
Cirrito, J. R., et al. (2005). Synaptic activity regulates interstitial fluid amyloid-beta levels in vivo. *Neuron*, *48*(6), 913–922, S0896-6273(05)00939-6 [pii]. https://doi.org/10.1016/j.neuron.2005.10.028.
Colwell, C. S. (2011). Linking neural activity and molecular oscillations in the SCN. *Nature Reviews. Neuroscience*, *12*(10), 553–569, nrn3086 [pii]. https://doi.org/10.1038/nrn3086.
Cronin, P., et al. (2017). Circadian alterations during early stages of Alzheimer's disease are associated with aberrant cycles of DNA methylation in BMAL1. *Alzheimers Dement*, *13*(6), 689–700, S1552-5260(16)33058-8 [pii]. https://doi.org/10.1016/j.jalz.2016.10.003.
Crosby, P., et al. (2019). Insulin/IGF-1 drives PERIOD synthesis to entrain circadian rhythms with feeding time. *Cell*, *177*, 896–909.e20. https://doi.org/10.1016/j.cell.2019.02.017.
De Pablo-Fernandez, E., et al. (2018). A histologic study of the circadian system in Parkinson disease, multiple system atrophy, and progressive supranuclear palsy. *JAMA Neurology*, *75*(8), 1008–1012. https://doi.org/10.1001/jamaneurol.2018.0640.
Di Meco, A., Joshi, Y. B., & Pratico, D. (2014). Sleep deprivation impairs memory, tau metabolism, and synaptic integrity of a mouse model of Alzheimer's disease with plaques and tangles. *Neurobiology of Aging*, *35*(8), 1813–1820, S0197-4580(14)00203-6 [pii]. https://doi.org/10.1016/j.neurobiolaging.2014.02.011.
Diem, S. J., et al. (2016). Measures of sleep-wake patterns and risk of mild cognitive impairment or dementia in older women. *The American Journal of Geriatric Psychiatry*, *24*(3), 248–258. https://doi.org/10.1016/j.jagp.2015.12.002.
Eckel-Mahan, K., & Sassone-Corsi, P. (2013). Metabolism and the circadian clock converge. *Physiological Reviews*, *93*, 107–135. https://doi.org/10.1152/physrev.00016.2012.
Foteinou, P. T., et al. (2018). Computational and experimental insights into the circadian effects of SIRT1. *Proceedings of the National Academy of Sciences of the United States of America*, *115*(45), 11643–11648. https://doi.org/10.1073/pnas.1803410115.
Fu, L., et al. (2002). The circadian gene Period2 plays an important role in tumor suppression and DNA damage response in vivo. *Cell*, *111*(1), 41–50. Available at: http://www.ncbi.nlm.nih.gov/entrez/query.fcgi?cmd=Retrieve&db=PubMed&dopt=Citation&list_uids=12372299.

Fultz, N. E., et al. (2019). Coupled electrophysiological, hemodynamic, and cerebrospinal fluid oscillations in human sleep. *Science*, *366*, 628–631. https://doi.org/10.1126/science.aax5440.

Gehrman, P., et al. (2005). The relationship between dementia severity and rest/activity circadian rhythms. *Neuropsychiatric Disease and Treatment*, *1*(2), 155–163.

Gooley, J. J., et al. (2001). Melanopsin in cells of origin of the retinohypothalamic tract. *Nature Neuroscience*, *4*, 1165. https://doi.org/10.1038/nn768.

Gu, Z., et al. (2015). Association of ARNTL and PER1 genes with Parkinson's disease: A case-control study of Han Chinese. *Scientific Reports*, *5*, 15891. https://doi.org/10.1038/srep15891.

Hahn, E. A., et al. (2014). A change in sleep pattern may predict Alzheimer disease. *The American Journal of Geriatric Psychiatry*, *22*(11), 1262–1271, S1064-7481(13)00233-9 [pii]. https://doi.org/10.1016/j.jagp.2013.04.015.

Harper, D. G., et al. (2008). Dorsomedial SCN neuronal subpopulations subserve different functions in human dementia. *Brain*, *131*(Pt. 6), 1609–1617. https://doi.org/10.1093/brain/awn049.

Harrison, D. E., et al. (2009). Rapamycin fed late in life extends lifespan in genetically heterogeneous mice. *Nature*, 460, 392–395. https://doi.org/10.1038/nature08221.

Hatfield, C. F., et al. (2004). Disrupted daily activity/rest cycles in relation to daily cortisol rhythms of home-dwelling patients with early Alzheimer's dementia. *Brain*, *127*(Pt. 5), 1061–1074, awh129 [pii]. https://doi.org/10.1093/brain/awh129.

Hattar, S., et al. (2002). Melanopsin-containing retinal ganglion cells: Architecture, projections, and intrinsic photosensitivity. *Science*, *295*, 1065–1070. https://doi.org/10.1126/science.1069609.

Herzog, E. D., et al. (2017). Regulating the suprachiasmatic nucleus (SCN) circadian clockwork: Interplay between cell-autonomous and circuit-level mechanisms. *Cold Spring Harbor Perspectives in Biology*, *9*(1), a027706, 9/1/a027706 [pii]. https://doi.org/10.1101/cshperspect.a027706.

Holth, J. K., et al. (2019). The sleep-wake cycle regulates brain interstitial fluid tau in mice and CSF tau in humans. *Science*, *363*(6429), 880–884, science.aav2546 [pii]. https://doi.org/10.1126/science.aav2546.

Hor, C. N., et al. (2019). Sleep–wake-driven and circadian contributions to daily rhythms in gene expression and chromatin accessibility in the murine cortex. *Proceedings of the National Academy of Sciences*, *116*, 201910590. https://doi.org/10.1073/pnas.1910590116.

Houtkooper, R. H., Pirinen, E., & Auwerx, J. (2012). Sirtuins as regulators of metabolism and healthspan. *Nature Reviews Molecular Cell Biology*, *13*(4), 225–238, Nature Publishing Group. https://doi.org/10.1038/nrm3293.

Hsu, P. K., Ptacek, L. J., & Fu, Y. H. (2015). Genetics of human sleep behavioral phenotypes. *Methods in Enzymology*, *552*, 309–324, S0076-6879(14)00047-0 [pii]. https://doi.org/10.1016/bs.mie.2014.10.046.

Huang, Y., et al. (2012). Effects of age and amyloid deposition on Abeta dynamics in the human central nervous system. *Archives of Neurology*, *69*(1), 51–58, archneurol.2011.235 [pii]. https://doi.org/10.1001/archneurol.2011.235.

Huang, G., et al. (2016). The circadian clock regulates autophagy directly through the nuclear hormone receptor Nr1d1/Rev-erbalpha and indirectly via Cebpb/(C/ebpbeta) in zebrafish. *Autophagy*, *12*(8), 1292–1309. https://doi.org/10.1080/15548627.2016.1183843.

Huang, Z., et al. (2018). Circadian rhythm dysfunction accelerates disease progression in a mouse model with amyotrophic lateral sclerosis. *Frontiers in Neurology*, *9*, 218. https://doi.org/10.3389/fneur.2018.00218.

Huber, A. L., et al. (2016). CRY2 and FBXL3 cooperatively degrade c-MYC. *Molecular Cell*, *64*, 774–789. https://doi.org/10.1016/j.molcel.2016.10.012.

Imai, S. i., & Guarente, L. (2014). NAD+ and sirtuins in aging and disease. *Trends in Cell Biology*, *24*(8), 464–471. https://doi.org/10.1016/j.tcb.2014.04.002.

Imai, S. i., & Johnson, S. (2018). NAD+ biosynthesis, aging, and disease. *F1000Research*, 7, 1–10. https://doi.org/10.12688/f1000research.12120.1.

Jacobi, D., et al. (2015). Hepatic Bmal1 regulates rhythmic mitochondrial dynamics and promotes metabolic fitness. *Cell Metabolism*, *22*(4), 709–720, Elsevier Inc. https://doi.org/10.1016/j.cmet.2015.08.006.

Jiang, M., et al. (2011). Neuroprotective role of SIRT1 in mammalian models of Huntington's disease through activation of multiple SIRT1 targets. *Nature Medicine*, *18*(1), 153–158. https://doi.org/10.1038/nm.2558.Neuroprotective.

Jiang, X., et al. (2018). Neurodegeneration-associated FUS is a novel regulator of circadian gene expression. *Translational Neurodegeneration*, 7, 24. https://doi.org/10.1186/s40035-018-0131-y.

Ju, Y. E., et al. (2013). Sleep quality and preclinical Alzheimer disease. *JAMA Neurology*, *70*(5), 587–593, 1663363 [pii]. https://doi.org/10.1001/jamaneurol.2013.2334.

Ju, Y. E. S., et al. (2017). Slow wave sleep disruption increases cerebrospinal fluid amyloid-β levels. *Brain*, *140*, 2104–2111. https://doi.org/10.1093/brain/awx148.

Kalliolia, E., et al. (2014). Plasma melatonin is reduced in Huntington's disease. *Movement Disorders*, *29*(12), 1511–1515. John Wiley and Sons Inc. https://doi.org/10.1002/mds.26003.

Kang, J. E., et al. (2009). Amyloid-beta dynamics are regulated by orexin and the sleep-wake cycle. *Science*, *326*(5955), 1005–1007. Available at: http://www.ncbi.nlm.nih.gov/entrez/query.fcgi?cmd=Retrieve&db=PubMed&dopt=Citation&list_uids=19779148.

Kim, D., et al. (2007). SIRT1 deacetylase protects against neurodegeneration in models for Alzheimer's disease and amyotrophic lateral sclerosis. *EMBO Journal*, *26*(13), 3169–3179. https://doi.org/10.1038/sj.emboj.7601758.

Kress, G. J., et al. (2018). Regulation of amyloid-β dynamics and pathology by the circadian clock. *Journal of Experimental Medicine*, *215*(4). jem.20172347https://doi.org/10.1084/jem.20172347.

Kudo, T., et al. (2011). Dysfunctions in circadian behavior and physiology in mouse models of Huntington's disease. *Experimental Neurology*, *228*(1), 80–90, S0014-4886(10)00450-4 [pii]. https://doi.org/10.1016/j.expneurol.2010.12.011 .

La Morgia, C., et al. (2016). Melanopsin retinal ganglion cell loss in Alzheimer disease. *Annals of Neurology*, *79*(1), 90–109. https://doi.org/10.1002/ana.24548.

Lamia, K. A., et al. (2009). AMPK regulates the circadian clock by cryptochrome phosphorylation and degradation. *Science*, *326*(5951), 437–440. Available at: http://www.ncbi.nlm.nih.gov/entrez/query.fcgi?cmd=Retrieve&db=PubMed&dopt=Citation&list_uids=19833968.

Lananna, B. V., et al. (2018). Cell-autonomous regulation of astrocyte activation by the circadian clock protein BMAL1. *Cell Reports*, *25*(1), 1–9.e5. https://doi.org/10.1016/j.celrep.2018.09.015.

Laposky, A., et al. (2005). Deletion of the mammalian circadian clock gene BMAL1/Mop3 alters baseline sleep architecture and the response to sleep deprivation. *Sleep*, *28*(4), 395–409. Available at: http://www.ncbi.nlm.nih.gov/entrez/query.fcgi?cmd=Retrieve&db=PubMed&dopt=Citation&list_uids=16171284.

Lauretti, E., et al. (2017). Circadian rhythm dysfunction: A novel environmental risk factor for Parkinson's disease. *Molecular Psychiatry*, *22*, 280–286. https://doi.org/10.1038/mp.2016.47.

Lazar, A. S., et al. (2015). Sleep deficits but no metabolic deficits in premanifest Huntington's disease. *Annals of Neurology*, *78*(4), 630–648. https://doi.org/10.1002/ana.24495.

Lech, K., et al. (2016). Dissecting daily and circadian expression rhythms of clock-controlled genes in human blood. *Journal of Biological Rhythms*, *31*, 68–81. https://doi.org/10.1177/0748730415611761.

Leng, Y., et al. (2019). Association between circadian rhythms and neurodegenerative diseases. *Lancet Neurology*, *18*(3), 307–318. https://doi.org/10.1016/S1474-4422(18)30461-7.

Lim, A. S., et al. (2013). Sleep fragmentation and the risk of incident Alzheimer's disease and cognitive decline in older persons. *Sleep*, *36*(7), 1027–1032. https://doi.org/10.5665/sleep.2802.

Lim, A. S., et al. (2014). Sleep is related to neuron numbers in the ventrolateral preoptic/intermediate nucleus in older adults with and without Alzheimer's disease. *Brain*, *137*(Pt. 10), 2847–2861, awu222 [pii]. https://doi.org/10.1093/brain/awu222.

Lipton, J. O., et al. (2015). The circadian protein BMAL1 regulates translation in response to S6K1-mediated phosphorylation. *Cell*, *161*, 1138–1151. https://doi.org/10.1016/j.cell.2015.04.002.

Liu, G. Y., & Sabatini, D. M. (2020). mTOR at the nexus of nutrition, growth, ageing and disease. *Nature Reviews Molecular Cell Biology*, 1–21. https://doi.org/10.1038/s41580-019-0199-y.

Liu, D., et al. (2018). mTOR signaling in VIP neurons regulates circadian clock synchrony and olfaction. *Proceedings of the National Academy of Sciences of the United States of America*, *115*, E3296–E3304. https://doi.org/10.1073/pnas.1721578115.

Lucey, B. P., et al. (2017). Effect of sleep on overnight CSF amyloid-beta kinetics. *Annals of Neurology*, *83*, 197–204. https://doi.org/10.1002/ana.25117.

Marpegan, L., et al. (2011). Circadian regulation of ATP release in astrocytes. *The Journal of Neuroscience*, *31*(23), 8342–8350. Available at: http://www.ncbi.nlm.nih.gov/entrez/query.fcgi?cmd=Retrieve&db=PubMed&dopt=Citation&list_uids=21653839.

Martin, P. R., et al. (1986). Sleep EEG in Korsakoff's psychosis and Alzheimer's disease. *Neurology*, *36*(3), 411–414. Available at: http://www.ncbi.nlm.nih.gov/pubmed/3951711.

Masri, S., et al. (2013). Circadian acetylome reveals regulation of mitochondrial metabolic pathways. *Proceedings of the National Academy of Sciences of the United States of America*, *110*(9), 3339–3344. https://doi.org/10.1073/pnas.1217632110.

Menzies, F. M., Fleming, A., & Rubinsztein, D. C. (2015). Compromised autophagy and neurodegenerative diseases. *Nature Reviews Neuroscience*, *16*(6), 345–357, Nature Publishing Group. https://doi.org/10.1038/nrn3961.

Mishima, K., et al. (1999). Melatonin secretion rhythm disorders in patients with senile dementia of Alzheimer's type with disturbed sleep-waking. *Biological Psychiatry*, *45*(4), 417–421.

Mohawk, J. A., Green, C. B., & Takahashi, J. S. (2012). Central and peripheral circadian clocks in mammals. *Annual Review of Neuroscience*, *35*, 445–462. Available at: http://www.ncbi.nlm.nih.gov/entrez/query.fcgi?cmd=Retrieve&db=PubMed&dopt=Citation&list_uids=22483041.

Möller-Levet, C. S., et al. (2013). Effects of insufficient sleep on circadian rhythmicity and expression amplitude of the human blood transcriptome. *Proceedings of the National Academy of Sciences of the United States of America*, *110*, E1132–E1141. https://doi.org/10.1073/pnas.1217154110.

Mongrain, V., et al. (2011). Sleep loss reduces the DNA-binding of BMAL1, CLOCK, and NPAS2 to specific clock genes in the mouse cerebral cortex. *PLoS One*, *6*(10), e26622. https://doi.org/10.1371/journal.pone.0026622.

Morton, A. J., et al. (2005). Disintegration of the sleep-wake cycle and circadian timing in Huntington's disease. *The Journal of Neuroscience*, *25*(1), 157–163. Available at: http://www.ncbi.nlm.nih.gov/entrez/query.fcgi?cmd=Retrieve&db=PubMed&dopt=Citation&list_uids=15634777.

Musiek, E. S., et al. (2013). Circadian clock proteins regulate neuronal redox homeostasis and neurodegeneration. *The Journal of Clinical Investigation*, *123*(12), 5389–5400. https://doi.org/10.1172/JCI70317.In.

Musiek, E. S., et al. (2018). Circadian rest-activity pattern changes in aging and preclinical Alzheimer disease. *JAMA Neurology*, *75*(5), 582–590. https://doi.org/10.1001/jamaneurol.2017.4719.

Nakahata, Y., et al. (2008). The NAD+-dependent deacetylase SIRT1 modulates CLOCK-mediated chromatin remodeling and circadian control. *Cell*, *134*(2), 329–340. https://doi.org/10.1016/j.cell.2008.07.002.

Nakahata, Y., et al. (2009). Circadian control of the NAD+ salvage pathway by CLOCK-SIRT1. *Science*, *324*(5927), 654–657. https://doi.org/10.1126/science.1170803.

Neufeld-Cohen, A., et al. (2016). Circadian control of oscillations in mitochondrial rate-limiting enzymes and nutrient utilization by PERIOD proteins. *Proceedings of the National Academy of Sciences of the United States of America*, *113*(12), E1673–E1682. https://doi.org/10.1073/pnas.1519650113.

Noya, S. B., et al. (2019). The forebrain synaptic transcriptome is organized by clocks but its proteome is driven by sleep. *Science*, *366*(6462), eaav2642, American Association for the Advancement of Science. https://doi.org/10.1126/science.aav2642.

Obeso, J. A., et al. (2017). Past, present, and future of Parkinson's disease: A special essay on the 200th Anniversary of the Shaking Palsy. *Movement Disorders*, *32*, 1264–1310. https://doi.org/10.1002/mds.27115.

Oh, J., et al. (2019). Profound degeneration of wake-promoting neurons in Alzheimer's disease. *Alzheimer's & Dementia*, *15*, 1253–1263. https://doi.org/10.1016/j.jalz.2019.06.3916.

Ooms, S., et al. (2014). Effect of 1 night of total sleep deprivation on cerebrospinal fluid beta-amyloid 42 in healthy middle-aged men: A randomized clinical trial. *JAMA Neurology*, *71*(8), 971–977, 1875833 [pii]. https://doi.org/10.1001/jamaneurol.2014.1173.

Pallier, P. N., et al. (2007). Pharmacological imposition of sleep slows cognitive decline and reverses dysregulation of circadian gene expression in a transgenic mouse model of Huntington's disease. *The Journal of Neuroscience*, *27*(29), 7869–7878. Available at: http://www.ncbi.nlm.nih.gov/entrez/query.fcgi?cmd=Retrieve&db=PubMed&dopt=Citation&list_uids=17634381.

Panda, S. (2016). Circadian physiology of metabolism. *Science*, *354*, 1008–1015. https://doi.org/10.1126/science.aah4967.

Peek, C. B., et al. (2013). Circadian clock NAD+ cycle drives mitochondrial oxidative metabolism in mice. *Science*, *342*(6158), 1243417. https://doi.org/10.1126/science.1243417.

Pekovic-Vaughan, V., et al. (2014). The circadian clock regulates rhythmic activation of the NRF2/glutathione-mediated antioxidant defense pathway to modulate pulmonary fibrosis. *Genes & Development*, *28*(6), 548–560. https://doi.org/10.1101/gad.237081.113.

Peter-Derex, L., et al. (2014). Sleep and Alzheimer's disease. *Sleep Medicine Reviews*, *19*, 29–38, S1087-0792(14)00040-9 [pii]. https://doi.org/10.1016/j.smrv.2014.03.007.

Ramanathan, C., et al. (2018). mTOR signaling regulates central and peripheral circadian clock function. *PLoS Genetics*, *14*, e1007369. https://doi.org/10.1371/journal.pgen.1007369.

Ramsey, K. M., et al. (2009). Circadian clock feedback cycle through NAMPT-mediated NAD+ biosynthesis. *Science*, *324*(5927), 651–654. https://doi.org/10.1126/science.1171641.

Ravikumar, B., et al. (2004). Inhibition of mTOR induces autophagy and reduces toxicity of polyglutamine expansions in fly and mouse models of Huntington disease. *Nature Genetics*, *36*, 585–595. https://doi.org/10.1038/ng1362.

Rey, G., et al. (2016). The pentose phosphate pathway regulates the circadian clock. *Cell Metabolism*, *24*, 462–473. https://doi.org/10.1016/j.cmet.2016.07.024.

Roh, J. H., et al. (2014). Potential role of orexin and sleep modulation in the pathogenesis of Alzheimer's disease. *The Journal of Experimental Medicine*, *211*(13), 2487–2496. https://doi.org/10.1084/jem.20141788.

Roos, R. A. C. (2010). Huntington's disease: A clinical review. *Orphanet Journal of Rare Diseases*, *5*, 40. https://doi.org/10.1186/1750-1172-5-40.

Rothman, S. M., et al. (2013). Chronic mild sleep restriction accentuates contextual memory impairments, and accumulations of cortical Abeta and pTau in a mouse model of Alzheimer's disease. *Brain Research*, *1529*, 200–208, S0006-8993(13)00975-X [pii]. https://doi.org/10.1016/j.brainres.2013.07.010.

Satlin, A., et al. (1995). Circadian locomotor activity and core-body temperature rhythms in Alzheimer's disease. *Neurobiology of Aging*, *16*(5), 765–771.

Schenck, C. H., Boeve, B. F., & Mahowald, M. W. (2013). Delayed emergence of a parkinsonian disorder or dementia in 81% of older men initially diagnosed with idiopathic rapid eye movement sleep behavior disorder: A 16-year update on a previously reported series. *Sleep Medicine*, *14*(8), 744–748. https://doi.org/10.1016/j.sleep.2012.10.009.

Schernhammer, E. S., et al. (2015). Occupational history of night shift work and Parkinson's disease in Denmark. *Scandinavian Journal of Work, Environment & Health*, *41*, 377–383. https://doi.org/10.5271/sjweh.3502.

Shokri-Kojori, E., et al. (2018). Beta-amyloid accumulation in the human brain after one night of sleep deprivation. *Proceedings of the National Academy of Sciences of the United States of America*, *115*(17), 4483–4488, 721694115 [pii]. https://doi.org/10.1073/pnas.17216941151.

Skene, D. J., & Swaab, D. F. (2003). Melatonin rhythmicity: Effect of age and Alzheimer's disease. *Experimental Gerontology*, *38*(1–2), 199–206.

Spira, A. P., et al. (2013). Self-reported sleep and beta-amyloid deposition in community-dwelling older adults. *JAMA Neurology*, *70*(12), 1537–1543, 1757019 [pii]. https://doi.org/10.1001/jamaneurol.2013.4258.

Sprecher, K. E., et al. (2017). Poor sleep is associated with CSF biomarkers of amyloid pathology in cognitively normal adults. *Neurology*, *89*(5), 445–453, WNL.0000000000004171 [pii]. https://doi.org/10.1212/WNL.0000000000004171.

St Louis, E. K., & Boeve, B. F. (2017). REM sleep behavior disorder: Diagnosis, clinical implications, and future directions. *Mayo Clinic Proceedings*, *92*, 1723–1736. https://doi.org/10.1016/j.mayocp.2017.09.007.

Stopa, E. G., et al. (1999). Pathologic evaluation of the human suprachiasmatic nucleus in severe dementia. *Journal of Neuropathology and Experimental Neurology*, *58*(1), 29–39.

St-Pierre, J., et al. (2006). Suppression of reactive oxygen species and neurodegeneration by the PGC-1 transcriptional coactivators. *Cell*, *127*(2), 397–408. https://doi.org/10.1016/j.cell.2006.09.024.

Swaab, D. F., Fliers, E., & Partiman, T. S. (1985). The suprachiasmatic nucleus of the human brain in relation to sex, age and senile dementia. *Brain Research*, *342*(1), 37–44.

Tranah, G. J., et al. (2011). Circadian activity rhythms and risk of incident dementia and mild cognitive impairment in older women. *Annals of Neurology*, *70*(5), 722–732. Available at: http://www.ncbi.nlm.nih.gov/entrez/query.fcgi?cmd=Retrieve&db=PubMed&dopt=Citation&list_uids=22162057.

Tsvetkov, A. S., et al. (2010). A small-molecule scaffold induces autophagy in primary neurons and protects against toxicity in a Huntington disease model. *Proceedings of the National Academy of Sciences of the United States of America*, *107*, 16982–16987. https://doi.org/10.1073/pnas.1004498107.

Uchida, K., et al. (1996). Daily rhythm of serum melatonin in patients with dementia of the degenerate type. *Brain Research*, *717*(1–2), 154–159.
Videnovic, A., & Golombek, D. (2013). Circadian and sleep disorders in Parkinson's disease. *Experimental Neurology*, *243*, 45–56. https://doi.org/10.1016/j.expneurol.2012.08.018.
Videnovic, A., et al. (2014a). The clocks that time us'—Circadian rhythms in neurodegenerative disorders. *Nature Reviews. Neurology*, *10*(12), 683–693. https://doi.org/10.1038/nrneurol.2014.206.
Videnovic, A., et al. (2014b). Circadian melatonin rhythm and excessive daytime sleepiness in Parkinson disease. *JAMA Neurology*, *71*, 463–469. https://doi.org/10.1001/jamaneurol.2013.6239.
Vitiello, M. V., & Prinz, P. N. (1989). Alzheimer's disease. Sleep and sleep/wake patterns. *Clinics in Geriatric Medicine*, *5*(2), 289–299.
Volicer, L., et al. (2001). Sundowning and circadian rhythms in Alzheimer's disease. *The American Journal of Psychiatry*, *158*(5), 704–711. https://doi.org/10.1176/appi.ajp.158.5.704.
Wang, J. L., et al. (2015). Suprachiasmatic neuron numbers and rest-activity circadian rhythms in older humans. *Annals of Neurology*, *78*(2), 317–322. https://doi.org/10.1002/ana.24432.
Wang, H. B., et al. (2017). Blue light therapy improves circadian dysfunction as well as motor symptoms in two mouse models of Huntington's disease. *Neurobiology of Sleep and Circadian Rhythms*, *2*, 39–52, Elsevier Inc. https://doi.org/10.1016/j.nbscr.2016.12.002.
Whitehead, D. L., et al. (2008). Circadian rest-activity rhythm is altered in Parkinson's disease patients with hallucinations. *Movement Disorders*, *23*(8), 1137–1145. Available at: http://www.ncbi.nlm.nih.gov/entrez/query.fcgi?cmd=Retrieve&db=PubMed&dopt=Citation&list_uids=18442142.
Whittaker, D. S., et al. (2018). Circadian-based treatment strategy effective in the BACHD mouse model of Huntington's disease. *Journal of Biological Rhythms*, *33*(5), 535–554, SAGE Publications Inc. https://doi.org/10.1177/0748730418790401.
Woldt, E., et al. (2013). Rev-erb-α modulates skeletal muscle oxidative capacity by regulating mitochondrial biogenesis and autophagy. *Nature Medicine*, *19*(8), 1039–1046. https://doi.org/10.1038/nm.3213.
Xie, L., et al. (2013). Sleep drives metabolite clearance from the adult brain. *Science*, *342*(6156), 373–377, 342/6156/373 [pii]. https://doi.org/10.1126/science.1241224.
Yamada, K., et al. (2014). Neuronal activity regulates extracellular tau in vivo. *The Journal of Experimental Medicine*, *211*(3), 387–393, jem.20131685 [pii]. https://doi.org/10.1084/jem.20131685.
Yang, Y. K., et al. (2013). The polymorphism of CLOCK gene 3111T/C C>T is associated with susceptibility of Alzheimer disease in Chinese population. *Journal of Investigative Medicine*, *61*(7), 1084–1087. https://doi.org/10.2310/JIM.0b013e31829f91c0.
Yang, G., et al. (2016). Timing of expression of the core clock gene Bmal1 influences its effects on aging and survival. *Science Translational Medicine*, *8*(324), 324ra16. https://doi.org/10.1126/scitranslmed.aad3305.
Yu, E. A., & Weaver, D. R. (2011). Disrupting the circadian clock: Gene-specific effects on aging, cancer, and other phenotypes. *Aging (Albany NY)*, *3*(5), 479–493. Available at: http://www.ncbi.nlm.nih.gov/entrez/query.fcgi?cmd=Retrieve&db=PubMed&dopt=Citation&list_uids=21566258.
Zhang, R., et al. (2014). A circadian gene expression atlas in mammals: Implications for biology and medicine. *Proceedings of the National Academy of Sciences of the United States of America*, *111*(45), 16219–16224. https://doi.org/10.1073/pnas.1408886111.

Zhang, T., et al. (2018). Sleep and circadian abnormalities precede cognitive deficits in R521C FUS knockin rats. *Neurobiology of Aging*, *72*, 159–170. https://doi.org/10.1016/j.neurobiolaging.2018.08.025.
Zhou, J. N., Hofman, M. A., & Swaab, D. F. (1995). VIP neurons in the human SCN in relation to sex, age, and Alzheimer's disease. *Neurobiology of Aging*, *16*(4), 571–576.
Zhu, Y., et al. (2018). Chronic sleep disruption advances the temporal progression of tauopathy in P301S mutant mice. *Journal of Neuroscience*, *38*, 10255–10270. https://doi.org/10.1523/JNEUROSCI.0275-18.2018.

CHAPTER FIFTEEN

The impact of neurovascular, blood-brain barrier, and glymphatic dysfunction in neurodegenerative and metabolic diseases

Molly Braun[a,b], **Jeffrey J. Iliff**[a,b,c,*]
[a]Department of Psychiatry and Behavioral Sciences, University of Washington School of Medicine, Seattle, WA, United States
[b]VISN 20 Mental Illness Research, Education and Clinical Center (MIRECC), VA Puget Sound Health Care System, Seattle, WA, United States
[c]Department of Neurology, University of Washington School of Medicine, Seattle, WA, United States
*Corresponding author: e-mail address: jiliff@uw.edu

Contents

1. Introduction 415
2. The cerebral vasculature and blood-brain barrier 415
   2.1 Elements of the neurovascular unit 416
   2.2 Cerebral blood flow regulation 418
   2.3 Blood-brain barrier function 418
3. Cerebrospinal fluid barriers 418
   3.1 Choroid plexus and CSF secretion 419
   3.2 Classical routes of cerebral spinal fluid reabsorption 420
4. Perivascular exchange and the glymphatic system 420
   4.1 Mechanisms of glymphatic influx and efflux 421
   4.2 Sleep-wake regulation and other regulation 422
   4.3 Meningeal lymphatics 423
5. Aging 424
6. Neurodegenerative diseases 425
   6.1 Alzheimer's disease and vascular dementia 425
   6.2 Parkinson's and Huntington's diseases 426
7. Metabolic disease 428
8. Conclusion 429
References 430

International Review of Neurobiology, Volume 154
ISSN 0074-7742
https://doi.org/10.1016/bs.irn.2020.02.006

Abstract

The cerebral vasculature serves as the crossroads of the CNS, supporting exchange of nutrients, metabolic wastes, solutes and cells between the compartments of the brain, including the blood, brain interstitium, and cerebrospinal fluid (CSF). The blood-brain barrier (BBB) regulates the entry and efflux of molecules into brain tissue. The cells of the neurovascular unit regulate cerebral blood flow, matching local metabolic demand to blood supply. The blood-CSF barrier at the choroid plexus secretes CSF, which supports the brain and provides a sink for interstitial solutes not cleared across the BBB. Recent studies have characterized the glymphatic system, a brain-wide network of perivascular spaces that supports CSF and interstitial fluid exchange and the clearance of interstitial solutes to the CSF. The critical role that these structures play in maintaining brain homeostasis is illustrated by the established and emerging roles that their dysfunctions play in the development of neurodegenerative diseases, such as Alzheimer's disease (AD). Loss of BBB and blood-CSF barrier function is reported both in rodent models of AD, and in human AD subjects. Cerebrovascular dysfunction and ischemic injury are well established contributors to both vascular dementia and to a large proportion of cases of sporadic AD. In animal models, the slowed glymphatic clearance of interstitial proteins, such as amyloid β or tau, are proposed to contribute to the development of neurodegenerative diseases, including AD. In total, these findings suggest that cellular and molecular changes occurring within and around the cerebral vasculature are among the key drivers of neurodegenerative disease pathogenesis.

Abbreviations

Aβ	amyloid beta
AD	Alzheimer's disease
AQP1	aquaporin-1
AQP4	aquaporin-4
BBB	blood-brain barrier
BCSFB	blood-cerebrospinal fluid barrier
CBF	cerebral blood flow
CNS	central nervous system
CP	choroid plexus
CSF	cerebral spinal fluid
ECs	brain endothelial cells
HD	Huntington's disease
ISF	interstitial fluid
LRP1	low density lipoprotein receptor 1
LRRK2	leucine-rich repeat kinase 2
MRI	magnetic resonance imaging
NVU	neurovascular unit
P-gp	P-glycoprotein
PCC	posterior cingulate cortex
PD	Parkinson's disease
T2DM	type 2 diabetes mellitus

1. Introduction

The aggregation of misfolded proteins is the hallmark of many neurodegenerative diseases, including Alzheimer's disease (AD), Parkinson's disease (PD), and others (Cai et al., 2012; Wanker, 2000). This common feature suggests that impairment in proteins and waste clearance from the brain may be one common mechanism underlying these otherwise disparate conditions. Recent studies have described the glymphatic system, a brain-wide network of perivascular spaces that supports exchange of cerebrospinal fluid (CSF) and interstitial fluid (ISF), facilitating the clearance of interstitial proteins and wastes including amyloid β (Aβ) (Iliff et al., 2012, 2013a). Glymphatic function is inextricably linked with the cerebral vasculature and blood-brain barrier (BBB), which become dysfunctional in the setting of aging and neurodegenerative disease (Kress et al., 2014). Our developing understanding of the glymphatic system and its interactions with the cerebral vasculature and BBB, are rapidly changing how we understand both normal physiological processes in the central nervous system, but also the pathological mechanisms underlying neurodegenerative and metabolic diseases.

2. The cerebral vasculature and blood-brain barrier

The cerebral vasculature is principally organized around two overlapping anatomical and physiological concepts: the neurovascular unit (NVU) and the BBB. The BBB separates the blood compartment from the brain extracellular and intracellular fluid compartments, and at its most basic level is a physical and chemical barrier consisting of specialized endothelial cells lining the brain vasculature linked by tight junctions that limits the free movement of fluid and solutes into and out of the brain (Jessen et al., 2015). The NVU consists of the endothelial and mural cells (vascular smooth muscle cells and pericytes) that make up the brain vasculature, as well as the surrounding astrocytes, neurons, microglia and immune cells that each contribute to the regulation of integrated cerebrovascular function, including BBB permeability, cerebral blood flow (CBF) regulation, nutrient delivery, and glymphatic function (McConnell et al., 2017).

2.1 Elements of the neurovascular unit

2.1.1 *Endothelial cells*

Specialized brain endothelial cells (ECs) line the cerebral vasculature and are connected via tight junctions that restrict the transport of fluid, ions, proteins and cells into the brain parenchyma (McConnell et al., 2017; Redzic, 2011). Brain endothelial cells express polarized ion receptors, ion channels and solute transporters on their luminal and abluminal membranes that mediate the delivery of nutrients, including glucose and amino acids, to the cells of the central nervous system (CNS) and mediate the efflux waste products and xenobiotics from the brain to the blood (McConnell et al., 2017; Redzic, 2011; Sanchez-Covarrubias et al., 2014). Capillary ECs are surrounded by the basal lamina made up of extracellular matrix proteins including collagen and laminin. These basement membrane proteins support and anchor the vessel, act as a mechanical barrier to macromolecular and cellular movement, and play a role in signal transduction (Yurchenco and Patton, 2009).

2.1.2 *Smooth muscle cells and pericytes*

Mural cells, which include vascular smooth muscle cells and pericytes, surround the endothelial cell layer of cerebral blood vessels from the resistance arteries and draining veins at the brain surface, to penetrating arterioles and ascending venules, to the terminal capillary beds where the majority of nutrient exchange occurs (Armulik et al., 2011). Vascular smooth muscle cells surround arteries and arterials, providing contractile force to produce vasodilation and vasoconstriction, regulating cerebrovascular resistance and blood flow to the downstream capillary beds (Hill et al., 2015).

More recently, a key role of pericytes has emerged in the regulation of the cerebral vasculature. Although these cells, embedded within the basal lamina and extending their long processes along capillaries, are found throughout the body, they are particularly important in the brain due to its high metabolic demands and need for tight regulation of blood flow and BBB permeability (Brown et al., 2019). The cerebral vasculature has the densest pericyte coverage of any organ (Armulik et al., 2011), and pericytes contribute to the development, maintenance and stability of the vasculature, to the regulation of cerebral blood flow (Hall et al., 2014), to BBB function (Armulik et al., 2010), and may possess immune modulatory and stem cell properties (Armulik et al., 2011). Pericyte degeneration leads to neurovascular uncoupling and decreased oxygenation of the brain (Kisler et al., 2017b).

2.1.3 Astrocytes

Astrocytes are the space-filling glial cell of the brain, providing structural, biochemical, and metabolic support to neurons and to other cells (Bélanger and Magistretti, 2009). They serve as key physiological integrators, recycling glutamate released by excitatory neurons, shuttling metabolic substrates in the form of lactate from endothelial cells to neurons, dispersing K^+ and water away from active synapses, and regulating cerebral blood flow to match metabolic demand to local blood supply (McConnell et al., 2017). Astrocytes extend processes to surround synapses, as well as long endfoot processes that surround the cerebral vasculature and are connected to each other via gap junctions (Iadecola and Nedergaard, 2007). This gap junctional coupling permits long-range communication between astrocytes, as well as the integration of neuronal and vascular processes across large distances (Fields et al., 2015).

Perivascular astrocytic endfeet ensheathing the cerebral vasculature express large amounts of the astroglial water channel aquaporin-4 (AQP4), which plays an important role in osmotically balancing K^+ uptake and redistributing K^+ from active synapses throughout the astroglial network via gap junctions (Bélanger and Magistretti, 2009; Jin et al., 2013), as well as in supporting glymphatic exchange of CSF and ISF along perivascular spaces (Iliff et al., 2012). Astroglial endfeet also form the glial limitans, a functional and structural barrier at the pial surface of the brain separating the subarachnoid CSF compartment from the brain parenchyma. An invagination of the glial limitans also surrounds the initial segments of penetrating vessels as they enter the brain from the cortical surface (Mastorakos and McGavern, 2019; McConnell et al., 2017).

2.1.4 Perivascular spaces

Pial vessels penetrate into the brain parenchyma, forming penetrating arterioles and branching to form the terminal capillary bed. Penetrating vessels are surrounded by leptomeningeal cells and the glial limitans, forming CSF-filled perivascular or Virchow-Robin spaces (Cipolla, 2009; Jessen et al., 2015). As arterioles penetrate further into the brain and transition into capillaries, this space shrinks and eventually disappears, as the microcirculation is surrounded only by basal lamina and perivascular astrocytic endfeet (Jessen et al., 2015). These perivascular spaces form a critical exchange route for the glymphatic system, supporting the movement of CSF into brain tissue from the subarachnoid space and the efflux of interstitial solutes out of the brain (Iliff et al., 2012).

2.2 Cerebral blood flow regulation

Cerebral blood flow is regulated at the level of the NVU to match local blood supply to metabolic demands driven by local neural activity (Nippert et al., 2018). This occurs at the level of penetrating arterioles, which dilate in response to increases in local neuronal activity to increase downstream blood supply, glucose and oxygen delivery (McConnell et al., 2017). This process, termed neurovascular coupling or functional hyperemia, is mediated either by the release of vasoactive metabolites, such as K^+, adenosine, or nitric oxide from nearby active neurons, or through the release of vasoactive substances from astrocytes following stimulation with synaptically released glutamate (Nippert et al., 2018). Uncoupling of this process can lead to vascular and BBB dysfunction, neural injury, and likely contributes to the pathogenesis of many neurological and neurodegenerative conditions (Kisler et al., 2017a).

2.3 Blood-brain barrier function

The BBB is critical for maintaining ionic and cellular homeostasis in the brain, and for protecting the brain from peripheral insults including changes in blood composition, pathogens, and toxins (Daneman and Prat, 2015). Blood-brain barrier disruption can occur acutely, in response to a discrete insult (Bramlett and Dietrich, 2004), such as cerebral ischemia, traumatic brain injury, or infection. In these cases, BBB breakdown leads to the movement of plasma proteins into brain tissue, leading to the formation of edema, and the infiltration of peripheral immune cells, leading to the development of secondary inflammatory injury (Chodobski et al., 2011). Blood-brain barrier disruption can also be more subtle, developing chronically in the setting of aging, neuroinflammatory diseases, or neurodegenerative diseases, including AD and PD (Farrall and Wardlaw, 2009).

3. Cerebrospinal fluid barriers

The cerebrospinal fluid compartment is classically considered distinct from the brain parenchyma, including the intracellular and extracellular spaces, and from the blood compartment. Comprised of the lateral, third and fourth ventricles internally, and externally of the cisterns and subarachnoid space surrounding the brain, the CSF compartment is in intimate communication with the brain interstitial compartment (Brinker et al., 2014). Ependymal cells lining the cerebral ventricles form the barrier between these CSF compartments and the surrounding brain tissue. Believed to lack tight junctions, but possessing adherens junctions, this barrier permits the movement of small molecules between the CSF and ISF, but restricts the movement of macromolecules and cells (Del Bigio, 2010). The subarachnoid CSF compartment is separated from the brain parenchyma by the leptomeningeal pia mater and glial limitans, a barrier that similarly restricts the movement of macromolecules and cells, but permits the movement of small molecules (Mastorakos and McGavern, 2019). Thus, while the CSF and brain interstitial compartments are classically regarded as separate, and are separated by respective barrier structures, these compartments are intimately connected to one another through extensive exchange across these barriers.

The subarachnoid CSF compartment is segregated from the periphery by the tight junction coupled arachnoid barrier, while the ventricular CSF compartment is separated from the periphery by the blood-cerebrospinal fluid barrier (BCSFB) of the choroid plexus (CP) (Laterra et al., 1999). Although not as widely discussed as the BBB, the BCSFB is a key regulator of bidirectional transport between the blood and CSF.

3.1 Choroid plexus and CSF secretion

The CP is located in lateral, third and fourth ventricles and is composed of three layered elements. Fenestrated capillaries surrounded by a stromal space allow for the free filtration of plasma and immune cells from the blood into the CP (Lun et al., 2015). A layer of polarized cuboidal epithelial cells, connected by tight junctions, separates the CP stromal space from the ventricular CSF compartment (Lun et al., 2015). The epithelial cells of the CP express numerous ion channels, solute transporters, as well as Na/K-ATPases and the water channel aquaporin-1 (AQP1), polarized to the apical and basolateral membranes facing the CSF and CP stroma, respectively (Laterra et al., 1999; Praetorius and Nielsen, 2006). Through the activity of these pumps, transporters and channels, CSF is actively secreted into the ventricles.

CSF has many important functions, including serving as a physical cushion to the brain against injury, providing for the delivery of nutrients across the CP to the rest of the brain, and serving as a sink for solutes and wastes from the brain interstitium that cannot readily be cleared across the BBB

(Laterra et al., 1999). The transport of solutes across the CP is not unidirectional from the blood to the CSF. Transporters within the CP appear to also support the clearance of solutes, including Aβ, from the CSF to the blood as well. Lastly, in addition to solute exchange and CSF secretion, the CP also serves as an important gateway for peripheral immune cell trafficking into the CSF (Prinz and Priller, 2017).

3.2 Classical routes of cerebral spinal fluid reabsorption

Cerebral spinal fluid is secreted primarily by the CPs within the cerebral ventricles, moves outward by pulsatile bulk flow through the ventricular system, then fills the cisterns and subarachnoid spaces surrounding the brain and spinal cord (Hladky and Barrand, 2014; Iliff et al., 2012, 2013a). Cerebral spinal fluid, and the solutes and cells within it, are classically believed to be reabsorbed into the venous blood via arachnoid villi and granulations, valve-like outpouchings through the arachnoid barrier into venous sinuses of the dura (Hladky and Barrand, 2014). In addition to this route of reabsorption, CSF also flows outward along cranial and spinal nerve sheathes to reach the peripheral interstitium, where uptake by peripheral lymphatic vessels supports eventual return to the venous circulation (Aspelund et al., 2015; Johnston et al., 2004). For example, CSF solutes moving along the olfactory nerves cross the cribriform plate and are taken up into nasal lymphatic vessels, passing to the deep cervical lymphatic drainage (Johnston et al., 2004; Liu et al., 2012).

4. Perivascular exchange and the glymphatic system

The glymphatic system is a recently defined network of perivascular pathways the supports fluid and solute exchange between the CSF and interstitial compartments, allowing for the clearance of waste and metabolites from the brain interstitium (Jessen et al., 2015; Lundgaard et al., 2015; Rangroo Thrane et al., 2013). This work builds on prior studies describing perivascular solute movement through the brain that also linked these processes with aging and a potential role for the development of neurodegenerative diseases, such as AD (Carare et al., 2008; Hawkes et al., 2011; Rennels et al., 1985; Weller et al., 2008). The term "glymphatic" referred to the dependence of this interstitial protein clearance (a lymphatic function) upon the perivascular astroglial water channel AQP4 (glial-lymphatic, glymphatic). It highlights the macroscopic scale of this waste clearance system, which is elaborated across the entire cerebral vasculature, and emphasizes the role of glial cells and AQP4 in this process.

Glymphatic exchange consists of the influx of CSF from the subarachnoid space along perivascular spaces surrounding penetrating arteries and its exchange into the surrounding interstitium (Iliff et al., 2012). Interstitial solutes in turn, are cleared to the sinus-associated cisternal compartments along white matter tracks and large caliber veins of the deep venous drainage (Iliff et al., 2012, 2014). The influx of CSF and the efflux of interstitial solutes is dependent upon AQP4, as *Aqp4* gene deletion slows both processes (Iliff et al., 2012; Mestre et al., 2018a, 2018b). Important to pathophysiological processes (detailed below), the perivascular localization of AQP4 is critical to its role in glymphatic exchange, as glymphatic function is suppressed in transgenic mice that express normal levels of AQP4, but lack perivascular localization (Mestre et al., 2018a).

4.1 Mechanisms of glymphatic influx and efflux

The glymphatic system includes two exchange processes: subarachnoid CSF solute influx into the interstitium and ISF solute efflux to the cisternal CSF compartments. The initial descriptions of the pathway proposed that fluids and solutes move along perivascular spaces and through the interstitium by convective bulk flow (Iliff et al., 2012, 2013a, 2013b; Jessen et al., 2015; Mestre et al., 2018a; Rasmussen et al., 2018; Xie et al., 2013). However, more recent experimental and modeling studies have suggested that while fluid and solute movement along perivascular spaces and white matter tracks may occur by bulk flow, the movement of solutes through the interstitium in between these perivascular spaces may occur by local diffusion (Ray et al., 2019).

The influx of CSF along perivascular spaces surrounding cerebral arteries is likely driven by multiple mechanisms. Continuous production of CSF by the CP builds pressure that drives the flow of CSF from the ventricles to the subarachnoid space and along perivascular space. Cerebral spinal fluid movement through the ventricular system in humans is also driven in part by respiration (Yamada et al., 2013) and by cardiac pulsation (Kiviniemi et al., 2016; Rennels et al., 1985, 1990). Experimental models modulating arterial pulsation, either surgically or pharmacologically, demonstrated that arterial pulsation is likely a key driver of glymphatic exchange (Iliff et al., 2013b; Mestre et al., 2018b; Stoodley et al., 1997). Indeed, a recent CSF particle tracking study demonstrated that fluid movement along perivascular spaces at the brain surface entrains strongly to the cardiac cycle (Iliff et al., 2013b). Most recently, a magnetic resonance imaging (MRI) study carried out in human subjects demonstrated that during sleep, slow-wave oscillations entrain low-frequency oscillations in cerebral blood volume, which in turn drive the bulk movement of CSF through the ventricular system (Fultz et al., 2019). This study supports the intriguing possibility that oscillations in neural activity couple to CSF flow dynamics through physio-mechanical coupling with the cerebral vasculature.

The mechanisms driving interstitial solute efflux are somewhat less clear. Several studies have demonstrated that rates of solute efflux cannot be explained solely by diffusion and have reported size-independent kinetics of interstitial solute clearance (Cserr et al., 1981; Iliff et al., 2012), which supports a role for bulk flow in interstitial solute efflux. A recent computational study combining simulations and experimental data found that diffusion dominates the movement of small molecules within the brain interstitium (<3 kDa), while convective transport is likely more important for the movement of larger molecules (>3 kDa), suggesting both diffusion and bulk flow contribute to interstitial solute efflux (Ray et al., 2019). Thus, a more complex picture of glymphatic clearance has emerged, with diffusion contributing to transport of smaller molecules over shorter distances, and with bulk flow, particularly along permissive anatomical routes such as white matter tracks and perivascular spaces, dominating the movement of macromolecules over longer anatomical distances (Iliff and Simon, 2019). Other factors, including solute chemistry, size, interactions with cellular environment including membranes, transporters and receptors, and physiological (and pathological) state all appear to influence interstitial solute efflux (Groothuis et al., 2007). Thus, our understanding of the drivers of glymphatic influx and efflux will surely continue to evolve.

4.2 Sleep-wake regulation and other regulation

One novel feature of glymphatic exchange is the observation that it is active primarily during sleep. In experimental rodent studies, CSF tracer influx and the clearance of interstitial solutes, including Aβ, were approximately two-fold higher during sleep than during wakefulness. Supporting sleep-active glymphatic function, during sleep or under anesthesia, the volume of the interstitial space expands by 60%, supporting more rapid movement of CSF and interstitial solutes (Xie et al., 2013). Studies in rodents and in human subjects demonstrate that interstitial and CSF Aβ levels increase during waking and decline during sleep (Lim et al., 2014; Roh et al., 2012; Shokri-Kojori et al., 2018). In human subjects, a single night of sleep deprivation results in reduced CSF Aβ levels and increased parenchymal Aβ burden measured by positron emission tomography (Shokri-Kojori et al., 2018). These studies are consistent with a role for sleep-active glymphatic function in the clearance of Aβ from the brain.

One of the mechanisms regulating sleep-active glymphatic function appears to be central noradrenergic signaling. Central administration of noradrenergic antagonists increased the extracellular volume and accelerated rates of CSF influx and interstitial solute efflux during waking to levels observed during sleep (Xie et al., 2013), suggesting that wake-active locus coeruleus output likely suppresses glymphatic function during waking. More recently, as described above, a study in human subjects described coupling between slow-wave, hemodynamic, and CSF oscillations primarily during deep (slow-wave) sleep (Fultz et al., 2019).

In addition to levels of consciousness, body position also affects glymphatic exchange with lateral positioning being optimal for tracer clearance vs prone or supine, which demonstrated slower clearance (Lee et al., 2015). Given that this is a common position for sleeping, body positioning during sleep likely enhances the increased glymphatic activity observed. In addition, exercise promoted glymphatic clearance of Aβ and reduced astrogliosis in aged mice, suggesting that exercise also modulates glymphatic activity (He et al., 2017). As the glymphatic system continues to be studied, more and more regulatory and modulatory mechanisms will come to light.

4.3 Meningeal lymphatics

The recent characterization of the meningeal lymphatic vessels has shed new light on the mechanisms underlying solute efflux from the brain interstitium and CSF compartments (Da Mesquita et al., 2018a). The meninges are made up of three distinct layers: dura mater, arachnoid mater, and pia mater. Meningeal lymphatic vessels have been characterized within the dura, the thickest and outermost meningeal layer, where they are associated with dural sinuses and also reside within the dura of the skull base. Below the dura is the arachnoid layer, which forms a tight junction barrier between the subarachnoid CSF and the dura, classically understood to be crossed by arachnoid villi and granulations (Rua and McGavern, 2018). Recent experimental studies defining meningeal lymphatic structure and function have demonstrated that these vessels play an important role in the efflux of solutes and cells from the CSF compartment into the cervical lymphatic drainage (Da Mesquita et al., 2018a). Although glymphatic exchange and meningeal lymphatic drainage have primarily been described separately, it is likely that they function in concert to support the efflux of interstitial solutes out of the cranium: glymphatic exchange supporting the efflux of interstitial solutes to the subarachnoid CSF in cisterns associated with dural sinuses and meningeal lymphatic vessels supporting the uptake of those solutes from the subarachnoid CSF to the lymphatic drainage (Louveau et al., 2017).

5. Aging

It is important to note that the greatest risk factor for the development of neurodegenerative diseases, including AD and PD, is aging. This suggests that changes occur in the aging brain that may promote neurodegenerative conditions. Aging in the absence of other co-morbidities is associated with neurovascular dysfunction and reduced overall CBF (Duncombe et al., 2017), factors that are worsened in the setting of AD. Aged mice exhibit neurovascular uncoupling as a result of breakdown across the neurovascular unit, including degeneration of pericytes (Kisler et al., 2017b), astrogliosis, microglial activation, and loss of contact between astrocytic endfeet with the vasculature (Duncombe et al., 2017). This neurovascular uncoupling worsens progressively with increasing age and is proposed to contribute to ongoing metabolic stress and injury (Duncombe et al., 2017).

In addition to changes in CBF regulation, perivascular glymphatic exchange is impaired in the aging brain. In a study in healthy aging mice, glymphatic function was reduced 80–90% in aged compared to young mice (Kress et al., 2014), age-related impairment that has been confirmed across several subsequent reports (Benveniste et al., 2019; He et al., 2017). Several factors likely contribute to age-related decline in glymphatic exchange. For example, aging is associated with loss of perivascular AQP4 localization (Kress et al., 2014), which given the role of AQP4 in glymphatic clearance, could contribute to age-related glymphatic decline. Glymphatic exchange is also driven in part by arterial pulsations (Iliff et al., 2013b; Kiviniemi et al., 2016; Stoodley et al., 1997); therefore, arterial and arteriolar stiffening and decreased arterial pulsatility that occur in aging may also contribute to impairment of glymphatic function. Downstream of perivascular glymphatic exchange, aged mice also exhibit reduced CSF solute efflux into the deep cervical lymph nodes (Ma et al., 2017), as well as significant meningeal lymphatic dysfunction (Da Mesquita et al., 2018b). This may result from slowed CSF production (Attier-Zmudka et al., 2019) and reduced CSF pressure (Fleischman et al., 2012; Iliff et al., 2013b), which decline in the setting of aging. Lastly, age-related changes in sleep architecture, including declining sleep duration and efficiency, and decreased slow-wave sleep, observed in adults over the age of 60 (Redline et al., 2004), may impair the overall sleep-active glymphatic function.

Thus, numerous physiological changes, including neurovascular dysfunction, glymphatic pathway impairment, changes in CSF secretion and reabsorption, and alterations in sleep architecture occur in the aging brain and may confer vulnerability to age-related neurovascular and neurodegenerative disease.

6. Neurodegenerative diseases

Decline of glymphatic system function in normal aging may contribute to the development of neurodegenerative diseases, such as AD (and other dementias), PD, and Huntington's disease (HD) due to decreased clearance of waste and accumulation of aggregated proteins, which are hallmarks of neurodegenerative diseases. Numerous studies suggest that neurovascular, BBB, and glymphatic dysfunction contribute to the development and pathogenesis of AD and other neurodegenerative diseases.

6.1 Alzheimer's disease and vascular dementia

Alzheimer's disease, the most common cause of dementia, is a neurodegenerative disease associated with progressive cognitive impairment and memory loss (Matthews et al., 2019). It is characterized histopathologically by the accumulation of extracellular Aβ in the form of insoluble plaques, aggregation of intracellular neurofibrillary tangles composed of hyperphosphorylated tau, synaptic dysfunction, and neuronal loss. In aging human subjects, the clearance of Aβ slows, suggesting that it is the impairment of Aβ clearance that underlies plaque accumulation (Mawuenyega et al., 2010; Weller et al., 2008).

The efflux of Aβ from the brain is mediated in part through efflux across the BBB, and is mediated by the activity of the endocytic receptor low density lipoprotein receptor-related protein 1 (LRP1) and the efflux membrane transporter P-glycoprotein (P-gp) (Storck et al., 2018). Blood-brain barrier efflux of Aβ is impaired in the aging brain. In addition to BBB efflux, several studies suggest that perivascular glymphatic exchange contributes to the clearance of Aβ from the brain interstitium. Amyloid beta is cleared more rapidly from the sleeping than the waking brain (Xie et al., 2013). Glymphatic function is slowed in mouse models of AD (Peng et al., 2016), with impairment preceding the development of Aβ plaque pathology. In *Aqp4* knockout mice, glymphatic clearance of Aβ is dramatically reduced (Iliff et al., 2012), and when the *Aqp4* gene is knocked out in a mouse model of AD, Aβ plaque formation and cognitive decline are accelerated (Xu et al., 2015).

In human subjects, post mortem histopathological studies in the frontal cortex demonstrate that loss of perivascular AQP4 localization predicts AD status (Zeppenfeld et al., 2017) and is associated with AD pathology. A single night of sleep deprivation increases accumulation of Aβ (Shokri-Kojori et al., 2018), suggesting that declining sleep quality could contribute to glymphatic decline and decreased Aβ clearance. These studies suggest that impairment of Aβ efflux, both across the BBB and via perivascular glymphatic exchange, may underlie the development of AD pathology, and highlights the possibility that these pathways could be therapeutically targeted in early stages of the disease (Peng et al., 2016).

Vascular dementia, the second leading cause of dementia following AD, results from accumulating cerebrovascular injury linked to ischemic injury and cerebrovascular dysfunction (Buffington et al., 2013). While cerebrovascular dysfunction, including loss of neurovascular coupling and BBB disruption, are well established both in human subjects suffering from vascular dementia and in rodent models of cerebrovascular disease, the role of glymphatic pathway impairment in vascular dementia is not as well established. In a mouse model of microinfarcts, diffuse ischemic injury was associated with loss of perivascular AQP4 localization (Wang et al., 2012) and impairment of glymphatic pathway function (Wang et al., 2017). In a similar rat model of multiple microinfarcts, cerebrovascular pathology impaired glymphatic function and was associated with loss of perivascular AQP4 localization (Venkat et al., 2017). These early studies suggest that cerebrovascular disease and pathology may impair glymphatic function, which may contribute to the development of vascular and mixed dementia.

6.2 Parkinson's and Huntington's diseases

Blood-brain barrier and CBF dysfunction contribute to PD and HD pathophysiology both in distinct and in similar ways. A subset of PD cases are associated with a leucine-rich repeat kinase 2 (LRRK2) missense mutation, the most common of which is G2019S, a gain of function mutation that has been shown to increase monocyte adhesion to endothelial cells (Hongge et al., 2015). Mutations in the MDR1 gene locus, which encodes the aforementioned P-gp, are also associated with PD. P-glycoprotein efflux transporters are highly expressed on the BBB endothelium, but this expression is reduced in PD patients (Kortekaas et al., 2005). These PD-associated mutations suggest a role for dysfunctional vascular and BBB integrity in PD pathophysiology. Magnetic resonance imaging and histological analysis revealed BBB breakdown throughout the basal ganglia of PD patient brains compared to controls (Sweeney et al., 2019). Interestingly, a recent MRI study reported worse motor symptoms were associated with higher CBF throughout the basal ganglia and thalamus, possibly due to aberrant angiogenesis that has been previously reported in PD (Pelizzari et al., 2019). Huntington's disease is associated with increased BBB permeability and cerebrovascular dysfunction as evidenced by the breakdown of endothelial tight junctions and decreased blood vessel diameters, as well as the presence of mutant huntingtin protein in all major components of the NVU (Drouin-Ouellet et al., 2015).

Parkinson's disease and HD have been far less studied than AD in the context of the glymphatic function; however, dysfunctional glymphatic clearance may contribute to the accumulation of α-synuclein and huntingtin aggregates in PD and HD, respectively, or in the process of α-synuclein and huntingtin aggregate propagation that may underlie the neuroanatomical progression of disease pathology (Abbott et al., 2018). Parkinson's disease is a neurodegenerative movement disorder that is characterized by substantial neuronal loss in the substantia nigra and accumulation of α-synuclein aggregates and ubiquitinated proteins in the cytoplasm of neurons, known as Lewy bodies (Wanker, 2000). A recent study reported decreased glymphatic influx, increased perivascular aggregation of α-synuclein, and loss of AQP4 perivascular localization in the substantia nigra in a PD-like transgenic mouse model (Zou et al., 2019). In addition, blockage of meningeal lymphatic drainage via cervical lymphatic ligation resulted in worsened PD pathology and increased aggregation of α-synuclein (Zou et al., 2019). This study suggests that glymphatic function may be decreased in the setting of PD and may lead to accumulation of α-synuclein aggregates. It is also possible that neurodegenerative pathology impairs glymphatic function, creating a pathogenic feed-forward loop driving the formation and propagation of pathogenic aggregates.

7. Metabolic disease

Numerous neurodegenerative diseases including AD and PD, are associated with coexisting metabolic dysfunction (Cai et al., 2012). Whether these alterations in metabolism drive the development of neurodegenerative disease, whether they are a consequence of the diseases themselves, or a complex interplay of both, remains to be elucidated. Type 2 diabetes mellitus (T2DM), a common metabolic disease found in middle-aged and older populations, is associated with cognitive impairment and an increased risk of developing dementia in older patients. Insulin resistance and hyperglycemia associated with T2DM can impair memory function, damage neuronal cells, and cause neuroinflammation (Kim et al., 2018). This insulin deficiency and resistance in the brain may contribute to cognitive decline in AD (Cai et al., 2012).

Type 2 diabetes mellitus is associated with BBB dysfunction and T2DM-associated hyperglycemia damages the structural integrity and function of the CNS microvasculature by promoting the production of reactive oxygen species, increasing vascular permeability, thickening capillary walls, and disrupting tight junctions (Prasad et al., 2014). Metabolic disorders, such as T2DM and obesity, are also associated with systemic inflammation and hypertension which contribute to increased BBB permeability and neuroinflammation (Van Dyken and Lacoste, 2018). A recent MRI study examining CBF in patients with T2DM reported cerebral hypoperfusion in the posterior cingulate cortex (PCC) and precuneus which was strongly correlated with higher insulin resistance and cognitive deficits (Cui et al., 2017). Posterior cingulate cortex hypometabolism is characteristic of AD, further emphasizing the relationship between T2DM and AD.

In a mouse model of T2DM, glymphatic exchange was impaired in the hippocampus and hypothalamus. This impairment was highly correlated with learning and memory deficits, suggesting a possible role for glymphatic dysfunction in cognitive decline associated with T2DM and diabetes-induced dementia (Jiang et al., 2016). Accordingly, T2DM and obesity are significant risk factors for developing AD and increased tau phosphorylation is observed in mouse models of T2DM (Cai et al., 2012). Together, these studies suggest that glymphatic pathway impairment may be one mechanistic link between T2DM and metabolic dysfunction to the development of neurodegenerative disease and diseases of aging.

In addition to a role in T2DM, a recent report found impaired glymphatic exchange and decreased AQP4 expression in an experimental rodent model of hepatic encephalopathy. Hepatic encephalopathy is associated with cognitive dysfunction and results from the accumulation of toxic metabolic waste products in the brain due to chronic liver disease (Hadjihambi et al., 2019). The authors propose that impaired glymphatic flow results in the accumulation of metabolic waste products in the brain. However, this ignores the likely role of accumulating peripheral metabolic waste products in the blood in promoting this glymphatic dysfunction. Increased metabolic waste products likely affect the neurovasculature and BBB and may promote astrogliosis, which may result in a bidirectional feedback loop that worsens glymphatic function and promotes the accumulation of waste products in the brain. Further research is needed to elucidate the complex dynamics underlying the pathogenesis of glymphatic impairment in hepatic encephalopathy and other metabolic disease models.

8. Conclusion

Emerging evidence suggests that neurovascular, BBB, and glymphatic dysfunction are associated with the aging process and may be major drivers of neurodegenerative disease. The glymphatic system facilitates the exchange of CSF and ISF through the parenchyma, enabling waste clearance and the transport of non-waste compounds such as lipids (Rangroo Thrane et al., 2013) and glucose (Lundgaard et al., 2015). This system is anatomically and functionally interconnected with the brain vasculature, BBB, and BCSFB and together they maintain brain homeostasis and regulate the movement and exchange of fluid and solutes throughout the brain (Jessen et al., 2015). While glymphatic exchange and clearance rates are impaired in aging (Benveniste et al., 2019; Kress et al., 2014) and neurodegenerative disorders (Peng et al., 2016; Rasmussen et al., 2018; Weller et al., 2008; Xu et al., 2015), they are enhanced during sleep (Xie et al., 2013), after physical exercise (He et al., 2017), and in certain body posture positions (Lee et al., 2015). These accumulating data suggest that glymphatic dysfunction may be one of the key drivers of protein mis-aggregation that underlies various neurodegenerative diseases, including AD and PD. Ongoing research will provide a clearer understanding of the physiological regulators of glymphatic function, and how they become dysfunctional in the setting of human disease. These findings may provide a novel therapeutic approach to the prevention and treatment of neurodegenerative disease.

References

Abbott, N. J., Pizzo, M. E., Preston, J. E., Janigro, D., & Thorne, R. G. (2018). The role of brain barriers in fluid movement in the CNS: Is there a 'glymphatic' system? *Acta Neuropathologica, 135*(3), 387–407. https://doi.org/10.1007/s00401-018-1812-4.

Armulik, A., Genove, G., & Betsholtz, C. (2011). Pericytes: Developmental, physiological, and pathological perspectives, problems, and promises. *Developmental Cell, 21*(2), 193–215. https://doi.org/10.1016/j.devcel.2011.07.001.

Armulik, A., Genove, G., Mae, M., Nisancioglu, M. H., Wallgard, E., Niaudet, C., et al. (2010). Pericytes regulate the blood-brain barrier. *Nature, 468*(7323), 557–561. https://doi.org/10.1038/nature09522.

Aspelund, A., Antila, S., Proulx, S. T., Karlsen, T. V., Karaman, S., Detmar, M., et al. (2015). A dural lymphatic vascular system that drains brain interstitial fluid and macromolecules. *The Journal of Experimental Medicine, 212*(7), 991–999. https://doi.org/10.1084/jem.20142290.

Attier-Zmudka, J., Sérot, J.-M., Valluy, J., Saffarini, M., Macaret, A.-S., Diouf, M., et al. (2019). Decreased cerebrospinal fluid flow is associated with cognitive deficit in elderly patients. *Frontiers in Aging Neuroscience, 11*, 87. https://doi.org/10.3389/fnagi.2019.00087.

Bélanger, M., & Magistretti, P. J. (2009). The role of astroglia in neuroprotection. *Dialogues in Clinical Neuroscience, 11*(3), 281–295. Retrieved from https://www.ncbi.nlm.nih.gov/pubmed/19877496 https://www.ncbi.nlm.nih.gov/pmc/articles/PMC3181926/.

Benveniste, H., Liu, X., Koundal, S., Sanggaard, S., Lee, H., & Wardlaw, J. (2019). The glymphatic system and waste clearance with brain aging: A review. *Gerontology, 65*(2), 106–119. https://doi.org/10.1159/000490349.

Bramlett, H. M., & Dietrich, W. D. (2004). Pathophysiology of cerebral ischemia and brain trauma: Similarities and differences. *Journal of Cerebral Blood Flow and Metabolism, 24*(2), 133–150. https://doi.org/10.1097/01.Wcb.0000111614.19196.04.

Brinker, T., Stopa, E., Morrison, J., & Klinge, P. (2014). A new look at cerebrospinal fluid circulation. *Fluids and Barriers of the CNS, 11*, 10. https://doi.org/10.1186/2045-8118-11-10.

Brown, L. S., Foster, C. G., Courtney, J.-M., King, N. E., Howells, D. W., & Sutherland, B. A. (2019). Pericytes and neurovascular function in the healthy and diseased brain. *Frontiers in Cellular Neuroscience, 13*, 282. https://doi.org/10.3389/fncel.2019.00282.

Buffington, A. L. H., Lipski, D. M., & Westfall, E. (2013). Dementia: An evidence-based review of common presentations and family-based interventions. *The Journal of the American Osteopathic Association, 113*(10), 768–775. https://doi.org/10.7556/jaoa.2013.046.

Cai, H., Cong, W.-N., Ji, S., Rothman, S., Maudsley, S., & Martin, B. (2012). Metabolic dysfunction in Alzheimer's disease and related neurodegenerative disorders. *Current Alzheimer Research, 9*(1), 5–17. Retrieved from https://www.ncbi.nlm.nih.gov/pubmed/22329649; https://www.ncbi.nlm.nih.gov/pmc/articles/PMC4097094/.

Carare, R. O., Bernardes-Silva, M., Newman, T. A., Page, A. M., Nicoll, J. A., Perry, V. H., et al. (2008). Solutes, but not cells, drain from the brain parenchyma along basement membranes of capillaries and arteries: Significance for cerebral amyloid angiopathy and neuroimmunology. *Neuropathology and Applied Neurobiology, 34*(2), 131–144. https://doi.org/10.1111/j.1365-2990.2007.00926.x.

Chodobski, A., Zink, B. J., & Szmydynger-Chodobska, J. (2011). Blood-brain barrier pathophysiology in traumatic brain injury. *Translational Stroke Research*, *2*(4), 492–516. https://doi.org/10.1007/s12975-011-0125-x.

Cipolla, M. (2009). Chapter 2, Anatomy and ultrastructure. In D. N. Granger & J. Granger (Eds.), *The cerebral circulation*. San Rafael, CA: Morgan & Claypool Life Sciences.

Cserr, H. F., Cooper, D. N., Suri, P. K., & Patlak, C. S. (1981). Efflux of radiolabeled polyethylene glycols and albumin from rat brain. *The American Journal of Physiology*, *240*(4), F319–F328. https://doi.org/10.1152/ajprenal.1981.240.4.F319.

Cui, Y., Liang, X., Gu, H., Hu, Y., Zhao, Z., Yang, X.-Y., et al. (2017). Cerebral perfusion alterations in type 2 diabetes and its relation to insulin resistance and cognitive dysfunction. *Brain Imaging and Behavior*, *11*(5), 1248–1257. https://doi.org/10.1007/s11682-016-9583-9.

Da Mesquita, S., Fu, Z., & Kipnis, J. (2018a). The meningeal lymphatic system: A new player in neurophysiology. *Neuron*, *100*(2), 375–388. https://doi.org/10.1016/j.neuron.2018.09.022.

Da Mesquita, S., Louveau, A., Vaccari, A., Smirnov, I., Cornelison, R. C., Kingsmore, K. M., et al. (2018b). Functional aspects of meningeal lymphatics in ageing and Alzheimer's disease. *Nature*, *560*(7717), 185–191. https://doi.org/10.1038/s41586-018-0368-8.

Daneman, R., & Prat, A. (2015). The blood-brain barrier. *Cold Spring Harbor Perspectives in Biology*, 7(1), a020412. https://doi.org/10.1101/cshperspect.a020412.

Del Bigio, M. R. (2010). Ependymal cells: Biology and pathology. *Acta Neuropathologica*, *119*(1), 55–73. https://doi.org/10.1007/s00401-009-0624-y.

Drouin-Ouellet, J., Sawiak, S. J., Cisbani, G., Lagace, M., Kuan, W. L., Saint-Pierre, M., et al. (2015). Cerebrovascular and blood-brain barrier impairments in Huntington's disease: Potential implications for its pathophysiology. *Annals of Neurology*, *78*(2), 160–177. https://doi.org/10.1002/ana.24406.

Duncombe, J., Lennen, R. J., Jansen, M. A., Marshall, I., Wardlaw, J. M., & Horsburgh, K. (2017). Ageing causes prominent neurovascular dysfunction associated with loss of astrocytic contacts and gliosis. *Neuropathology and Applied Neurobiology*, *43*(6), 477–491. https://doi.org/10.1111/nan.12375.

Farrall, A. J., & Wardlaw, J. M. (2009). Blood–brain barrier: Ageing and microvascular disease—Systematic review and meta-analysis. *Neurobiology of Aging*, *30*(3), 337–352. https://doi.org/10.1016/j.neurobiolaging.2007.07.015.

Fields, R. D., Woo, D. H., & Basser, P. J. (2015). Glial regulation of the neuronal connectome through local and long-distant communication. *Neuron*, *86*(2), 374–386. https://doi.org/10.1016/j.neuron.2015.01.014.

Fleischman, D., Berdahl, J. P., Zaydlarova, J., Stinnett, S., Fautsch, M. P., & Allingham, R. R. (2012). Cerebrospinal fluid pressure decreases with older age. *PLoS One*, 7(12), e52664. https://doi.org/10.1371/journal.pone.0052664.

Fultz, N. E., Bonmassar, G., Setsompop, K., Stickgold, R. A., Rosen, B. R., Polimeni, J. R., et al. (2019). Coupled electrophysiological, hemodynamic, and cerebrospinal fluid oscillations in human sleep. *Science*, *366*(6465), 628. https://doi.org/10.1126/science.aax5440.

Groothuis, D. R., Vavra, M. W., Schlageter, K. E., Kang, E. W., Itskovich, A. C., Hertzler, S., et al. (2007). Efflux of drugs and solutes from brain: The interactive roles of diffusional transcapillary transport, bulk flow and capillary transporters. *Journal of Cerebral Blood Flow and Metabolism*, *27*(1), 43–56. https://doi.org/10.1038/sj.jcbfm.9600315.

Hadjihambi, A., Harrison, I. F., Costas-Rodriguez, M., Vanhaecke, F., Arias, N., Gallego-Duran, R., et al. (2019). Impaired brain glymphatic flow in experimental hepatic encephalopathy. *Journal of Hepatology*, *70*(1), 40–49. https://doi.org/10.1016/j.jhep.2018.08.021.

Hall, C. N., Reynell, C., Gesslein, B., Hamilton, N. B., Mishra, A., Sutherland, B. A., et al. (2014). Capillary pericytes regulate cerebral blood flow in health and disease. *Nature*, *508*(7494), 55–60. https://doi.org/10.1038/nature13165.

Hawkes, C. A., Hartig, W., Kacza, J., Schliebs, R., Weller, R. O., Nicoll, J. A., et al. (2011). Perivascular drainage of solutes is impaired in the ageing mouse brain and in the presence of cerebral amyloid angiopathy. *Acta Neuropathologica*, *121*(4), 431–443. https://doi.org/10.1007/s00401-011-0801-7.

He, X. F., Liu, D. X., Zhang, Q., Liang, F. Y., Dai, G. Y., Zeng, J. S., et al. (2017). Voluntary exercise promotes glymphatic clearance of amyloid beta and reduces the activation of astrocytes and microglia in aged mice. *Frontiers in Molecular Neuroscience*, *10*, 144. https://doi.org/10.3389/fnmol.2017.00144.

Hill, R. A., Tong, L., Yuan, P., Murikinati, S., Gupta, S., & Grutzendler, J. (2015). Regional blood flow in the normal and ischemic brain is controlled by arteriolar smooth muscle cell contractility and not by capillary pericytes. *Neuron*, *87*(1), 95–110. https://doi.org/10.1016/j.neuron.2015.06.001.

Hladky, S. B., & Barrand, M. A. (2014). Mechanisms of fluid movement into, through and out of the brain: Evaluation of the evidence. *Fluids and Barriers of the CNS*, *11*(1), 26. https://doi.org/10.1186/2045-8118-11-26.

Hongge, L., Kexin, G., Xiaojie, M., Nian, X., & Jinsha, H. (2015). The role of LRRK2 in the regulation of monocyte adhesion to endothelial cells. *Journal of Molecular Neuroscience*, *55*(1), 233–239. https://doi.org/10.1007/s12031-014-0312-9.

Iadecola, C., & Nedergaard, M. (2007). Glial regulation of the cerebral microvasculature. *Nature Neuroscience*, *10*(11), 1369–1376. https://doi.org/10.1038/nn2003.

Iliff, J. J., Chen, M. J., Plog, B. A., Zeppenfeld, D. M., Soltero, M., Yang, L., et al. (2014). Impairment of glymphatic pathway function promotes tau pathology after traumatic brain injury. *The Journal of Neuroscience*, *34*(49), 16180. https://doi.org/10.1523/JNEUROSCI.3020-14.2014.

Iliff, J. J., Lee, H., Yu, M., Feng, T., Logan, J., Nedergaard, M., et al. (2013a). Brain-wide pathway for waste clearance captured by contrast-enhanced MRI. *The Journal of Clinical Investigation*, *123*(3), 1299–1309. https://doi.org/10.1172/JCI67677.

Iliff, J., & Simon, M. (2019). CrossTalk proposal: The glymphatic system supports convective exchange of cerebrospinal fluid and brain interstitial fluid that is mediated by perivascular aquaporin-4. *The Journal of Physiology*, *597*(17), 4417–4419. https://doi.org/10.1113/jp277635.

Iliff, J. J., Wang, M., Liao, Y., Plogg, B. A., Peng, W., Gundersen, G. A., et al. (2012). A paravascular pathway facilitates CSF flow through the brain parenchyma and the clearance of interstitial solutes, including amyloid β. *Science Translational Medicine*, *4*(147), 147ra111. https://doi.org/10.1126/scitranslmed.3003748.

Iliff, J. J., Wang, M., Zeppenfeld, D. M., Venkataraman, A., Plog, B. A., Liao, Y., et al. (2013b). Cerebral arterial pulsation drives paravascular CSF-interstitial fluid exchange in the murine brain. *The Journal of Neuroscience*, *33*(46), 18190–18199. https://doi.org/10.1523/JNEUROSCI.1592-13.2013.

Jessen, N. A., Munk, A. S. F., Lundgaard, I., & Nedergaard, M. (2015). The glymphatic system: A beginner's guide. *Neurochemical Research*, *40*(12), 2583–2599. https://doi.org/10.1007/s11064-015-1581-6.

Jiang, Q., Zhang, L., Ding, G., Davoodi-Bojd, E., Li, Q., Li, L., et al. (2016). Impairment of the glymphatic system after diabetes. *Journal of Cerebral Blood Flow & Metabolism*, *37*(4), 1326–1337. https://doi.org/10.1177/0271678X16654702.

Jin, B. J., Zhang, H., Binder, D. K., & Verkman, A. S. (2013). Aquaporin-4-dependent K(+) and water transport modeled in brain extracellular space following neuroexcitation. *The Journal of General Physiology*, *141*(1), 119–132. https://doi.org/10.1085/jgp.201210883.

Johnston, M., Zakharov, A., Papaiconomou, C., Salmasi, G., & Armstrong, D. (2004). Evidence of connections between cerebrospinal fluid and nasal lymphatic vessels in humans, non-human primates and other mammalian species. *Cerebrospinal Fluid Research*, *1*(1), 2. https://doi.org/10.1186/1743-8454-1-2.

Kim, Y.-K., Nam, K. I., & Song, J. (2018). The glymphatic system in diabetes-induced dementia. *Frontiers in Neurology*, *9*, 867. Retrieved from https://www.frontiersin.org/article/10.3389/fneur.2018.00867.

Kisler, K., Nelson, A. R., Montagne, A., & Zlokovic, B. V. (2017a). Cerebral blood flow regulation and neurovascular dysfunction in Alzheimer disease. *Nature Reviews. Neuroscience*, *18*(7), 419–434. https://doi.org/10.1038/nrn.2017.48.

Kisler, K., Nelson, A. R., Rege, S. V., Ramanathan, A., Wang, Y., Ahuja, A., et al. (2017b). Pericyte degeneration leads to neurovascular uncoupling and limits oxygen supply to brain. *Nature Neuroscience*, *20*(3), 406–416. https://doi.org/10.1038/nn.4489.

Kiviniemi, V., Wang, X., Korhonen, V., Keinanen, T., Tuovinen, T., Autio, J., et al. (2016). Ultra-fast magnetic resonance encephalography of physiological brain activity—Glymphatic pulsation mechanisms? *Journal of Cerebral Blood Flow and Metabolism*, *36*(6), 1033–1045. https://doi.org/10.1177/0271678x15622047.

Kortekaas, R., Leenders, K. L., van Oostrom, J. C., Vaalburg, W., Bart, J , Willemsen, A. T., et al. (2005). Blood-brain barrier dysfunction in parkinsonian midbrain in vivo. *Annals of Neurology*, *57*(2), 176–179. https://doi.org/10.1002/ana.20369.

Kress, B. T., Iliff, J. J., Xia, M., Wang, M., Wei, H. S., Zeppenfeld, D., et al. (2014). Impairment of paravascular clearance pathways in the aging brain. *Annals of Neurology*, *76*(6), 845–861. https://doi.org/10.1002/ana.24271.

Laterra, J., Keep, R., Betz, L. A., et al. (1999). Blood—Cerebrospinal fluid barrier. In S. GJ & R. W. Alberset al.*Basic neurochemistry: Molecular, cellular and medical aspects*. (6th ed.). Philadelphia: Lippincott-Raven.

Lee, H., Xie, L., Yu, M., Kang, H., Feng, T., Deane, R., et al. (2015). The effect of body posture on brain glymphatic transport. *The Journal of Neuroscience*, *35*(31), 11034. https://doi.org/10.1523/JNEUROSCI.1625-15.2015.

Lim, M. M., Gerstner, J. R., & Holtzman, D. M. (2014). The sleep-wake cycle and Alzheimer's disease: What do we know? *Neurodegenerative Disease Management*, *4*(5), 351–362. https://doi.org/10.2217/nmt.14.33.

Liu, H., Ni, Z., Chen, Y., Wang, D., Qi, Y., Zhang, Q., et al. (2012). Olfactory route for cerebrospinal fluid drainage into the cervical lymphatic system in a rabbit experimental model. *Neural Regeneration Research*, 7(10), 766–771. https://doi.org/10.3969/j.issn.1673-5374.2012.10.009.

Louveau, A., Plog, B. A., Antila, S., Alitalo, K., Nedergaard, M., & Kipnis, J. (2017). Understanding the functions and relationships of the glymphatic system and meningeal lymphatics. *The Journal of Clinical Investigation*, *127*(9), 3210–3219. https://doi.org/10.1172/JCI90603.

Lun, M. P., Monuki, E. S., & Lehtinen, M. K. (2015). Development and functions of the choroid plexus-cerebrospinal fluid system. *Nature Reviews. Neuroscience*, *16*(8), 445–457. https://doi.org/10.1038/nrn3921.

Lundgaard, I., Li, B., Xie, L., Kang, H., Sanggaard, S., Haswell, J. D., et al. (2015). Direct neuronal glucose uptake heralds activity-dependent increases in cerebral metabolism. *Nature Communications*, *6*, 6807. https://doi.org/10.1038/ncomms7807.

Ma, Q., Ineichen, B. V., Detmar, M., & Proulx, S. T. (2017). Outflow of cerebrospinal fluid is predominantly through lymphatic vessels and is reduced in aged mice. *Nature Communications*, *8*(1), 1434. https://doi.org/10.1038/s41467-017-01484-6.

Mastorakos, P., & McGavern, D. (2019). The anatomy and immunology of vasculature in the central nervous system. *Science Immunology*, *4*(37), eaav0492. https://doi.org/10.1126/sciimmunol.aav0492.

Matthews, K. A., Xu, W., Gaglioti, A. H., Holt, J. B., Croft, J. B., Mack, D., et al. (2019). Racial and ethnic estimates of Alzheimer's disease and related dementias in the United States (2015–2060) in adults aged ≥65 years. *Alzheimers Dement, 15*(1), 17–24. https://doi.org/10.1016/j.jalz.2018.06.3063.

Mawuenyega, K. G., Sigurdson, W., Ovod, V., Munsell, L., Kasten, T., Morris, J. C., et al. (2010). Decreased clearance of CNS β-amyloid in Alzheimer's disease. *Science, 330*(6012), 1774. https://doi.org/10.1126/science.1197623.

McConnell, H. L., Kersch, C. N., Woltjer, R. L., & Neuwelt, E. A. (2017). The translational significance of the neurovascular unit. *The Journal of Biological Chemistry, 292*(3), 762–770. https://doi.org/10.1074/jbc.R116.760215.

Mestre, H., Hablitz, L. M., Xavier, A. L. R., Feng, W., Zou, W., Pu, T., et al. (2018a). Aquaporin-4-dependent glymphatic solute transport in the rodent brain. *eLife, 7*, e40070. https://doi.org/10.7554/eLife.40070.

Mestre, H., Tithof, J., Du, T., Song, W., Peng, W., Sweeney, A. M., et al. (2018b). Flow of cerebrospinal fluid is driven by arterial pulsations and is reduced in hypertension. *Nature Communications, 9*(1), 4878. https://doi.org/10.1038/s41467-018-07318-3.

Nippert, A. R., Biesecker, K. R., & Newman, E. A. (2018). Mechanisms mediating functional hyperemia in the brain. *The Neuroscientist: A Review Journal Bringing Neurobiology, Neurology and Psychiatry, 24*(1), 73–83. https://doi.org/10.1177/1073858417703033.

Pelizzari, L., Laganà, M. M., Rossetto, F., Bergsland, N., Galli, M., Baselli, G., et al. (2019). Cerebral blood flow and cerebrovascular reactivity correlate with severity of motor symptoms in Parkinson's disease. *Therapeutic Advances in Neurological Disorders, 12*, 1756286419838354. https://doi.org/10.1177/1756286419838354.

Peng, W., Achariyar, T. M., Li, B., Liao, Y., Mestre, H., Hitomi, E., et al. (2016). Suppression of glymphatic fluid transport in a mouse model of Alzheimer's disease. *Neurobiology of Disease, 93*, 215–225. https://doi.org/10.1016/j.nbd.2016.05.015.

Praetorius, J., & Nielsen, S. (2006). Distribution of sodium transporters and aquaporin-1 in the human choroid plexus. *American Journal of Physiology Cell Physiology, 291*(1), C59–C67. https://doi.org/10.1152/ajpcell.00433.2005.

Prasad, S., Sajja, R. K., Naik, P., & Cucullo, L. (2014). Diabetes mellitus and blood-brain barrier dysfunction: An overview. *Journal of Pharmacovigilance, 2*(2), 125. https://doi.org/10.4172/2329-6887.1000125.

Prinz, M., & Priller, J. (2017). The role of peripheral immune cells in the CNS in steady state and disease. *Nature Neuroscience, 20*, 136. https://doi.org/10.1038/nn.4475.

Rangroo Thrane, V., Thrane, A. S., Plog, B. A., Thiyagarajan, M., Iliff, J. J., Deane, R., et al. (2013). Paravascular microcirculation facilitates rapid lipid transport and astrocyte signaling in the brain. *Scientific Reports, 3*, 2582. https://doi.org/10.1038/srep02582.

Rasmussen, M. K., Mestre, H., & Nedergaard, M. (2018). The glymphatic pathway in neurological disorders. *The Lancet Neurology, 17*(11), 1016–1024. https://doi.org/10.1016/S1474-4422(18)30318-1.

Ray, L., Iliff, J. J., & Heys, J. J. (2019). Analysis of convective and diffusive transport in the brain interstitium. *Fluids and Barriers of the CNS, 16*(1), 6. https://doi.org/10.1186/s12987-019-0126-9.

Redline, S., Kirchner, H. L., Quan, S., Gottlieb, D., Kapur, V., & Newman, A. (2004). The effects of age, sex, ethnicity, and sleep-disordered breathing on sleep architecture. *Archives of Internal Medicine, 164*, 406–418. https://doi.org/10.1001/archinte.164.4.406.

Redzic, Z. (2011). Molecular biology of the blood-brain and the blood-cerebrospinal fluid barriers: Similarities and differences. *Fluids and Barriers of the CNS, 8*(1), 3. https://doi.org/10.1186/2045-8118-8-3.

Rennels, M. L., Blaumanis, O. R., & Grady, P. A. (1990). Rapid solute transport throughout the brain via paravascular fluid pathways. *Advances in Neurology, 52*, 431–439.

Rennels, M. L., Gregory, T. F., Blaumanis, O. R., Fujimoto, K., & Grady, P. A. (1985). Evidence for a 'paravascular' fluid circulation in the mammalian central nervous system, provided by the rapid distribution of tracer protein throughout the brain from the subarachnoid space. *Brain Research*, *326*(1), 47–63. https://doi.org/10.1016/0006-8993(85)91383-6.

Roh, J. H., Huang, Y., Bero, A. W., Kasten, T., Stewart, F. R., Bateman, R. J., et al. (2012). Disruption of the sleep-wake cycle and diurnal fluctuation of β-amyloid in mice with Alzheimer's disease pathology. *Science Translational Medicine*, *4*(150), 150ra122. https://doi.org/10.1126/scitranslmed.3004291.

Rua, R., & McGavern, D. B. (2018). Advances in meningeal immunity. *Trends in Molecular Medicine*, *24*(6), 542–559. https://doi.org/10.1016/j.molmed.2018.04.003.

Sanchez-Covarrubias, L., Slosky, L. M., Thompson, B. J., Davis, T. P., & Ronaldson, P. T. (2014). Transporters at CNS barrier sites: Obstacles or opportunities for drug delivery? *Current Pharmaceutical Design*, *20*(10), 1422–1449. https://doi.org/10.2174/13816128113199990463.

Shokri-Kojori, E., Wang, G. J., Wiers, C. E., Demiral, S. B., Guo, M., Kim, S. W., et al. (2018). Beta-amyloid accumulation in the human brain after one night of sleep deprivation. *Proceedings of the National Academy of Sciences of the United States of America*, *115*(17), 4483–4488. https://doi.org/10.1073/pnas.1721694115.

Stoodley, M. A., Brown, S. A., Brown, C. J., & Jones, N. R. (1997). Arterial pulsation-dependent perivascular cerebrospinal fluid flow into the central canal in the sheep spinal cord. *Journal of Neurosurgery*, *86*(4), 686–693. https://doi.org/10.3171/jns.1997.86.4.0686.

Storck, S. E., Hartz, A. M. S., Bernard, J., Wolf, A., Kachlmeier, A., Mahringer, A., et al. (2018). The concerted amyloid-beta clearance of LRP1 and ABCB1/P-gp across the blood-brain barrier is linked by PICALM. *Brain, Behavior, and Immunity*, *73*, 21–33. https://doi.org/10.1016/j.bbi.2018.07.017.

Sweeney, M. D., Zhao, Z., Montagne, A., Nelson, A. R., & Zlokovic, B. V. (2019). Blood-brain barrier: From physiology to disease and back. *Physiological Reviews*, *99*(1), 21–78. https://doi.org/10.1152/physrev.00050.2017.

Van Dyken, P., & Lacoste, B. (2018). Impact of metabolic syndrome on neuroinflammation and the blood-brain barrier. *Frontiers in Neuroscience*, *12*, 930. https://doi.org/10.3389/fnins.2018.00930.

Venkat, P., Chopp, M., Zacharek, A., Cui, C., Zhang, L., Li, Q., et al. (2017). White matter damage and glymphatic dysfunction in a model of vascular dementia in rats with no prior vascular pathologies. *Neurobiology of Aging*, *50*, 96–106. https://doi.org/10.1016/j.neurobiolaging.2016.11.002.

Wang, M., Ding, F., Deng, S., Guo, X., Wang, W., Iliff, J. J., et al. (2017). Focal solute trapping and global glymphatic pathway impairment in a murine model of multiple microinfarcts. *The Journal of Neuroscience*, *37*(11), 2870–2877. https://doi.org/10.1523/JNEUROSCI.2112-16.2017.

Wang, M., Iliff, J. J., Liao, Y., Chen, M. J., Shinseki, M. S., Venkataraman, A., et al. (2012). Cognitive deficits and delayed neuronal loss in a mouse model of multiple microinfarcts. *The Journal of Neuroscience*, *32*(50), 17948–17960. https://doi.org/10.1523/JNEUROSCI.1860-12.2012.

Wanker, E. E. (2000). Protein aggregation in Huntington's and Parkinson's disease: Implications for therapy. *Molecular Medicine Today*, *6*(10), 387–391.

Weller, R. O., Subash, M., Preston, S. D., Mazanti, I., & Carare, R. O. (2008). Perivascular drainage of amyloid-beta peptides from the brain and its failure in cerebral amyloid angiopathy and Alzheimer's disease. *Brain Pathology*, *18*(2), 253–266. https://doi.org/10.1111/j.1750-3639.2008.00133.x.

Xie, L., Kang, H., Xu, Q., Chen, M. J., Liao, Y., Thiyagarajan, M., et al. (2013). Sleep drives metabolite clearance from the adult brain. *Science*, *342*(6156), 373–377. https://doi.org/10.1126/science.1241224.

Xu, Z., Xiao, N., Chen, Y., Huang, H., Marshall, C., Gao, J., et al. (2015). Deletion of aquaporin-4 in APP/PS1 mice exacerbates brain Aβ accumulation and memory deficits. *Molecular Neurodegeneration*, *10*, 58. https://doi.org/10.1186/s13024-015-0056-1.

Yamada, S., Miyazaki, M., Yamashita, Y., Ouyang, C., Yui, M., Nakahashi, M., et al. (2013). Influence of respiration on cerebrospinal fluid movement using magnetic resonance spin labeling. *Fluids and Barriers of the CNS*, *10*(1), 36. https://doi.org/10.1186/2045-8118-10-36.

Yurchenco, P. D., & Patton, B. L. (2009). Developmental and pathogenic mechanisms of basement membrane assembly. *Current Pharmaceutical Design*, *15*(12), 1277–1294. https://doi.org/10.2174/138161209787846766.

Zeppenfeld, D. M., Simon, M., Haswell, J. D., D'Abreo, D., Murchison, C., Quinn, J. F., et al. (2017). Association of perivascular localization of aquaporin-4 with cognition and Alzheimer disease in aging brains. *JAMA Neurology*, *74*(1), 91–99. https://doi.org/10.1001/jamaneurol.2016.4370.

Zou, W., Pu, T., Feng, W., Lu, M., Zheng, Y., Du, R., et al. (2019). Blocking meningeal lymphatic drainage aggravates Parkinson's disease-like pathology in mice overexpressing mutated α-synuclein. *Translational Neurodegeneration*, *8*(1), 7. https://doi.org/10.1186/s40035-019-0147-y.

FSC
www.fsc.org
MIX
Papier | Fördert gute Waldnutzung
FSC® C083411

Kolibri 360 GmbH
Ferdinand-Jühlke-Straße 7
99095 Erfurt, Deutschland
produktsicherheit@kolibri360.de